Leukemia

Cytology and Cytochemistry

GOPAL PARIJA, M.D.

GOPAL C. PARIJA
M.B.B.S., M.D., F.C.A.P., M.I.A.C.

Leukemia
Cytology and Cytochemistry

Lawrence Kass, M.D.
Professor of Medicine, Pathology, and
Hematopathology
Case Western Reserve University
School of Medicine
Cleveland, Ohio

J. B. Lippincott Company
Philadelphia • Toronto

The author and publisher have exerted every effort to ensure that drug selection and dosage set forth in this text are in accord with current recommendations and practice at the time of publication. However, in view of ongoing research, changes in government regulations, and the constant flow of information relating to drug therapy and drug reactions, the reader is urged to check the package insert for each drug for any change in indications and dosage and for added warnings and precautions. This is particularly important when the recommended agent is a new or infrequently employed drug.

For information address J. B. Lippincott Company, East Washington Square, Philadelphia, Pennsylvania 19105.

6 5 4 3 2 1

Library of Congress Cataloging in Publication Data

Kass, Lawrence.
Leukemia, cytology and cytochemistry.

Bibliography
Includes index.
1. Leukemia—Diagnosis. 2. Cancer cells—Identification. 3. Cytochemistry. 4. Diagnosis, Cytologic.
I. Title. [DNLM: 1. Leukemia—Diagnosis. 2. Histocytochemistry. 3. Cytodiagnosis. QZ 350 K19L]
RC643.K36 616.99'41907582 80–29033
ISBN 0–397–50463–2

Printed in the United States of America

For my wife, Sara Fran

Contents

Preface

For most hematologists, pathologists, medical oncologists, and medical laboratory technologists, microscopic examination of panoptically-stained preparations of blood and bone marrow forms the foundation of the diagnosis of leukemia. In chronic leukemias, it is easy to distinguish one type from another. With acute leukemias, it may be more difficult.

Although morphologic criteria for distinguishing between typical myeloblasts and lymphoblasts have been known for many years, atypical cases still pose diagnostic problems. In these atypical cases, cytochemical stains can be helpful and have had increasing application and popularity. Cytochemistry represents biochemistry on a microscopic and submicroscopic level. When applied to blood cells, cytochemical techniques are most useful when they demonstrate unique properties found largely or exclusively in one cell type but not in another. Demonstration of the presence or absence of these properties, such as myeloperoxidase activity in granulocytes but not in lymphocytes, can be used to identify one cell type and distinguish it from others.

As such, cytochemistry amplifies conventional morphology and helps to circumvent some of the subjectiveness inherent in visual perceptive techniques such as light microscopy using traditional panoptic stains. Although the ideal cytochemical stains are yet to be devised, those that are available now can usually distinguish between lymphoid, granulocytic, and monocytic cells. These distinctions are less reliable when dealing with some primitive leukemic blast cells that have not manifested a pathway of differentiation that is recognizable by present techniques. In these instances, traditional cytochemical techniques fail because the cells in question show few, if any, distinctive properties that can be attributed to a specific cell line.

Although it is not difficult to diagnose most cases of acute and chronic leukemia using accepted morphologic and cytochemical criteria, difficulties often arise in trying to diagnose leukemia before these accepted criteria are met. In a group of panmyelotic disorders including preleukemia, Di Guglielmo syndrome, refractory anemia with excess blasts, and subacute myelomonocytic leukemia, some of the morphologic and cytochemical features of acute leukemia may be present. However, using accepted criteria, marrows from patients with these disorders contain less than 40% leukemic blasts, so they are not acute leukemia in the usual sense.

Because acute leukemia develops in some of these patients, their panmyeloses prior to the development of acute leukemia have been regarded as preleukemic disorders. These disorders are controversial. Some authors do not believe that they exist as such but rather represent leukemia in a slowly developing or indolent form. In view of the current interest in these preleukemic disorders, as well as the current efforts to identify leukemia and other diseases in an indolent and even preclinical form, this text emphasizes the morphologic and cytochemical features of the broad group of primary acquired panmyelotic disorders identified thus far that precede the development of acute leukemia.

In addition to the emphasis on preleukemic disorders, this book also emphasizes a number of evolutionary patterns of leukemia. Over the past several decades, these evolutionary patterns, from preleukemic disorder to acute leukemia and from lymphoma to leukemia, have been described and recognized as a part of the natural history of some of these and other disorders. Furthermore, acute leukemia can occur as a terminal event in a patient who has been treated for another neoplasm with chemotherapeutic drugs, particularly alkylating agents, procarbazine, and the nitrosoureas. In such a patient, the drugs may be instrumental in the development of acute leukemia. With greater understanding of these evolutionary patterns, it may be possible to gain insight into important pathogenetic factors in the development of acute leukemia. As time goes on, and with the perception of new morphologic criteria and the development of new and better cytochemical and immunologic tests, it may be possible to make a more accurate and reliable diagnosis of the type of leukemia than at present.

Lawrence Kass, M.D.

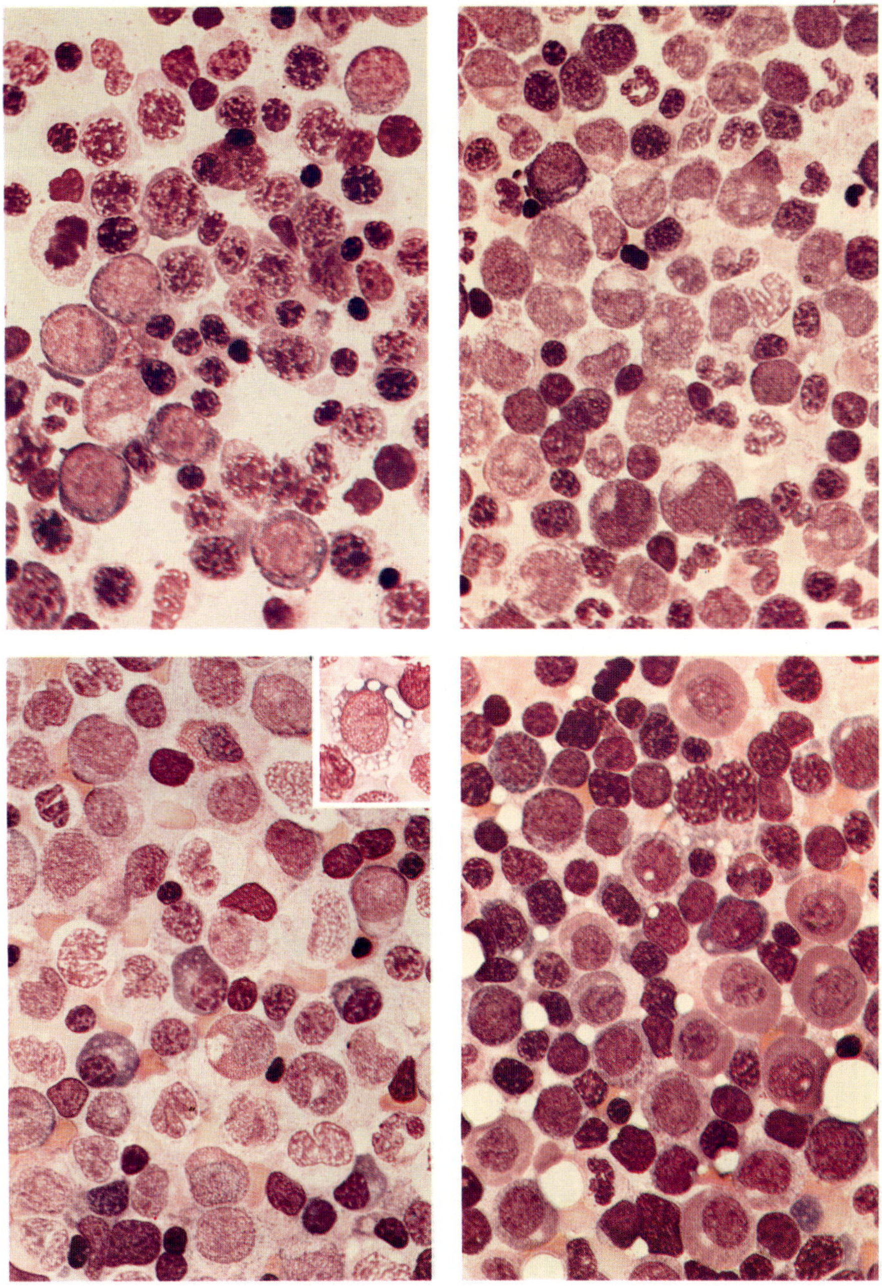

A B C D

Plate 1

(*A*) Bone marrow in chronic erythremic myelosis, demonstrating megaloblastoid erythroid hyperplasia and increased numbers of proerythroblasts. (*B*) Bone marrow in preleukemia, showing panmyelosis, megaloblastoid intermediate macronormoblasts, and increased numbers of myelocytes. (*C*) Bone marrow in preleukemia, depicting abnormalities of nuclear size and shape in granulocytic cells, plasmacytosis, and aberrant-appearing erythroblast with numerous cytoplasmic vacuoles (**inset**). (*D*) Bone marrow in smoldering acute leukemia (refractory anemia with excess blasts), illustrating increased numbers of myeloblasts, erythroblasts, and eosinophils.

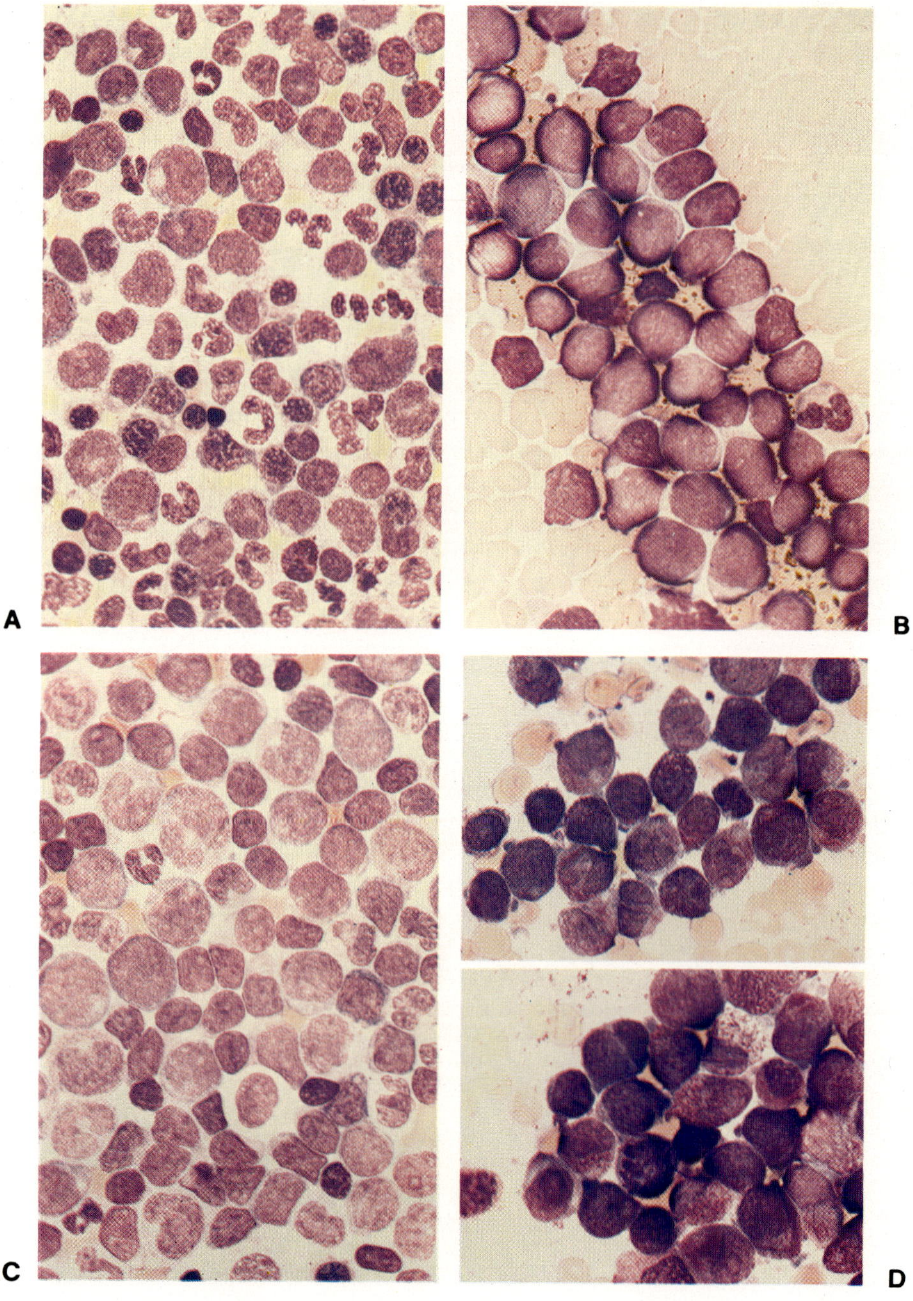

Plate 2 (*A*) Bone marrow in subacute myelomonocytic leukemia, demonstrating erythroid hyperplasia and increased numbers of monocytes. (*B*) Bone marrow in acute myeloblastic leukemia (M1). Little or no differentiation beyond the myeloblast stage is seen. (*C*) Bone marrow in acute myeloblastic leukemia (M2). Increased numbers of myeloblasts are seen, and there is differentiation beyond the promyelocytic stage. Some metamyelocytes have aberrant-appearing nuclei. (*D*) Bone marrow in acute promyelocytic leukemia (hypergranular promyelocytic leukemia, or M3). Some of the leukemic promyelocytes contain multiple Auer rods, in a "stack of twigs" configuration.

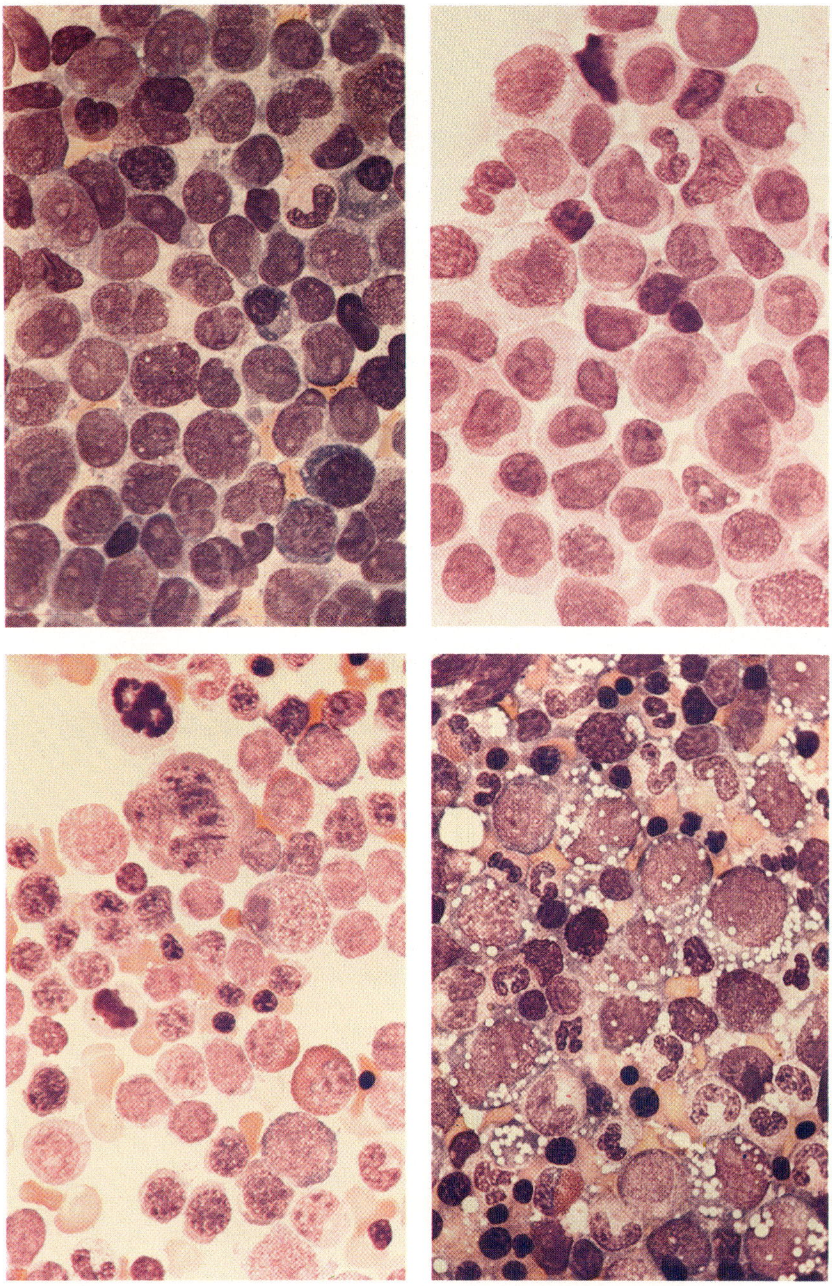

A B C D

Plate 3

(*A*) Bone marrow in acute myelomonocytic leukemia (M4). (*B*) Bone marrow in acute monocytic leukemia (M5). (*C*) Bone marrow in erythroleukemia (M6). (*D*) Bone marrow in leukemic phase of histiocytic lymphoma.

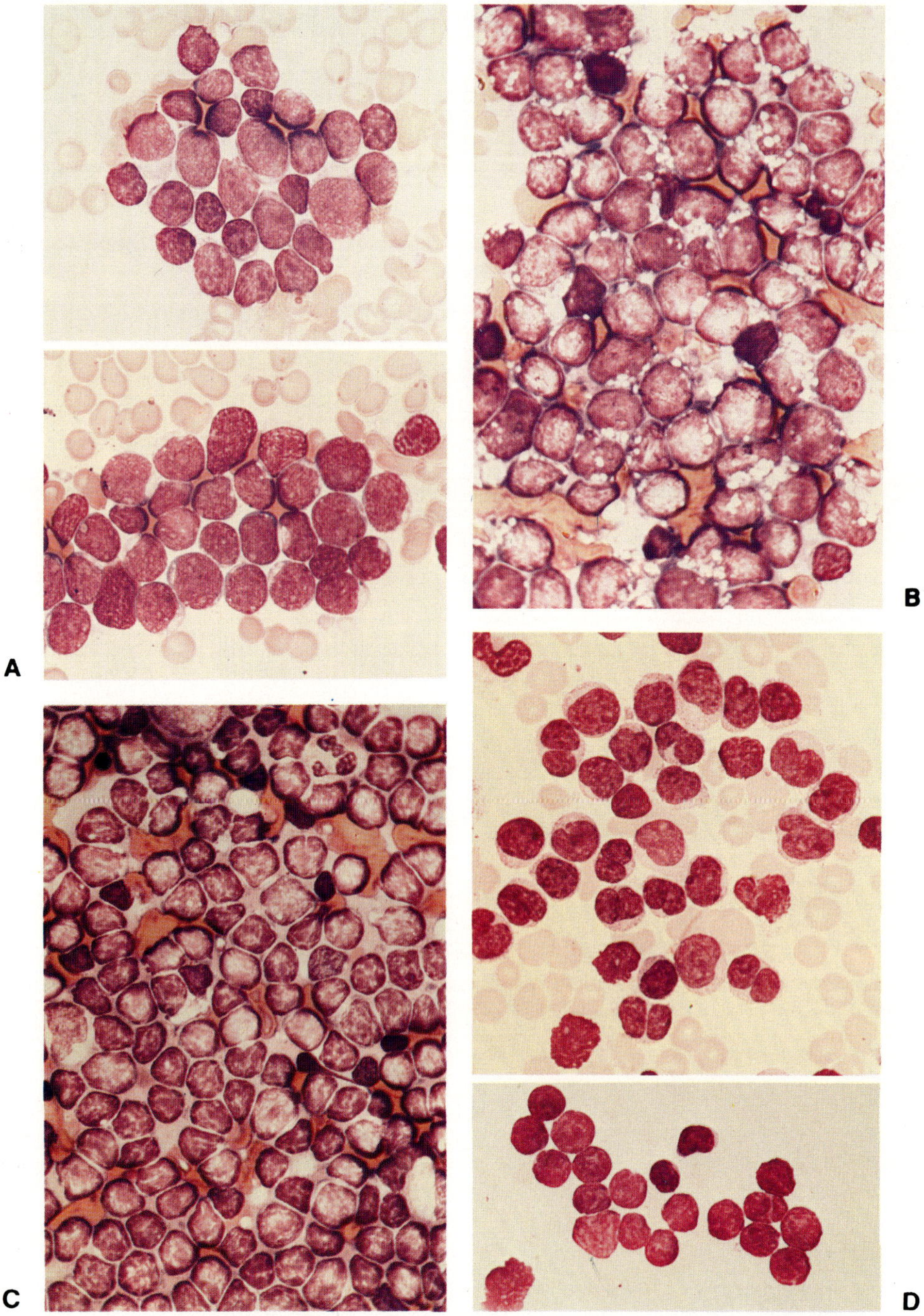

Plate 4 (*A*) Bone marrow in acute lymphoblastic leukemia (L1; **top**). Bone marrow in acute lymphoblastic leukemia (L2; **bottom**). (*B*) Bone marrow in acute lymphoblastic leukemia (Burkitt's type, or L3). (*C*) Bone marrow in chronic lymphocytic leukemia, well-differentiated type. (*D*) Bone marrow in chronic lymphocytic leukemia, moderately poorly differentiated type (**top**). Nuclear clefting and binuclearity are depicted. Bone marrow in chronic lymphocytic leukemia, poorly differentiated type (**bottom**). Nuclei of lymphocytes show multiple clefts and convolutions. Cytoplasm is scant.

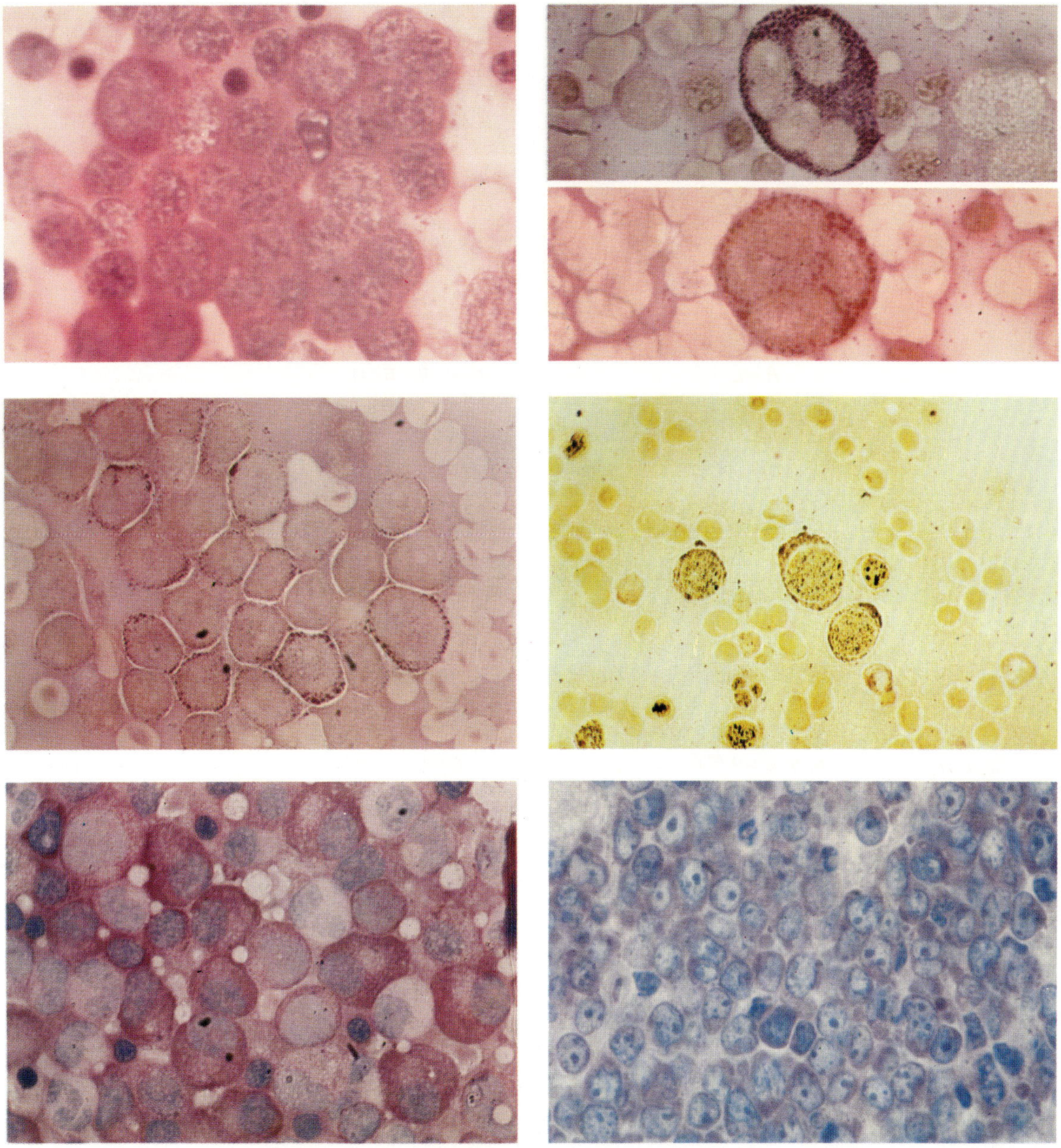

A B C D E F

Plate 5

(*A*) PAS stain, erythroblasts in chronic erythremic myelosis, illustrating punctate distribution of glycogen in proerythroblast, and diffuse cytoplasmic staining pattern in more mature erythroblasts. (*B*) PAS stain, erythroblasts in erythroleukemia, showing prominent, blocklike aggregates of glycogen in the cytoplasm. (*C*) PAS stain, lymphoblasts in acute lymphoblastic leukemia, demonstrating punctate pattern of glycogen distribution within the cytoplasm. (*D*) Ammoniacal silver stain in chronic erythremic myelosis, depicting black punctate distribution of arginine-rich histone in nuclei of proerythroblasts. (*E*) Methyl green pyronine stain, myeloma cells, showing intense cytoplasmic pyroninophilia. (*F*) Methyl green-pyronine stain, Burkitt's leukemia cells, illustrating strong pyroninophilia of cytoplasm in a 2-μ thick biopsy section.

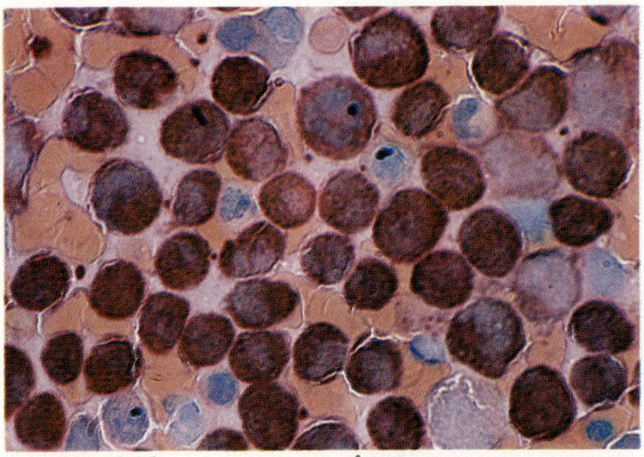

A

NON SP. EST. AML

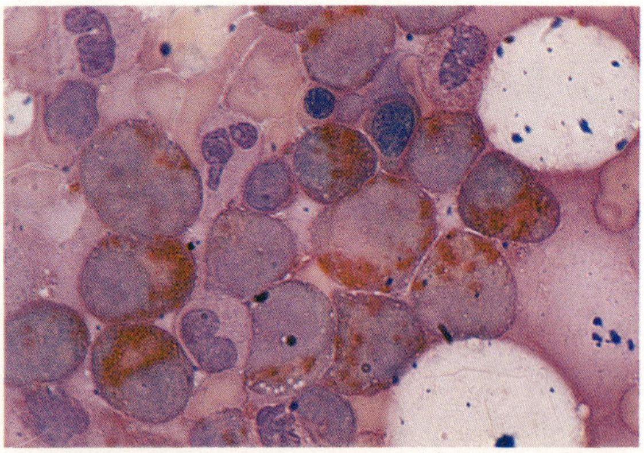

B

NON SP. EST. H.L.

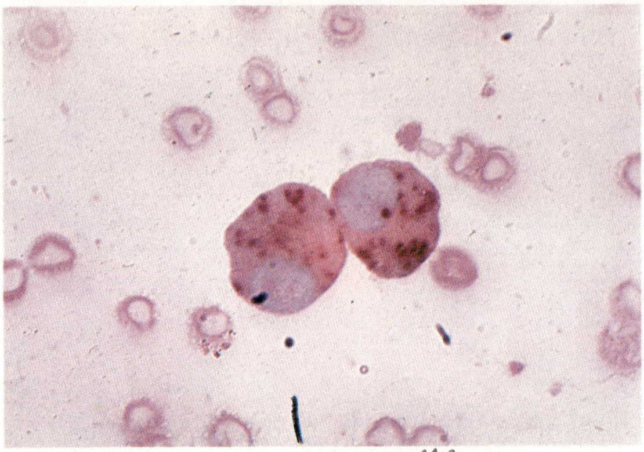

C

NONSP. EST. MYELOMA

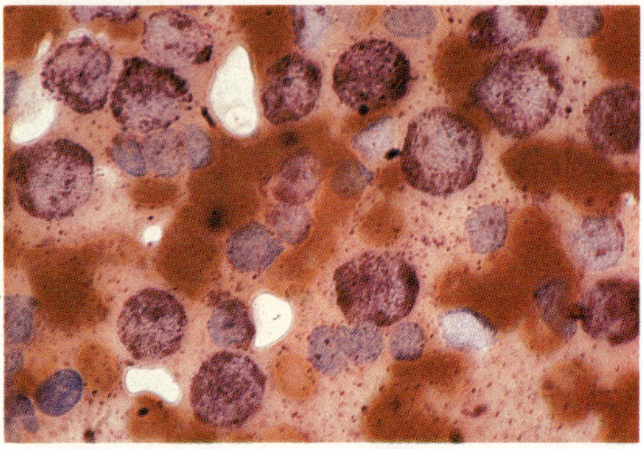

D

ACID PHOSPH. HAIRY CELL L.

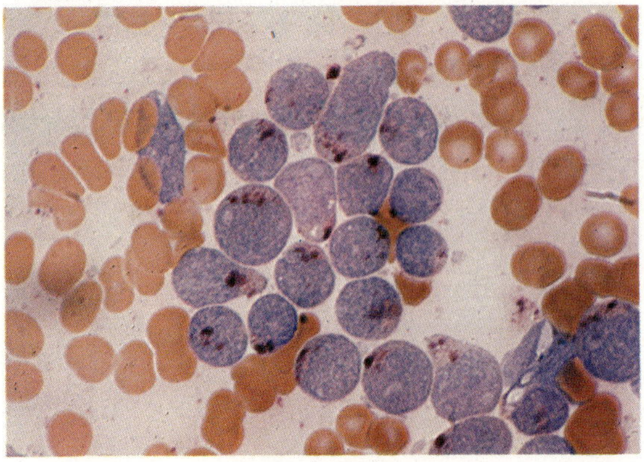

E

ACID PHOSP. T-CELL LEUK.

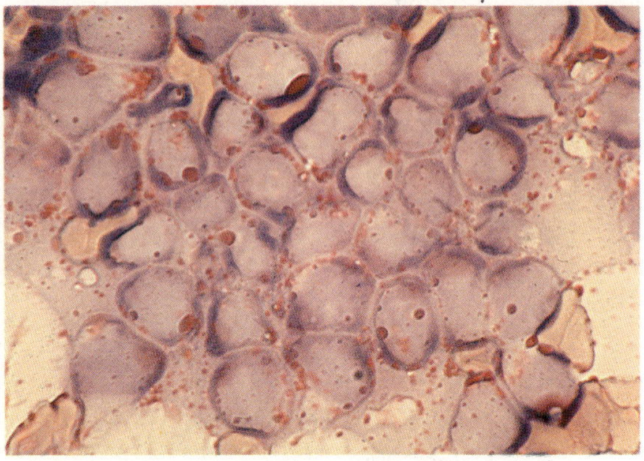

F

OIL RED 'O' BURKITT

Plate 6 (*A*) Nonspecific esterase reaction, leukemic monocytes, showing intense diffuse activity in the cytoplasm (alpha naphthyl acetate and hexazotized pararosaniline). (*B*) Nonspecific esterase reaction, leukemic histiocytic lymphoma cells, showing unipolar localization of the enzyme in a chunklike pattern. (*C*) Nonspecific esterase reaction, myeloma cells, showing large aggregates of reaction product representing distribution of enzyme activity. (*D*) Tartrate-resistant acid phosphatase activity, hairy cells, in hairy cell leukemia (naphthol AS-TR phosphoric acid and fast corinth V). (*E*) Acid phosphatase activity in unipolar distribution in T-cell convoluted leukemic lymphoblasts. (*F*) Oil red O reaction, Burkitt's leukemia cells, showing "necklace" of orange-stained vacuoles.

1

Preleukemic Disorders

For some time, various authors have recognized a group of primary acquired panmyelotic disorders that have defied categorization into the traditional divisions of acute and chronic leukemia. Described at first as examples of leukemias with unusually protracted clinical courses,[117,119,665,1096,1101] this heterogeneous group of panmyelopathies soon attracted the interest of a wider group of investigators interested in patterns of leukemic evolution. Although these disorders seemed to precede the development of acute leukemia in some patients, they were not the precursors of leukemia in all cases, suggesting the variability of the disease process even when similar morphologic criteria were met.

The story of these primary acquired panmyelopathies has its origin in the Italian school of hematology. About the turn of the century, Copelli,[213] Luzzato,[621] and subsequently Di Guglielmo[249–253] described and identified a group of predominantly erythroblastic disorders that seemed to have an unusually aggressive clinical course and to differ from what was at the time believed to be typical pernicious anemia. At about the same period, others, like von Leube,[1070] Parkes-Weber,[767] and Treadgold[1037] described cases of "leukanemia" in which blood specimens contained a mixture of erythroblasts and myeloblasts and resembled those described later by Di Guglielmo[249] as erythroleukemia. These early descriptions of acute erythremic myelosis and of erythroleukemia form the basis of the Di Guglielmo syndrome,[223–227] a group of predominantly erythroblastic disorders comprised of acute and chronic erythremic myelosis and erythroleukemia. In a variable number of patients with these disorders, acute myeloblastic or monocytic leukemia develops.

Currently it is believed that most if not all patients with acute myeloblastic or monocytic leukemia have passed through an erythremic phase, probably acute erythremic myelosis followed by erythroleukemia, prior to acute leukemic evolution. In most instances, this erythroblastic phase is unnoticed clinically and occurs during the prodromal phases of acute leukemia. Only those unusual patients who experience consequences of pancytopenia like hemorrhage or infection come to medical attention and undergo hematologic evaluation and subsequent diagnosis.

Viewed in this sense, the disorders included in the Di Guglielmo syndrome are the initial or erythroblastic expression of a disorder that at the outset is myeloblastic. In keeping with this concept, erythroblasts from patients with acute erythremic myelosis and erythroleukemia often contain properties of granulocytes, such as myeloperoxidase and specific esterase activity.[504,505] These cytochemical findings suggest that the abnormal erythroblasts have differentiated along two cell lines, erythroblastic and granulocytic. Only gradually, and at speeds which vary between patients, myeloblastosis or monocytosis of the marrow supervenes. We do not yet understand why the initial morphologic expression of acute leukemia in these persons is a disorder of erythropoiesis.

Various techniques have been used to predict the likelihood and the imminence of leukemic evolution in patients with the Di Guglielmo syndrome, but results have been inconsistent. Delayed colony formation in soft agar or liquid cultures, the occurrence of ringed sideroblasts, PAS-positive erythroblasts, the appearance of the pseudo-Pelger-Huët anomaly, and elevated serum and urinary lysozyme levels have all been suggested as possible unfavorable prognostic signs.[505,552]

In addition to the predominantly erythroblastic group of primary acquired panmyelotic disorders that precede the development of acute leukemia, another group has been identified over the past several decades. In this group, bone marrows are characterized predominantly by granulocytic hyperplasia and, in most instances, up to 20% myeloblasts or monocytes. Whether these predominantly myeloblastic or monocytic disorders had been preceded in their evolution by one of the disorders in the Di Guglielmo syndrome is unknown. Given different names by various authors, these predominantly granulocytic or monocytic preleukemic panmyelopathies have been called preleukemia,[98,318,596,597] smoldering acute leukemia or oligoblastic leukemia,[273,459,716,849] and subacute myelomonocytic leukemia.[69,169,946,1125,1126.]

In these disorders, the panmyelotic character of the bone marrow is often more apparent than in the preleukemic erythroblastic disorders. Bone marrows from patients with preleukemia frequently contain metamyelocytes with unusual nuclear configurations such as clubbing, doughnut-hole, and focal constrictions. Plasma cells and eosinophils are often increased, erythroblasts show megaloblastoid chromatin pattern, and megakaryocytes are often small with few or no nuclear lobulations, resembling micromegakaryocytes found in some cases of acute myeloblastic leukemia. Similar morphologic abnormalities reflecting disorders in all cell lines occur in bone marrows from patients with smoldering acute leukemia and subacute myelomonocytic leukemia.

Most of the disorders in this category are typified by slowly but steadily increasing myeloblastosis or monocytosis of the bone marrow and, in a variable number, by evolution into acute myeloblastic or monocytic leukemia. Interestingly, the Auer rod, which for over three quarters of a century has been a reliable marker for leukemic myeloblasts, is almost never identified in myeloblasts or monocytes from patients with the preleukemic disorders of granulopoiesis.[273,459,596,597,716,849] Occasionally, large, splinter-shaped granules resembling those in leukemic promyelocytes are found.

Cytochemically, these preleukemic disorders of granulopoiesis have few features that have been implicated as prognostic indicators of leukemic evolution. Although a myeloperoxidase-deficient variant of preleukemia has been identified,[162,163] it is currently unknown whether this enzyme deficiency confers a greater risk of acute leukemic evolution than otherwise.

Conceptual questions have been raised as to whether these preleukemic disorders are preleukemic in the literal sense or whether they represent acute leukemia in a slowly developing or indolent form. Most of the evidence to date suggests that the preleukemic disorders of granulopoiesis, like preleukemia and smoldering acute leukemia, represent leukemia at the outset but with a different degree of clinical aggressiveness and a different proportion of leukemic blasts than that found in florid acute leukemia. Substantiating the close relationships between preleukemia and leukemia, recent studies have shown that bone marrow from patients with preleukemia has growth characteristics *in vitro* that are identical to those in acute leukemia[379,945] and that cytogenetic abnormalities present during the preleukemic phase persist when evolution into acute leukemia occurs.[797] As with preleukemic disorders of erythropoiesis, the occurrence of an abrupt cytogenetic change and delayed growth of marrow in tissue culture are thought to indicate a high risk of leukemic transformation. However, high variability in the tempo of the leukemic process from case to case suggests that these parameters may not be as reliable as once thought.

In all probability, the conditions included under the broad term *preleukemic disorders* represent a continuum ranging from predominantly erythroblastic at one end to predominantly myeloblastic or monocytic at the other. Their close relationships can be documented in the evolutionary patterns identified in individual patients. For example, a case of acute or chronic erythremic myelosis may evolve into a disorder with marrow findings typical of smoldering acute leukemia. Similarly, a case of "preleukemia" may show evolution into erythroleukemia and then into acute myeloblastic leukemia. Usually, the general evolutionary direction is toward increasing myeloblastosis or monocytosis rather than increasing erythroblastosis. With further identification of morphologic, cytochemical, cytogenetic, and proliferative abnormalities of preleukemic panmyelopathies, it may be possible to identify these disorders with greater precision and with greater predictive value than at present.

Preleukemic Erythroblastic Disorders

Preleukemic erythroblastic disorders have been called the *Di Guglielmo syndrome.*[223,227] This includes acute and chronic erythremic myelosis and erythroleukemia.[223–227,249–258] Disorders within this spectrum are believed by some authors to precede most if not all cases of acute nonlymphoblastic leukemia[223–227] and may be present during the prodromal phases of acute leukemia prior to its clinical recognition.[223–227]

Acute Erythremic Myelosis

This rarely identified erythroblastic disorder is characterized by anemia, hypercellularity of the bone marrow and intense erythroblastic hyperplasia, and infiltration of visceral organs by atypical erythroblasts. Fever, hepatomegaly, and splenomegaly occur frequently. In most patients, acute erythremic myelosis is undetected. It may occur in most and perhaps all patients with acute nonlymphoblastic leukemia as a brief transitory phase preceding the evolution of acute leukemia. Rarely, patients are diagnosed as having acute erythremic myelosis as a part of evaluation of obscure fever, macrocytic anemia, or normoblastemia. Some patients with acute erthremic myelosis succumb to the consequences of pancytopenia before obvious leukemic evolution.[131,133,213,252,562,566,621,855,935,998]

Peripheral blood samples from patients with acute erythremic myelosis usually show pancytopenia with marked anisocytosis and poikilocytosis of erythrocytes. Giant macrocytes are seen often. Some erythrocytes demonstrate coarse basophilic stippling. Occasionally, atypical-appearing monocytes are present. Normoblasts appear in blood samples of most patients with acute erythremic myelosis. Some of the normoblasts have megaloblastoid nuclear chromatin patterns; others are multinuclear.

Bone marrow samples from patients with acute erythremic myelosis are hypercellular. Erythroblasts are increased markedly in number. Most of the erythroblasts are young and are recognizable as proerythroblasts (Figs. 1-1, 1-2). Gigantism of erythroblasts is seen often.[917,934] Some giant erythroblasts contain several atypical-appearing nuclear lobulations with megaloblastoid nuclear chromatin pattern (Fig. 1-3) and prominent nucleoli with perinucleolar condensation of chromatin (Fig. 1-4). Often erythroblasts are multinuclear (Figs. 1-5, 1-6). Nuclei vary in size and shape and show megaloblastic or megaloblastoid nuclear chromatin patterns. Other erythroblasts have prominent, multiple cytoplasmic vacuoles. Occasionally, gigantic erythrocytes containing Howell-Jolly bodies, Pappenheimer bodies, and Cabot's rings are found (Fig. 1-7). Some show nuclear aberrations such as accelerated pyknosis or abnormal karyorrhexis. Megakaryocytes and granulocytic cells are greatly reduced in number. Macrophages and erythroblastic "islands" are seen often.

Cytochemically, erythroblasts from patients with acute erythremic myelosis contain abundant glycogen demonstrable with the PAS (periodic acid-Schiff) reagent.[29,566,567,669] Glycogen may be detectable in an intense but diffuse staining pattern in some erythroblasts, and in a prominent, blocklike pattern in other erythroblasts (Fig. 1-8). Prussian blue stain for iron reveals numerous siderotic granules, representing ferric iron within the lamellae and cristae of mitochondria. In those erythroblasts that are unusually bizarre and multinuclear, siderotic granules are often numerous and coarse-appearing. Activity of nonspecific esterase is intense in these erythroblasts and is moderately resistant to inhibition by sodium fluoride.

Activities of specific esterase and peroxidase, both of which are properties of cells of granulocytic origin, are detectable in erythroblasts in acute erythremic myelosis. Demonstration of granulocytic properties in a cell that is morphologically erythroblastic and capable of synthesizing hemoglobin is in accord with the concept of the Di Guglielmo syndrome. In

(text continued on page 10)

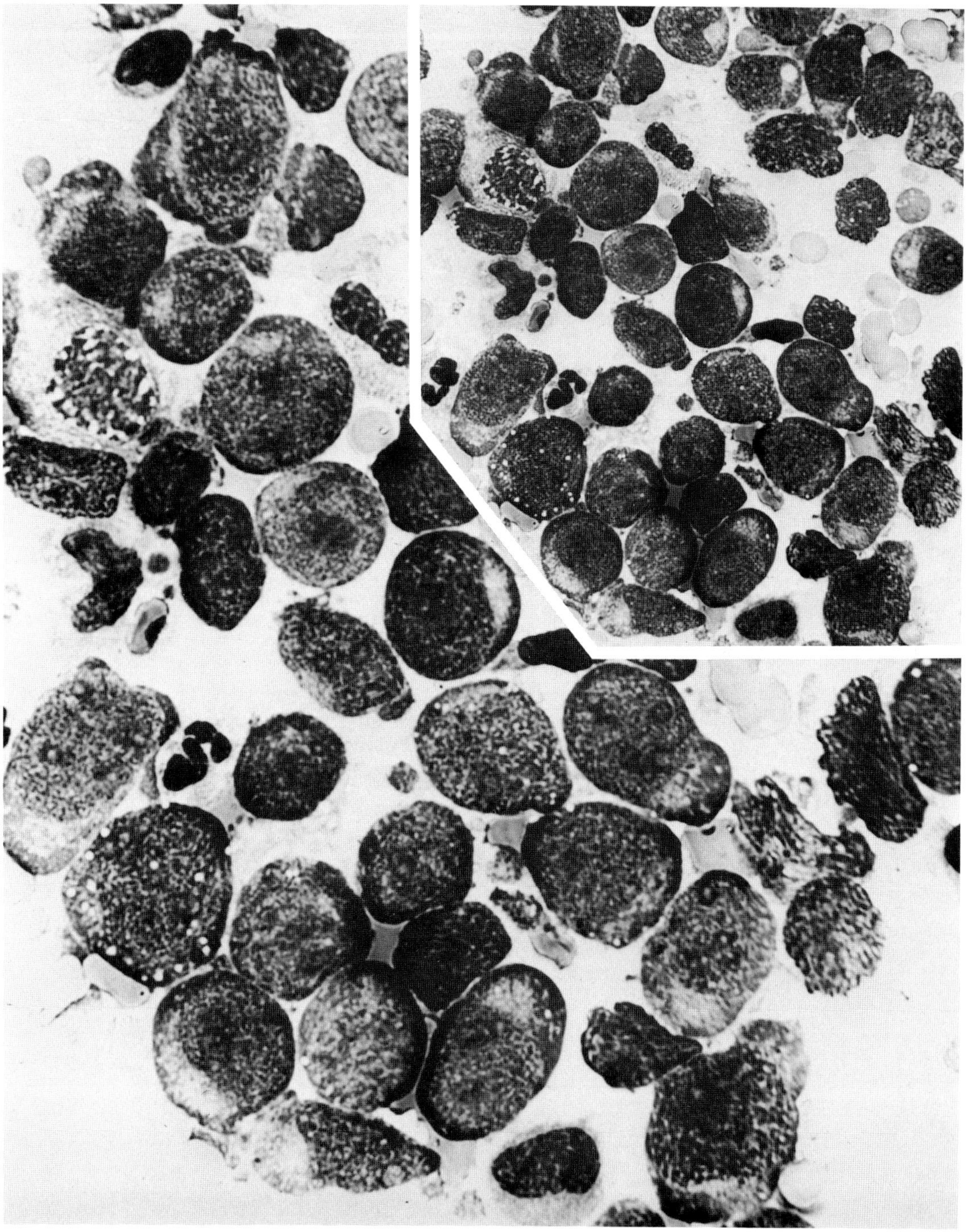

Figure 1-1. Bone marrow in acute erythremic myelosis, showing increased numbers of proerythroblasts, some of which contain cytoplasmic vacuoles.

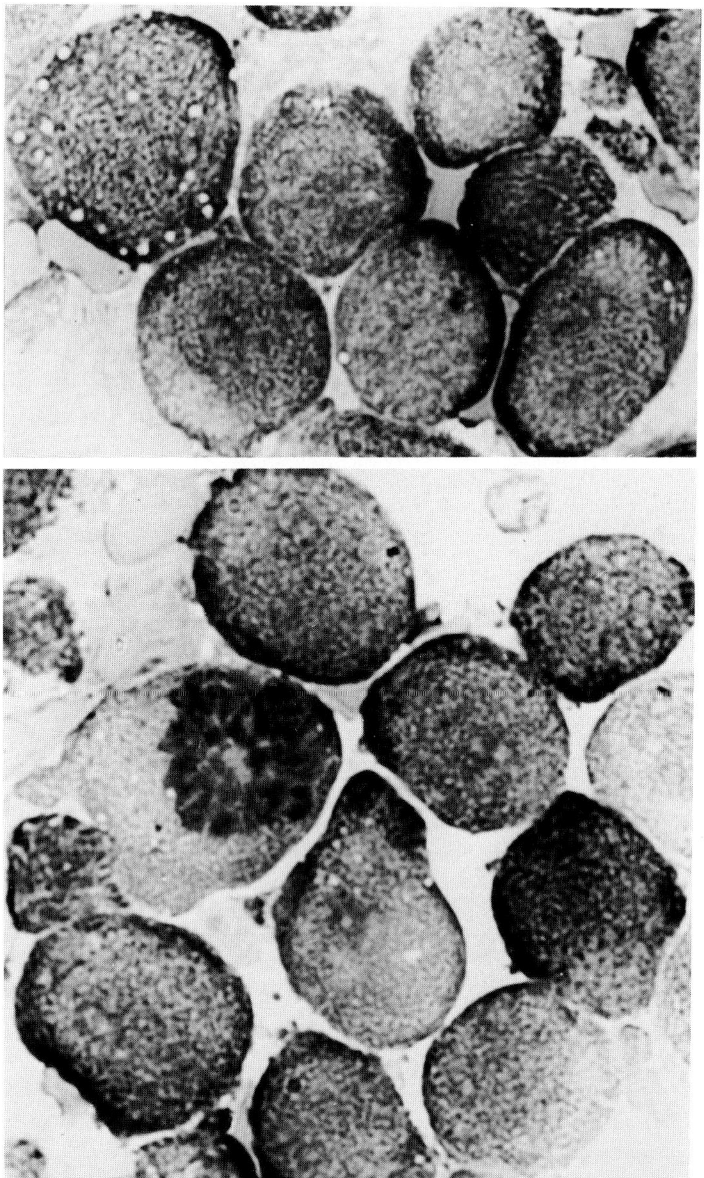

Figure 1-2. Proerythroblasts in acute erythremic myelosis. Proerythroblasts with cytoplasmic vacuoles **(top)**. Proerythroblasts and aberrant mitotic figures **(bottom).**

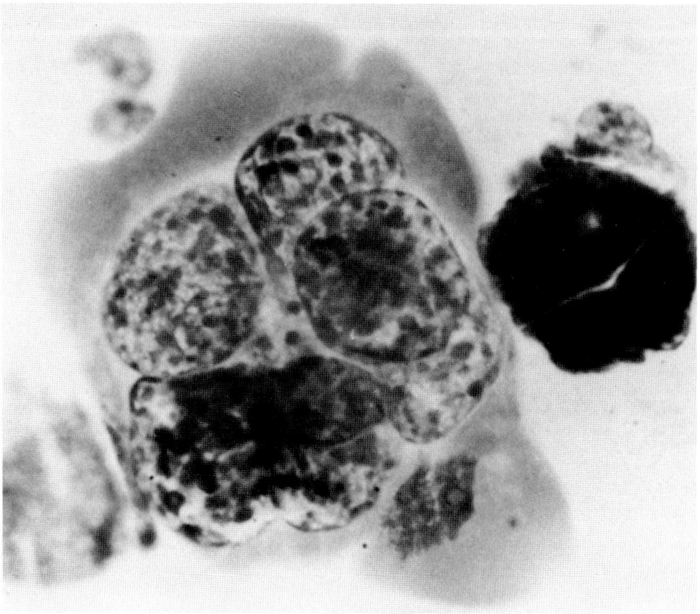

Figure 1-3. Gigantic megaloblastoid intermediate macronormoblast in acute erythremic myelosis, demonstrating multinuclearity and multiple nuclear lobulations.

Figure 1-4. Proerythroblasts in acute erythremic myelosis, showing prominent nucleoli with perinucleolar condensations of chromatin as well as multinuclearity and aberrant mitotic figures.

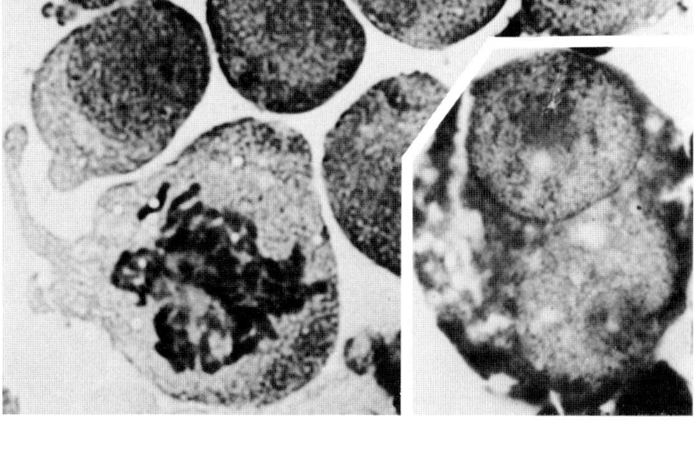

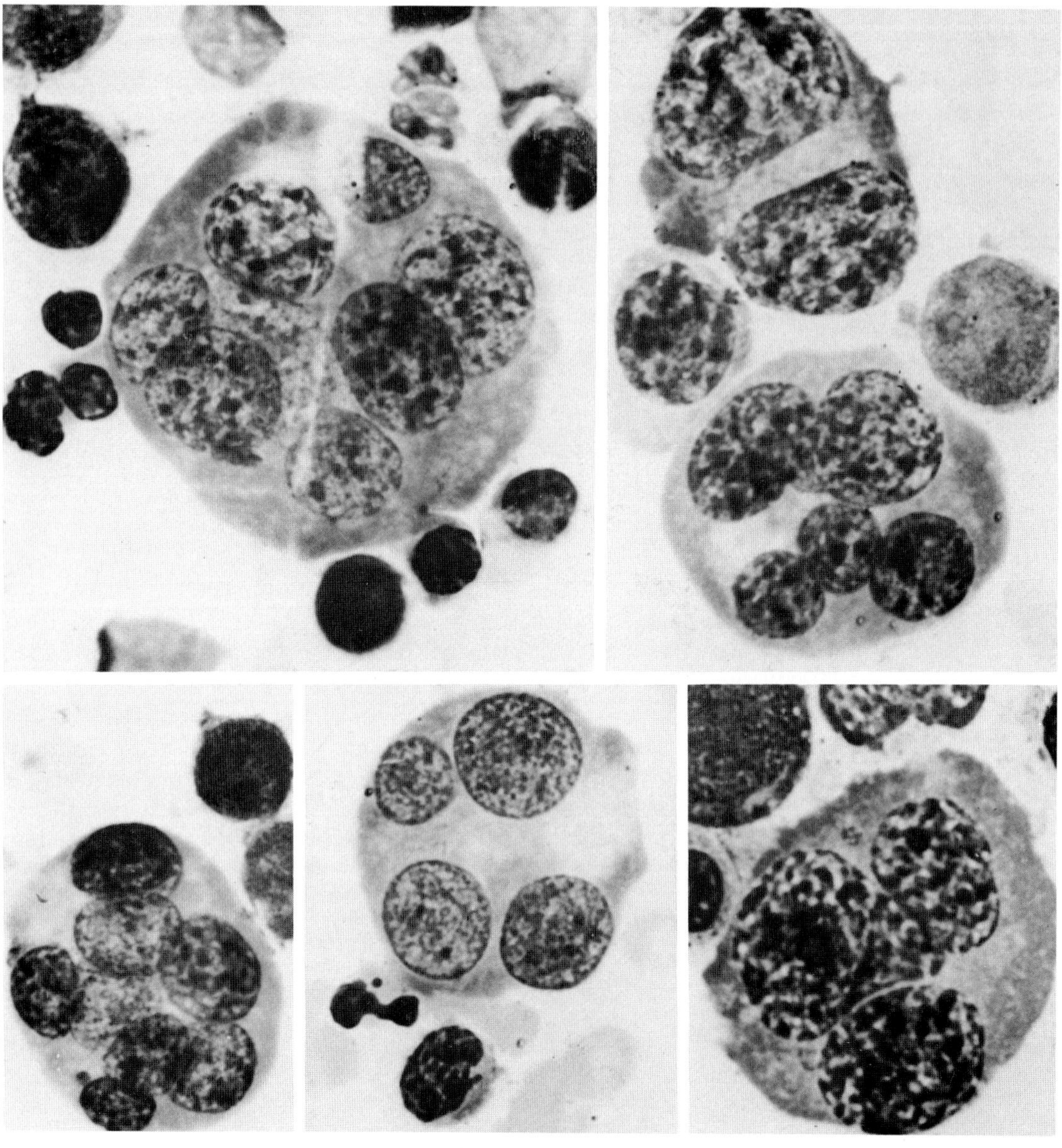

Figure 1-5. Giant multinuclear erythroblasts in acute erythremic myelosis, showing nuclei of varying sizes and shapes. Some of the nuclei demonstrate megaloblastoid nuclear chromatin patterns, whereas others show delicate nuclear chromatin patterns found in megaloblasts.

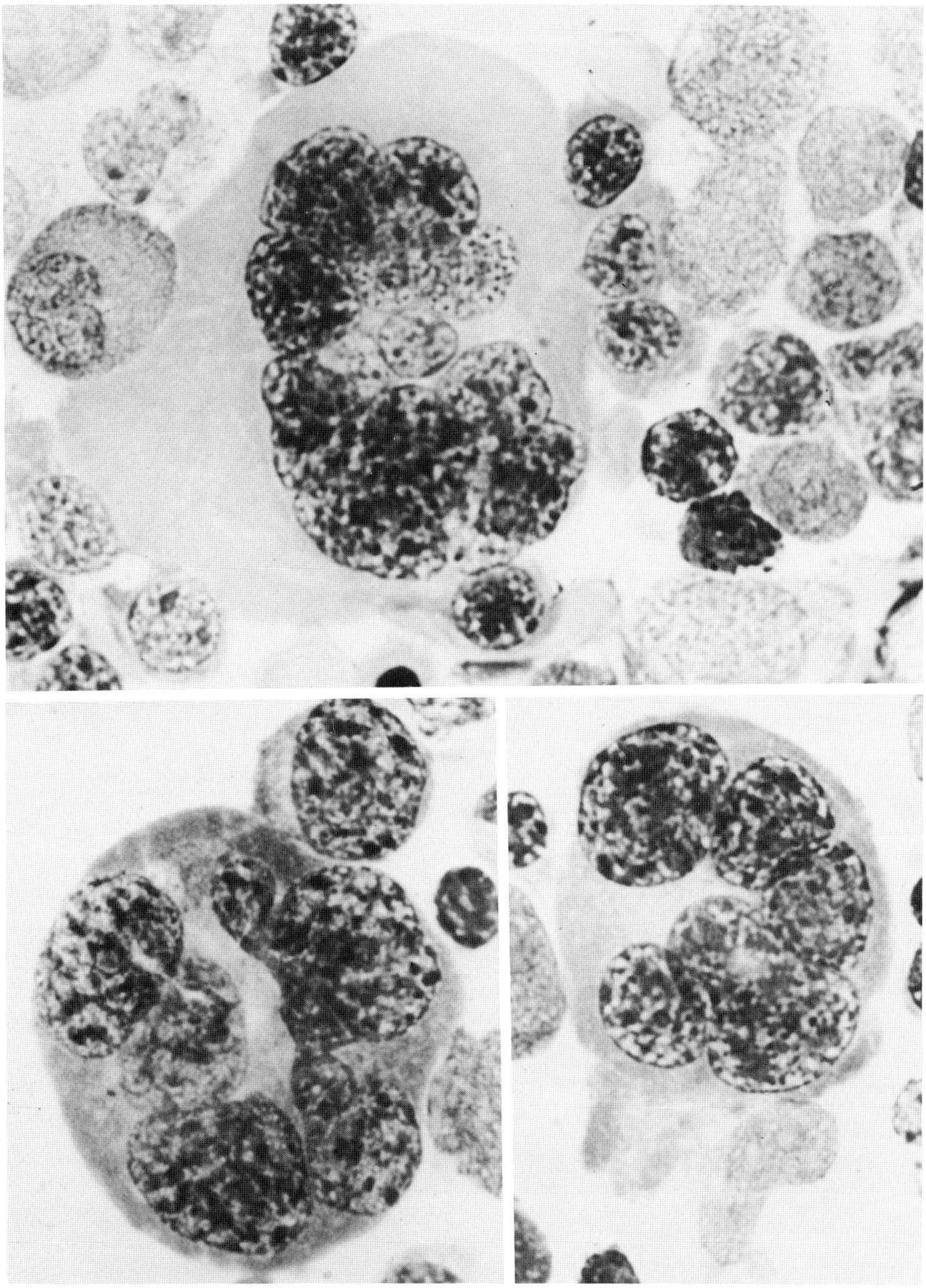

Giant multinuclear erythroblasts in acute erythremic myelosis. **Figure 1-6.**

this syndrome, a proliferation of abnormal erythroblasts may be the initial or erythroblastic expression of a disorder that is myeloblastic at the outset.[223,227]

H & E sections of biopsies from patients with acute erythremic myelosis show hypercellularity and increased numbers of cells with pale vesicular nuclei and prominent nucleoli. These probably represent immature erythroblasts, and on the basis of the biopsy alone they cannot be distinguished with certainty from malignant cells in histiocytic lymphoma. Few megakaryocytes or granulocytic cells are seen.

Chronic Erythremic Myelosis

Chronic erythremic myelosis is a disorder characterized by macrocytic anemia, usually refractory to present hematinics used alone or in combination, and by intense megaloblastoid erythroid hyperplasia of the bone marrow. Pancytopenia, monocytosis of the peripheral blood, normoblastemia, and splenomegaly occur often. Pathophysiologically, chronic erythremic myelosis is an example of ineffective erythropoiesis, in which erythroblasts proliferate in the bone marrow but die there prematurely and do not develop into circulating erythrocytes.

In some patients, this predominantly erythroblastic disorder precedes the development of acute leukemia. In others, obvious leukemic evolution does not seem to occur.[62,111,282,302,411,458,552,693,695,862,928,1036,1069] Some patients with chronic erythremic myelosis have a history of an antecedent hematologic disorder, such as polycythemia vera.[283] Evolution into acute leukemia may occur insidiously. It can become apparent by a sudden need for a blood transfusion in a patient who was stable previously, by steadily developing monocytosis of abnormal monocytes, or by worsening pancytopenia attributable to increasing myeloblastosis of the bone marrow.

Like erythroleukemia, some cases of chronic erythremic myelosis have been misdiagnosed as pernicious anemia and treated accordingly with vitamin B_{12}. Only after lack of response to this and other hematinics does the refractory character of the anemia become apparent. Usually it is possible to distinguish between chronic erythremic myelosis and untreated pernicious anemia or folate deficiency on the basis of assays of vitamin B_{12}, folate, and the morphologic properties of erythroblasts in the bone marrow. In deficiency of vitamin B_{12} or folate, serum levels of the vitamin are low in untreated and uncomplicated cases, erythropoiesis is megaloblastic, and giant atypical metamyelocytes are found.[506] In chronic erythremic myelosis, levels of vitamin B_{12} and folate may be unusually high, erythropoiesis is normoblastic but more often megaloblastoid, and giant atypical-appearing metamyelocytes are seen rarely or not at all.

Most cases of chronic erythremic myelosis are characterized by pancytopenia, macrocytosis of erythrocytes, and monocytosis.[84] In some patients, neutrophils in the peripheral blood show the pseudo-Pelger-Huët anomaly of hyposegmentation.[263,569] It has been suggested that this morphologic abnormality signifies an increased risk of leukemic evolution.

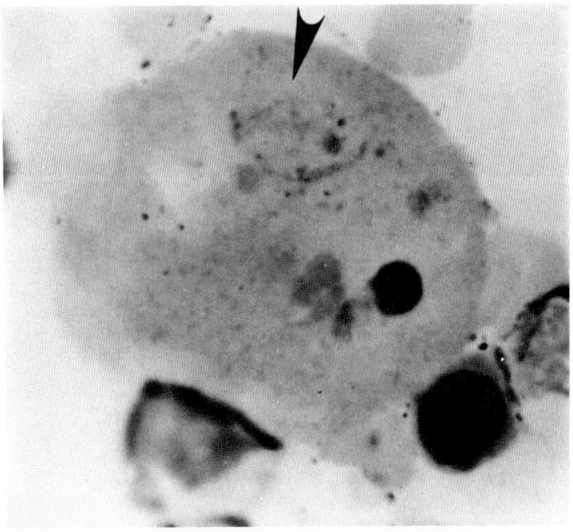

Figure 1-7. Gigantic erythrocyte in acute erythremic myelosis, showing prominent Howell-Jolly body and Cabot's ring (**arrow**).

Figure 1-8. Erythroblasts in acute erythremic myelosis, stained with the PAS reagent for glycogen. Cytoplasm of these cells stains intensely and diffusely.

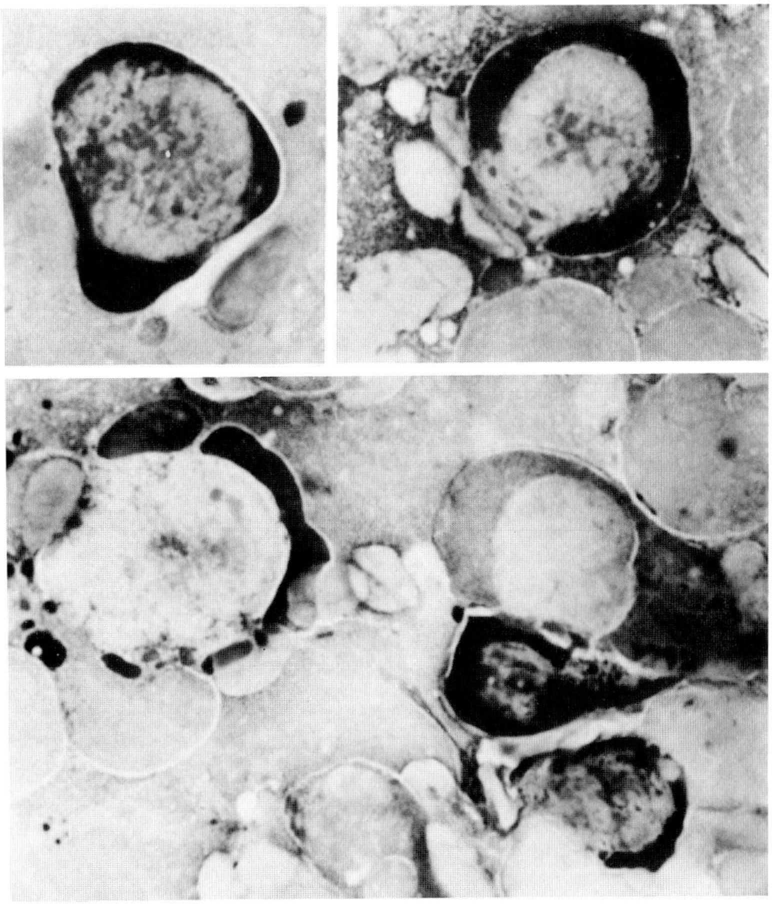

Macrocytosis of erythrocytes can be extreme, with mean corpuscular volume of erythrocytes as high as 130-140 cubic μ. In some erythrocytes, Pappenheimer bodies and Cabot's rings are found.

Morphologically, bone marrow specimens from patients with chronic erythremic myelosis are hypercellular.[84,193,519,594,928,1012,1062,1065,1066] Megakaryocytes are increased in number in some patients and normal or reduced in number in others. Sometimes these megakaryocytes appear normal, but more often they are micromegakaryocytes with a single, unlobulated nucleus, delicate nuclear chromatin pattern, and hypergranular cytoplasm.[12,154] Erythropoiesis is intense, and marrows from patients with chronic erythremic myelosis may contain up to 80% to 90% erythroblasts (Fig. 1-9 and Plate 1*A*). Proerythroblasts are increased in number, and some may be binuclear (Fig. 1-10). Clusters of macronormoblasts at a comparable maturational age occur frequently (Figs. 1-11 to 1-13). The maturational sequence of erythroblasts is abnormal, with increased numbers of erythroblasts at the early megaloblastoid intermediate macronormoblasts stage of development. At this stage, in which beginnings of hemoglobination can be seen in the form of cytoplasmic polychromasia, erythroblasts are thought to be in the G_2 phase of cell proliferation and to be unusually susceptible to intramedullary cell death.

Some erythroblasts are multinuclear (Figs. 1-10, 1-14) and may contain cytoplasmic vacuoles (Figs. 1-14, 1-15). Increased numbers of mitotic figures are seen in erythroblasts (Fig. 1-16). In some multinuclear erythroblasts, one nucleus shows megaloblastoid nuclear chromatin pattern, whereas an adjacent nucleus in the same cell demonstrates megaloblastoid chromatin pattern or dense nuclear chromatin, suggesting degeneration. In these atypical erythroblasts, asynchronism between cytoplasmic and nuclear maturation is often seen (Fig. 1-17). These nuclear and cytoplasmic abnormalities probably indicate disturbances of erythroblastic development and maturation in chronic erythremic myelosis.

Megaloblastoid nuclear chromatin patterns become recognizable first at the early megaloblastoid intermediate macronormoblast stage of development. At this stage, megaloblastoid chromatin patterns are defined most clearly and distinguishable from normoblastic and megaloblastic nuclear chromatin patterns.[499] In megaloblastoid erythropoiesis, nuclear chromatin strands appear disorganized and lack the usual circumferential pattern of euchromatin found in normoblasts. Furthermore, megaloblastoid erythropoiesis is characterized by large, blocklike aggregates of nuclear chromatin that vary in size and shape and are connected to each other by one or more thin strands of chromatin. These nuclear chromatin aggregates and their connections impart an open appearance to the nucleus.

Several morphologic abnormalities provide evidence for intramedullary cell death. These include accelerated pyknosis and abnormal karyorrhexis in erythroblasts, as well as increased numbers of macrophages (phagocytes) containing degenerating erythroblasts and mature

(*text continued on page 22*)

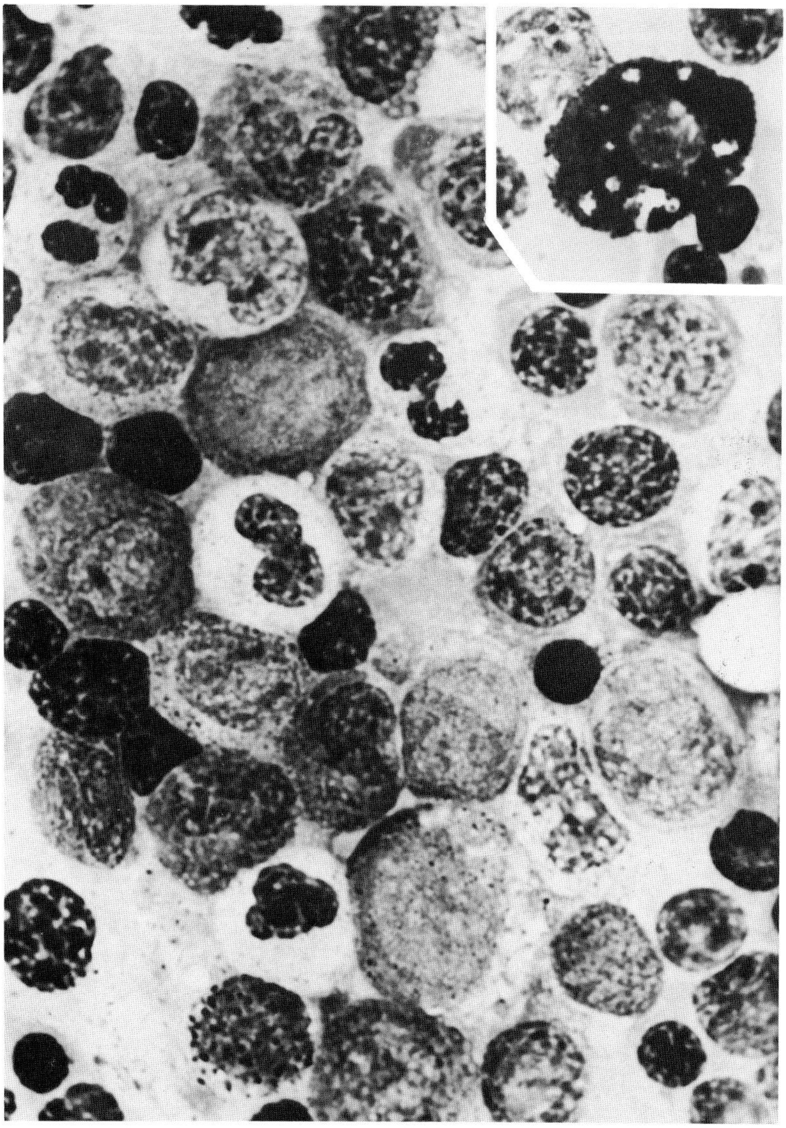

Figure 1-9. Bone marrow in chronic erythremic myelosis, showing increased numbers of promyelocytes and eosinophils as well as increased numbers of megaloblastoid intermediate macronormoblasts. Inset depicts a mast cell.

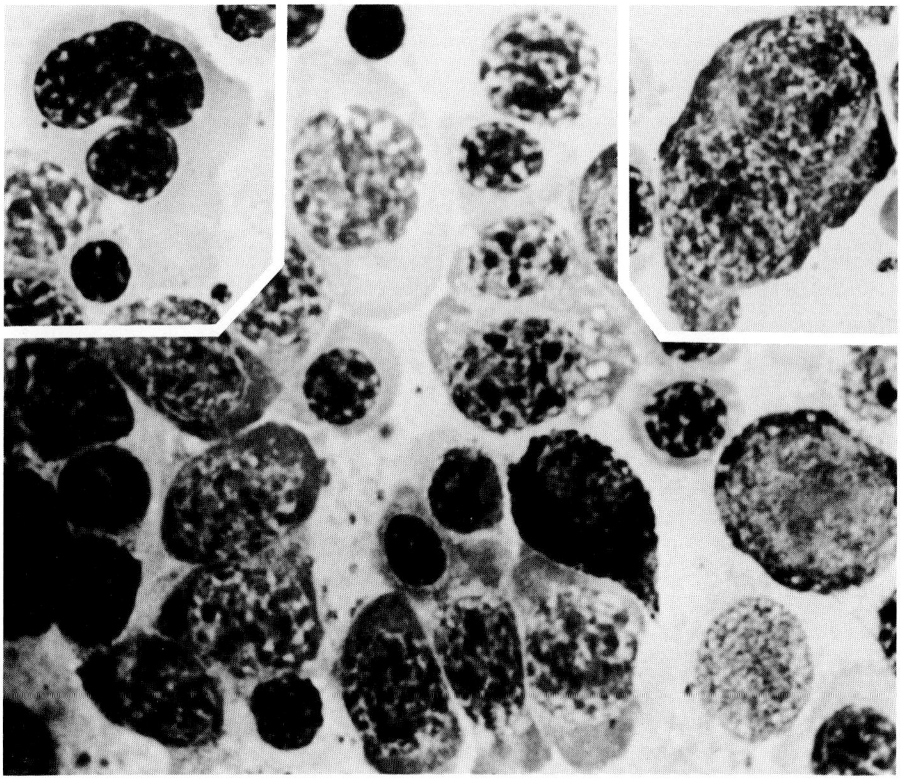

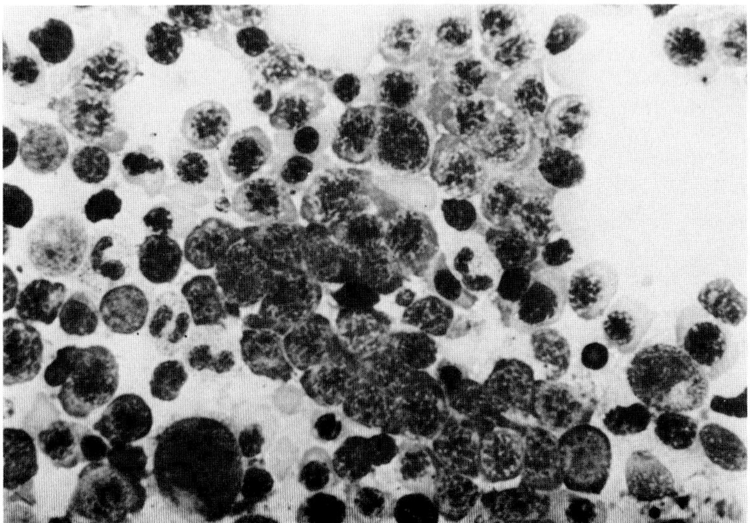

Figure 1-10. ← Bone marrow in chronic erythremic myelosis, showing a marked increase in megaloblastoid intermediate macronormoblasts with prominent, blocklike aggregates of nuclear chromatin, a mast cell (**center**), and multinuclear erythroblasts (**insets**).

Figure 1-11. ← Bone marrow in chronic erythremic myelosis, illustrating clumps of proerythroblasts (**center**) and megaloblastoid intermediate macronormoblasts, in a low-power view.

Figure 1-12. Megaloblastoid intermediate macronormoblasts in chronic erythremic myelosis. Inset shows degenerating erythroblast with darkly stained nucleus and coarse cytoplasmic basophilic stippling.

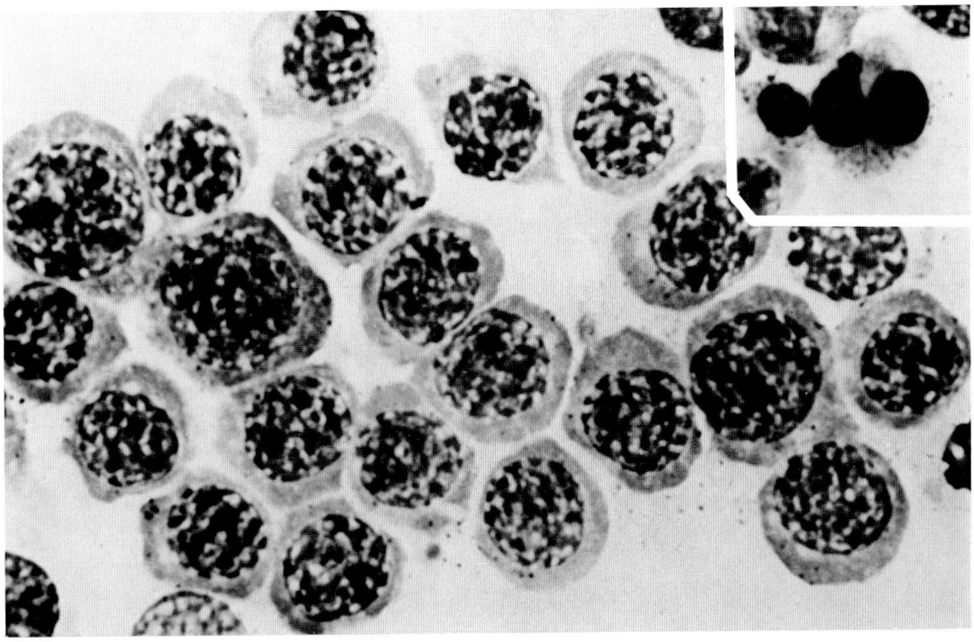

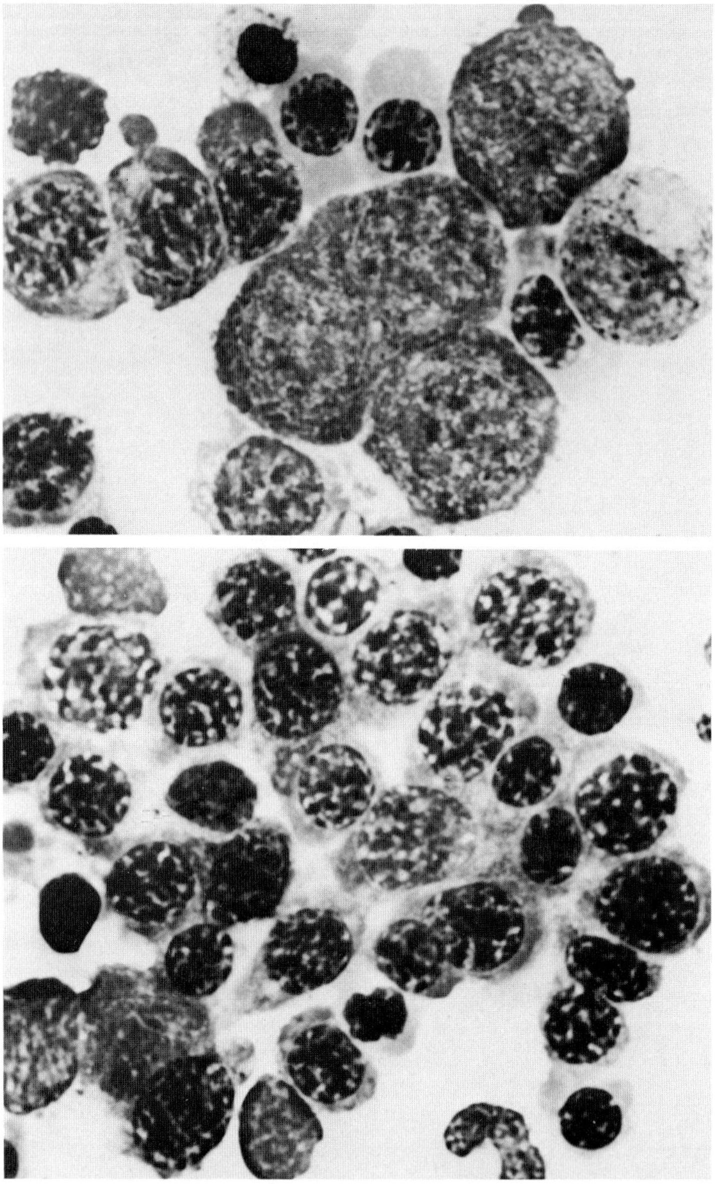

Figure 1-13. Bone marrow in chronic erythremic myelosis. Cluster of early intermediate normoblasts and proerythroblasts **(top)**. Cluster of megaloblastoid intermediate macronormoblasts **(bottom)**.

Figure 1-14. Bone marrow in chronic erythremic myelosis. Large, multinuclear erythroblasts and increased numbers of promyelocytes **(top).** Increased numbers of proerythroblasts, myelocytes, and erythroblasts with cytoplasmic vacuoles **(bottom, inset).**

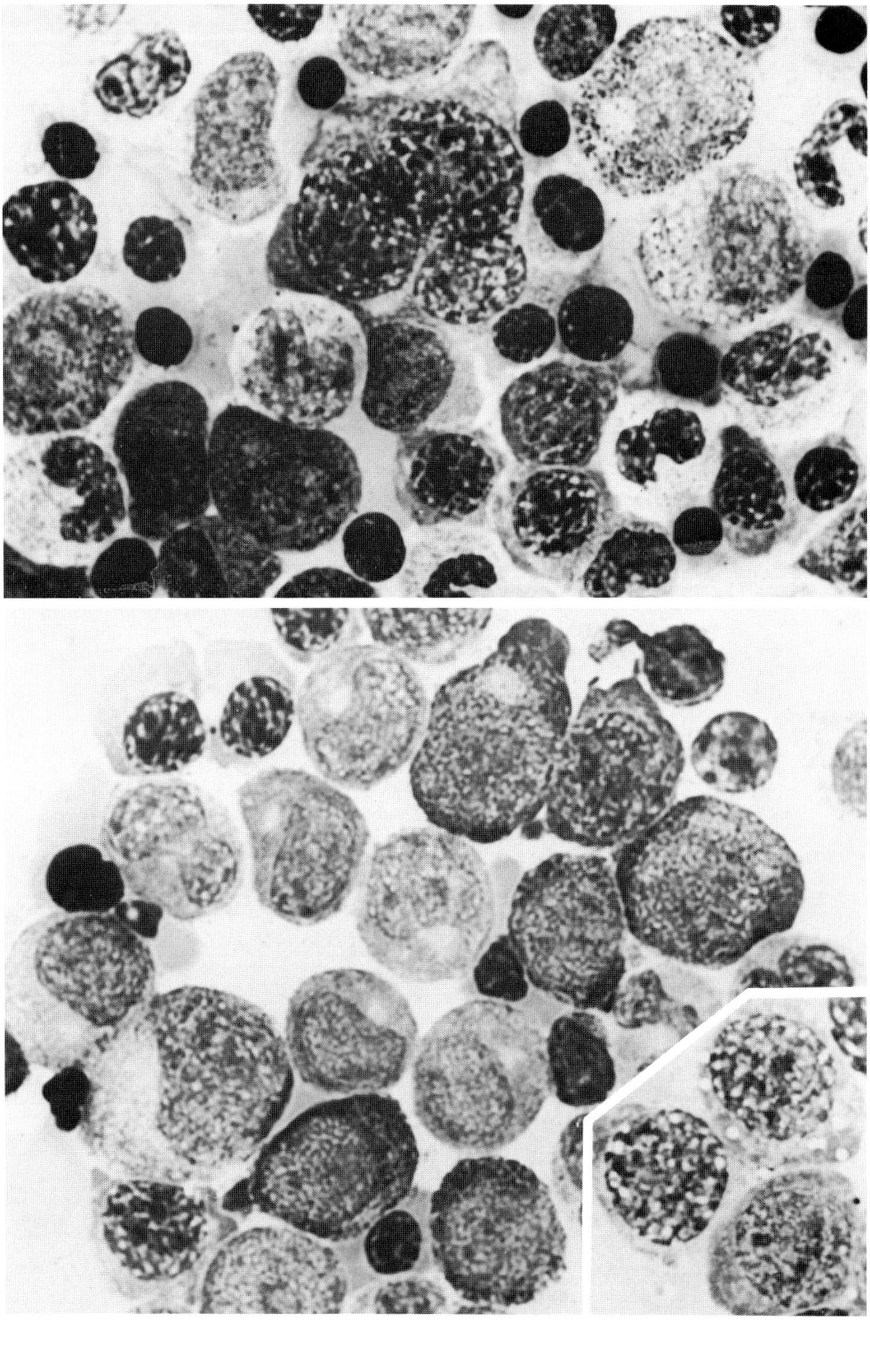

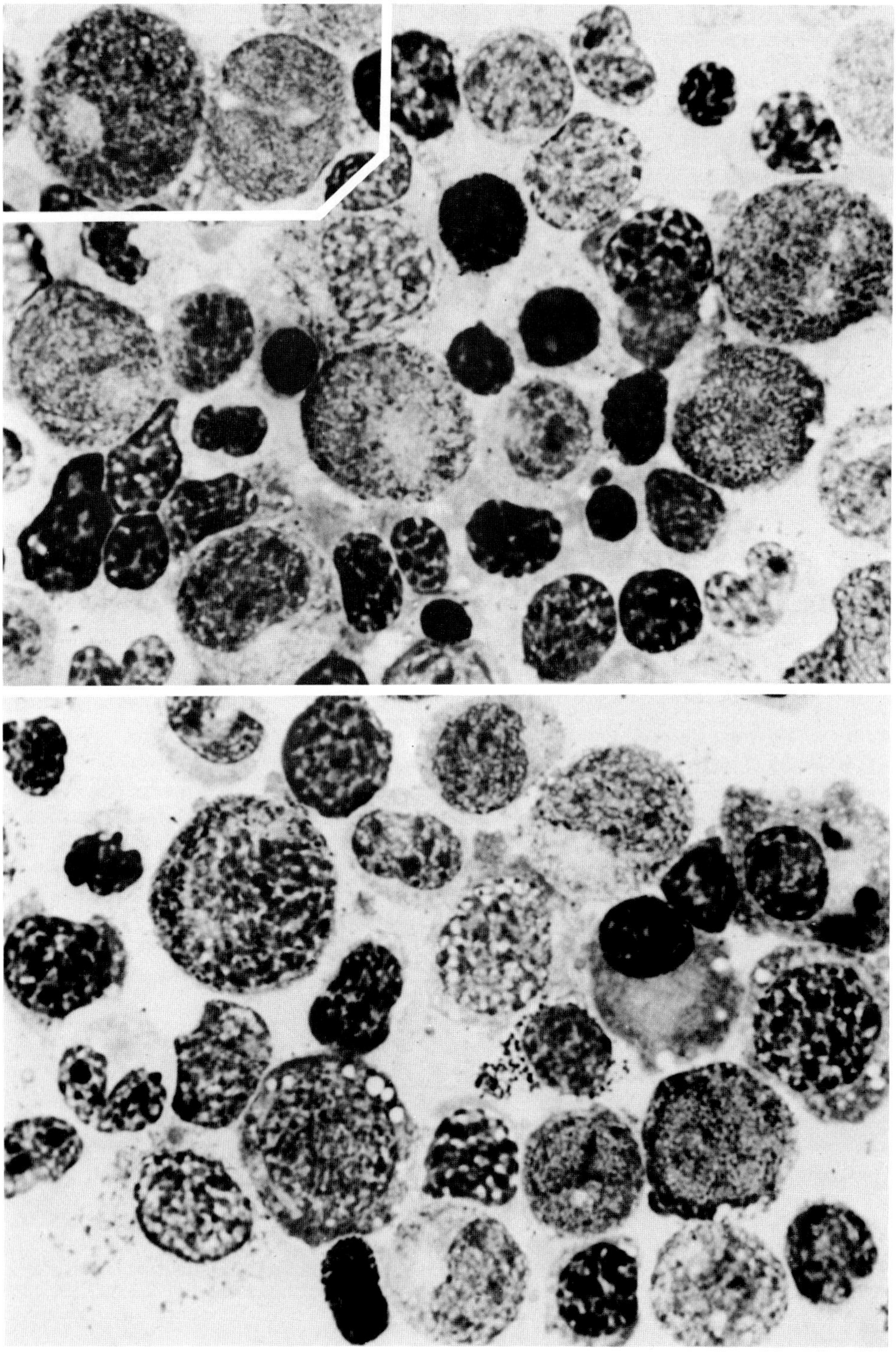

Figure 1-15. ← Bone marrow in chronic erythremic myelosis. Increased numbers of promyelocytes and myeloctyes (**top**) as well as atypical megakaryocytes (**inset**). Proerythroblasts with cytoplasmic vacuoles (**bottom**).

Figure 1-16. Bone marrow in chronic erythremic myelosis. Atypical mitotic figure in erythroblasts (**top**). Megaloblastoid intermediate macronormoblasts containing cytoplasmic vacuoles (**bottom**).

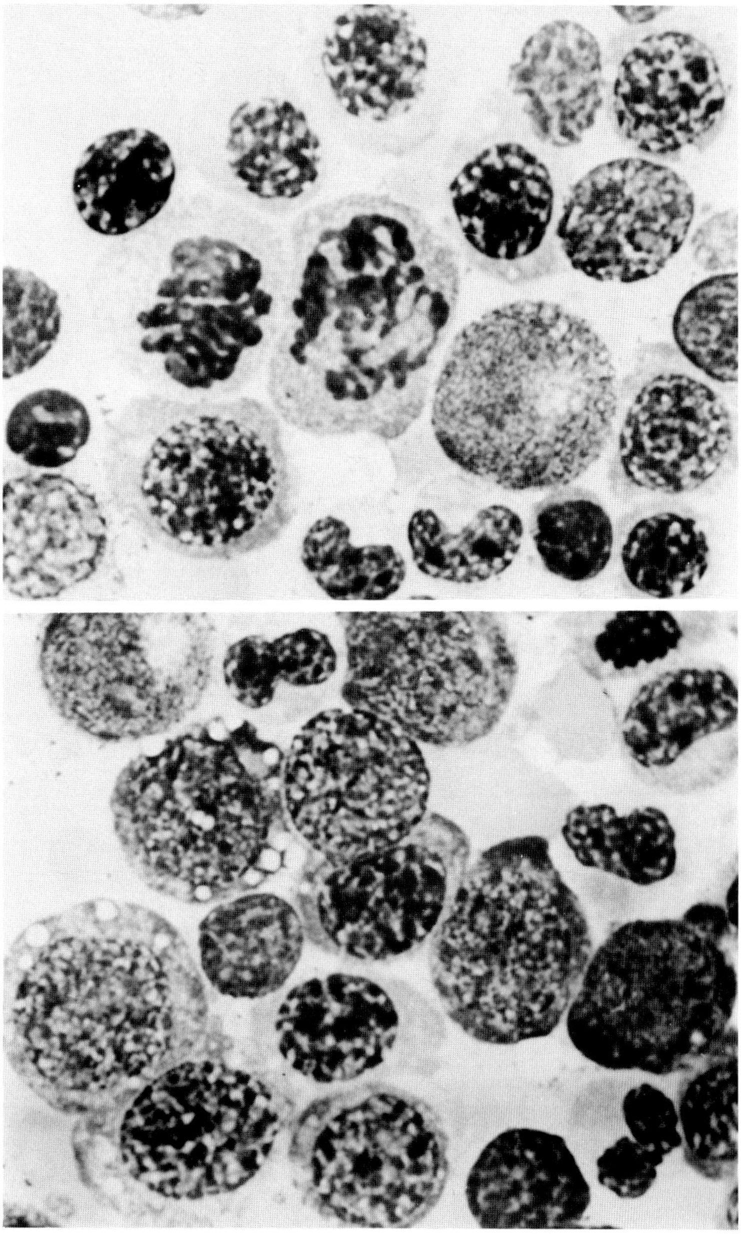

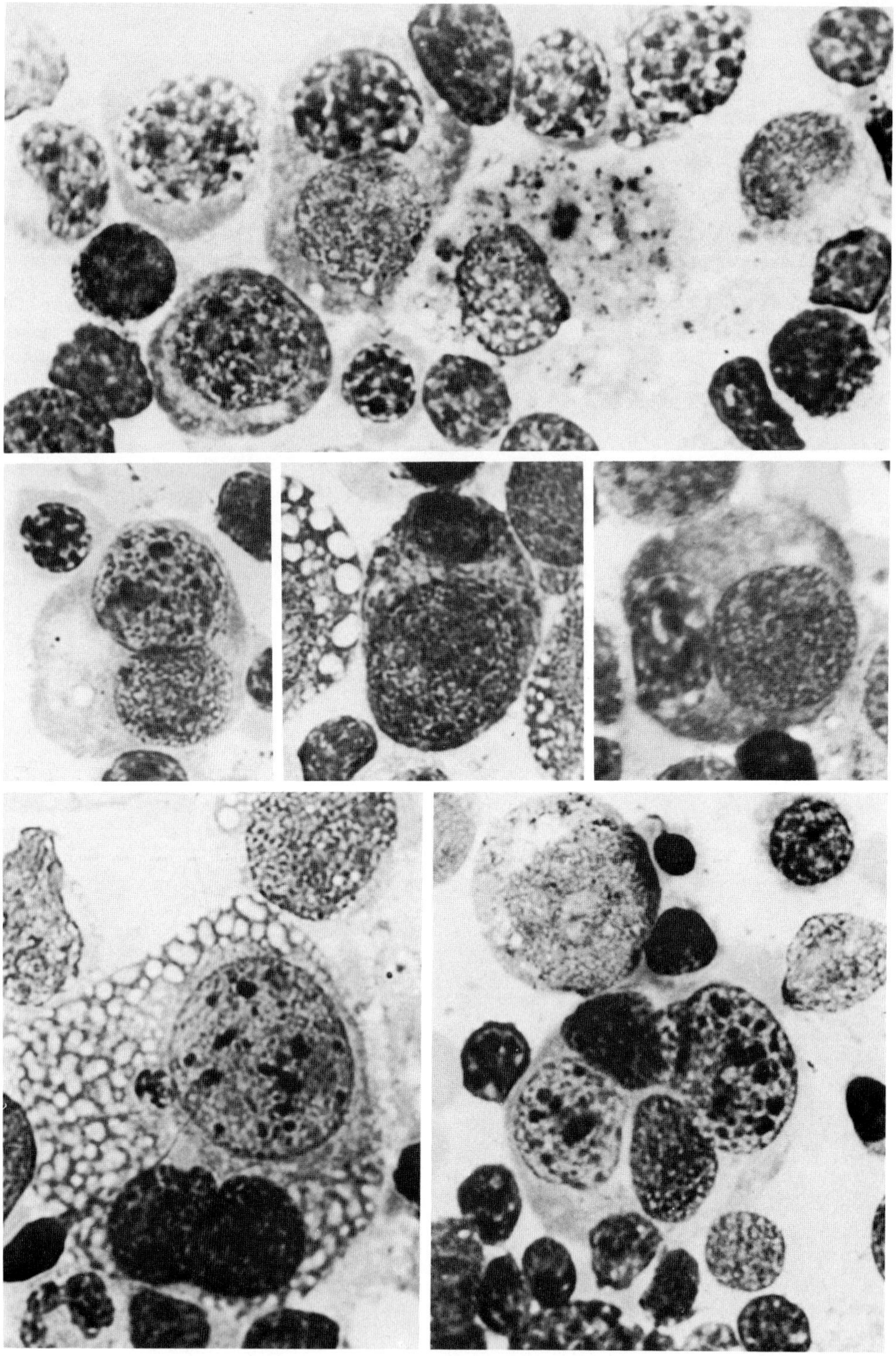

Figure 1-17. Bone marrow in chronic erythremic myelosis. Erythroblasts in the center of each inset show nuclei that vary in size and nuclear chromatin pattern. In some of the erythroblasts, nuclei are stained darkly, suggesting degeneration. Some of the nuclei show prominent, blocklike aggregates typical of megaloblastoid chromatin patterns. Others show a more delicate chromatin pattern and few, small aggregates of chromatin, as found in megaloblasts. In some cells, both megaloblastic-appearing nuclei and megaloblastoid-appearing nuclei are found in the same cell.

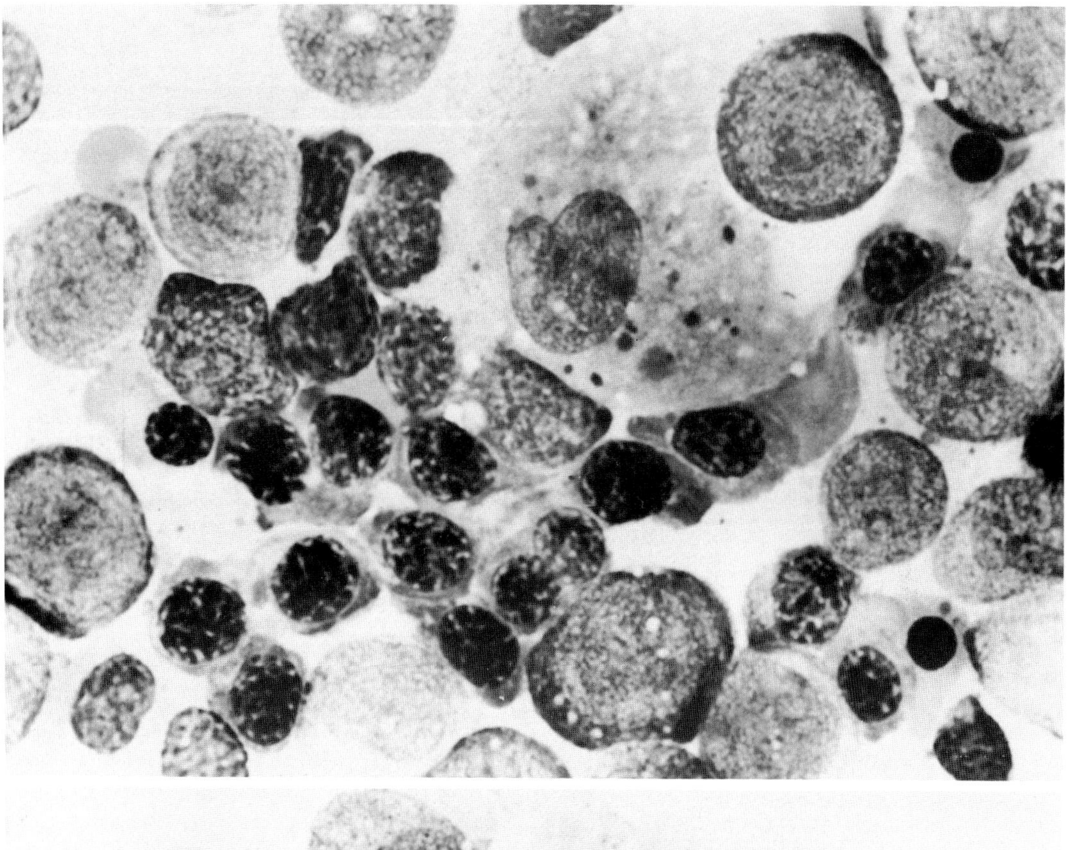

Figure 1-18. Bone marrow in chronic erythremic myelosis, showing large macrophages. Cytoplasm of these cells contains remnants of partially digested erythrocytes and degenerating cells. Intense hyperplasia of erythroblasts is seen.

erythrocytes (Figs. 1-17, 1-18). As an indicator of intramedullary cell death, erythrophagocytosis provides cytologic evidence for ineffective erythropoiesis. In this abnormality, erythroblasts proliferate actively but die within the bone marrow, and delivery of mature erythrocytes to the blood stream is greatly reduced compared to the intense erythroblastic proliferation in the marrow.

Granulocytic development is normal, and often there are increased numbers of promyelocytes and myelocytes (see Figs. 1-9, 1-14, 1-15). Some promyelocytes have prominent splinter-shaped granules (see Fig. 1-15). Myeloblasts are not identified in increased numbers and usually constitute less than 1% of the cells in a differential cell count of the bone marrow. Eosinophils, plasma cells, and mast cells may be increased (see Figs. 1-9, 1-10, 1-15). Eosinophils can have giant granules.[313]

Cytochemically, erythroblasts in marrows from patients with chronic erythremic myelosis show increased amounts of glycogen demonstrable with the PAS reagent,[30,404–409,442,546,671] as shown in Figure 1-19 and Plate 5*A*. In proerythroblasts the pattern of glycogen distribution is often punctate, whereas in more mature erythroblasts the pattern is diffuse. In other patients, glycogen may not be detectable in erythroblasts. Activity of phosphorylase is increased in erythroblasts that also contain abundant glycogen demonstrable with the PAS reagent. Pathophysiologically, increased phosphorylase activity may be induced as a compensatory catabolic mechanism by the increased glycogen. Alternatively, increased phosphorylase contributes to the biosynthesis of increased glycogen.[512] In some cases, granulocytic cells show reduced activity of myeloperoxidase.[580]

Alizarin red S,[498] and Prussian blue stains demonstrate increased iron stores in bone marrow samples from patients with chronic erythremic myelosis (Fig. 1-20). Many ringed sideroblasts occur in the marrows of some patients with chronic erythremic myelosis (Figs. 1-21, 1-22, 1-23), and few if any sideroblasts are found in the marrows of other patients. Iron is also demonstrable in the cytoplasm of large macrophages (Fig. 1-24). Activity of nonspecific esterase is intense in erythroblasts from some patients, and it may be moderately resistant to inhibition by sodium fluoride.[515] Activity of specific esterase and peroxidase are not detectable in erythroblasts in chronic erythremic myelosis. Activity of acid phosphatase is usually normal in erythroblasts from patients with chronic erythremic myelosis. As in normal erythroblasts, staining of the cytoplasm with pyronine is intense, reflecting increased amounts of cytoplasmic RNA.

Demonstration of increased amounts of several nuclear constituents in erythroblasts from patients with chronic erythremic myelosis may be of diagnostic value in distinguishing these erythroblasts from other types of erythroblasts. Using the ammoniacal silver stain to demonstrate arginine-rich histone and lysine-rich histone, nuclei of erythroblasts from patients with chronic erythremic myelosis contain large numbers of black to brown punctate deposits, representing increased amounts of arginine-

(text continued on page 26)

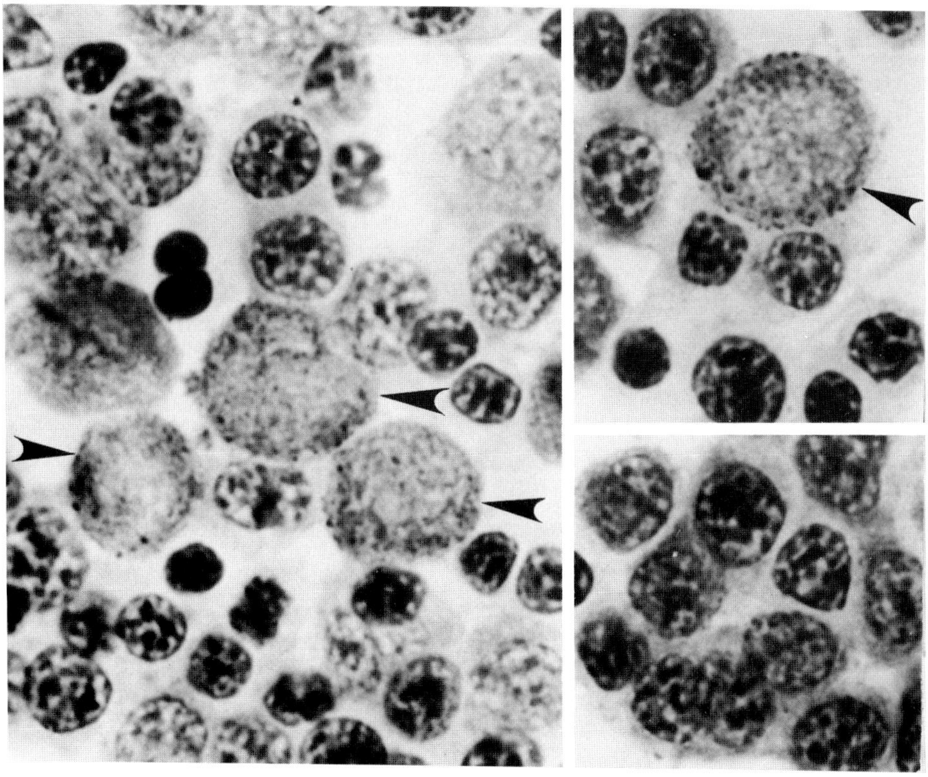

Figure 1-19. Bone marrow in chronic erythremic myelosis, stained with the PAS stain for glycogen. In proerythroblasts (**arrows**), glycogen appears as a punctate granular pattern in the cytoplasm. In more mature erythroblasts (**inset, lower right**), glycogen may appear in a diffuse staining pattern.

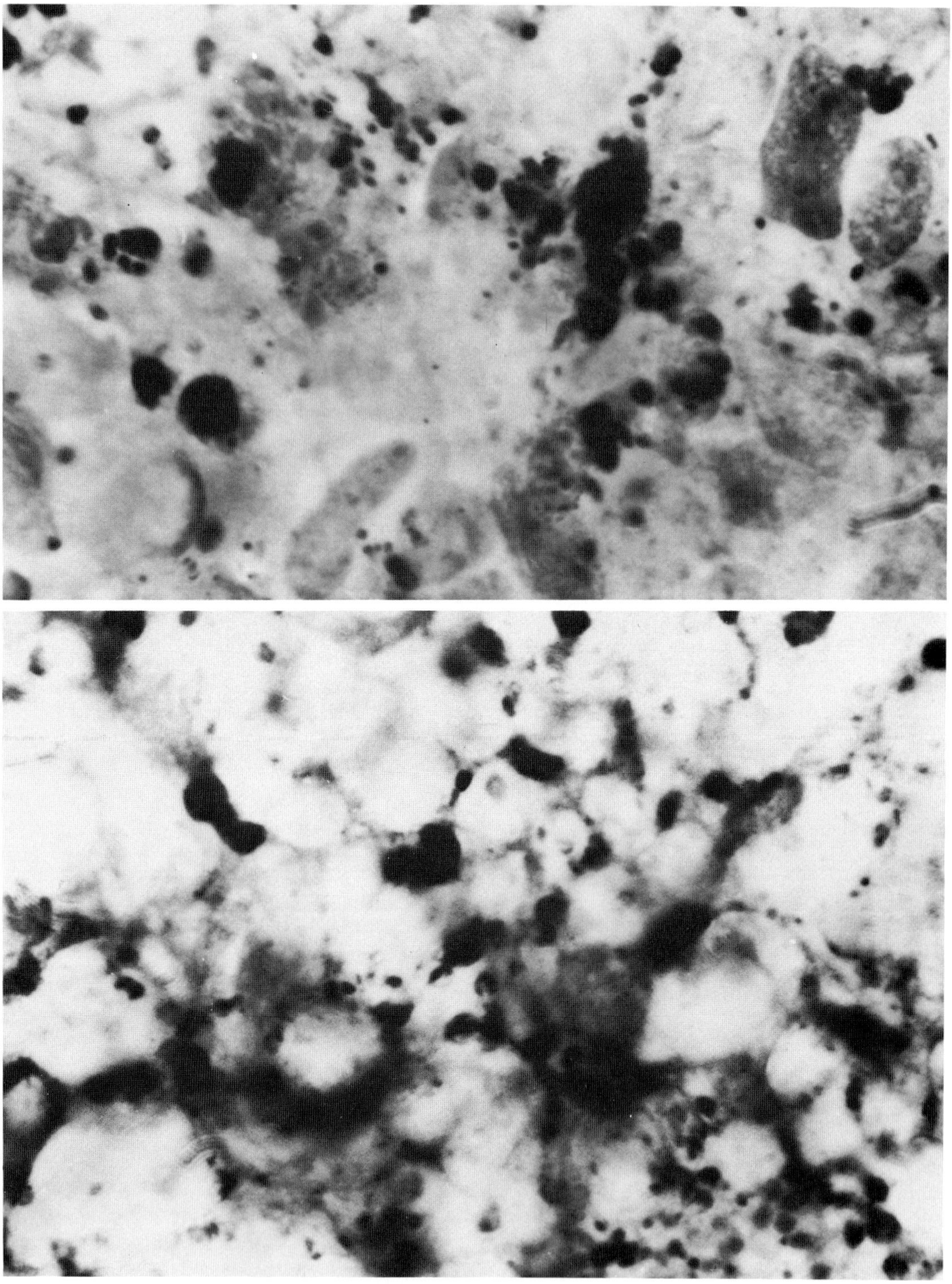

Figure 1-20. Bone marrow in chronic erythremic myelosis, stained with the Prussian blue stain for iron. Large agglomerates of ferric iron are seen, reflecting reduced turnover of iron and reduced incorporation into hemoglobin.

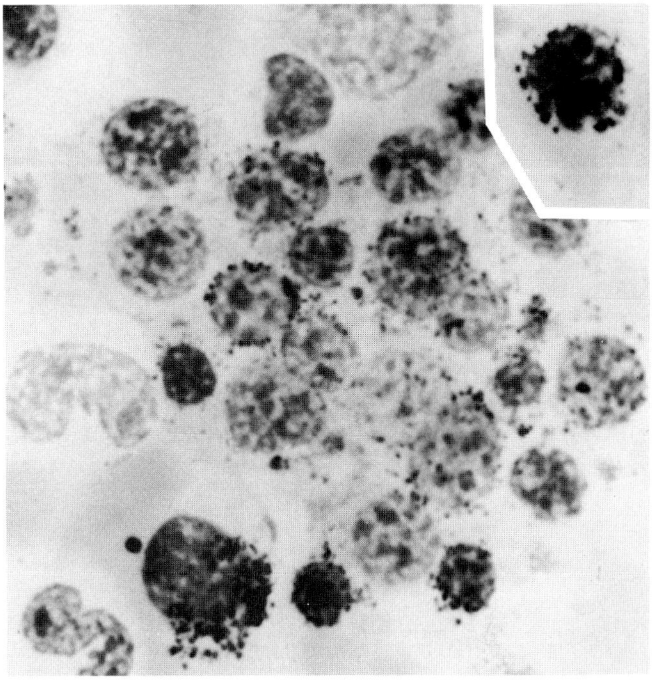

Figure 1-21. Bone marrow in chronic erythremic myelosis, stained with the Prussian stain for ferric iron. Many of the erythroblasts are sideroblasts, showing the darkly staining granular pattern of iron in a circumferential arrangement around the nucleus. The iron is contained within the lamellae and cristae of the mitochondria.

Figure 1-22. Bone marrow in chronic erythremic myelosis, stained with the alizarin red S stain for iron. Ferric iron appears as a purple perinuclear halo as well as coccoid-shaped structures in the cytoplasm of cells that appear as sideroblasts when stained with conventional Prussian blue stain. (Kass L: Bone Marrow Interpretation. Philadelphia, J.B. Lippincott, 1979)

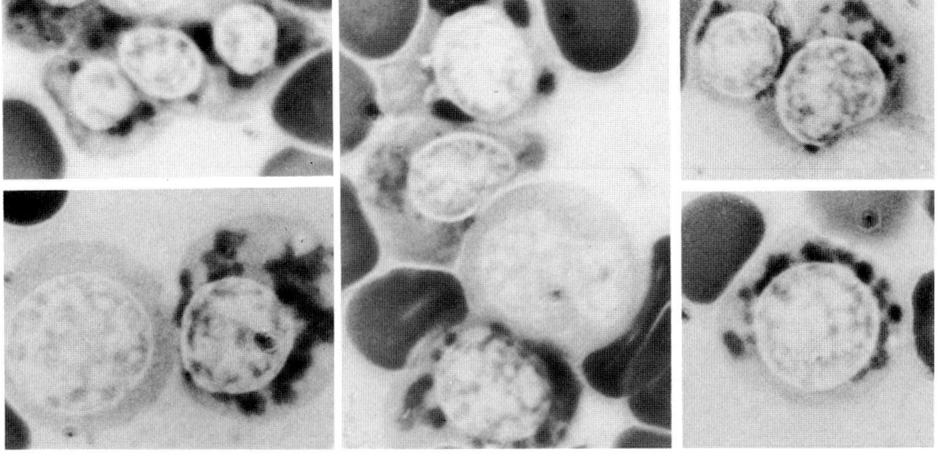

rich histone, as seen in Figure 1-25.[496] Ultrastructurally, these deposits of arginine-rich histone are localized in areas of heterochromatin or metabolically repressed DNA.[511]

Using the blue thiazine dye azure A, nuclei of erythroblasts from patients with chronic erythremic myelosis demonstrate an unusual pink metachromasia (Fig. 1-26). This staining property has been attributed to increased amounts of methylated arginine in the histones of these erythroblasts.[497] Homocysteine[500] is also increased in the cytoplasm of erythroblasts from patients with chronic erythremic myelosis, and a unique black-speckled nuclear staining is obtained with the use of gold chloride.[509]

H & E sections of biopsies from patients with chronic erythremic myelosis show hypercellularity and a marked increase in number of erythroblasts (Fig. 1-27). Immature erythroblasts, like proerythroblasts, demonstrate large, pale vesicular nuclei with prominent nucleoli and nuclear chromatin aggregates that seem to adhere to the nuclear membrane. Cytoplasmic staining is prominent. More mature erythroblasts show darkly-staining nuclei and a checkerboard chromatin pattern in which aggregates of nuclear chromatin appear within the central portion of the cell and few are adherent to the nuclear membrane (Figs. 1-27, 1-28). Megakaryocytes are increased in number in some instances, and normal or reduced in others. Numbers of granulocytic cells are reduced.

Erythroleukemia

Erythroleukemia is a disorder characterized by proliferation of myeloblasts and abnormal-appearing erythroblasts in the bone marrow and other visceral organs. In keeping with the concept of the Di Guglielmo syndrome, erythroleukemia may be a transitory phase in the development of acute myeloblastic or monocytic leukemia. Perhaps most if not all patients with acute myeloblastic or monocytic leukemia pass through an erythroleukemic phase that may be clinically inapparent, variable in duration, and present during the prodromal phases of acute leukemia. Some patients with erythroleukemia have a history of an antecedent hematologic disorder, such as chronic granulocytic leukemia, polycythemia vera,[794,1052] paroxysmal nocturnal hemoglobinemia,[186] and even lymphoproliferative disorders,[184,277,319,787,836,991] and histiocytic medullary reticulosis.[746] Others relate a history of exposure to volatile solvents such as benzene[330,537,879] that have been implicated as leukemogens.

Originally described by Di Guglielmo as a sudden myeloblastosis developing in a patient who had an erythroblastic disorder,[249] erythroleukemia as seen today usually presents as pancytopenia with increased numbers of macrocytic erythrocytes, some gigantic erythrocytes, and few atypical erythroblasts and myeloblasts in the blood.[384,385,414,457,519,628,631,657,733,741,767,938,941,947,958,1058] Only rarely does the peripheral blood contain large numbers of myeloblasts and normoblasts.[457,673,1037,1070] Evolution into acute

(*text continued on page 30*)

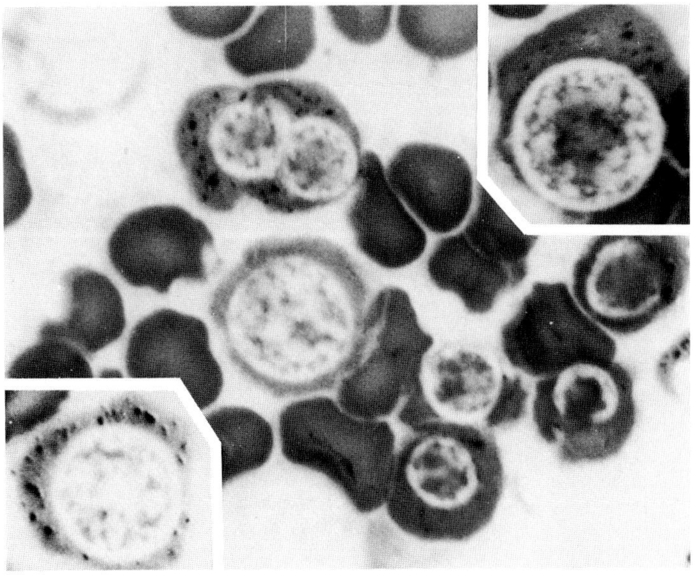

Figure 1-23. Bone marrow in chronic erythremic myelosis, stained with the bromochlorophenol blue stain for ferric iron. Iron appears as black and purple particulate structures within the cytoplasm of the erythroblasts. (Kass L: Bone Marrow Interpretation. Philadelphia, J.B. Lippincott, 1979)

Figure 1-24. Bone marrow in chronic erythremic myelosis, showing large macrophages containing prominent agglomerates of iron visualized with the Prussian blue stain (**inset also**). Several erythroblasts surround the central macrophage as in an erythroblastic island.

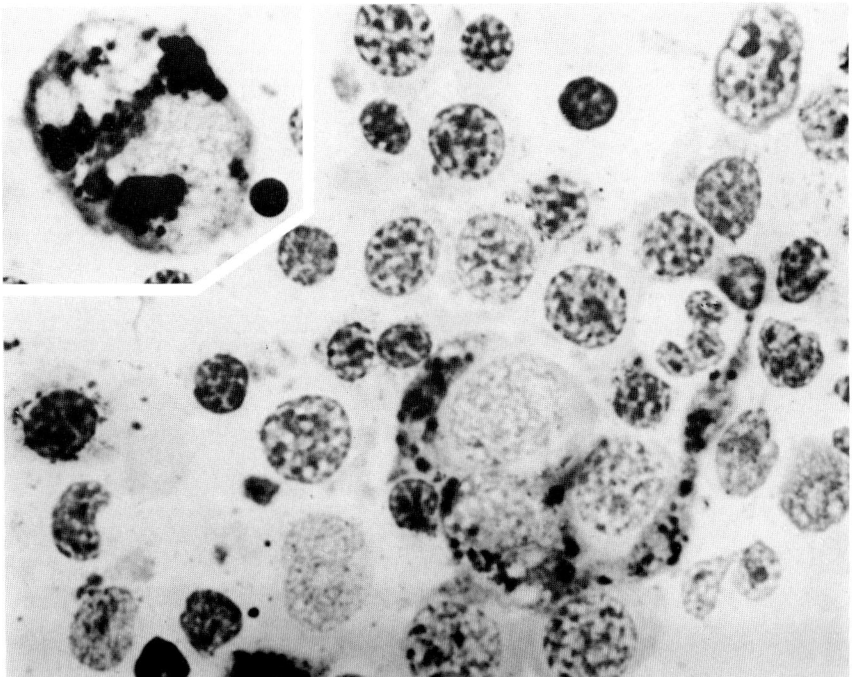

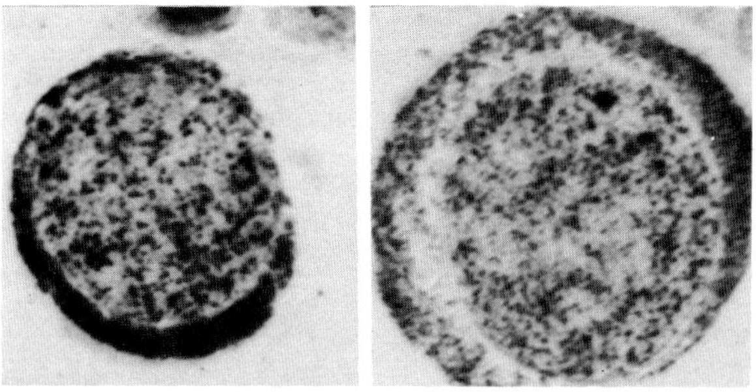

Figure 1-25. Bone marrow in chronic erythremic myelosis, stained with the ammoniacal silver stain. Using this reagent, histone rich in lysine stains yellow, whereas histone rich in arginine stains brown to black. In chronic erythremic myelosis and erythroleukemia, the nuclei of proerythroblasts as shown in the figure contain many black to brown punctate structures in a speckled pattern, reflecting abnormalities in the composition and distribution of arginine-rich histone.

Figure 1-26. Bone marrow in chronic erythremic myelosis, stained with azure A after removal of nucleic acids by acid digestion. Using this blue stain, nuclear proteins, largely histones, stain metachromatically pink. This metachromasia is thought to be partially due to increased content of methylated arginine-rich histone in the nuclei of these cells. (Kass L: Bone Marrow Interpretation. Philadelphia, J. B. Lippincott, 1979)

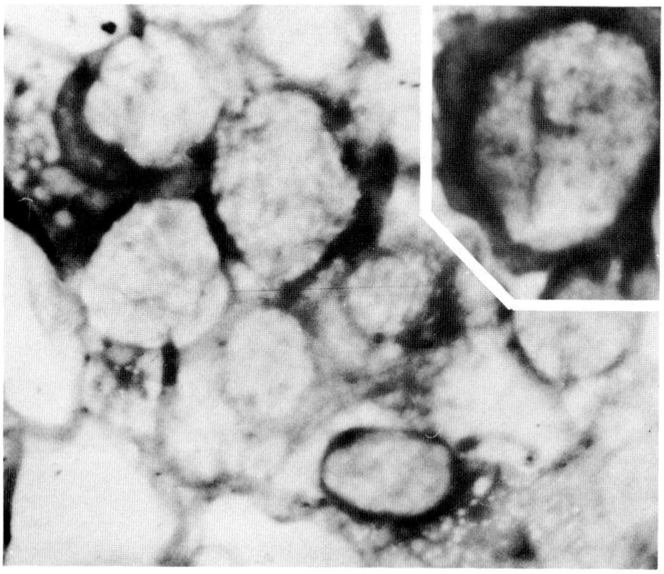

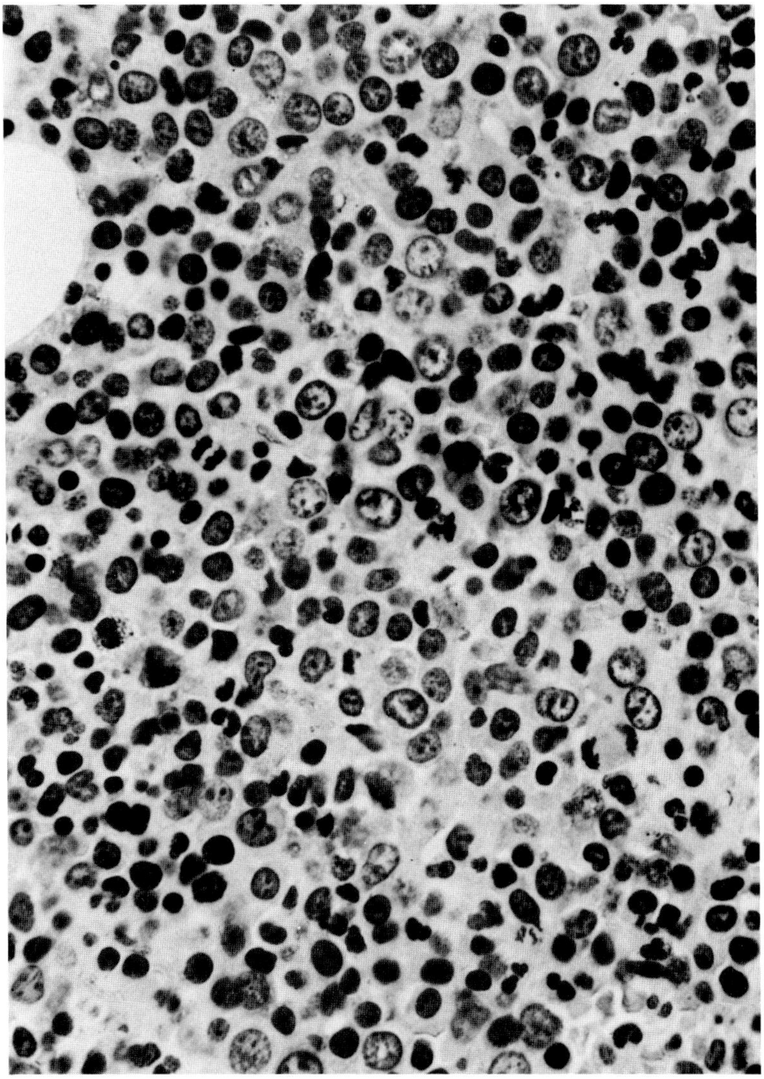

Figure 1-27. H & E section of bone marrow biopsy in chronic erythremic myelosis, showing intense hyperplasia of mature erythroblasts with darkly staining nuclei and increased numbers of proerythroblasts with pale vesicular nuclei.

leukemia may occur weeks to months after the initial diagnosis of erythroleukemia, and it is usually heralded by increasing numbers of blasts in the blood and bone marrow as well as abrupt cytogenetic changes.[185] In some cases, evolution into acute leukemia may not seem to occur within the lifetime of the patient, and death results from consequences of pancytopenia like hemorrhage or infection.

Marrows from patients with erythroleukemia are usually hypercellular, although in some instances they may be normocellular or even hypocellular. In all cases, there is a roughly even mixture of myeloblasts and erythroblasts, each comprising approximately 40% of the bone marrow (Fig. 1-29 and Plate 3C). When the proportion of myeloblasts or leukemic monocytes exceeds 40%, it may be more appropriate to call the disorder *acute myeloblastic* or *monocytic leukemia*, particularly if the tempo of the disease accelerates in conjunction with the myeloblastic or monocytic proliferation in the bone marrow.

Myeloblasts or leukemic monocytes in erythroleukemia may contain one or more Auer rods, as may megakaryocytes and histiocytic-ap-

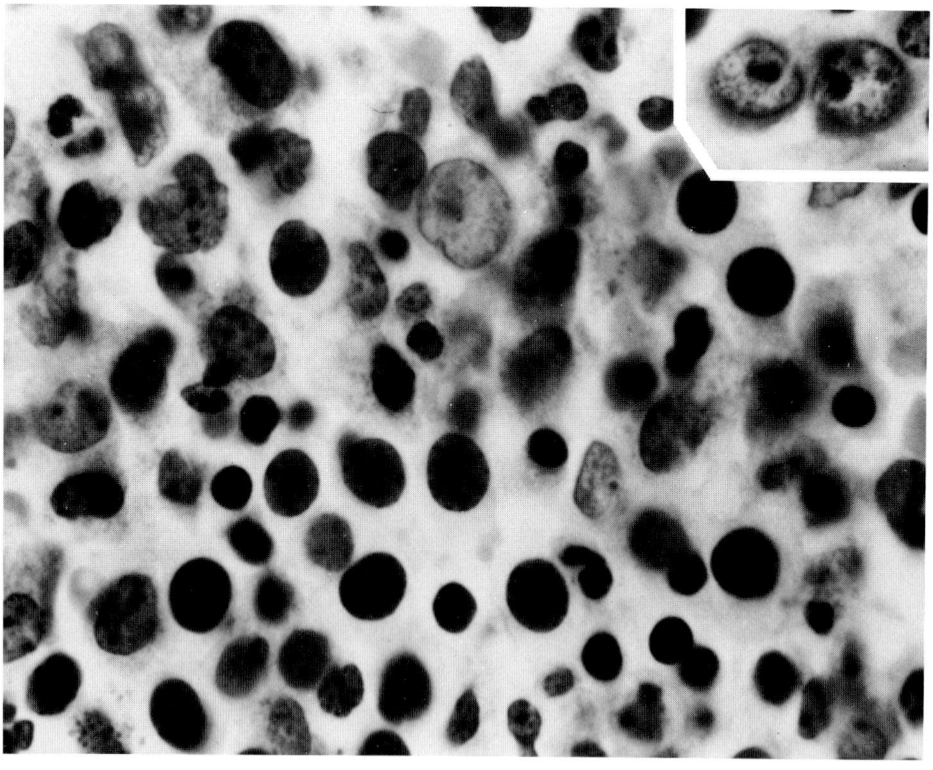

Figure 1-28. H & E section of bone marrow biopsy in chronic erythremic myelosis, showing increased numbers of late intermediate macronormoblasts with darkly staining nuclei, and increased numbers of proerythroblasts with prominent, darkly staining nucleoli **(inset)**.

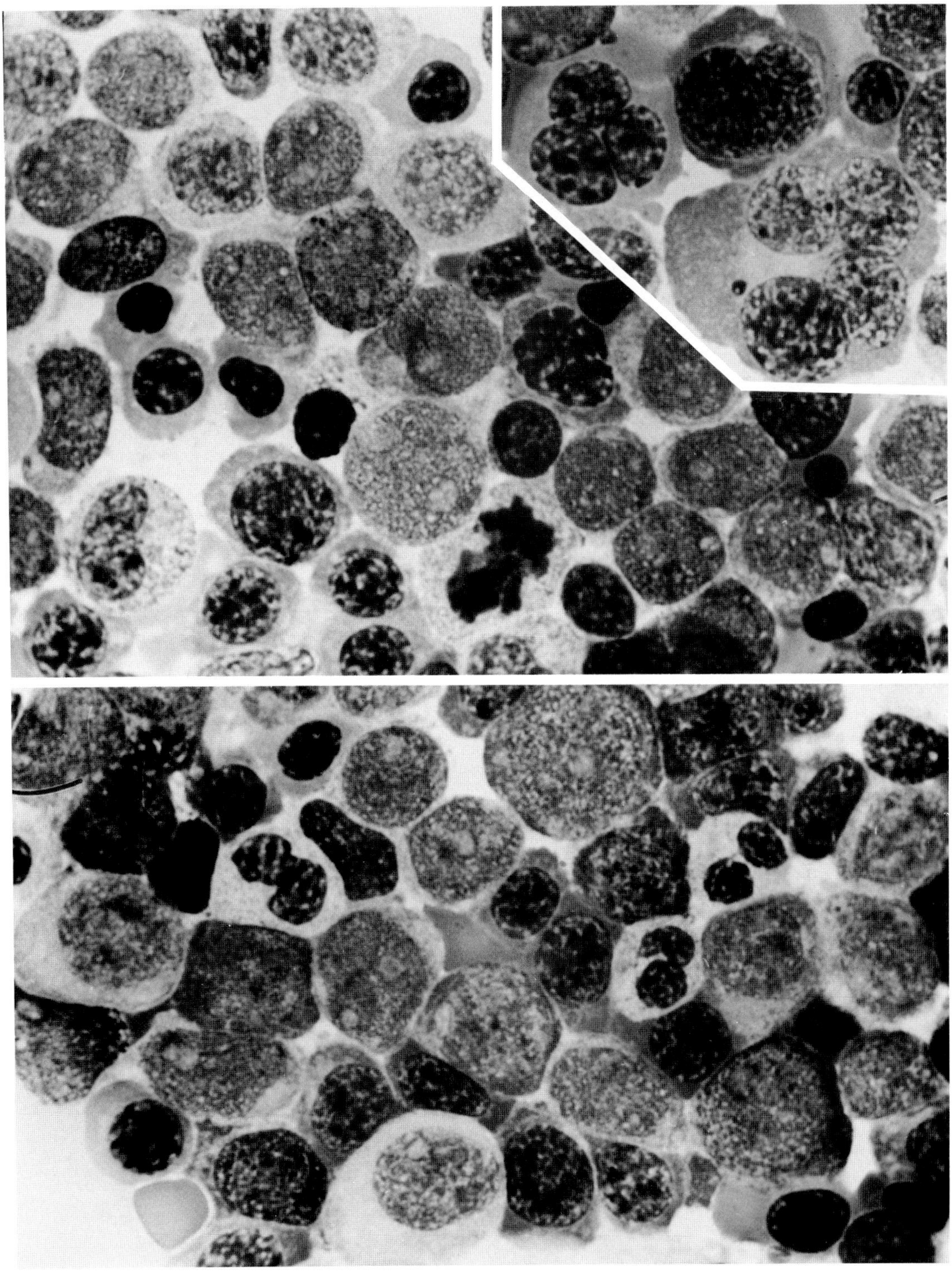

Figure 1-29. Bone marrow in erythroleukemia, showing myeloblasts with prominent nucleoli, and increased numbers of megaloblastoid intermediate macronormoblasts. Inset shows three megaloblastoid intermediate macronormoblasts with multiple nuclei and Howell-Jolly bodies.

pearing cells. Myeloblasts are usually normal in size, although a small population (usually less than 20%) of micromyeloblasts is usually found. Increased numbers of promyelocytes with atypical splinter-shaped granules occur. Mature granulocytic cells are reduced in number. Some neutrophils have ring-shaped nuclei.[993] Others may show pseudo-Pelger-Huët nuclear abnormalities.[795,989] In some granulocytic cells, both eosinophilic and basophilic granules are found in the same cell, reflecting disturbances in differentiation (Fig. 1-30). Increased numbers of macrophages demonstrating phagocytosis of mature erythrocytes and immature erythroblasts are present. Megakaryocytes are reduced in number, and some may be micromegakaryocytes.

Erythroblasts usually show megaloblastoid nuclear chromatin pattern (Fig. 1-31). Proerythroblasts are increased in number, and many are multinuclear with prominent nucleoli (Fig. 1-32*A*). Some large megaloblastoid intermediate macronormoblasts contain multiple nuclear lobulations (Fig. 1-32*B*). In some erythroblasts, small accessory nuclei are attached to the main nuclei by one or more thin stalks of chromatin (Fig. 1-33). Other gigantic erythroblasts show outpouchings of the nuclear membrane and Howell-Jolly bodies (Fig. 1-34). Abnormal karyorrhexis and accelerated pyknosis frequently occur (Figs. 1-35, 1-36, 1-37), as do atypical multipolar mitotic figures in erythroblasts (Figs. 1-38, 1-39).

In the nuclei of some erythroblasts, one or more strands of chromatin stain diffusely pale, in a flat, ribbonlike manner with routine panoptic stains. This altered staining property of DNA in these cells could be the result of nuclear degeneration (Fig. 1-40) and may reflect intramedullary cell death and ineffective erythropoiesis. In other erythroblasts, marked asynchronism between nuclear development and cytoplasmic maturation is found, and gigantic well-hemoglobinized intermediate macronormoblasts[917] with multiple cytoplasmic vacuoles[690] are often seen (Fig. 1-41). It is unclear as yet whether erythroblasts in erythroleukemia appear bizarre because they are themselves neoplastic or whether they appear bizarre as an epiphenomenon resulting from their proliferation in a microenvironment of myeloblastic leukemia.

Atypical megaloblasts are found in most cases of erythroleukemia.[499,505,506,843] These megaloblasts have centrally placed nuclei with stellate well-hemoglobinized cytoplasm and unusually delicate-appearing strands of nuclear chromatin. Aggregates of nuclear chromatin are rare, and nucleoli are not found (Fig. 1-42). Compared to megaloblasts from patients with untreated deficiency of vitamin B_{12} or folate,[499,506] megaloblasts in erythroleukemia have more delicate nuclear chromatin and fewer aggregates of chromatin. As in deficiency of vitamin B_{12} or folate, these chromatin abnormalities probably reflect biochemical aberrations in DNA and perhaps histone biosynthesis.[499] These atypical megaloblasts occur in varying numbers in marrows from patients with erythroleukemia, ranging from 1% to as high as 10% or 15% of the total number of erythroblasts. Furthermore, some of these atypical megaloblasts demonstrate unusual folding or twisting of the nuclei, resembling a monocytoid configuration.

(*text continued on page 42*)

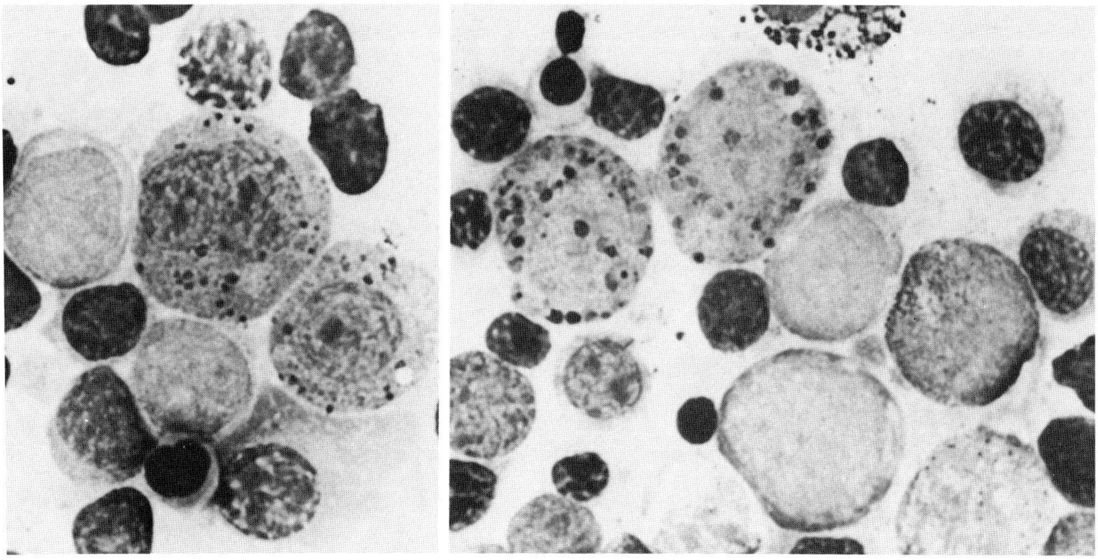

Figure 1-30. Bone marrow in erythroleukemia, showing promyelocytes containing both eosinophilic and basophilic granules in the same cell.

Figure 1-31. Bone marrow in erythroleukemia. Two early megaloblastoid intermediate macronormoblasts separated by cluster of large platelets (**left**). Megaloblastoid intermediate macronormoblasts in erythroleukemia (**right**). Chromatin in this cell demonstrates a blocklike pattern, and aggregates are connective to each other and to the nuclear membrane by delicate strands of chromatin. Cytoplasm is well hemoglobinized.

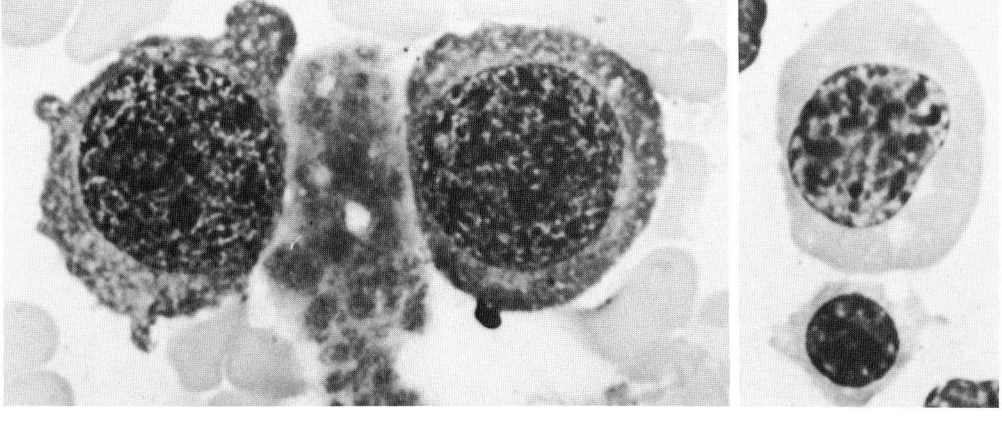

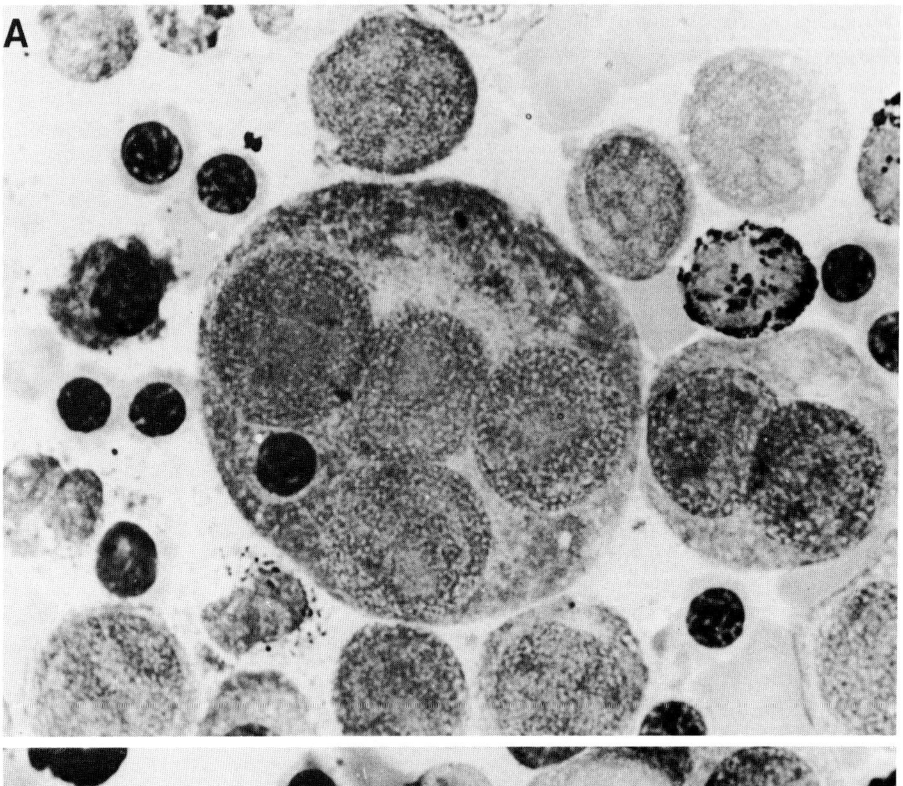
A

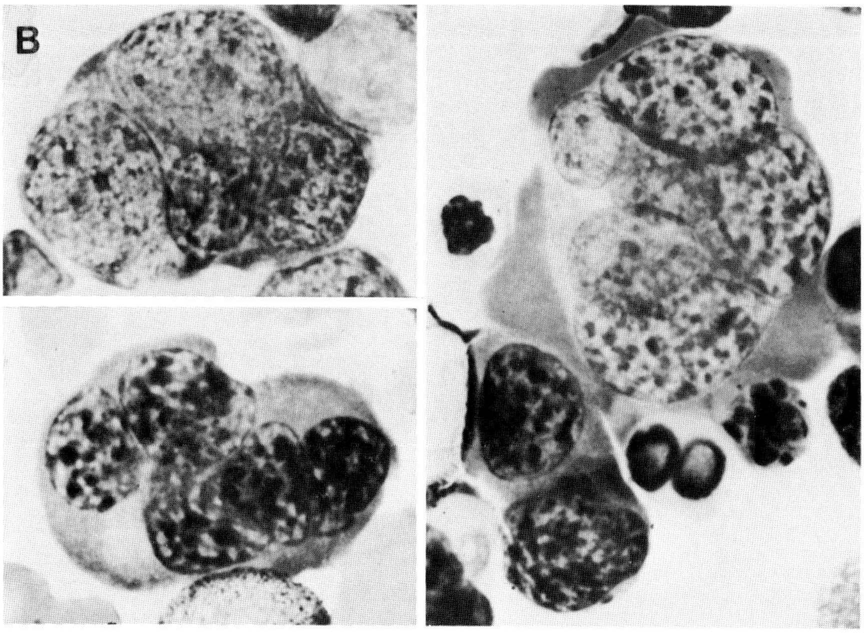

Figure 1-32. ← (**A**) Giant erythroblasts in erythroleukemia, showing multinuclearity, gigantic nucleoli, megaloblastoid nuclear chromatin pattern, and well-hemoglobinized cytoplasm containing vacuoles (**bottom**). (**B**) Atypical erythroblasts in erythroleukemia, illustrating megaloblastoid nuclear chromatin pattern, multinuclearity and multiple nuclear lobulations, and well-hemoglobinized cytoplasm indicating nucleocytoplasmic asynchronism.

Figure 1-33. Erythroblasts in erythroleukemia, depicting accessory nuclei and Howell-Jolly bodies connected to the main nucleus by "stalks" of chromatin.

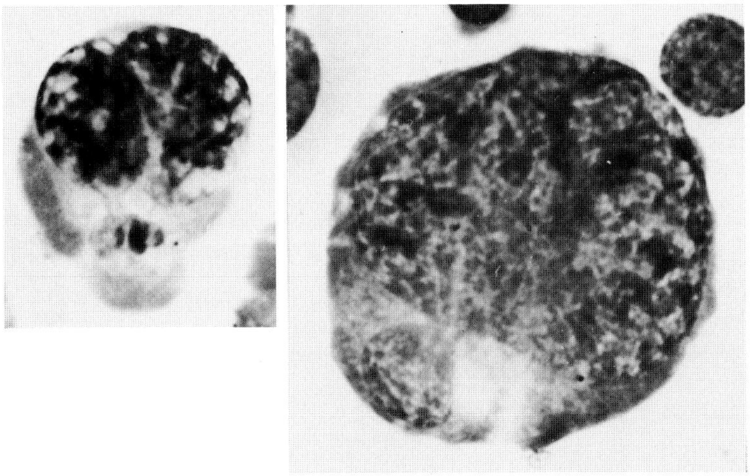

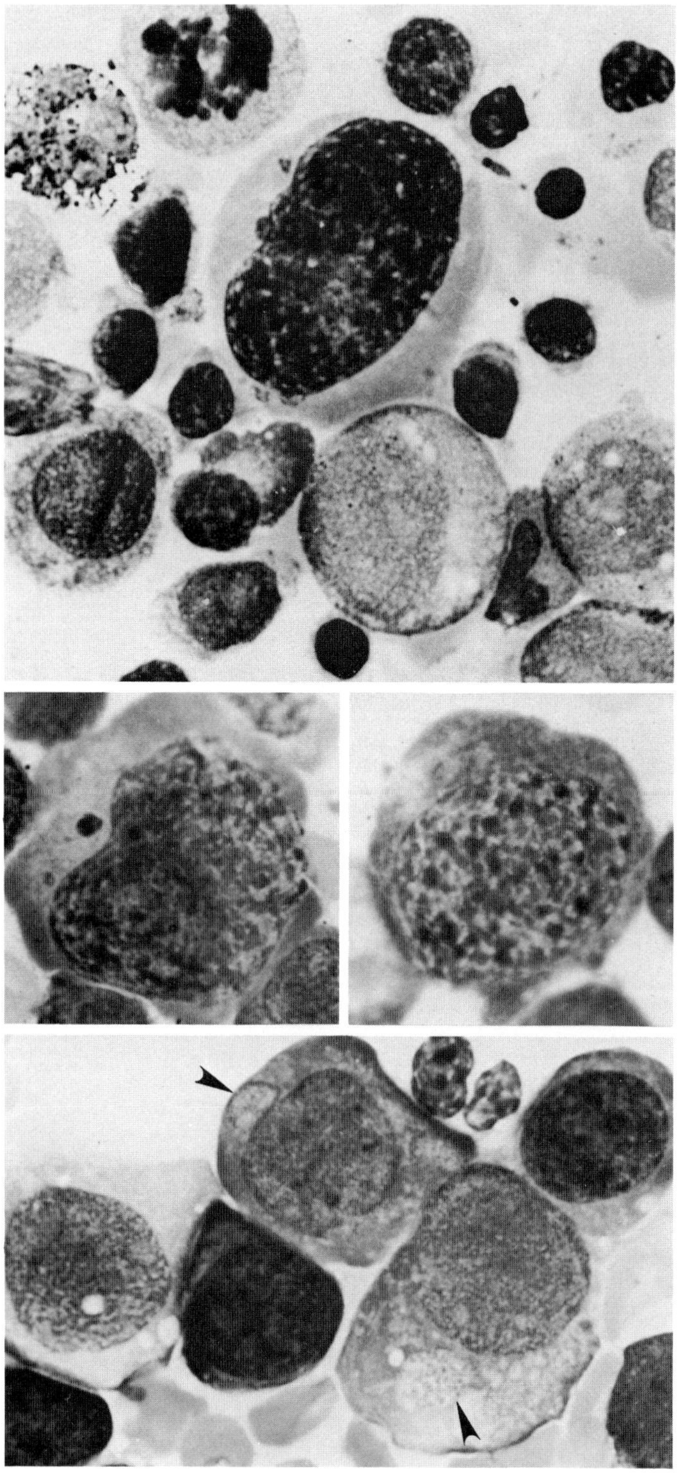

Figure 1-34. Erythroblasts in erythroleukemia, showing nuclei with irregular nuclear configurations, multiple large aggregates of nuclear chromatin in a megaloblastoid pattern, and Howell-Jolly bodies; also erythroblasts with accessory nuclei **(bottom, arrows)**.

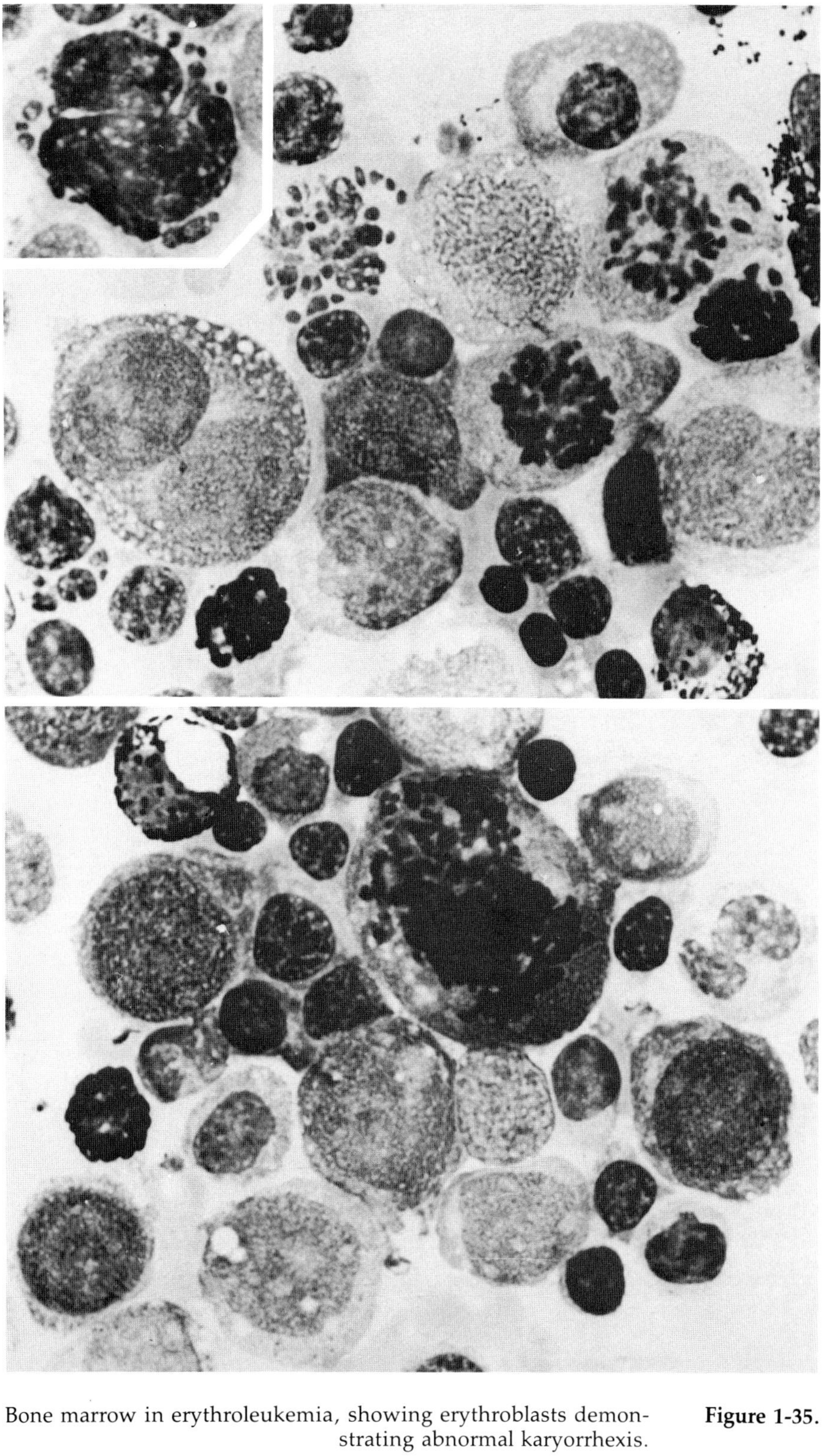

Bone marrow in erythroleukemia, showing erythroblasts demonstrating abnormal karyorrhexis. **Figure 1-35.**

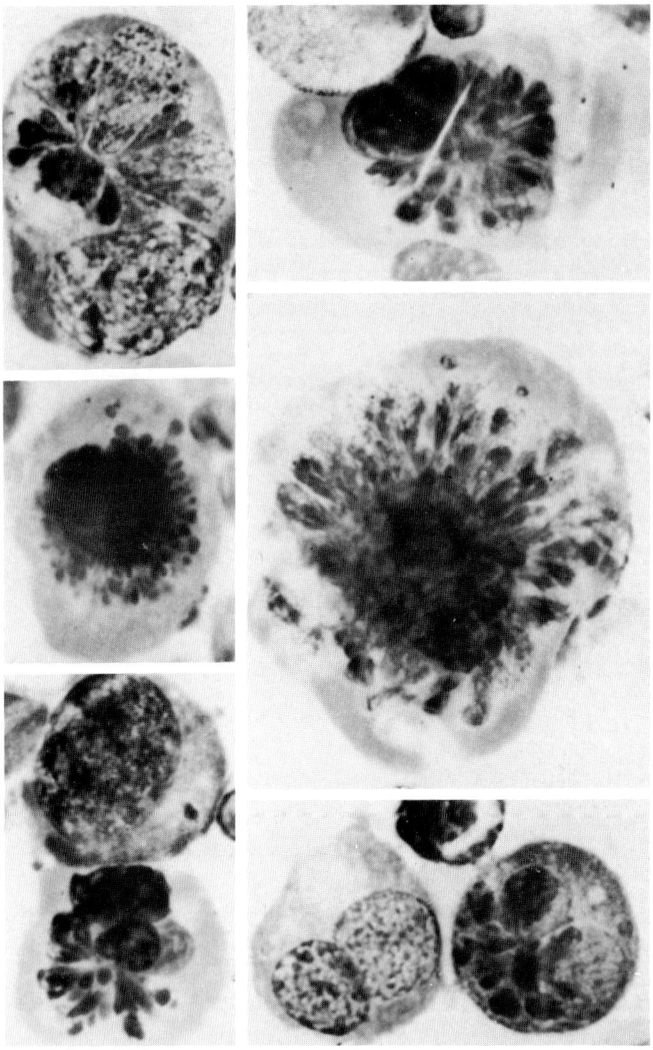

Figure 1-36. Bone marrow in erythroleukemia, showing erythroblasts with intense accelerated pyknosis.

Figure 1-37. Erythroblasts in erythroleukemia, showing accelerated pyknosis.

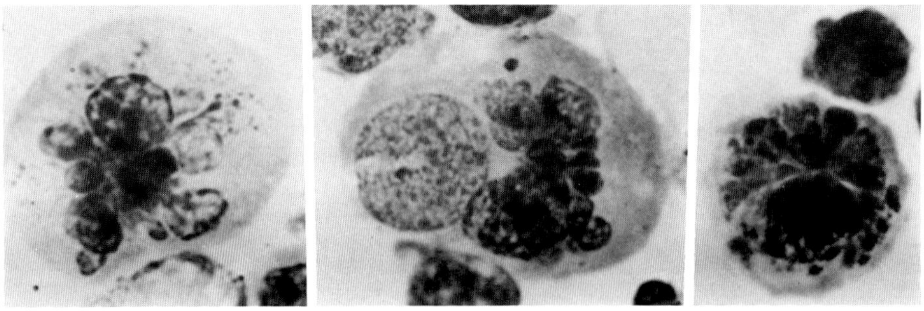

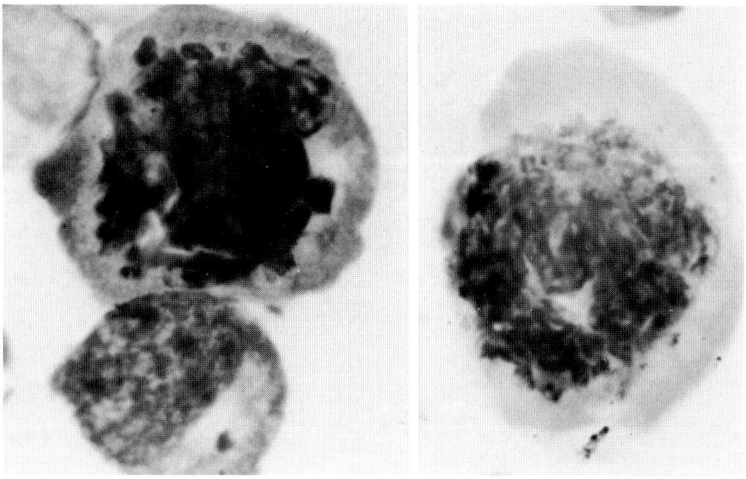

Figure 1-38. Abnormal mitotic figures in erythroblasts in erythroleukemia.

Figure 1-39. Abnormal, multipolar mitotic figures in erythroblasts in erythroleukemia.

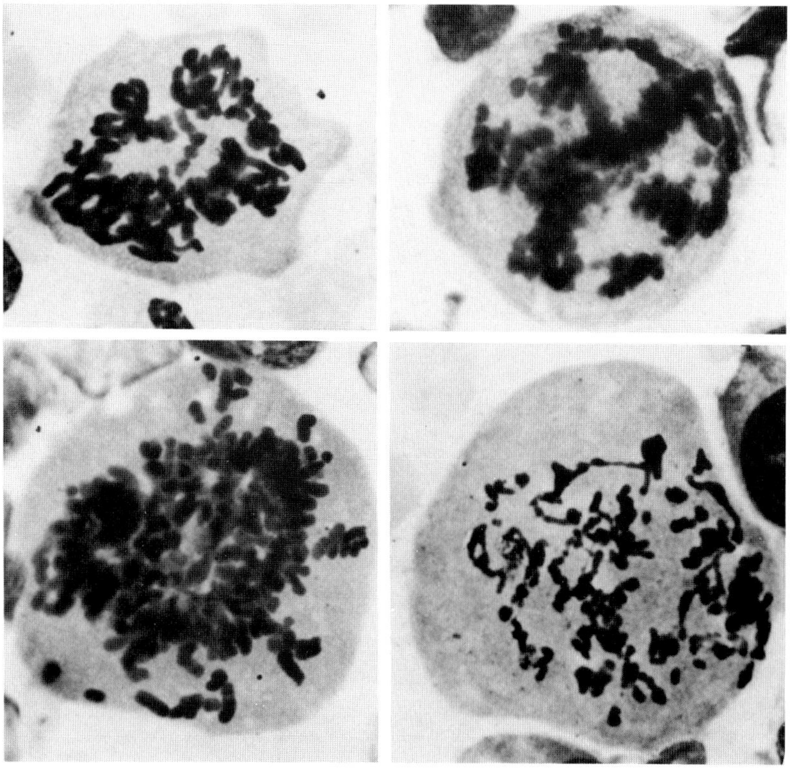

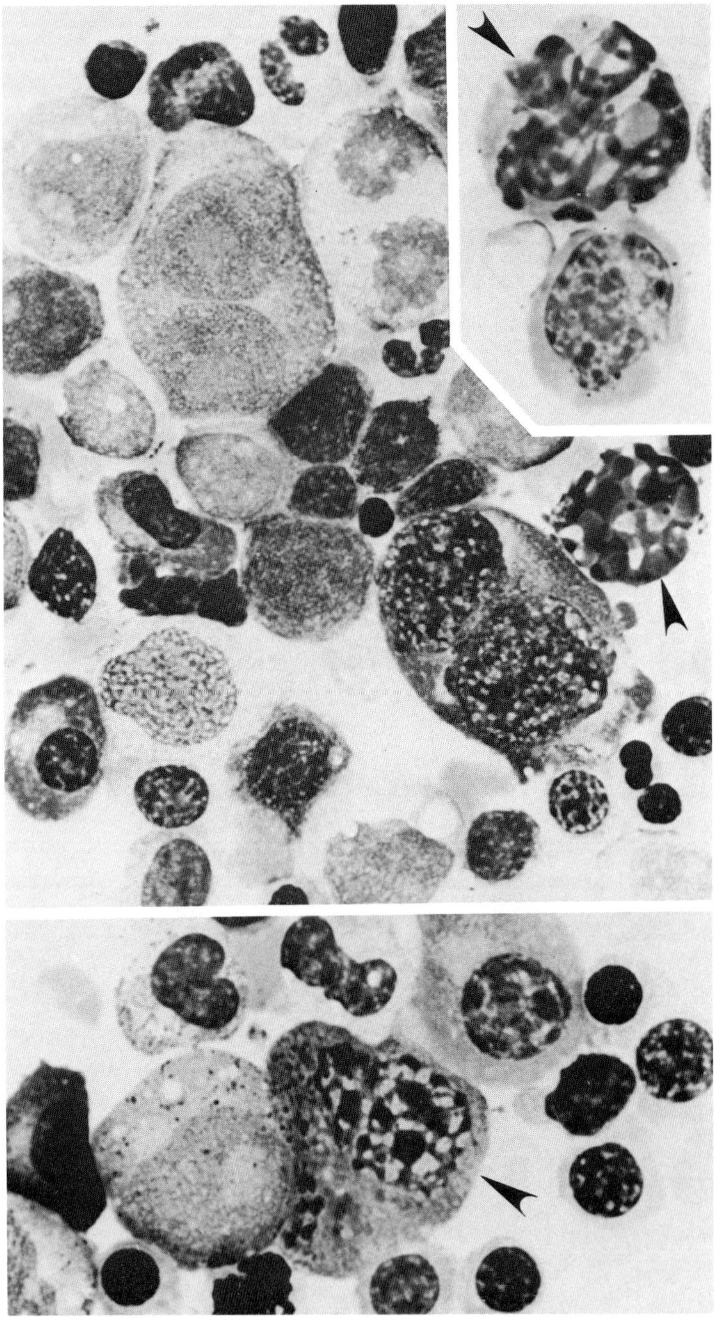

Figure 1-40. Areas of abnormally staining, ribbonlike threads of nuclear chromatin in erythroblasts in erythroleukemia **(arrows).** These may be degenerating erythroblasts.

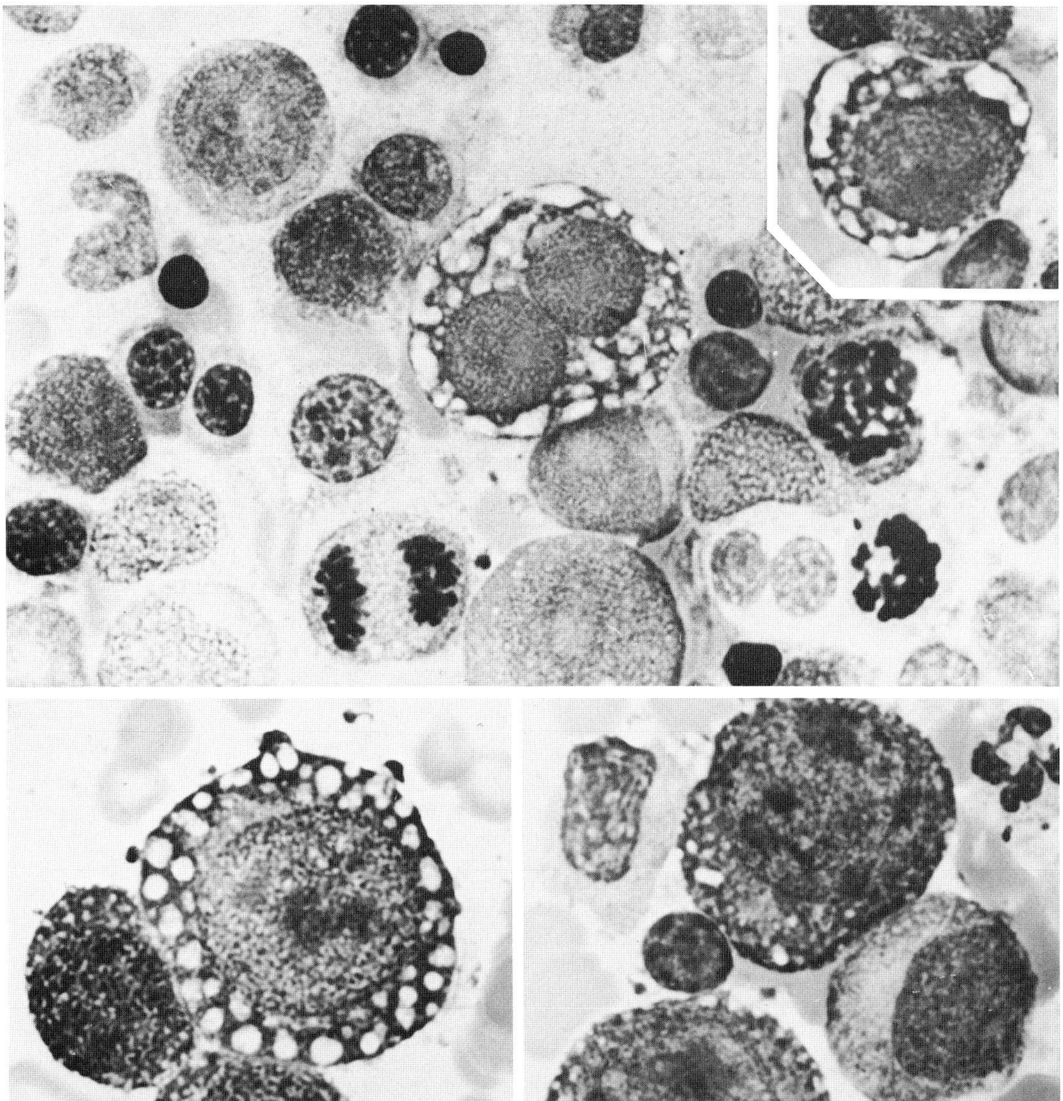

Erythroblasts in erythroleukemia, showing gigantism and multiple cytoplasmic vacuoles of varying sizes and shapes. **Figure 1-41.**

Cytochemically, myeloblasts from patients with erythroleukemia contain scant activity of specific esterase and myeloperoxidase, and weak activity of nonspecific esterase. Glycogen can usually be demonstrated in the cytoplasm of erythroblasts with the use of the PAS reagent.[8,29,825,973] Typically, erythroblasts in erythroleukemia contain large, blocklike aggregates of glycogen (Figs. 1-43, 1-44, and Plate 5*B*). This pattern of glycogen accumulation is most apparent in large, multinuclear erythroblasts. In more mature erythroblasts, the pattern of glycogen distribution may be blocklike, but more often it is intense and diffuse (Fig. 1-45). In other erythroblasts, both a diffuse and a blocklike pattern are present within the same cell. Although these patterns of glycogen distribution are sufficiently typical in erythroblasts in erythroleukemia to be useful diagnostically, their pathologic significance and prognostic implications are as yet unknown.

In those erythroblasts with abundant glycogen, activity of phosphorylase is increased, as seen in Figure 1-46.[504] Activity of glycolytic enzymes like LDH[126,243,300,366,504,711,826,875,876,878,1008–1011] are also increased in erythroblasts from patients with erythroleukemia (see Fig. 1-46). Nevertheless, increased activity of LDH and other glycolytic enzymes does not itself signify that the erythroblasts are malignant, as has been proposed in the case of other cells that show high levels of LDH.[5,57,63,105,136,291,302,391,561,776,886,1075,1076,1097] Activity of acid phosphatase is also increased and may occur in a perinuclear or unipolar distribution, suggesting a cluster of lysosomes in this area.

Using the ammoniacal silver stain for arginine-rich and lysine-rich histone, erythroblasts from patients with untreated erythroleukemia show abundant arginine-rich histone in a punctate pattern (Fig. 1-47 and Plate 5*D*). Activity of nonspecific esterase is unusually intense (Fig. 1-48) and resistant to inhibition by sodium fluoride.[501] In erythroleukemia, activities of myeloperoxidase and specific esterase can be demonstrated in many erythroblasts (see Fig. 1-48), suggesting that these cells contain properties of both granulocytic cells and erythroblasts. Myeloblasts from patients with erythroleukemia contain scant activity of specific esterase and myeloperoxidase and weak activity of nonspecific esterase that is resistant to inhibition by sodium fluoride.

With the Prussian blue stain for iron, bone marrows from patients with erythroleukemia often contain ringed sideroblasts (Fig. 1-49). As in chronic erythremic myelosis, those erythroblasts in erythroleukemia that are unusually large and multinuclear often contain the largest number of siderotic granules. Furthermore, siderotic granules in these giant atypical erythroblasts often appear coarser than in more mature erythroblasts.

In most instances, H & E sections of biopsies from patients with erythroleukemia show hypercellularity, but occasionally are normocellular or hypocellular (Fig. 1-50). Megakaryocytes are reduced in number. Large numbers of erythroblasts occur, and some are gigantic and multinuclear. Similarly, increased numbers of mitotic figures are present. Granulocytic cells, presumably myeloblasts, have diffuse-appearing nuclear chromatin

(text continued on page 46)

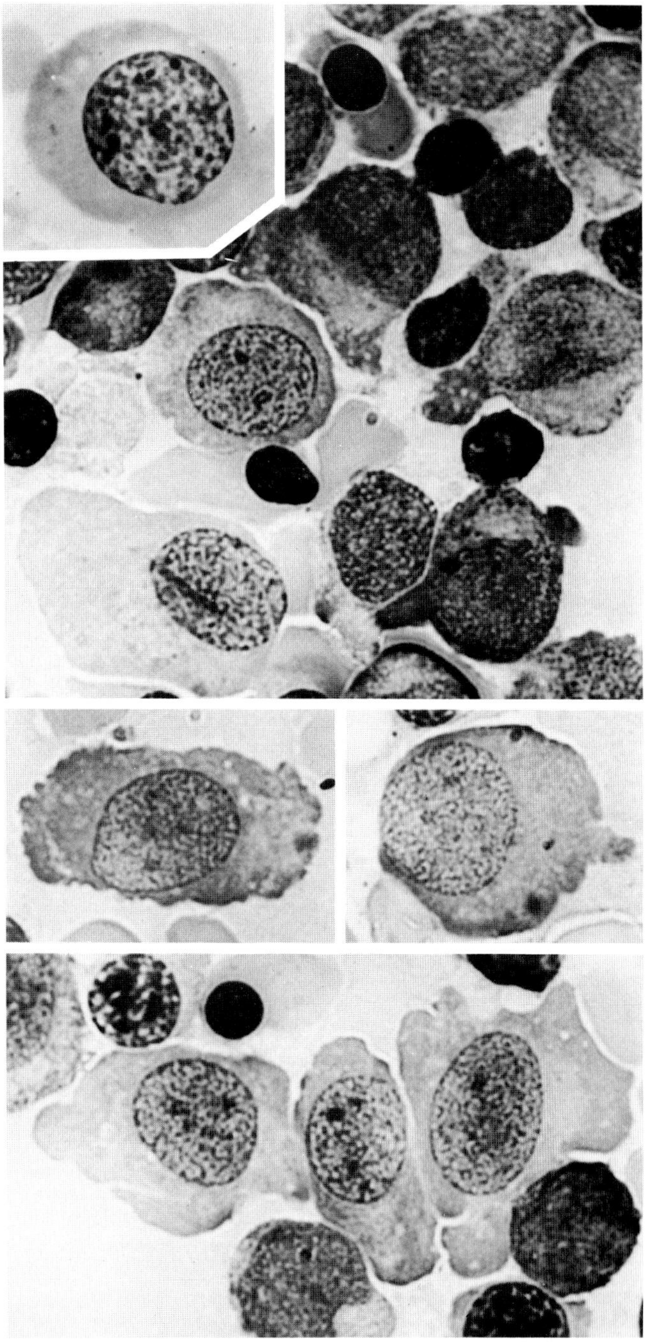

Figure 1-42. Atypical megaloblasts in erythroleukemia. In these cells, nuclei occupy a central position, imparting a fried-egg appearance to the cell. Nuclear chromatin is unusually attenuated, with few or no aggregates of chromatin.

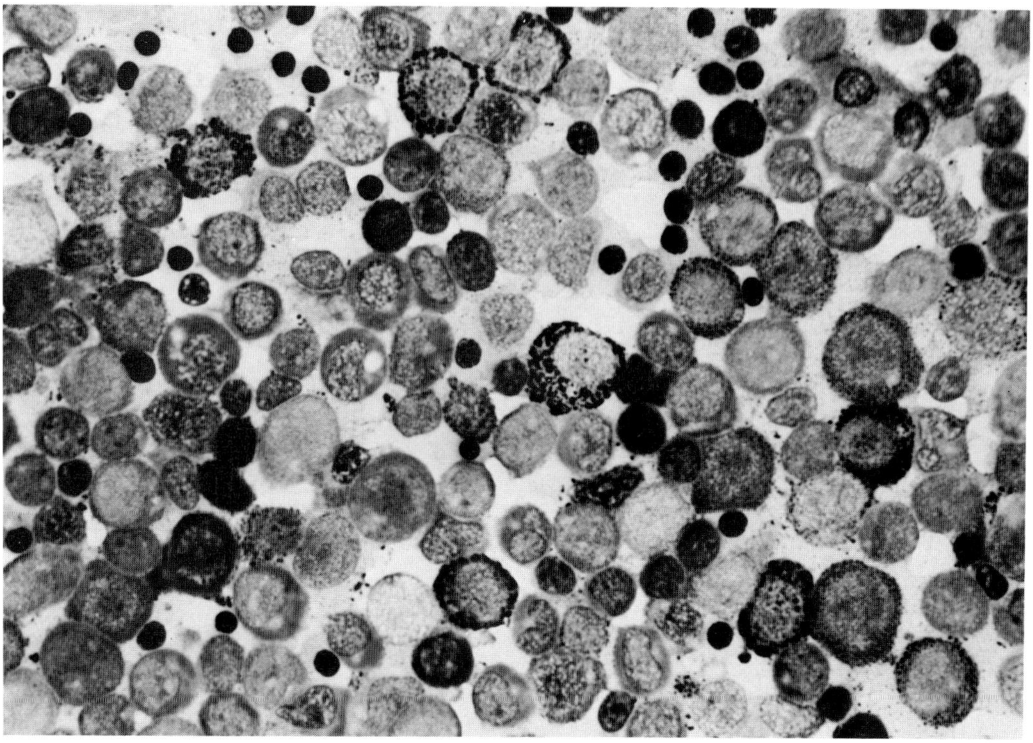

Figure 1-43. Bone marrow in erythroleukemia, stained with the PAS stain for glycogen. In atypical erythroblasts, glycogen appears as darkly staining material in the cytoplasm. In proerythroblasts, glycogen appears as dark punctate structures, as in this low-power view.

Figure 1-44. Erythroblasts in erythroleukemia, stained with the PAS stain for glycogen. In these cells, glycogen often appears in a blocklike pattern in the cytoplasm (**right;** after Kass L: Enzymatic abnormalities in erythroleukemia. Acta Haematol 59: 302, 1978. Reproduced by permission of S. Karger AG, Basel, Switzerland). ➔

Figure 1-45. Erythroblasts in erythroleukemia, stained with the PAS stain for glycogen. In these multinuclear erythroblasts, glycogen appears as an intense but diffuse staining of the cytoplasm. ➔

Figure 1-46. Erythroblasts in erythroleukemia. Intense activity of phosphorylase (**left;** black diffuse). Increased activity of lactic dehydrogenase (**right;** black punctate. After Kass L: Enzymatic abnormalities in erythroleukemia. Acta Haematol 59: 302, 1978. Reproduced by permission of S. Karger AG, Basel, Switzerland). ➔

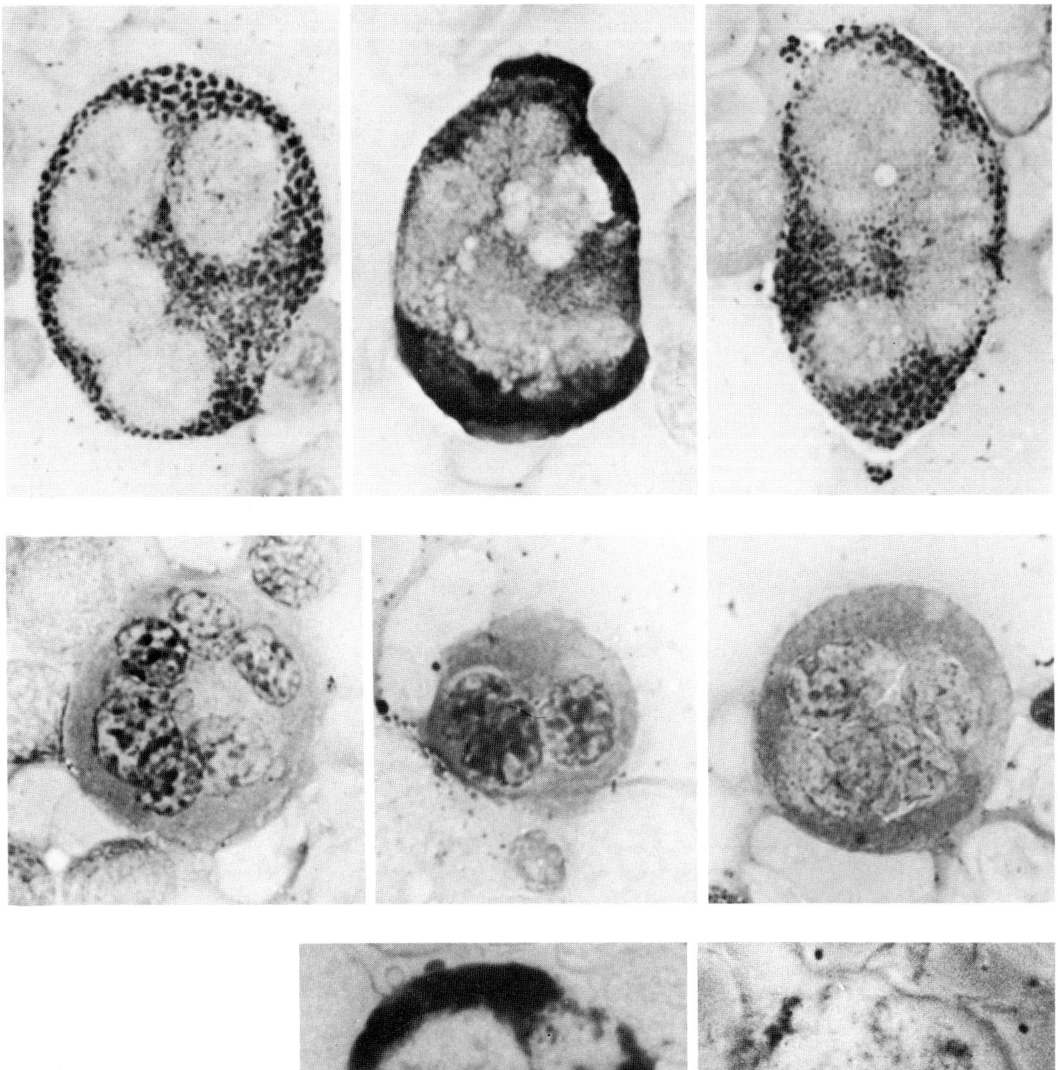

with dustlike aggregates evenly dispersed throughout the nucleus. Immature granulocytic cells such as promyelocytes and myeloblasts predominate, and mature neutrophils are rarely seen. Some contain a prominent darkly-staining nucleolus. On the basis of an H & E stained biopsy section, it is not possible to accurately distinguish myeloblasts from proerythroblasts or early megaloblastoid intermediate macronormoblasts.

Preleukemic Myeloblastic and Monocytic Disorders

A second group of preleukemic disorders is characterized by increased numbers of myeloblasts or monocytes in the bone marrow.[98,314,318,459,505,508,596,597,684,1099,1119] This group contrasts to the predominantly erythroblastic proliferation of the marrow seen in patients with Di Guglielmo syndrome. Because of the proliferation of myeloblasts or monocytes as well as abnormal growth characteristics in liquid culture, questions have been raised as to whether these disorders are preleukemic in the literal sense or represent leukemia itself in a slowly developing or indolent form. Transitions between the different aspects of the preleukemic erythroblastic and preleukemic myeloblastic or monocytic disorders occur, suggesting their close evolutionary relationships. Furthermore, evolutions have been identified within the erythroblastic group alone, as can be seen in transitions between acute or chronic erythremic myelosis and erythroleukemia. Evolutions have also been identified within the preleukemic myeloblastic or monocytic group alone, such as development of subacute myelomonocytic leukemia in a patient who had preleukemia, or the development of smoldering acute leukemia in a patient who had preleukemia.

In some instances, it is possible to document transitions between erythroblastic and myeloblastic disorders in a single patient. An example of such a transition is an evolution from preleukemia into erythroleukemia and finally into acute myeloblastic or monocytic leukemia. Instances such as these suggest close relationships between the two major types of preleukemic disorders. Although the time of evolution into acute leukemia is uncertain in a given individual, growth characteristics in liquid culture as well as cytogenetic analyses have proven to be of considerable diagnostic and prognostic value.[132,379,797,945] In most instances, these tests provide more information regarding the likelihood than the imminence of leukemic evolution.

Preleukemia

Preleukemia is a panmyelotic disorder that is believed to precede the development of acute myeloblastic or monocytic leukemia by months to years. Hematologically, preleukemia is characterized by macrocytic anemia, monocytosis, normoblastemia, and marked hypercellularity of the

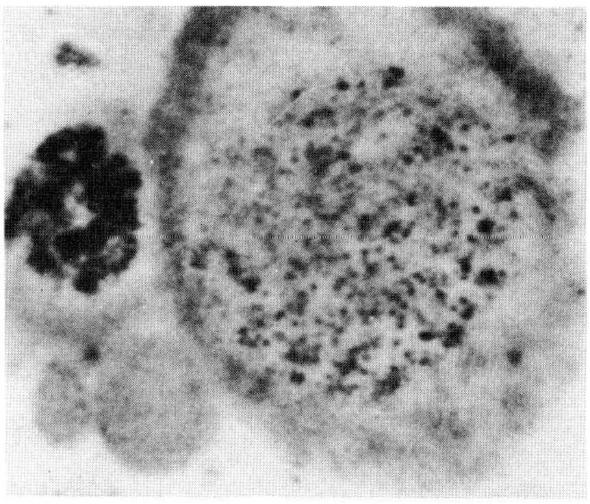

Figure 1-47. Erythroblast in erythroleukemia, stained with the ammoniacal silver stain. In the nucleus, black punctate material appears in a speckled pattern, representing abnormalities in the amount and distribution of arginine-rich histone.

Figure 1-48. Erythroblasts in erythroleukemia, stained with a combined stain for specific and nonspecific esterase. Gray color represents diffuse staining of the cytoplasm of these cells, representing increased activity of nonspecific esterase. Black punctate material represents increased activity of specific esterase, an enzyme that is not detectable in normal erythroblasts. Its presence in erythroblasts in erythroleukemia suggests that these abnormal cells have differentiated along both erythroblastic and granulocytic lines.

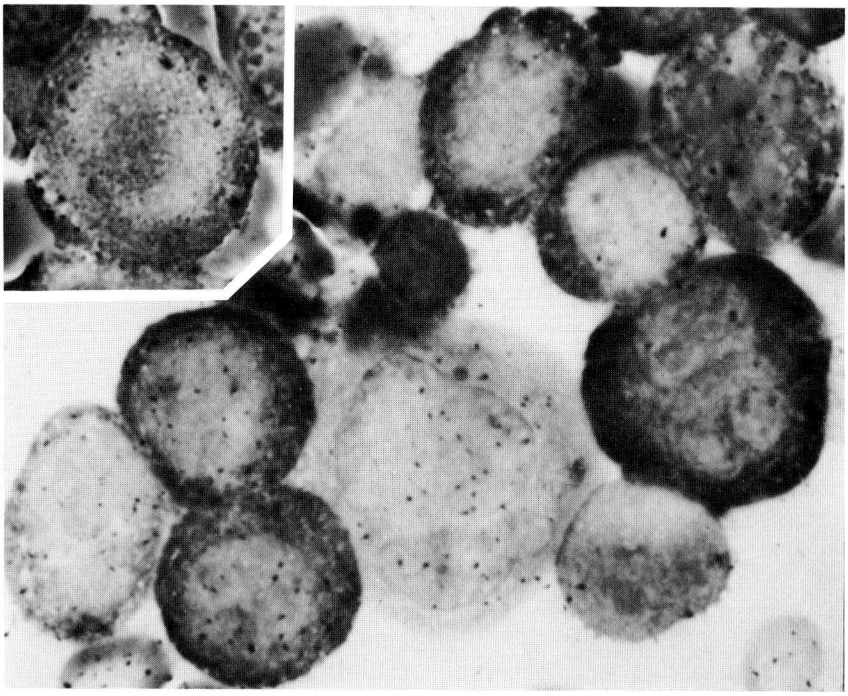

bone marrow. Clinically, patients with preleukemia often have fever, hepatomegaly, and splenomegaly. The marrow typically shows increased numbers of all cell types and individual morphologic abnormalities within these cell types, indicating that it is a panmyelosis. In a variable number of patients with preleukemia, acute myeloblastic or monocytic leukemia occurs as a terminal event.

Described under a variety of names over the years,[98,119,127,314,319,396,417,420,445,459,596,597,665,668,674,684,797,860,891,897,1022,1043,1096] preleukemia is probably best conceptualized now as a disorder in which the blood and bone marrow do not reveal typical findings of acute leukemia, namely, infiltration of the bone marrow by substantial numbers of leukemic blasts. Nevertheless, morphologic abnormalities in many of the cells in the blood and bone marrow are distinctive enough to suggest the likelihood of leukemic transformation. In many instances, the abnormalities resemble those in leukemic cells, raising the possibility that preleukemia as an entity may not exist as such but rather represents acute leukemia in an indolent or slowly developing form.[508,527]

In other patients, panmyelosis without distinctive morphologic abnormalities occur, making it difficult to diagnose preleukemia. In these instances, it may be more appropriate to describe the panmyelotic character of the marrow and suggest that continued observation over a period of time will be necessary to ascertain whether leukemia will develop. Since obvious leukemic evolution does not seem to occur in other patients with preleukemia who have panmyelotic marrows with the distinctive morphologic abnormalities, continued observation and sequential bone marrow examination is essential in all instances in which the diagnosis of preleukemia is made.

Peripheral blood samples from most patients with preleukemia show pancytopenia with macrocytosis of erythrocytes, occasional normoblasts, pseudo-Pelger-Huët anomaly of neutrophils, and monocytosis. Sometimes monocytes are increased in number and appear abnormal with multiple nuclear lobulations, cytoplasmic pseudopodia, and increased numbers of granules clustered in a perinuclear distribution. Platelets may be large and hypergranular or pale and hypogranular, and often demonstrate functional abnormalities.[885]

Marrows from patients with preleukemia demonstrate hypercellularity with markedly increased numbers of granulocytic cells (Fig. 1-51 and Plate 1*B*). Promyelocytes are increased in number and may show numerous splinter-shaped granules (Fig. 1-52). Myelocytes are also increased in number (Figs. 1-52, 1-53). Metamyelocytes and bands often exhibit atypical nuclear configurations such as twistings and convolutions, dumbbell-shape, or hole in the center of the nucleus (Figs. 1-54, 1-55, 1-56, 1-57, and Plate 1C). Nuclei of other metamyelocytes show clubbed ends. Cytoplasm of granulocytic cells may show increased numbers of primary and secondary granules. Bands and mature neutrophils often demonstrate aberrant nuclear configurations, including hypersegmentation (see Figs. 1-54 through 1-57). In contrast to deficiency in vitamin B_{12} or folate, multiple

(text continued on page 56)

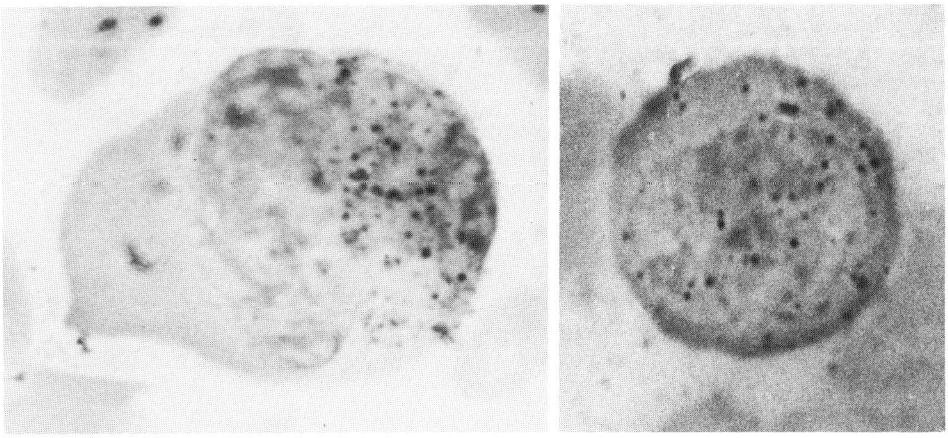

Figure 1-49. Erythroblasts in erythroleukemia, stained with the Prussian blue stain for iron. These cells are sideroblasts and contain multiple, coarse-appearing structures that stain positively for ferric iron.

Figure 1-50. H & E section of bone marrow biopsy in erythroleukemia, showing increased numbers of primitive-appearing cells that may be both erythroblasts and myeloblasts, as well as multinuclear cells that are probably erythroblasts and bizarre-appearing mitotic figures.

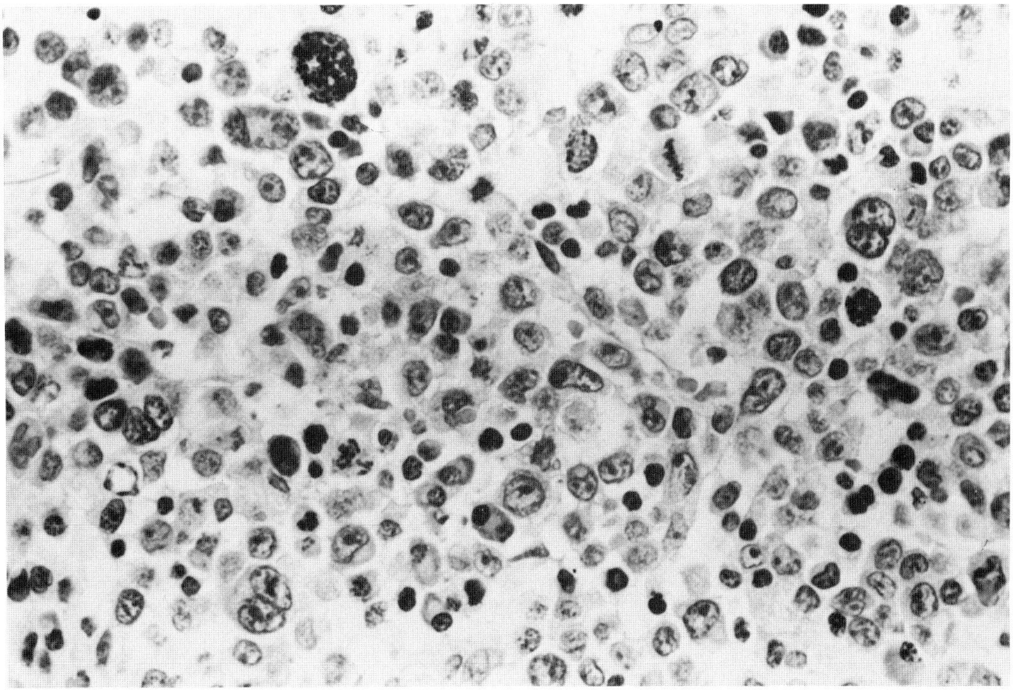

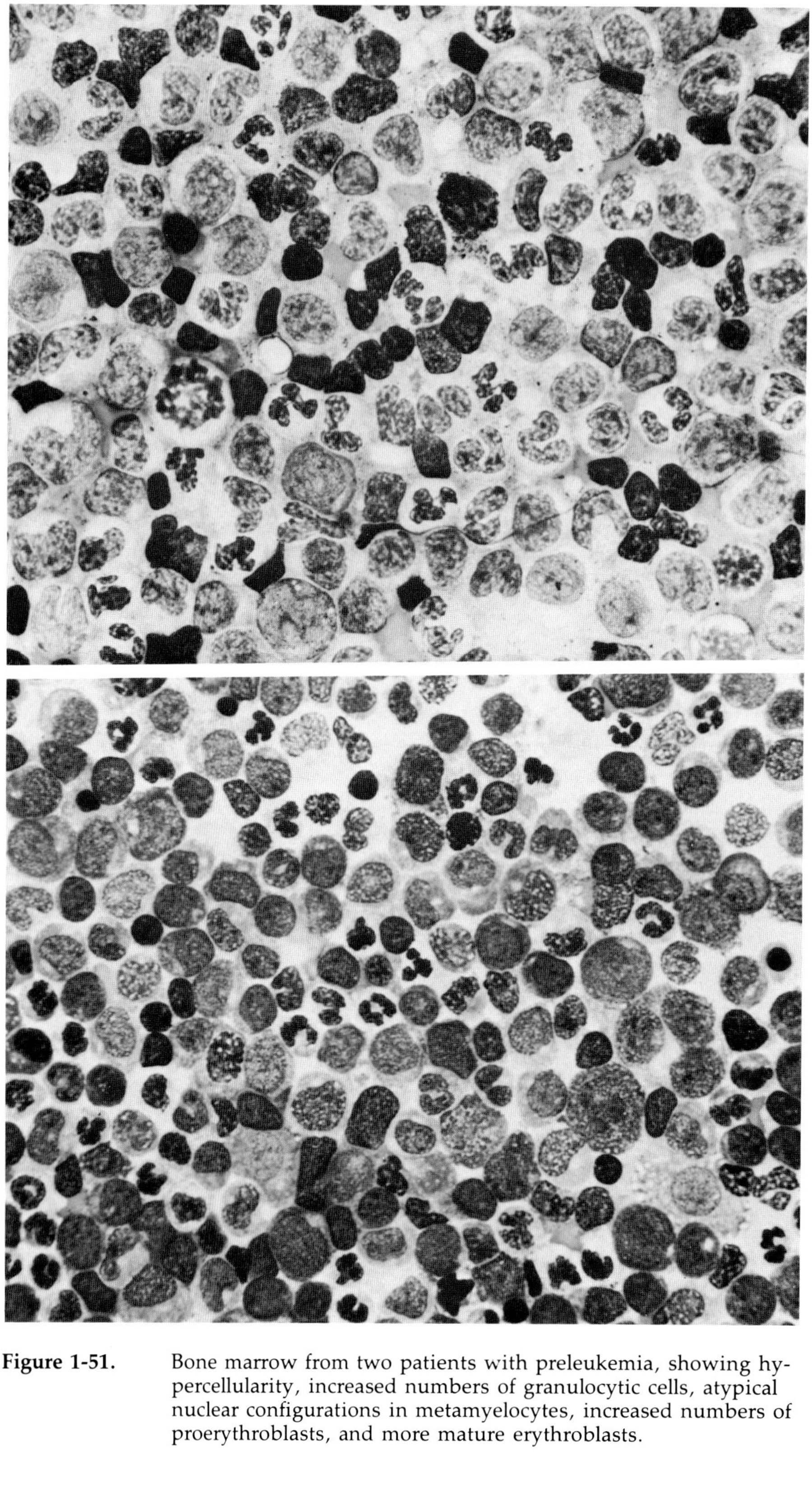

Figure 1-51. Bone marrow from two patients with preleukemia, showing hypercellularity, increased numbers of granulocytic cells, atypical nuclear configurations in metamyelocytes, increased numbers of proerythroblasts, and more mature erythroblasts.

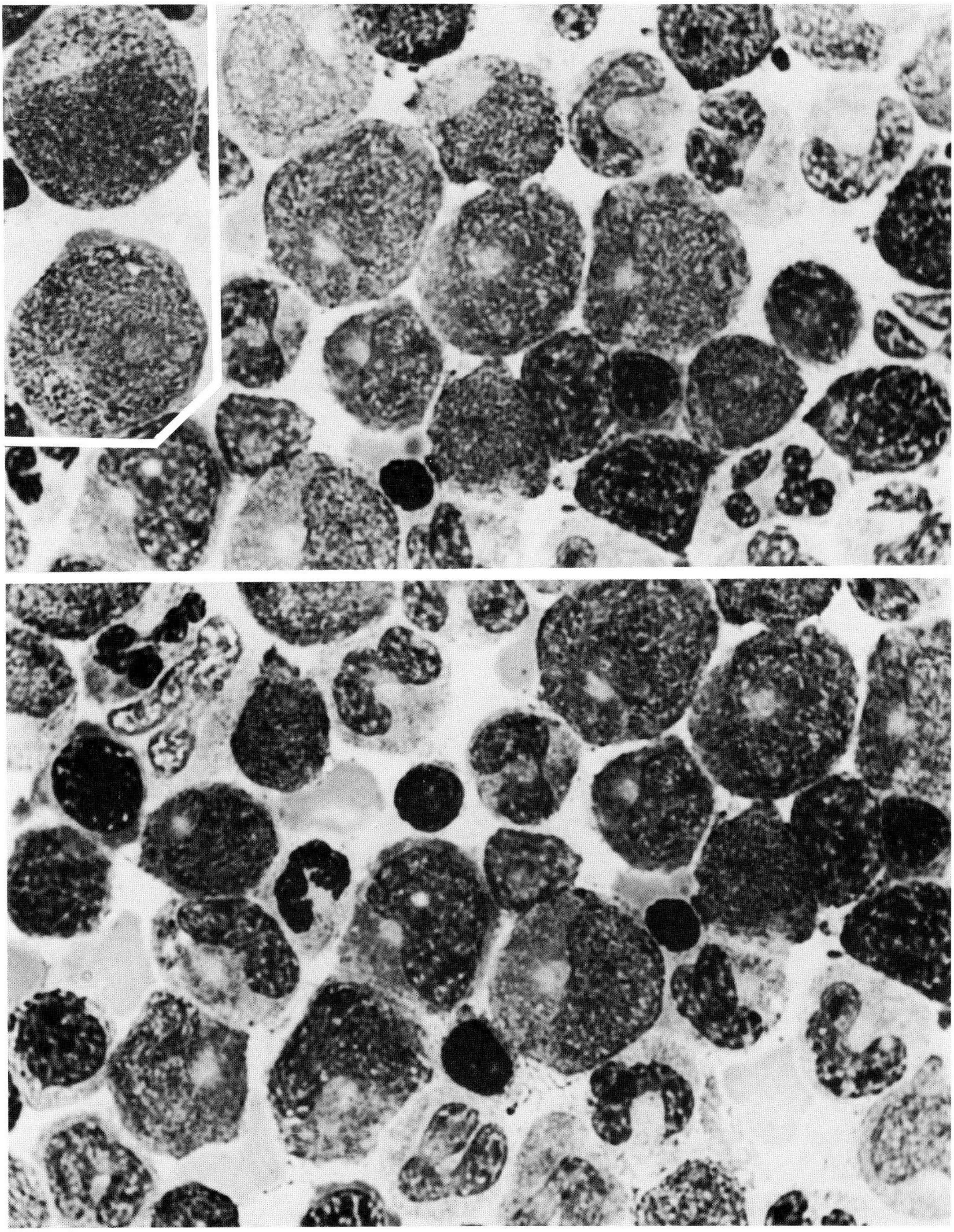

Bone marrow in preleukemia, showing increased numbers of promyelocytes as well as increased numbers of band forms, some of which have atypical nuclear configuration. **Figure 1-52.**

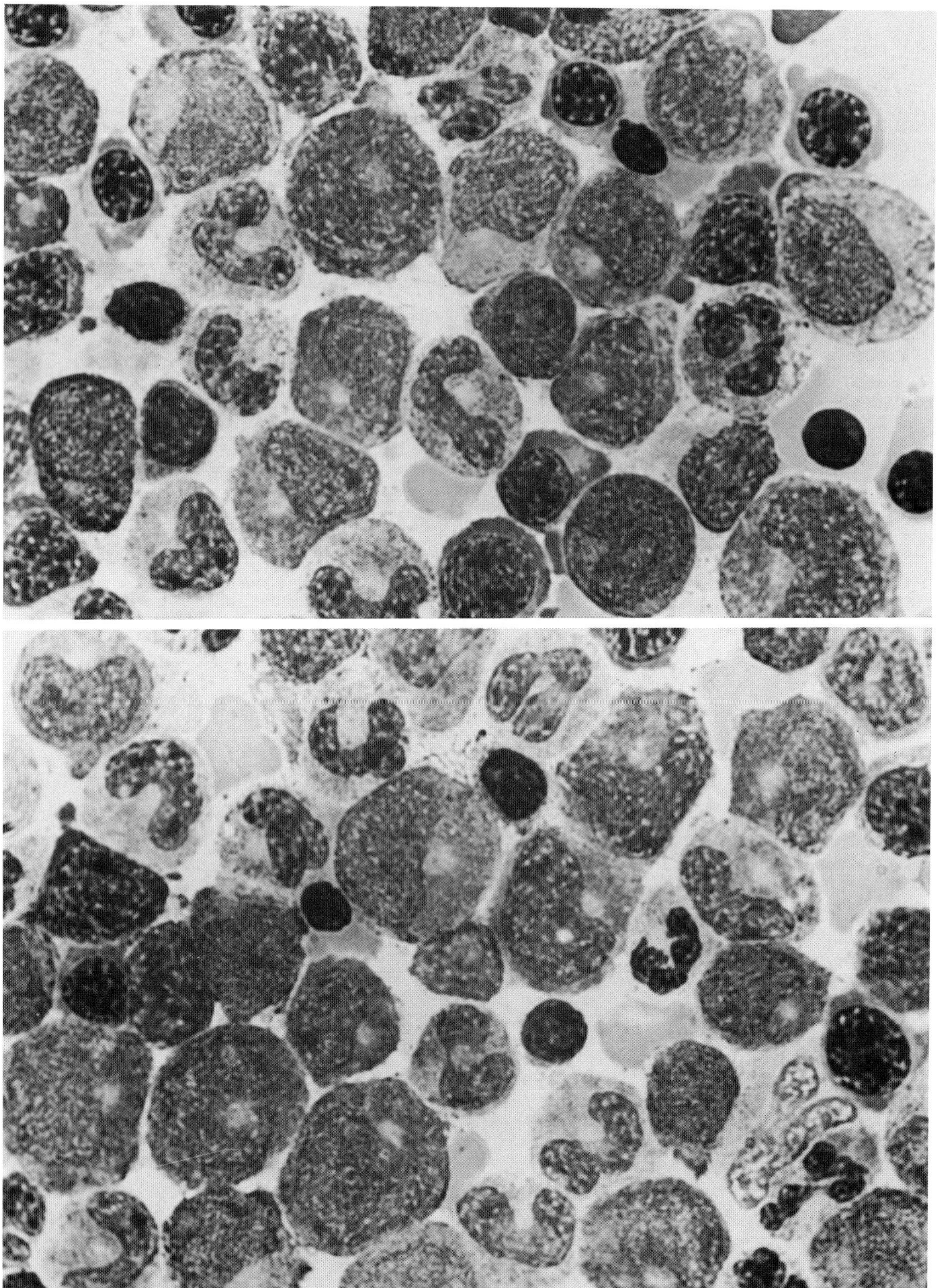

Figure 1-53. Bone marrow in preleukemia, showing increased numbers of myelocytes and atypical-appearing metamyeloctyes and band cells.

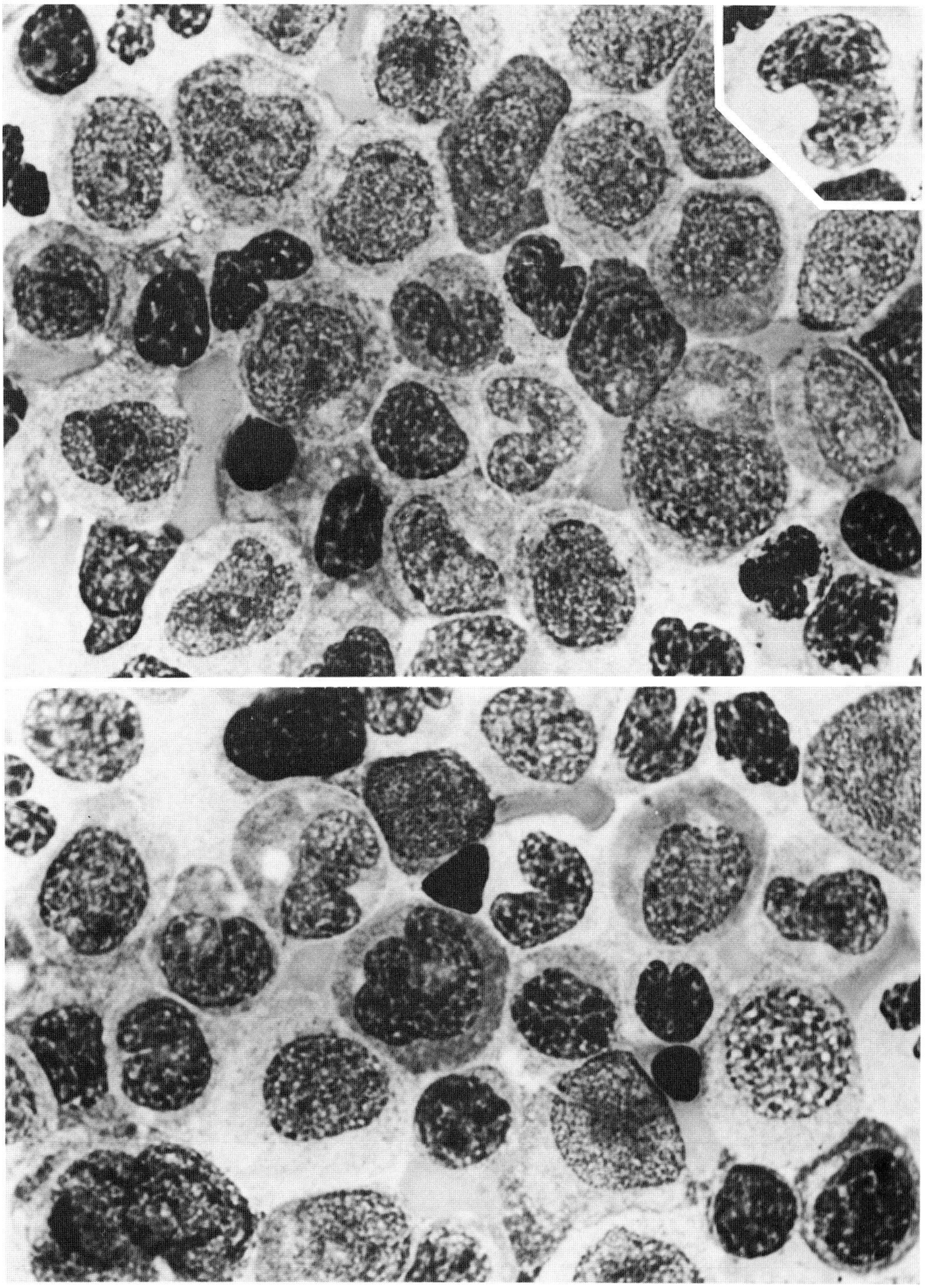

Bone marrow in preleukemia, depicting increased numbers of atypical-appearing metamyelocytes and bands. **Figure 1-54.**

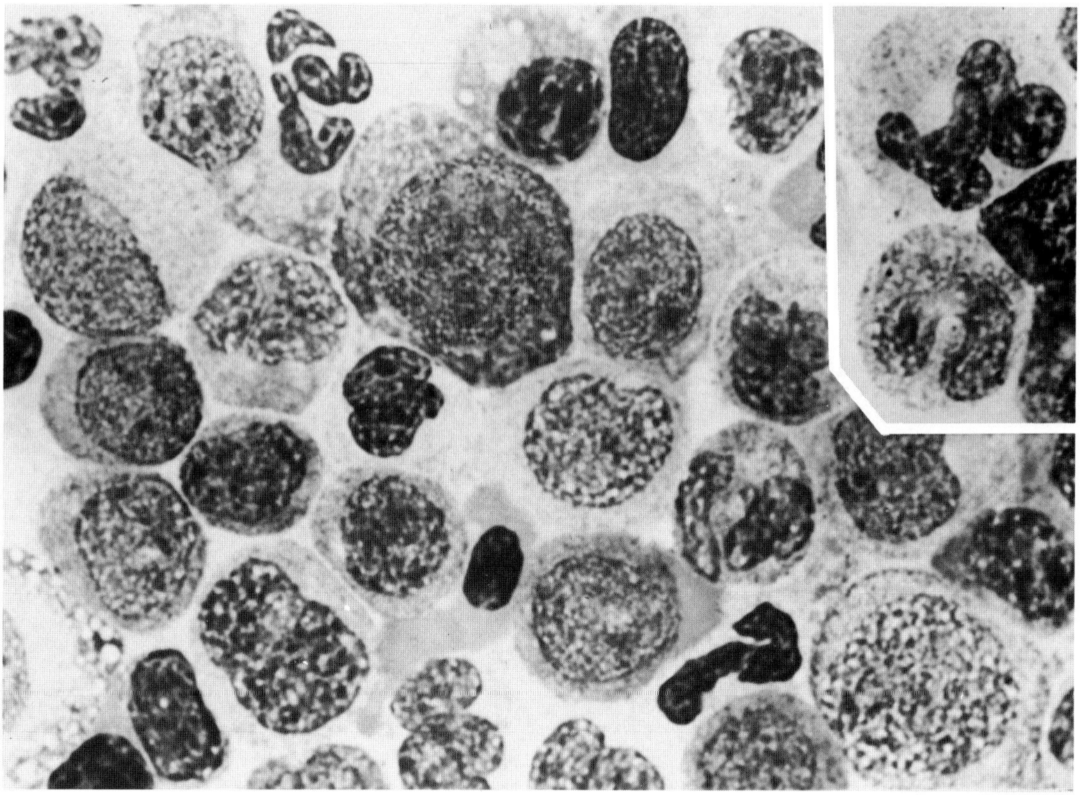

Figure 1-55. Bone marrow in preleukemia, showing increased numbers of atypical-appearing bands and neutrophils, with nuclear convolutions and bulbous protrusions.

Figure 1-56. Bone marrow in preleukemia, showing increased numbers of ➡ myelocytes and atypical appearing metamyelocytes.

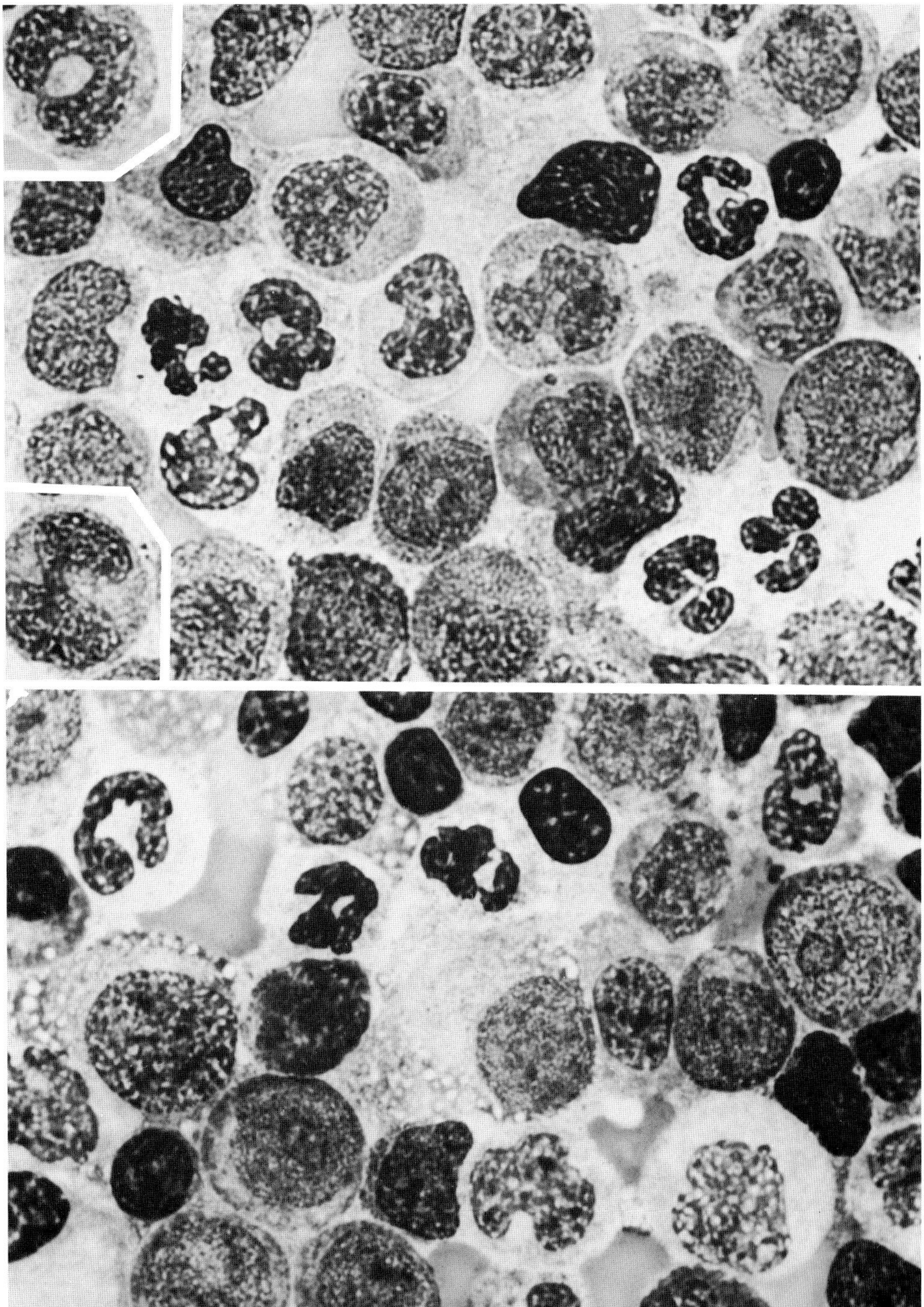

connections between adjacent lobes of neutrophils do not usually occur in the hypersegmented neutrophils of preleukemia. Eosinophil myeloctyes and mature eosinophils are increased in number in preleukemia (Fig. 1-58). Similarly, mature-appearing plasma cells are increased (Fig. 1-59 and Plate 1C). Megakaryocytes are increased in number. Some are large with multiple nuclear lobulations (Figs. 1-60, 1-61), and many are micromegakaryocytes[154,829,978] with a single unlobulated nucleus and richly granular cytoplasm (Figs. 1-62, 1-63).

Erythroblasts are increased in number in marrows from patients with preleukemia. Proerythroblasts are numerous, and megaloblastoid intermediate macronormoblasts occur in clusters (Fig. 1-64 and Plate 1*B*).

Recently, several morphologic abnormalities have been observed in erythroblasts in bone marrows of patients with preleukemia who have subsequently demonstrated evolution into acute myeloblastic or monocytic leukemia. Although these atypical erythroblasts occur in small numbers (less than 2% of the total number of erythroblasts in these marrows), their occurrence and appearance are sufficiently distinctive to make them potentially valuable as cytologic markers for leukemic evolution. First, atypical megaloblasts are found. These cells have unusually delicate nuclear chromatin with few aggregates of chromatin and centrally-placed location of the nucleus, giving the cell a fried-egg appearance (Fig. 1-65). Cytoplasm of these cells is often stellate in configuration and stains diffusely with the PAS reagent (Fig. 1-66, inset). Second, atypical megaloblasts with monocytic-appearing nuclear foldings occur. These erythroblasts demonstrate nucleocytoplasmic asynchronism, and their cytoplasm stains diffusely positive for glycogen with the PAS reagent (Fig. 1-67). Third, erythroblasts with prominent nucleoli, delicate nuclear chromatin pattern, and multiple cytoplasmic vacuoles of varying sizes can be identified (Figs. 1-68, 1-69, and Plate 1C). Using the PAS stain, these vacuoles reveal abundant amounts of glycogen, and they can be identified easily under the low-power lens of the microscope (Fig. 1-70).

Fourth, proerythroblasts in patients with preleukemia who subsequently demonstrate evolution into acute leukemia often contain multiple cytoplasmic pseudopodia (Fig. 1-71). In some instances, these pseudopodia appear detached from the main portion of the cytoplasm, suggesting increased cytoplasmic movements or increased fragility of the cytoplasm when making the coverslip film of bone marrow particles (Fig. 1-71). Fifth, some erythroblasts in erythroleukemia show nuclear chromatin strands that appear more smudged than adjacent erythroblasts (Fig. 1-72). Additionally, the area between the chromatin strands in these atypical erythroblasts contains globular-shaped deposits that stain orange with panoptic reagents like Wright's or Giemsa stains (Fig. 1-72). With continued experience and identification of these atypical erythroblasts in marrows of patients diagnosed as having preleukemia who ultimately demonstrate evolution into acute leukemia, it may be possible to utilize them along with other morphologic and cytogenetic information to predict the likelihood of leukemic evolution.

(text continued on page 72)

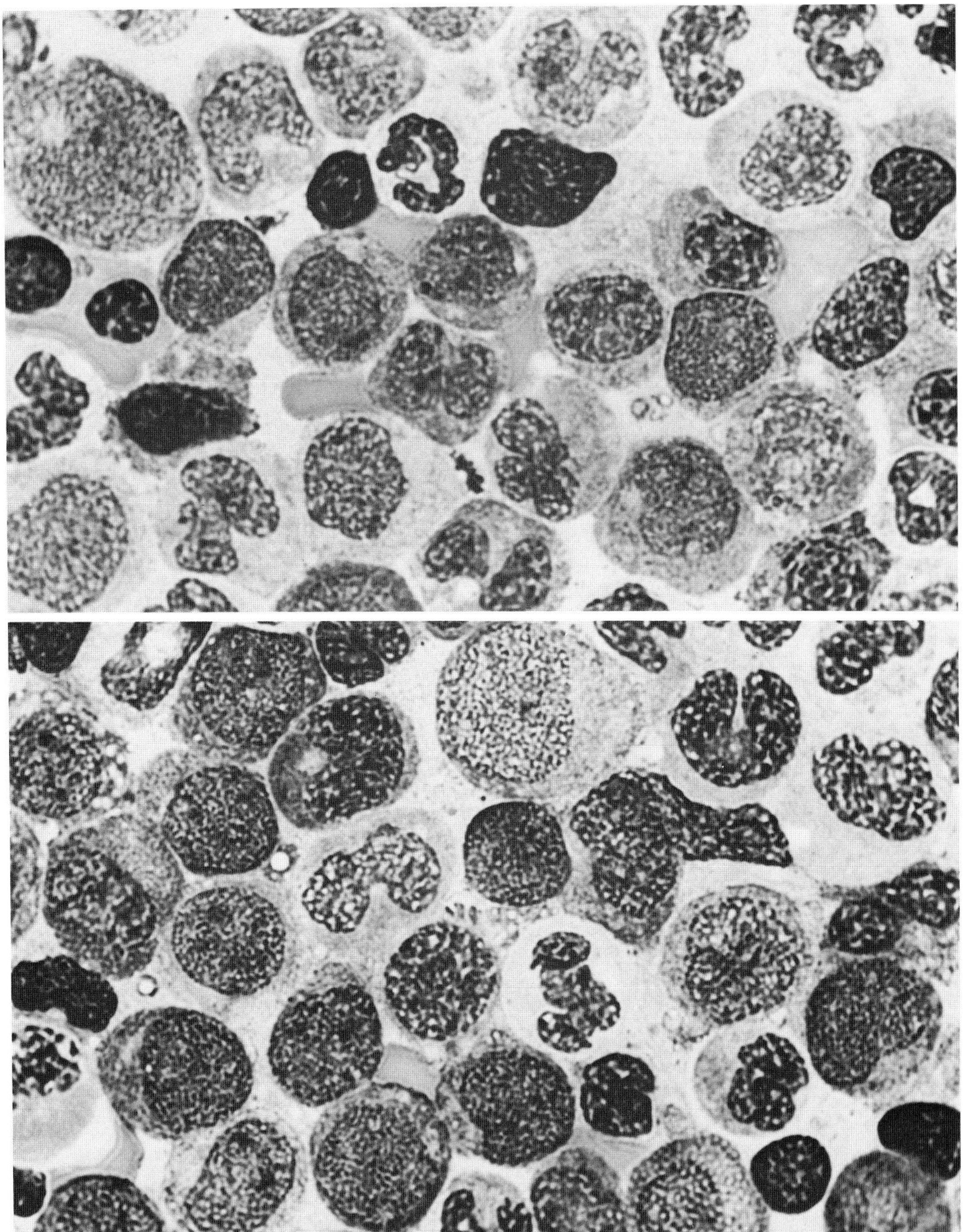

Figure 1-57. Bone marrow in preleukemia, illustrating increased numbers of myelocytes, eosinophils, and metamyelocytes, and bands with atypical nuclear configurations.

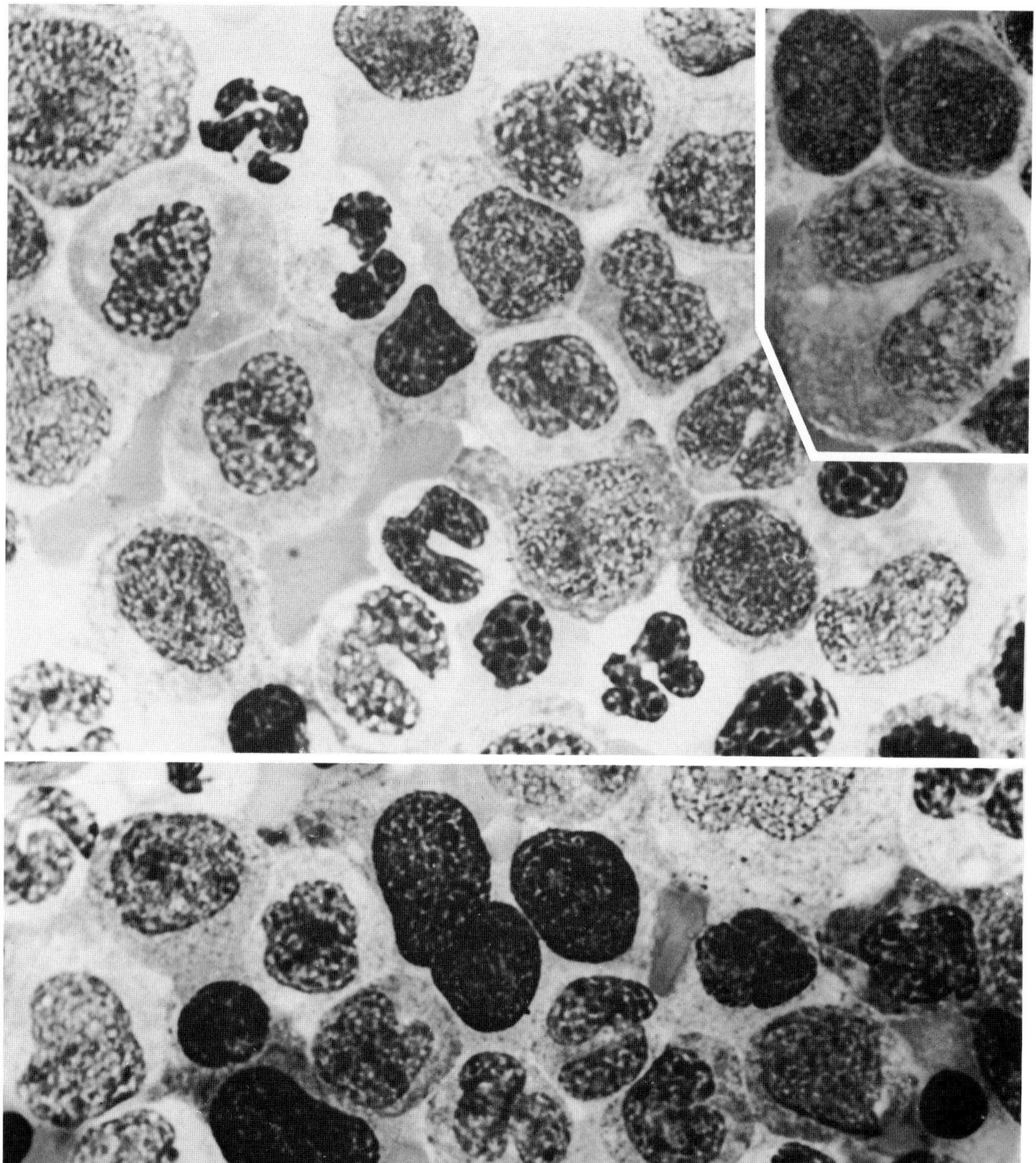

Figure 1-58. ← Bone marrow in preleukemia, showing increased numbers of eosinophils, bands with atypical nuclear configurations, and a multinuclear eosinophil and several myeloblasts (**inset**).

Figure 1-59. Bone marrow in preleukemia, demonstrating increased numbers of mature-appearing plasma cells.

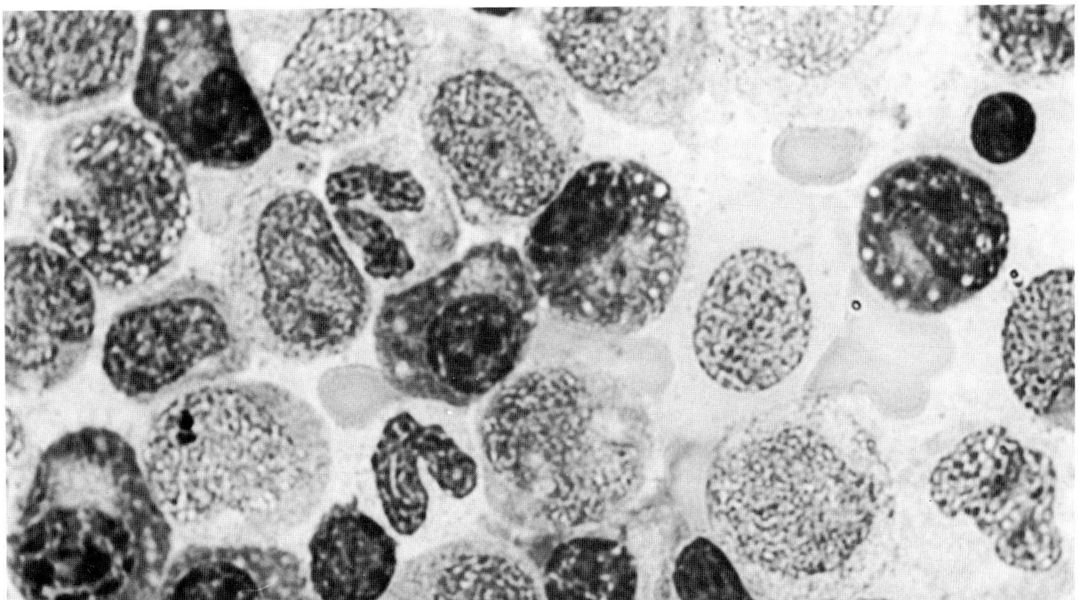

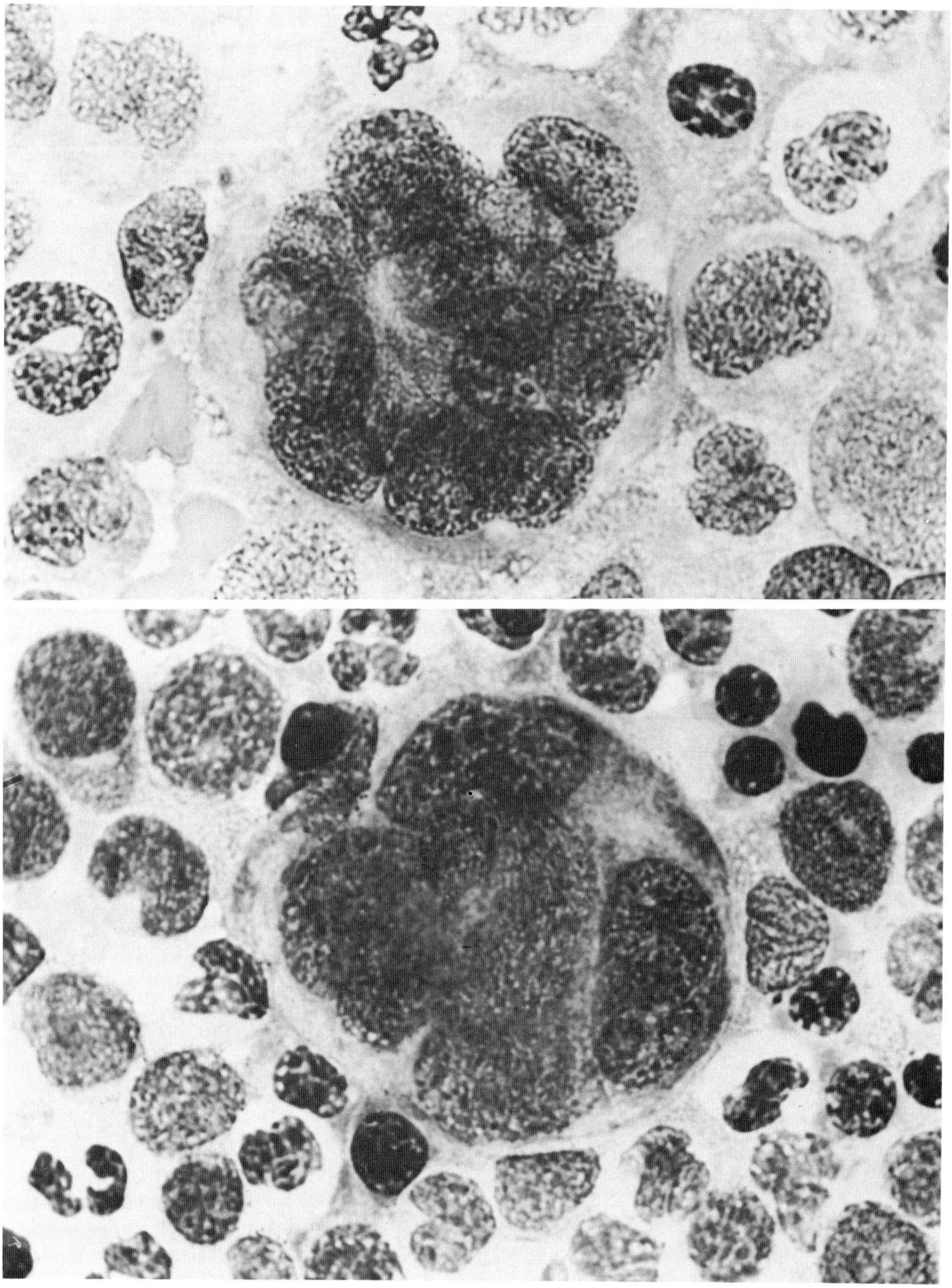

Figure 1-60. Bone marrow in preleukemia. Megakaryocytes are unusually large, with multiple nuclear lobulations and hypogranular cytoplasm.

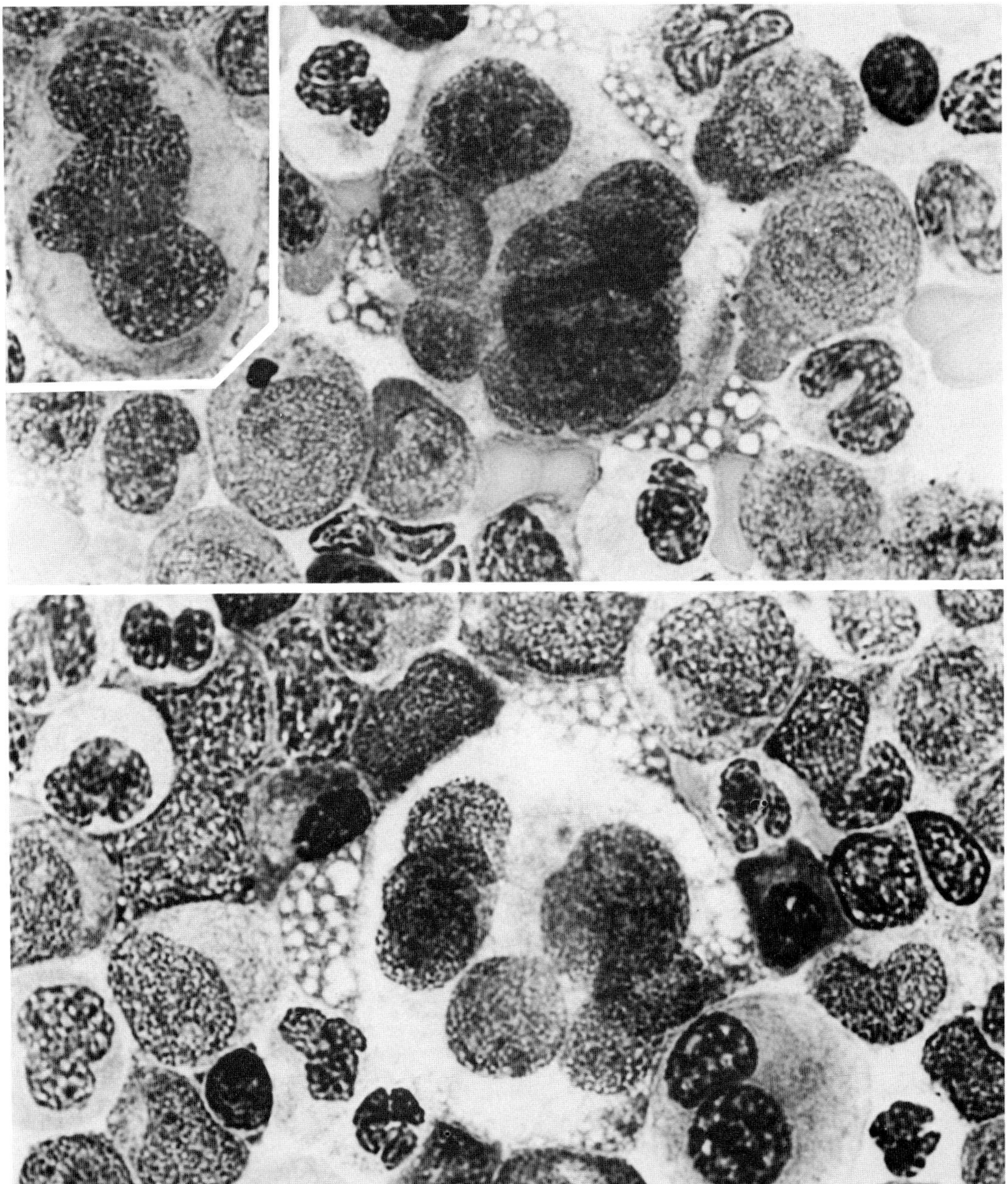

Figure 1-61. Bone marrow in preleukemia, showing megakaryocytes with multiple nuclear lobulations and hypogranular, vacuolated cytoplasm.

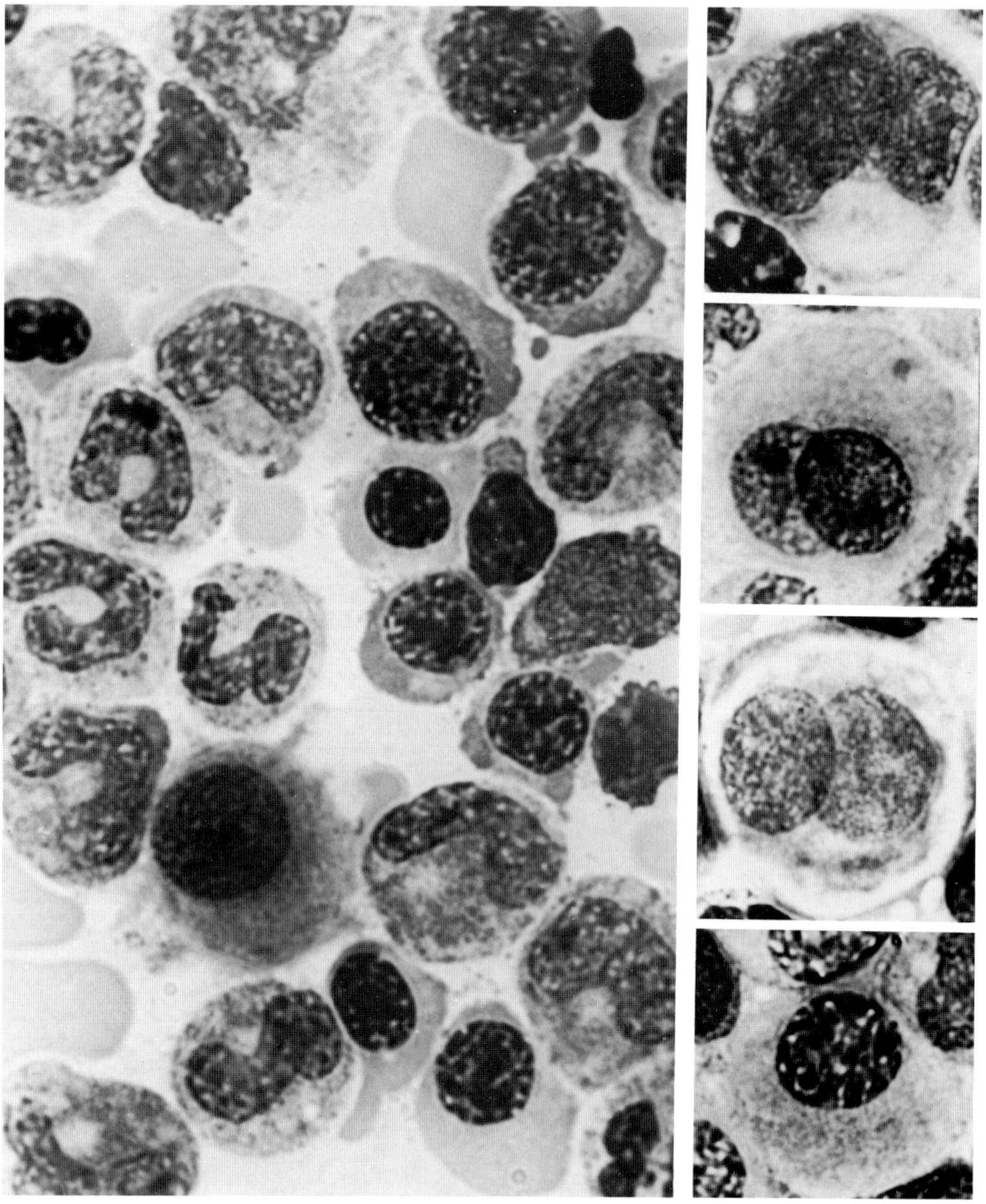

Figure 1-62. Bone marrow in preleukemia, depicting micromegakaryocytes with few or no nuclear lobulations and hypergranular cytoplasm. In some megakaryocytes, cytoplasm is pale and hypogranular.

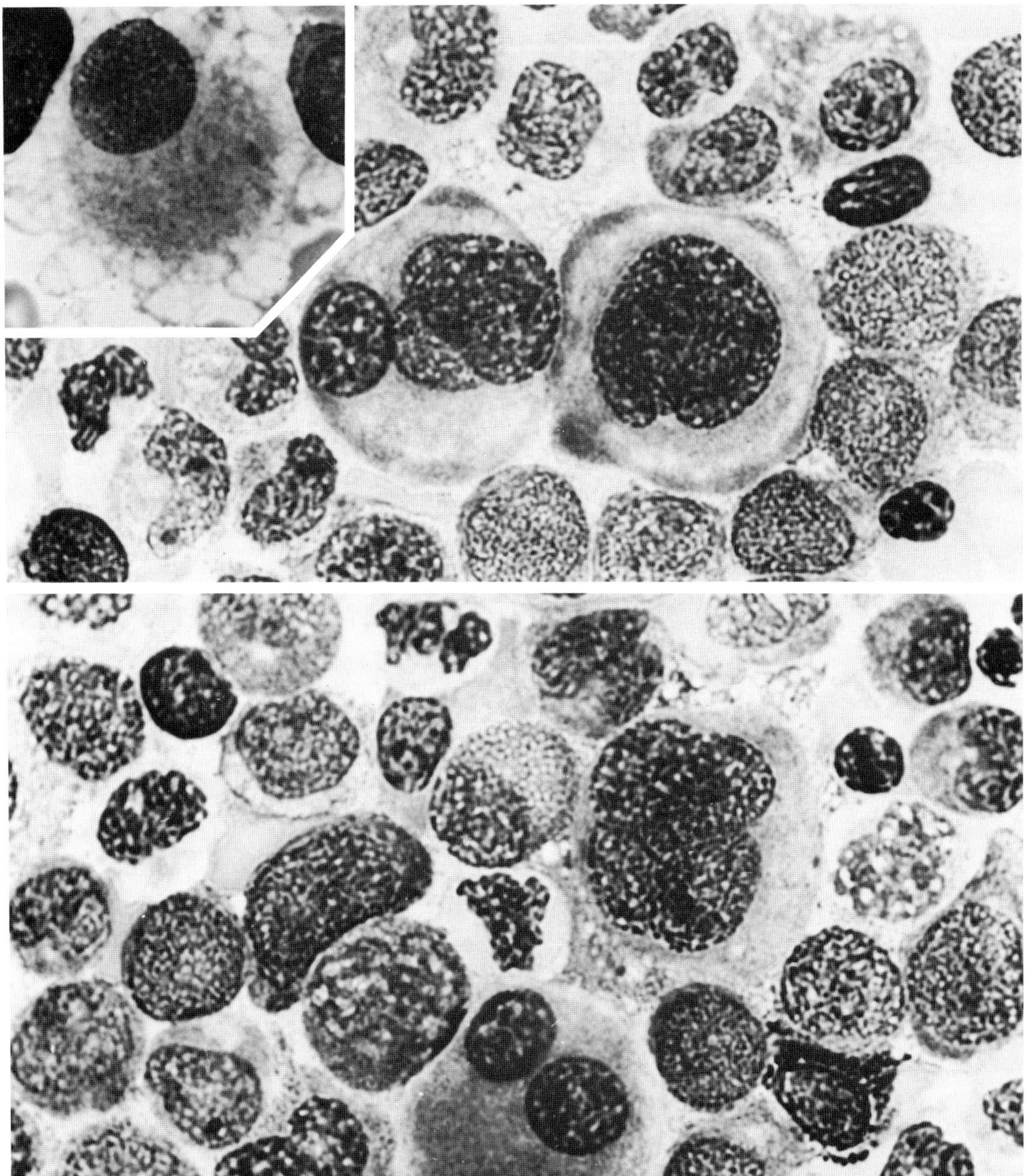

Bone marrow in preleukemia, showing increased numbers of eosinophils, plasma cells, and micromegakaryocytes. **Figure 1-63.**

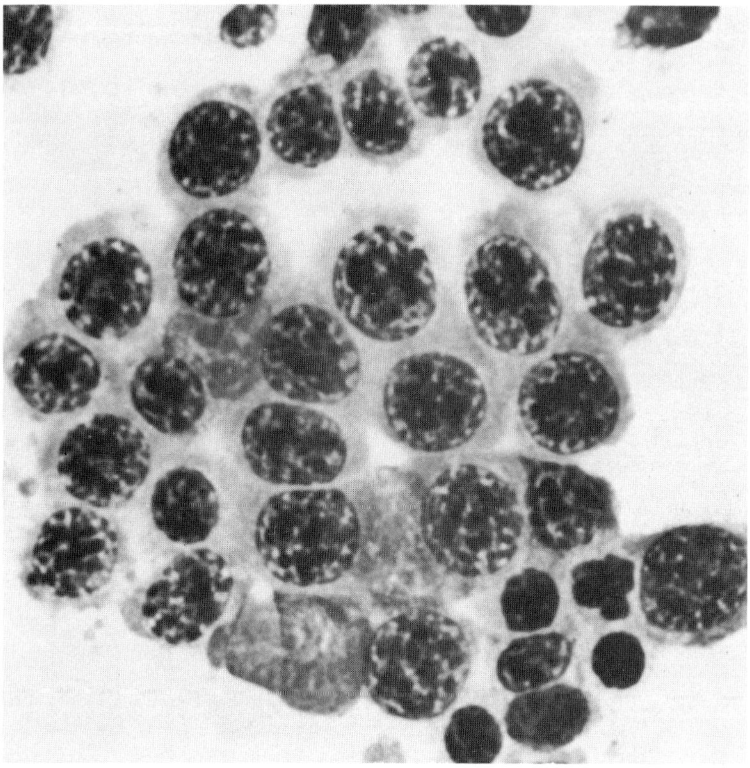

Figure 1-64. Bone marrow in preleukemia, showing cluster of megaloblastoid intermediate macronormoblasts.

Figure 1-65. Atypical megaloblasts in preleukemia **(top and arrow)**. These cells have centrally placed nuclei and delicate nuclear chromatin with few or no aggregates of chromatin. ➜

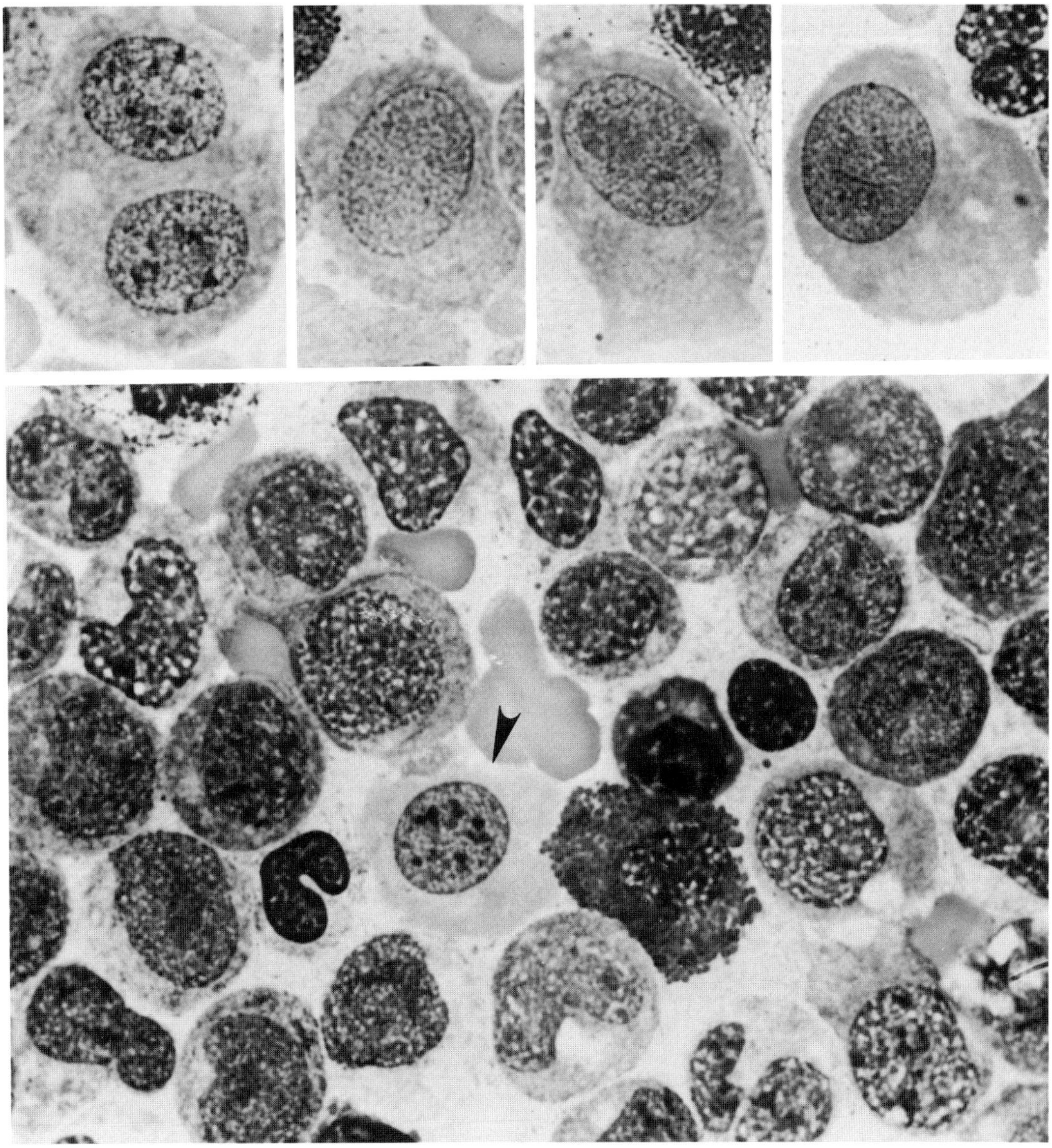

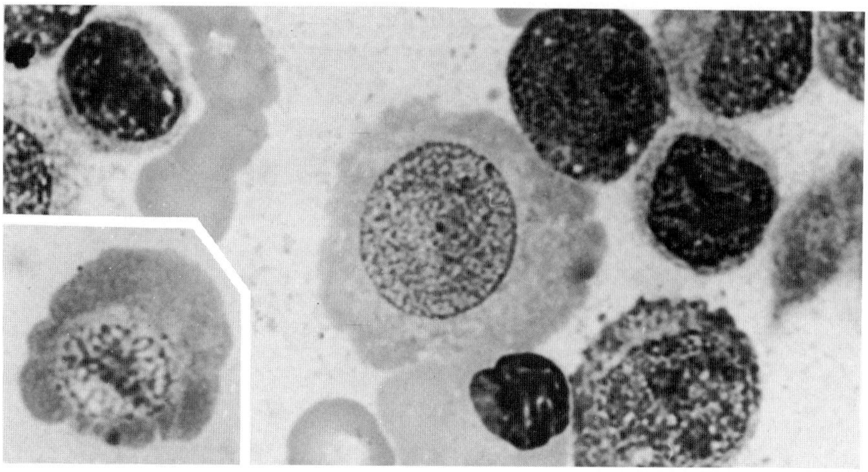

Figure 1-66. Atypical megaloblast in preleukemia. Inset shows diffuse staining of cytoplasm of atypical megaloblast by the PAS reagent.

Figure 1-67. Bone marrow in preleukemia, showing atypical megaloblasts with monocytoid-type foldings and convolutions of the nucleus **(top)**. These atypical megaloblasts often contain substantial amounts of glycogen, demonstrable with the PAS reagent **(bottom, arrow)**.

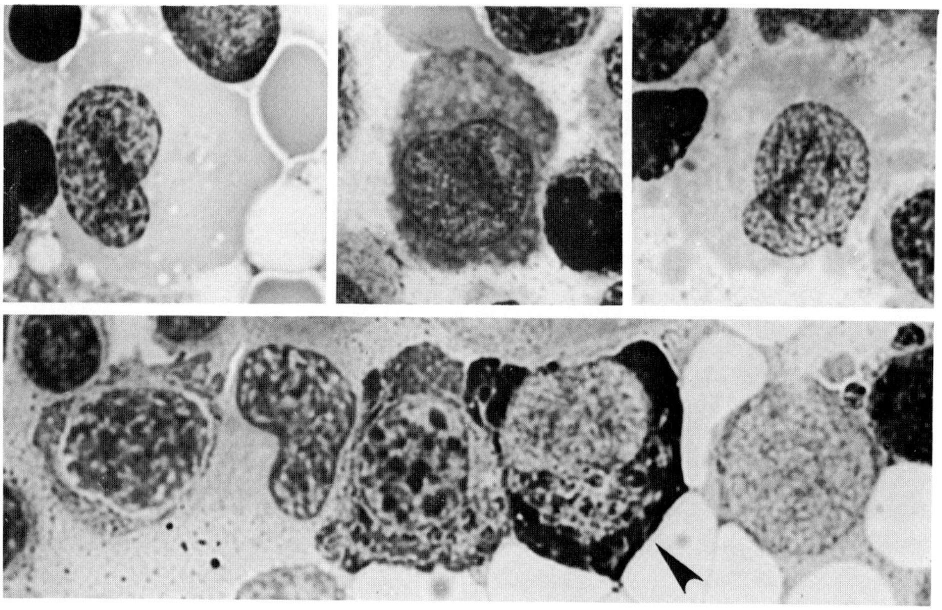

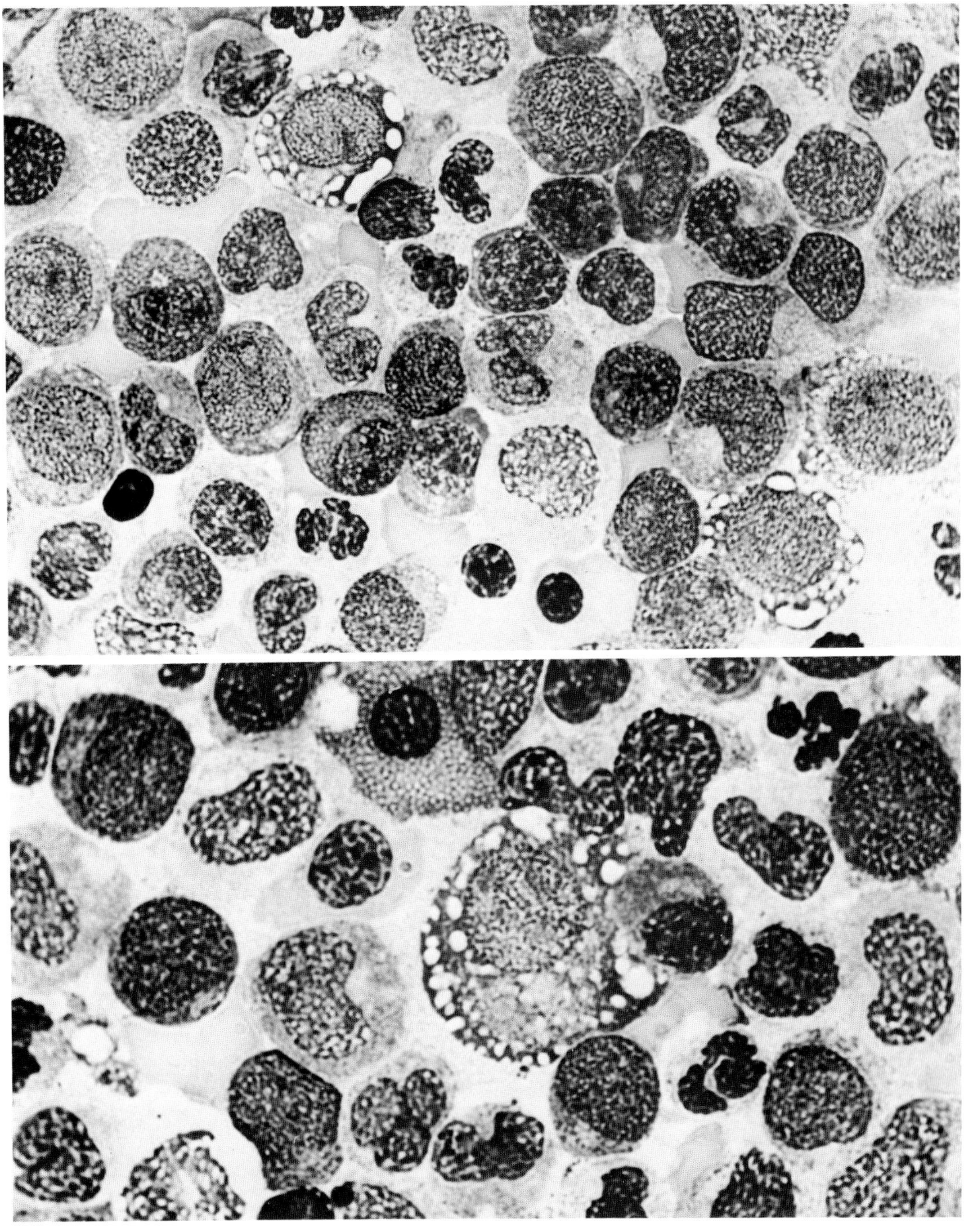

Figure 1-68. Bone marrow in preleukemia, demonstrating low- and higher-power views of atypical erythroblasts containing delicate megaloblastic-type nuclear chromatin and multiple cytoplasmic vacuoles.

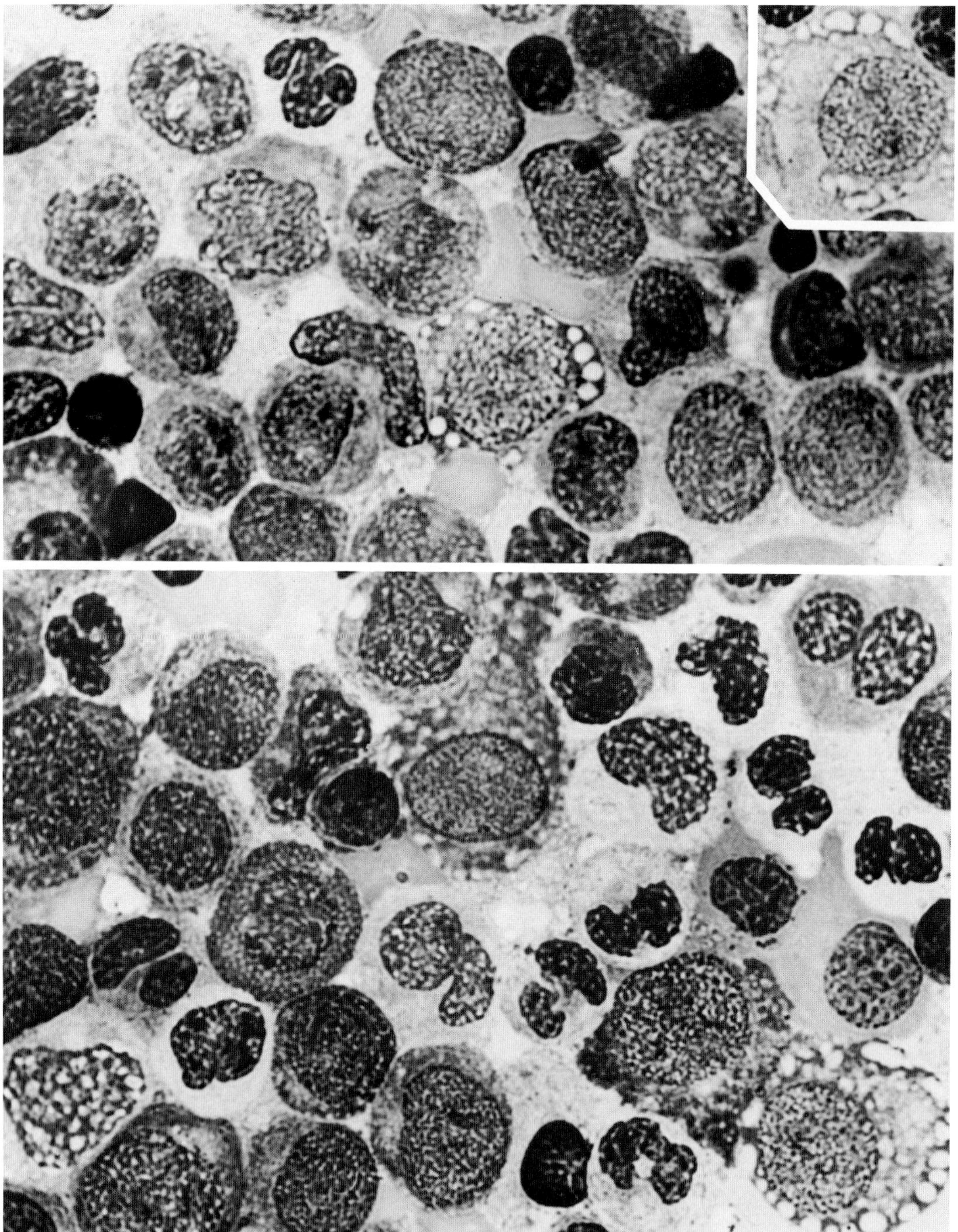

Figure 1-69. ← Bone marrow in preleukemia, showing atypical erythroblasts with delicate nuclear chromatin pattern and cytoplasmic vacuoles.

Figure 1-70. Erythroblasts in preleukemia, stained with Wright's stain, demonstrating prominent nucleolus, delicate nuclear chromatin, and multiple cytoplasmic vacuoles. Insets show atypical erythroblasts stained with the PAS reagent for glycogen, depicting globules of glycogen which correspond in location to the large vacuoles seen on the Wright-stained specimen.

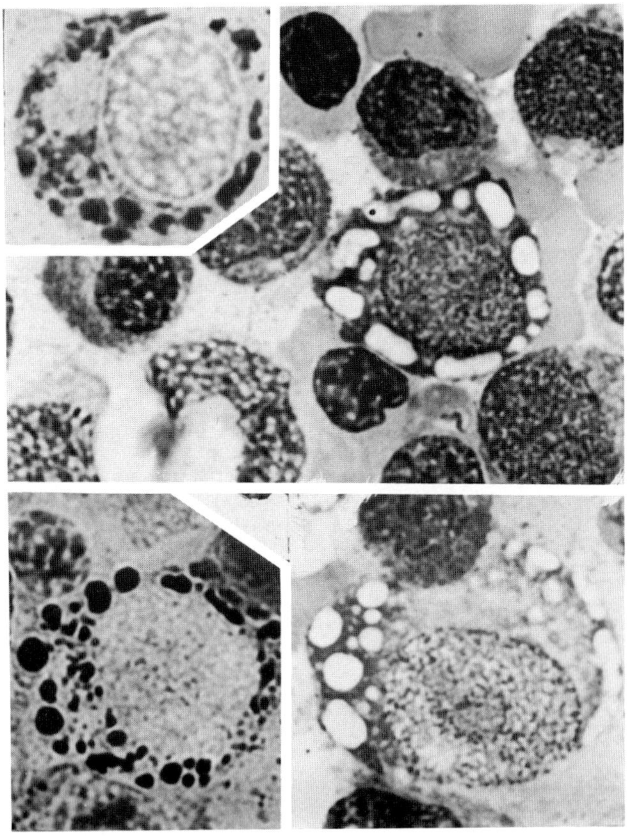

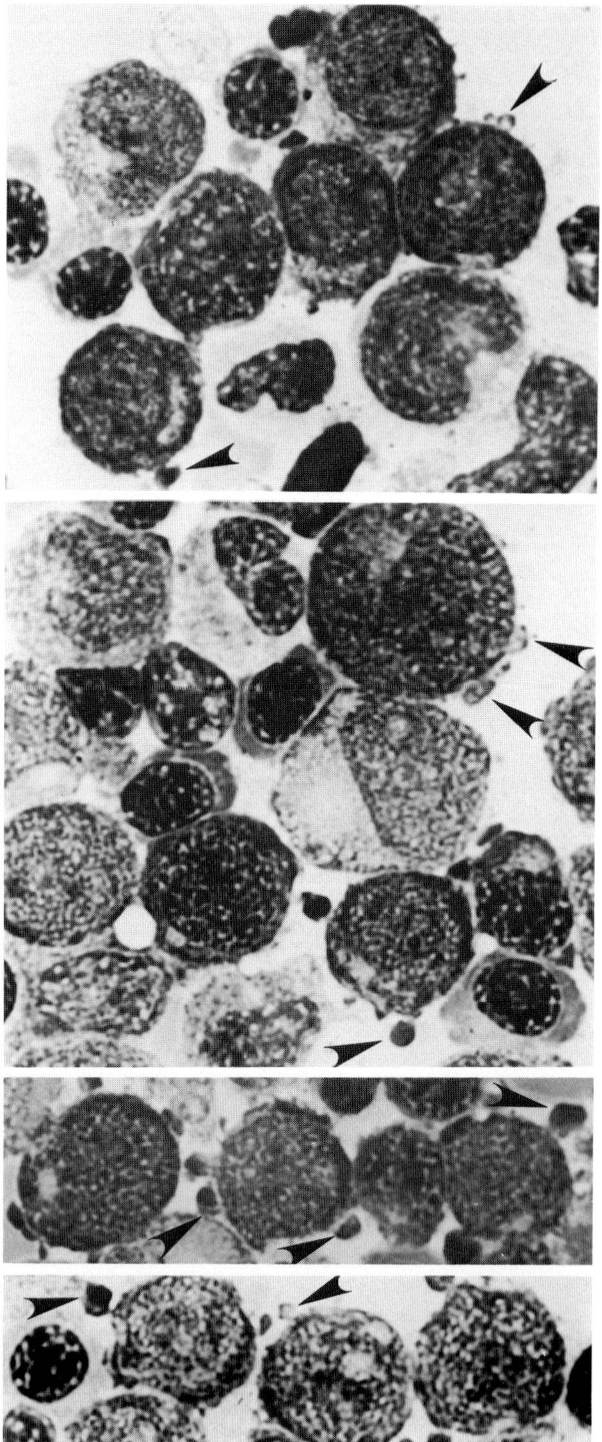

Figure 1-71. ← Proerythroblasts in preleukemia, demonstrating multiple cytoplasmic pseudopodia and cytoplasmic shedding (**arrows**).

Figure 1-72. Erythroblasts in preleukemia, illustrating areas of orange staining material (**arrows**) within the interstices of nuclear chromatin, presumably in areas of heterochromatin or metabolically repressed DNA.

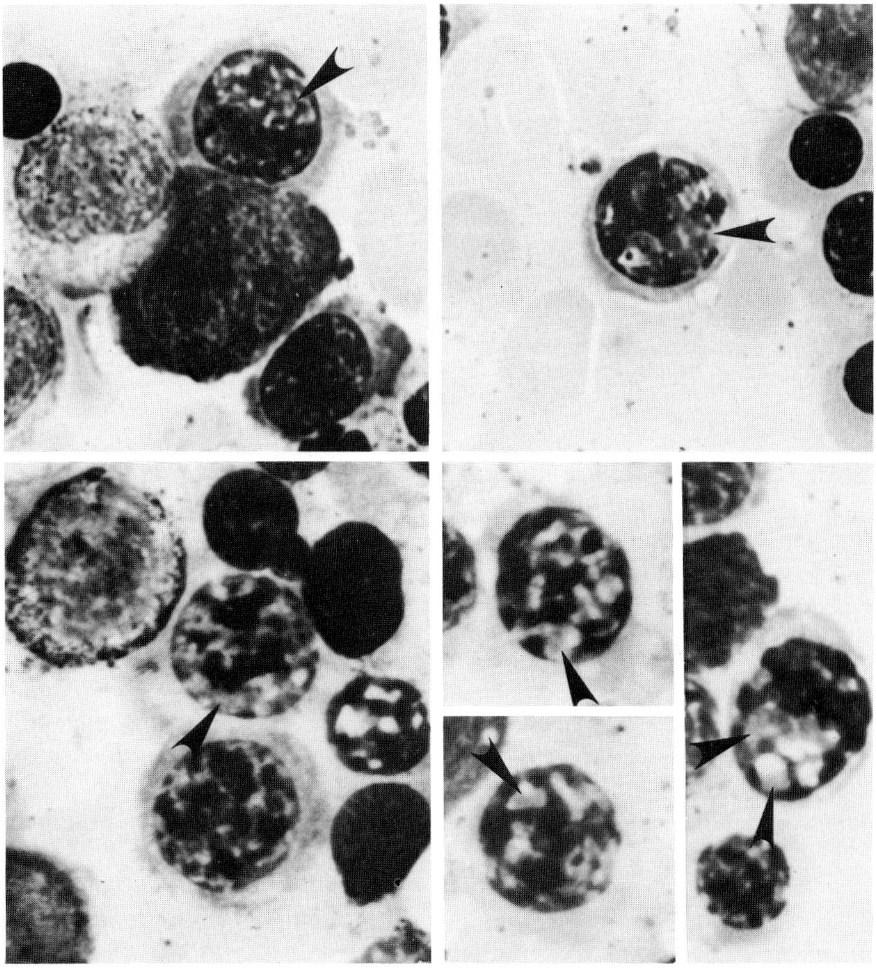

In another group of patients with preleukemic panmyelosis of the group labeled preleukemia, peripheral blood samples show pancytopenia. However, bone marrows show unusually intense granulopoiesis with increased numbers of granulocytic cells at all maturational stages, and bands and polymorphonuclear leukocytes are numerous. Characteristically, the cytoplasm of these bands and neutrophils is hypogranular, and contains a few pale-staining granules, as seen in Figure 1-73.[154] Megakaryocytes and erythroblasts are virtually absent. On H & E section of marrow biopsies from patients with this type of preleukemic panmyelosis, large numbers of mature and immature granulocytes are found, and megakaryocytes and erythroblasts are virtually absent. Cytochemically, the granulocytic cells in this type of preleukemic panmyelosis contain virtually no activity of specific esterase or myeloperoxidase. This variant may represent a peroxidase deficient type of preleukemia as described by others.[151–153,157,158,236]

H & E stained biopsies of bone marrow from patients with preleukemia show hypercellularity with increased numbers of both granulocytic cells and erythroblasts, as well as increased numbers of megakaryocytes, eosinophils, and plasma cells (Figs. 1-74, 1-75). Erythroblasts are increased in number and may occur in clusters (Fig. 1-75).

Cytochemically, granulocytes from marrows of most patients with preleukemia contain substantial activities of myeloperoxidase and specific esterase.[418–420] Nonspecific esterase activity is weak in these cells and resistant to inhibition by sodium fluoride. In some instances, scant amounts of glycogen can be demonstrated in erythroblasts using the PAS reagent. In other instances, erythroblasts do not contain glycogen in demonstrable amounts. With the Prussian blue reagent, ringed sideroblasts can be detected in the marrows of some patients with preleukemia; these are believed to represent a prognostic sign suggesting the likelihood of leukemic evolution. In other patients, ringed sideroblasts are not detectable. Activity of alkaline phosphatase is variable in granulocytic cells in preleukemia. Using the PAS stain for glycogen, the cytoplasm of granulocytic cells stains diffusely positive.

Smoldering Acute Leukemia

Smoldering acute leukemia is a disorder typified by macrocytic anemia and the occurrence of a few myeloblasts, normoblasts, atypical moncytes in the blood, and bone marrow that can vary in cellularity. Clinically, patients with smoldering acute leukemia often have fever and splenomegaly. In the bone marrow, approximately 5% to 20% myeloblasts are found, and there may be abnormalities of erythroblasts and more mature granulocytic cells, reflecting the panmyelotic disturbance.

Described under various names by different authors,[117,166,207,269–271,273,534,716,844,848,849,1015] smoldering acute leukemia, or refractory anemia with excess blasts, probably represents myeloblastic leukemia in a slowly developing or indolent form. In some cases, evolution into acute leukemia

(text continued on page 76)

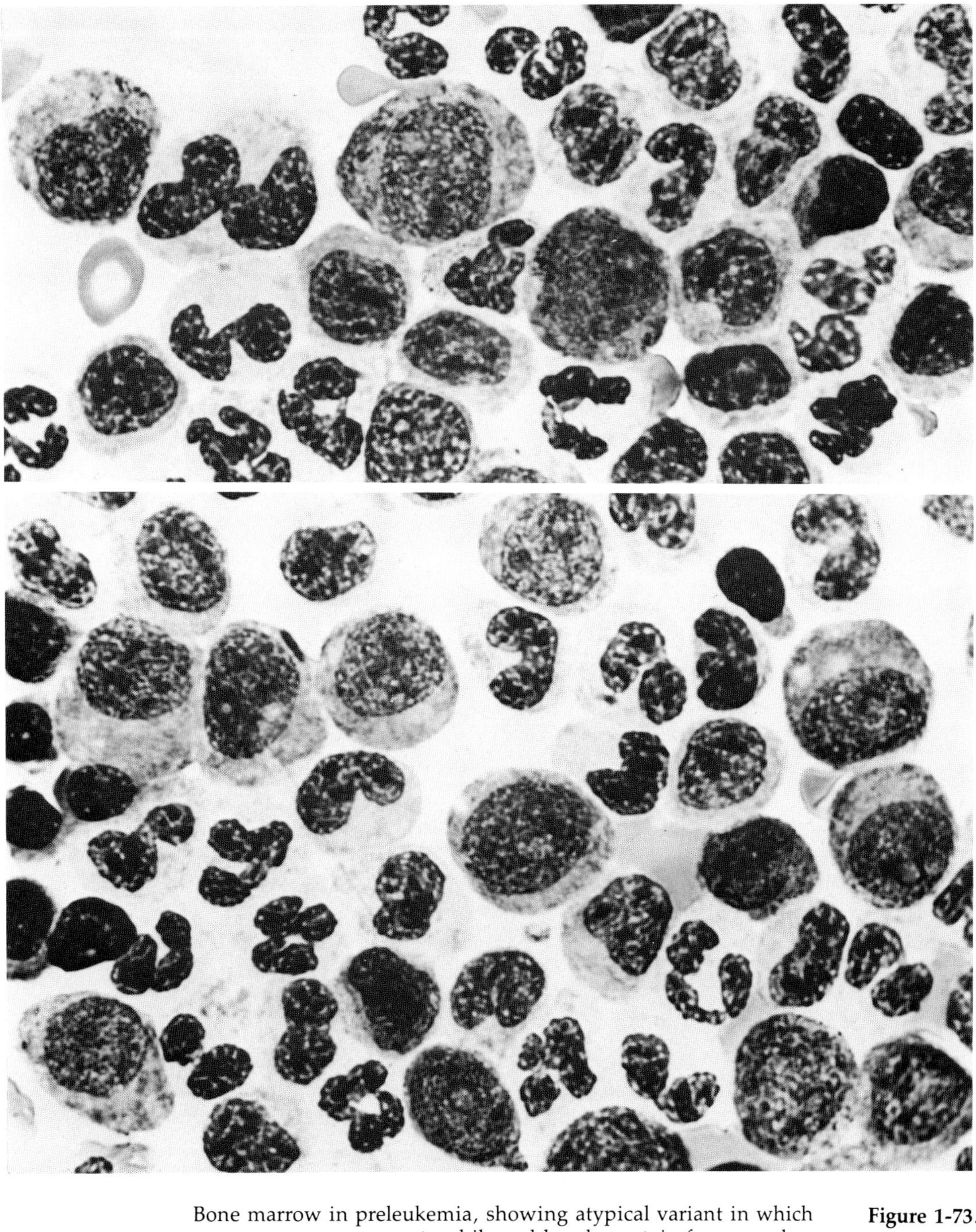

Bone marrow in preleukemia, showing atypical variant in which neutrophils and bands contain few granules.

Figure 1-73.

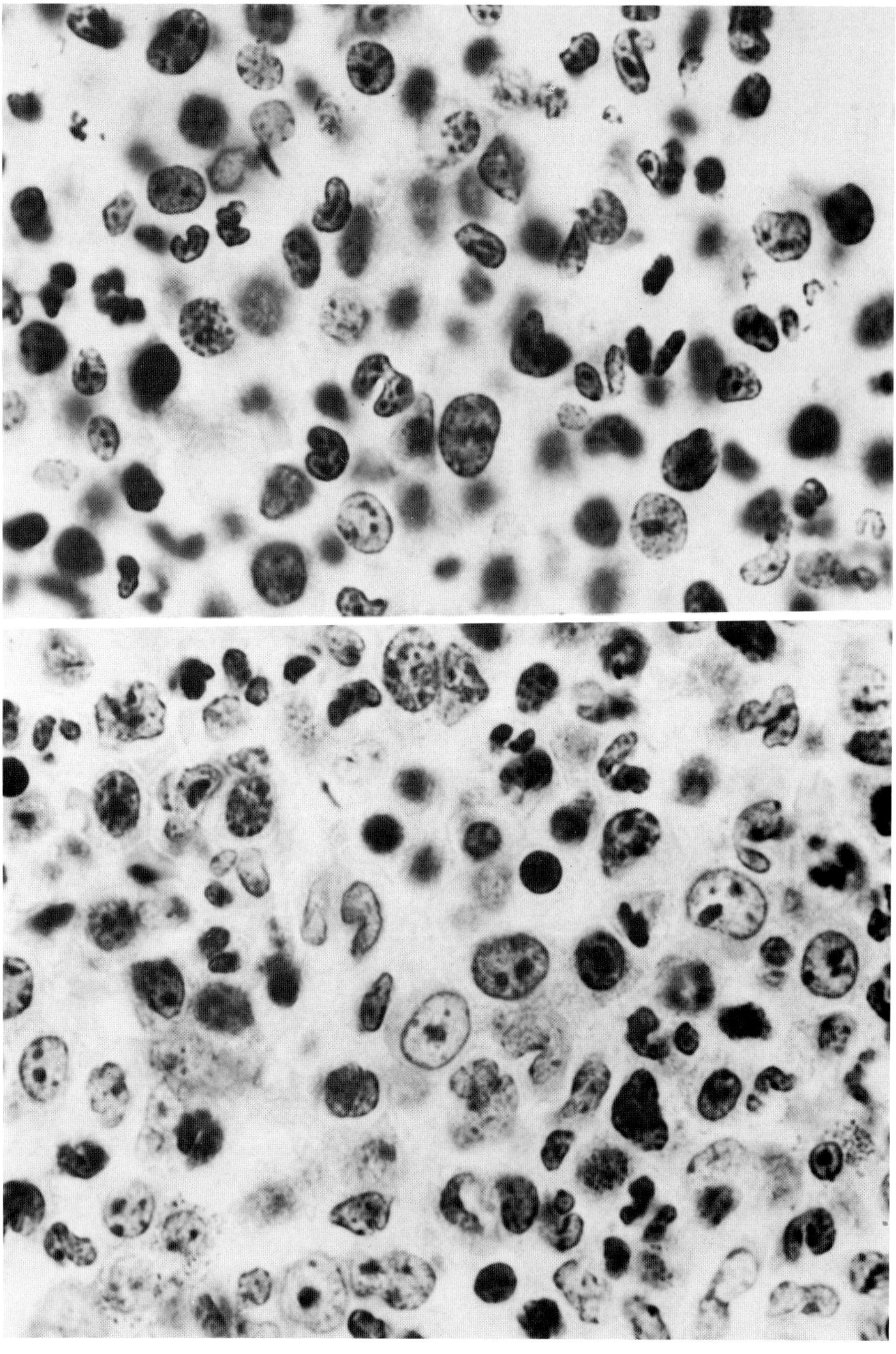

Figure 1-74. H & E section of bone marrow biopsy in preleukemia, demonstrating increased numbers of granulocytic precursors and mature neutrophils.

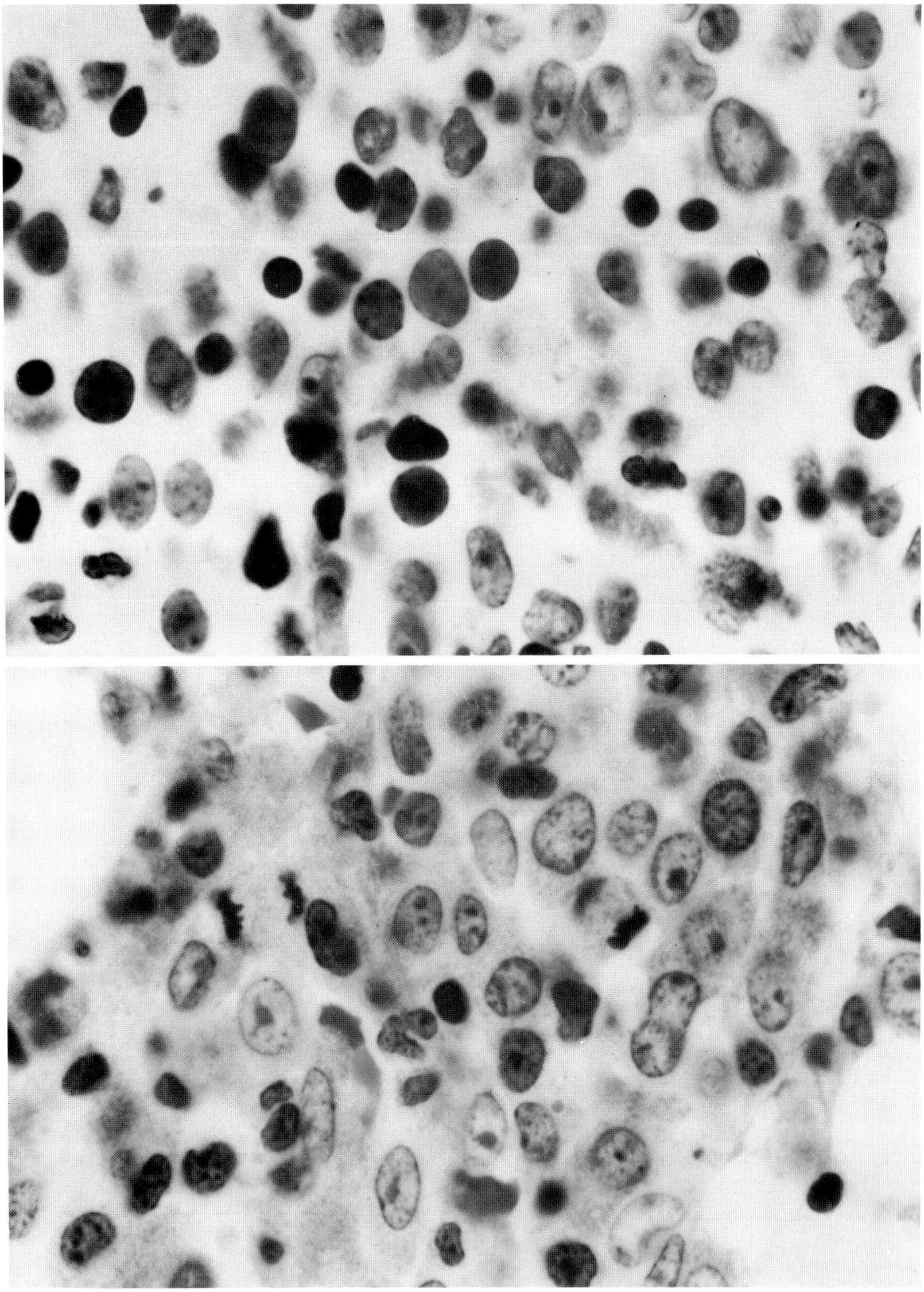

Figure 1-75. H & E section of bone marrow biopsy in preleukemia. Increased numbers of erythroblasts with darkly staining nuclei (**top**). Increased numbers of immature granulocytic cells, probably myelocytes and promyelocytes with pale-appearing nuclei containing dustlike nuclear chromatin (**bottom**). Several mitotic figures are seen.

does not occur, and patients succumb to consequences of pancytopenia. In other patients, the tempo of myeloblastic or monocytic proliferation is variable, and absolute numbers of myeloblasts or monocytes increase slowly or rapidly over a period of time.

Peripheral blood samples from most patients with smoldering acute leukemia demonstrate pancytopenia. Occasionally, several myeloblasts, atypical neutrophils, monocytes, giant erythrocytes, and normoblasts are present. Marrows from patients with smoldering acute leukemia are usually hypercellular (Fig. 1-76 and Plate 1*D*) but can be normocellular or hypocellular. Myeloblasts constitute approximately 20% or less of the bone marrow differential cell count, and range from 5% to 20%. In most instances, Auer rods are not detectable in the myeloblasts. Both normal-sized myeloblasts and micromyeloblasts occur (Figs. 1-77, 1-78, 1-79).

(text continued on page 80)

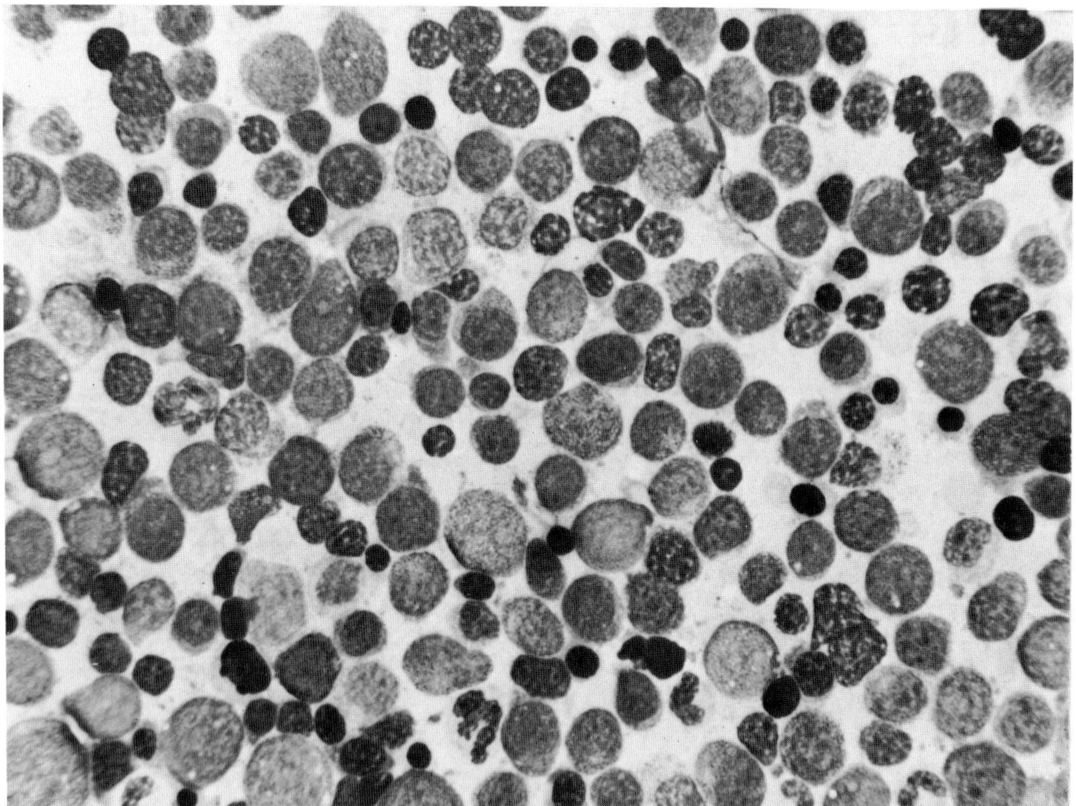

Figure 1-76. Bone marrow in smoldering acute leukemia, showing increased numbers of myeloblasts and promyelocytes having prominent nucleoli, at low magnification.

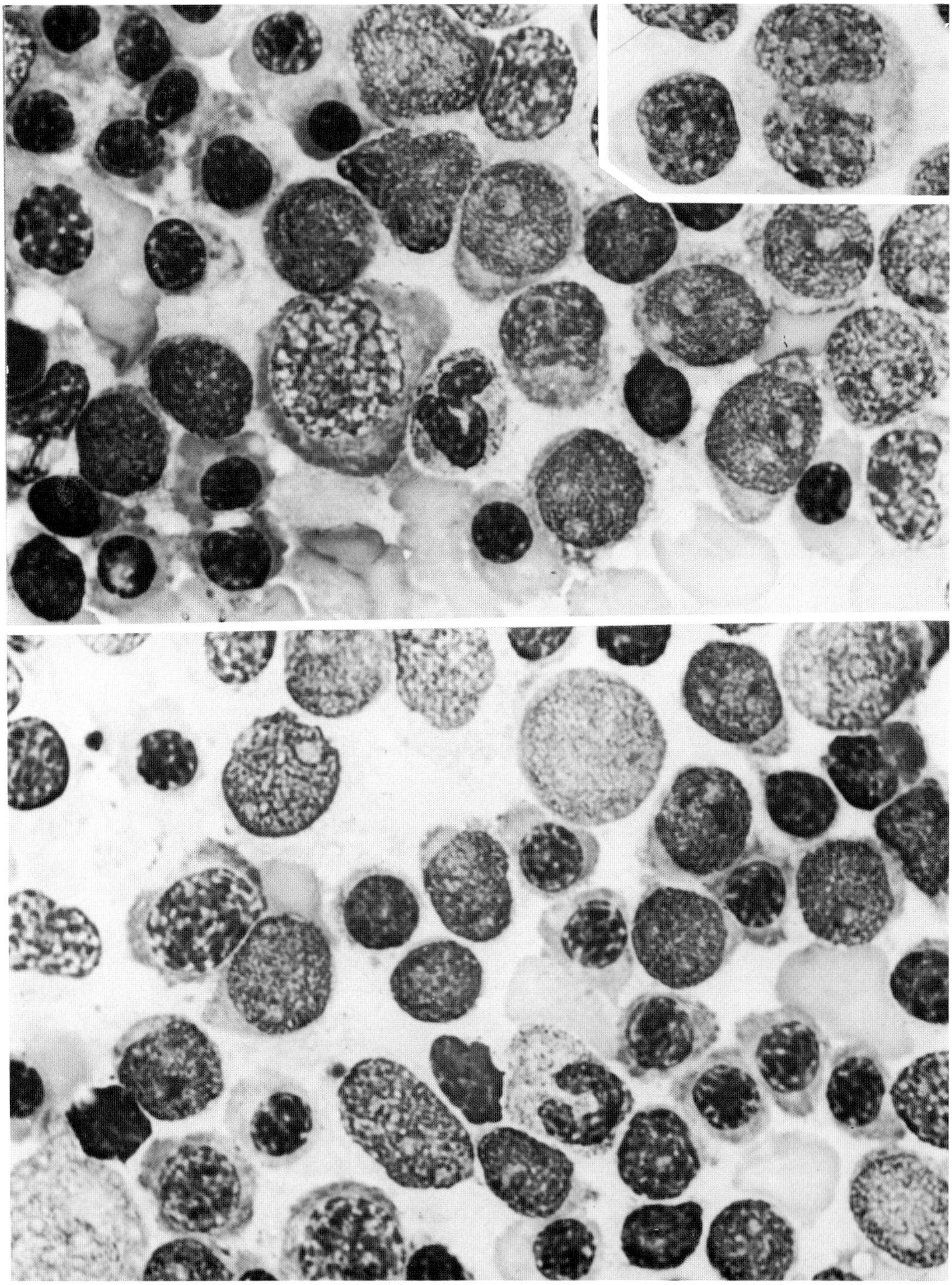

Figure 1-77. Bone marrow in smoldering acute leukemia, showing increased numbers of myeloblasts with prominent nucleoli, as well as increased numbers of megaloblastoid intermediate macronormoblasts. Inset shows gigantic metamyelocyte with aberrant nuclear configuration, and a small myeloblast.

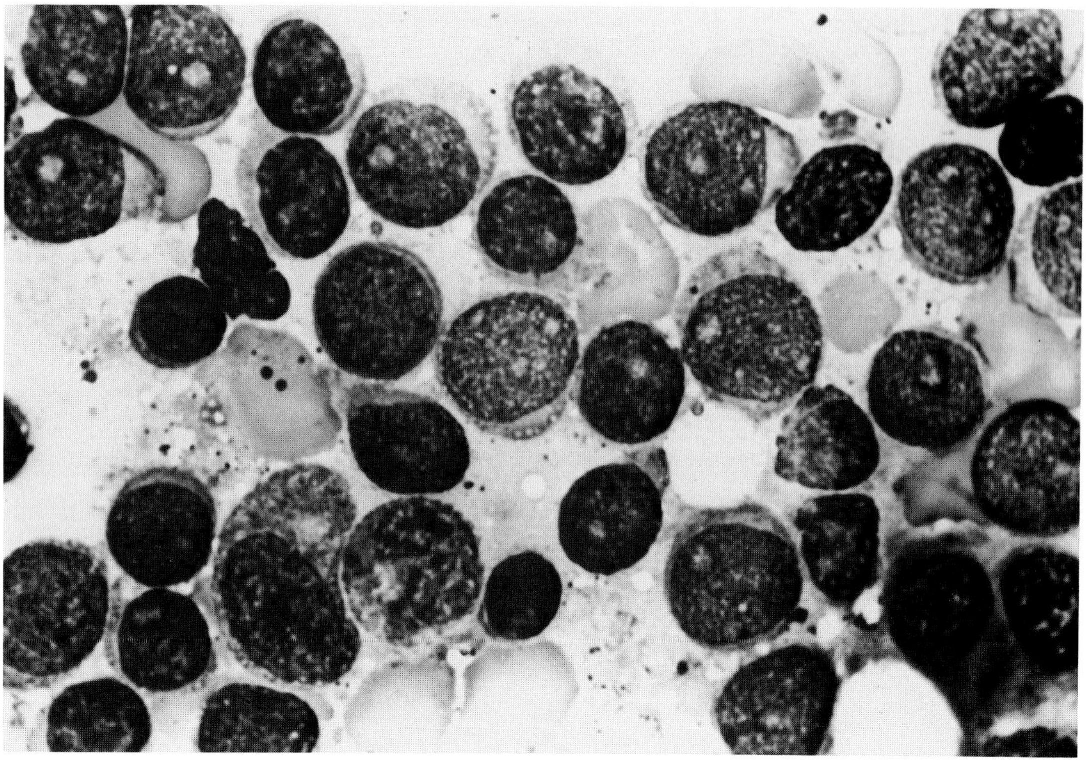

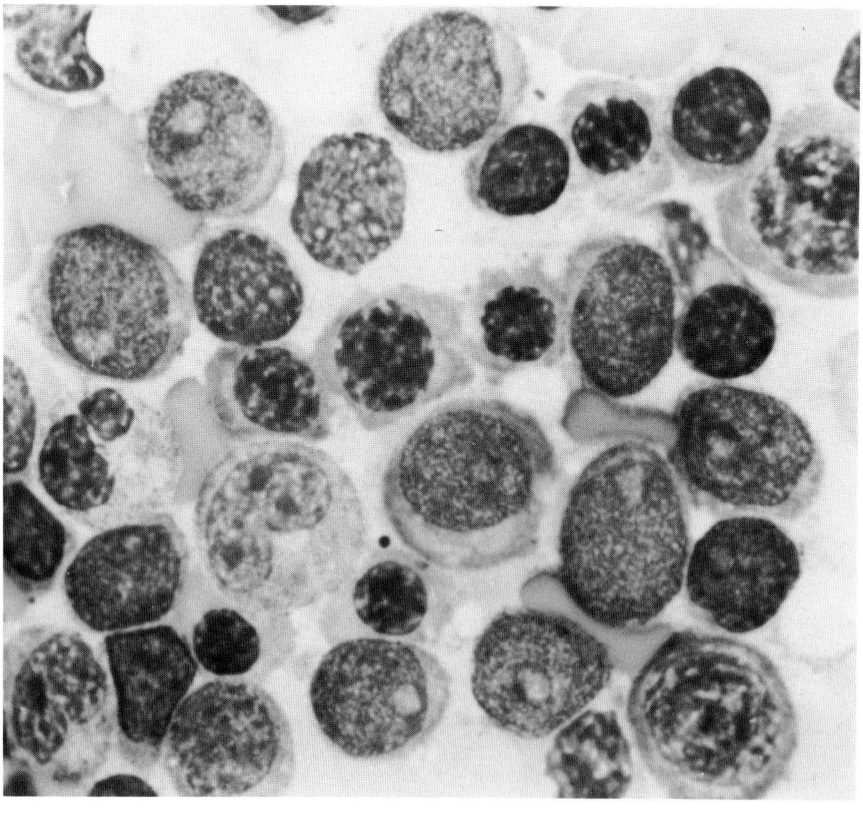

Figure 1-78. ← Bone marrow in smoldering acute leukemia, depicting increased numbers of small myeloblasts.

Figure 1-79. ← Bone marrow in smoldering acute leukemia, depicting increased numbers of myeloblasts with prominent nucleoli, and several megaloblastoid intermediate macronormoblasts.

Figure 1-80. Bone marrow in smoldering acute leukemia. Several myeloblasts and promyelocytes are shown along with megaloblastoid intermediate macronormoblasts and band forms having aberrant nuclear configurations.

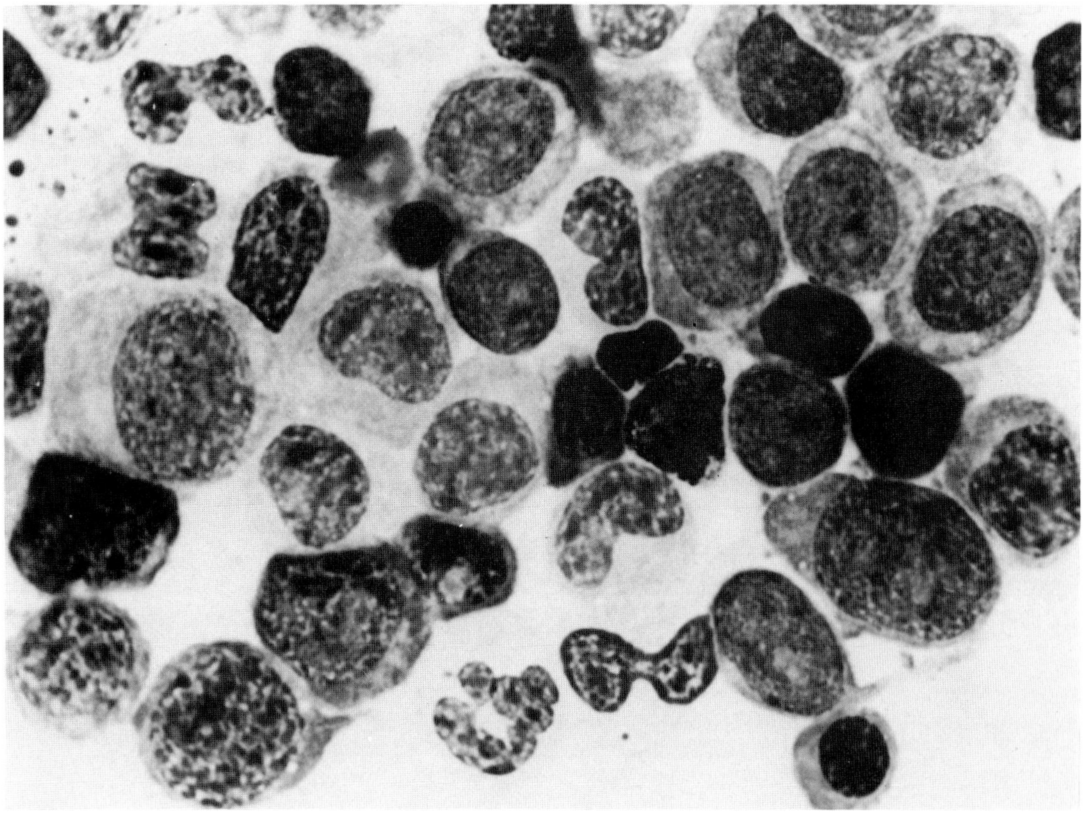

Since micromyeloblasts are believed to be dormant, the slow evolutionary tempo of many instances of smoldering acute leukemia may be due in part to predominance of micromyeloblasts. Promyelocytes are also increased (Figs. 1-80, 1-81) and metamyelocytes and bands can have atypical nuclear configurations (Figs. 1-77, 1-82, 1-83, 1-88).

Erythropoiesis is usually megaloblastoid, with increased numbers of proerythroblasts and megaloblastoid intermediate macronormoblasts (Figs. 1-84, 1-85). Gigantic, bizarre-appearing erythroblasts are seen (Figs. 1-86, 1-87). Aberrant mitotic figures occur in erythroblasts (Fig. 1-88), and some mature erythroblasts contain coarse basophilic stippling and aberrant nuclear configurations. Eosinophils and plasma cells may be increased in number (Fig. 1-89). Megakaryocytes are usually reduced in number, and may appear normal or unusually small (micromegakaryocytes). Other megakaryocytes are large and show multiple nuclear lobulations.

(text continued on page 88)

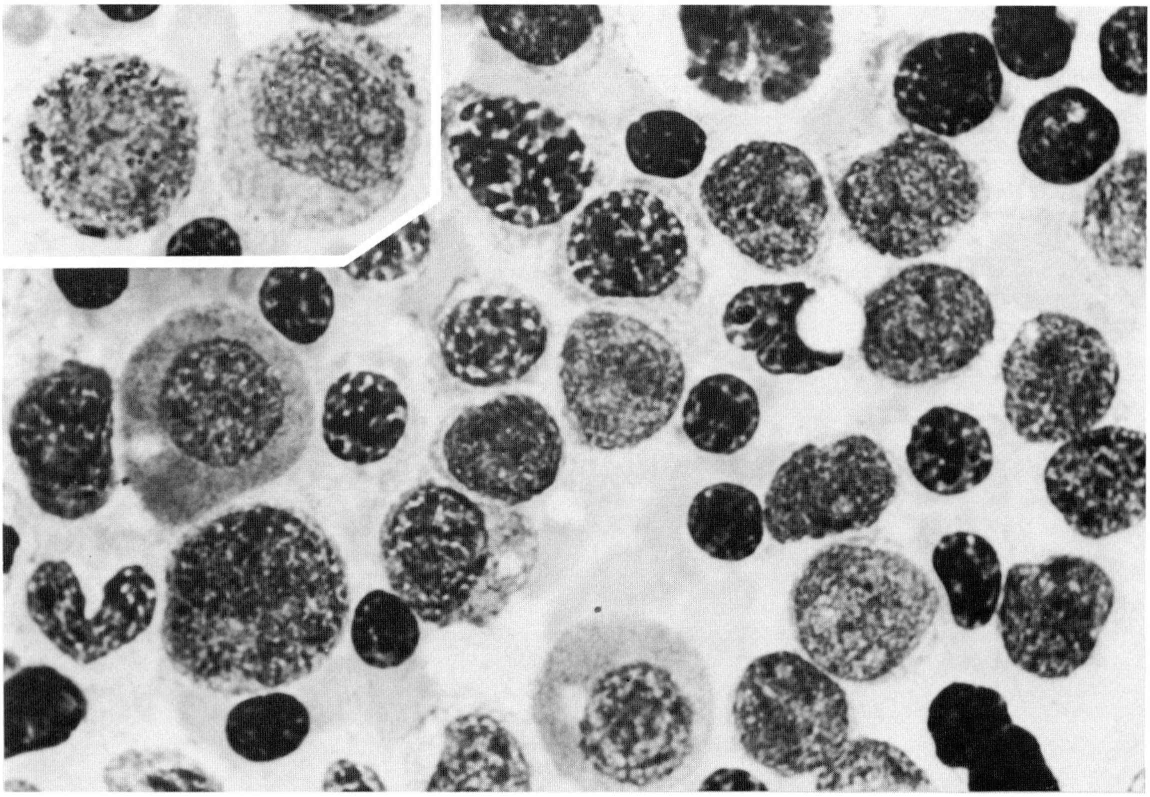

Figure 1-81. Bone marrow in smoldering acute leukemia, showing increased numbers of eosinophils, myelocytes, increased myeloblasts, and megaloblastoid intermediate macronormoblasts. Inset shows promyelocytes with prominent splinter-shaped granules and large nucleoli.

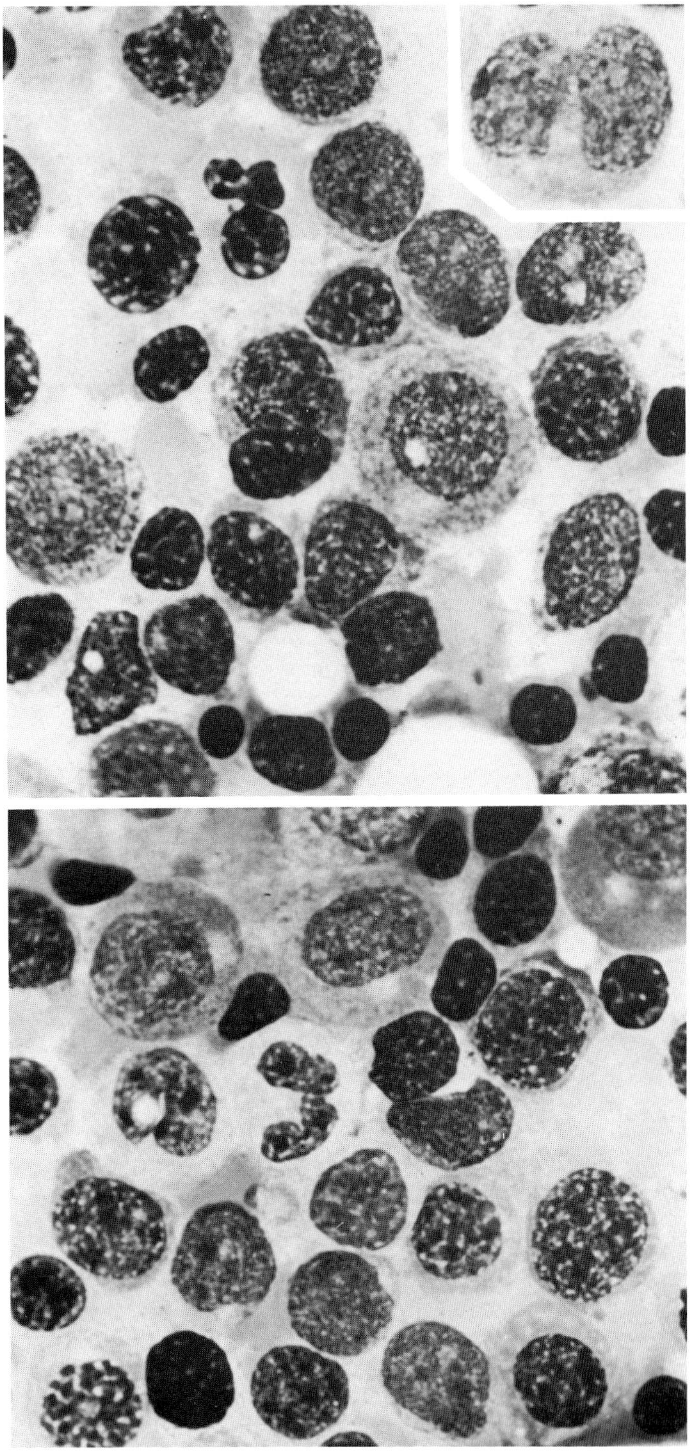

Figure 1-82.

Bone marrow in smoldering acute leukemia, illustrating erythroblasts having atypical nuclear configurations (**top**), along with myeloblasts and giant atypical metamyelocytes showing pseudo-Pelger-Huët anomaly (**inset**). Increased numbers of myeloblasts (**bottom**), metamyelocyte with atypical pseudo-Pelger-Huët nuclear configuration, and several megaloblastoid intermediate macronormoblasts.

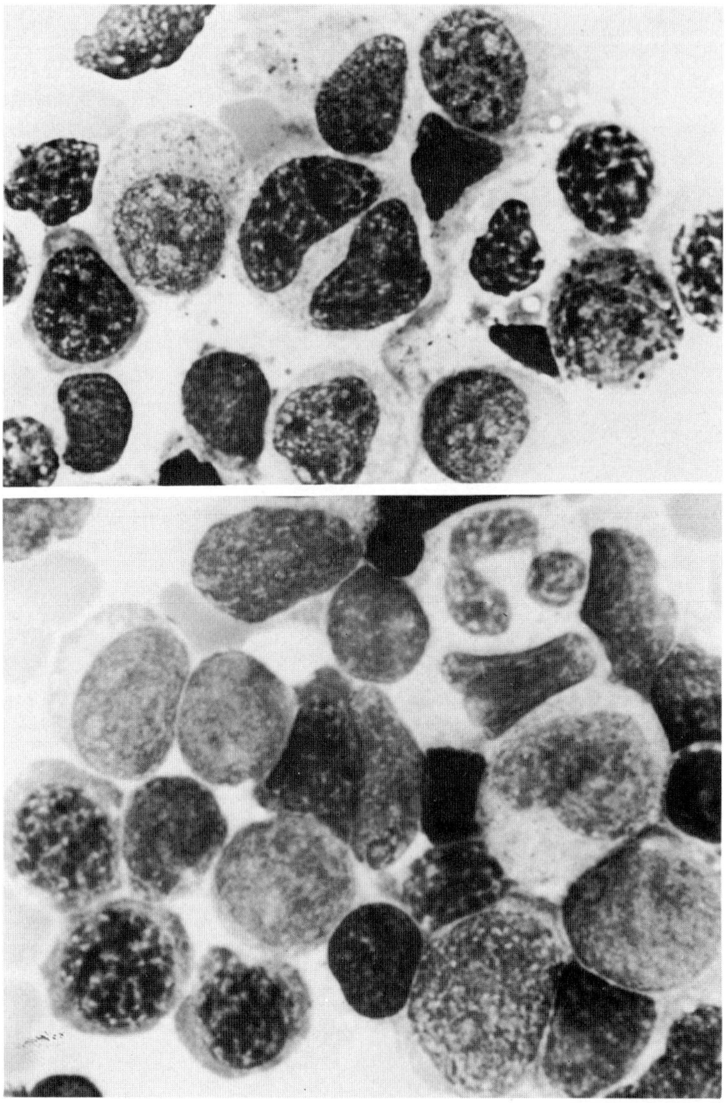

Figure 1-83. Bone marrow in smoldering acute leukemia, demonstrating giant metamyelocyte with atypical nuclear configuration, along with several promyelocytes **(top)**. Increased numbers of myeloblasts and megaloblastoid intermediate macronormoblasts **(bottom)**.

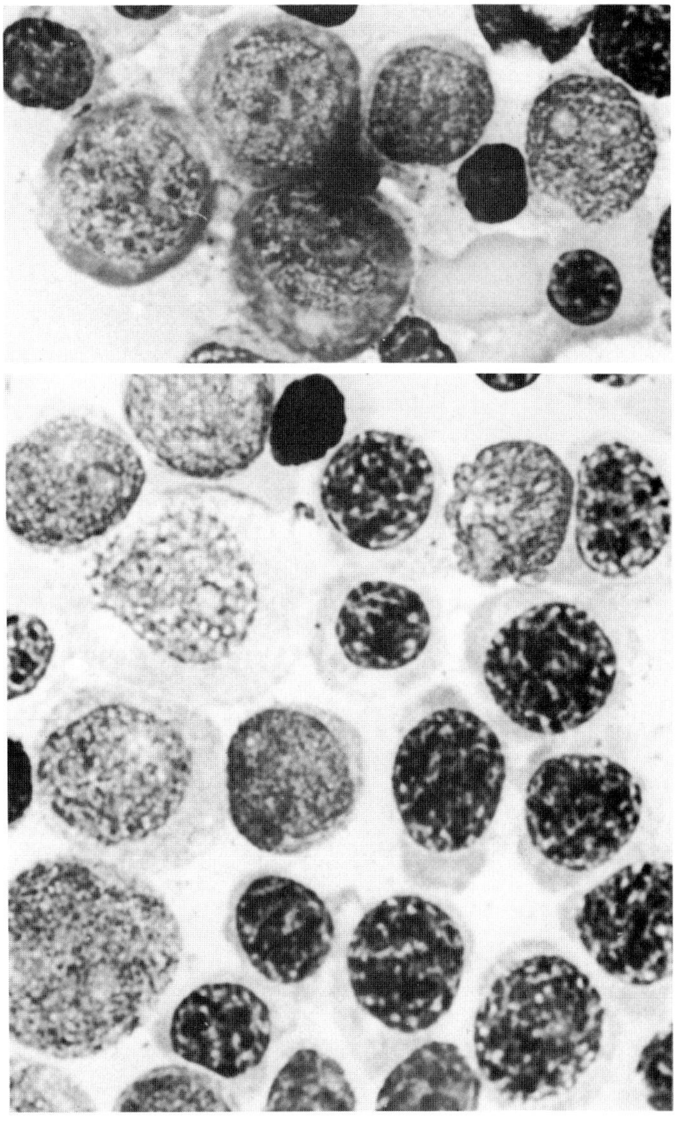

Figure 1-84. Bone marrow in smoldering acute leukemia, depicting cluster of proerythroblasts (**top**). Cluster of megaloblastoid intermediate macronormoblasts and several myeloblasts and promyelocytes (**bottom**).

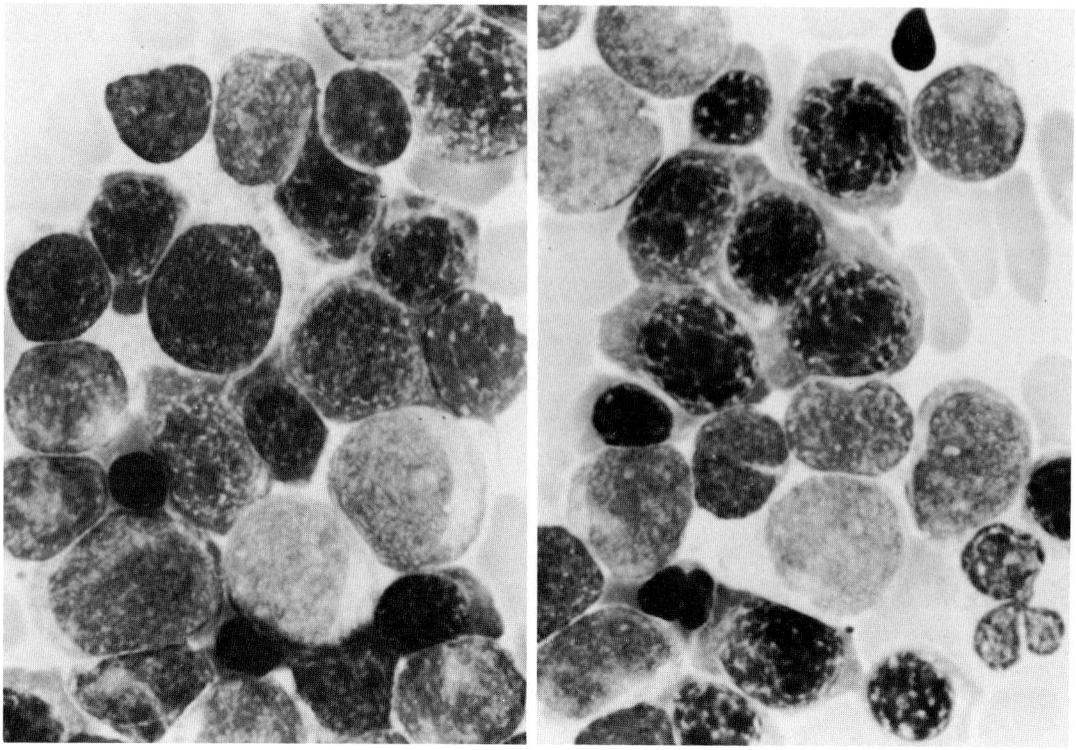

Figure 1-85. Bone marrow in smoldering acute leukemia, showing increased numbers of myeloblasts and megaloblastoid intermediate macronormoblasts.

Figure 1-86. Giant, atypical erythroblasts in smoldering acute leukemia. These cells show marked nucleocytoplasmic asynchronism as well as megaloblastic nuclear chromatin pattern with few aggregates of chromatin and prominent, darkly staining nucleoli **(left)**. Vacuoles may be found in the cytoplasm **(right)**.

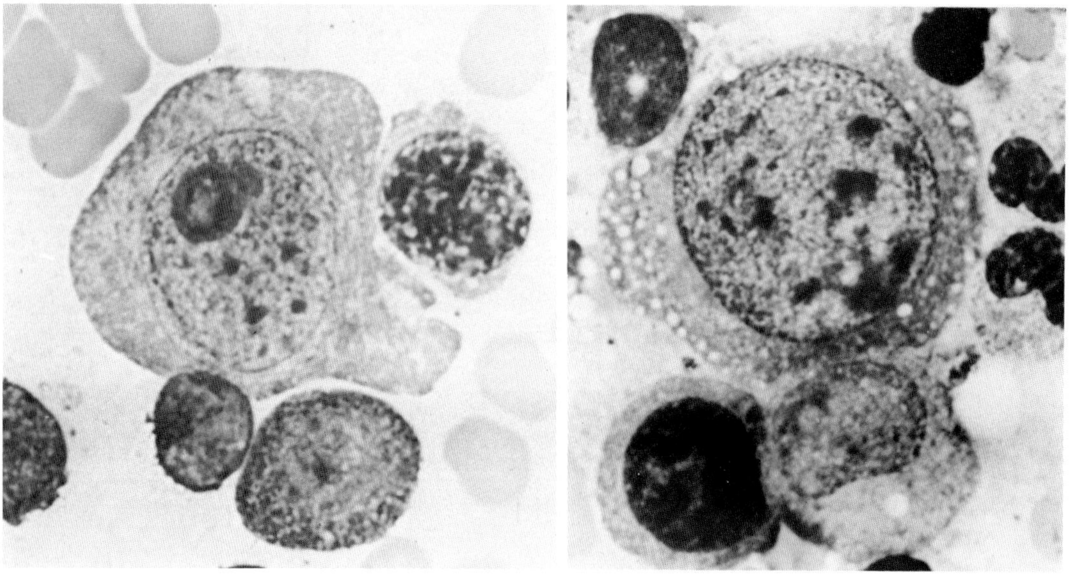

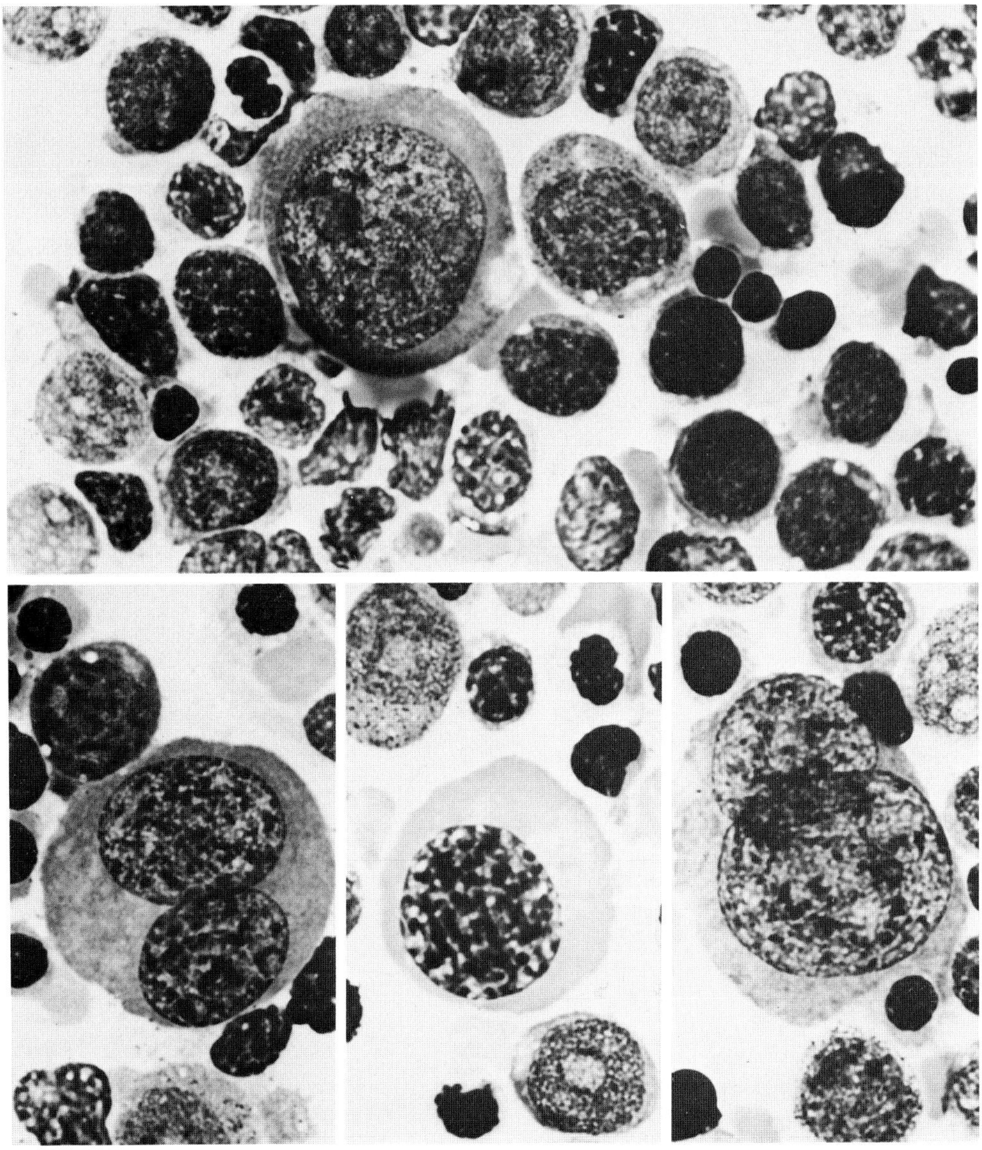

Figure 1-87. Giant erythroblasts in smoldering acute leukemia. These cells show marked nucleocytoplasmic asynchronism along with megaloblastoid or megaloblastic nuclear chromatin patterns and multinuclearity in some instances.

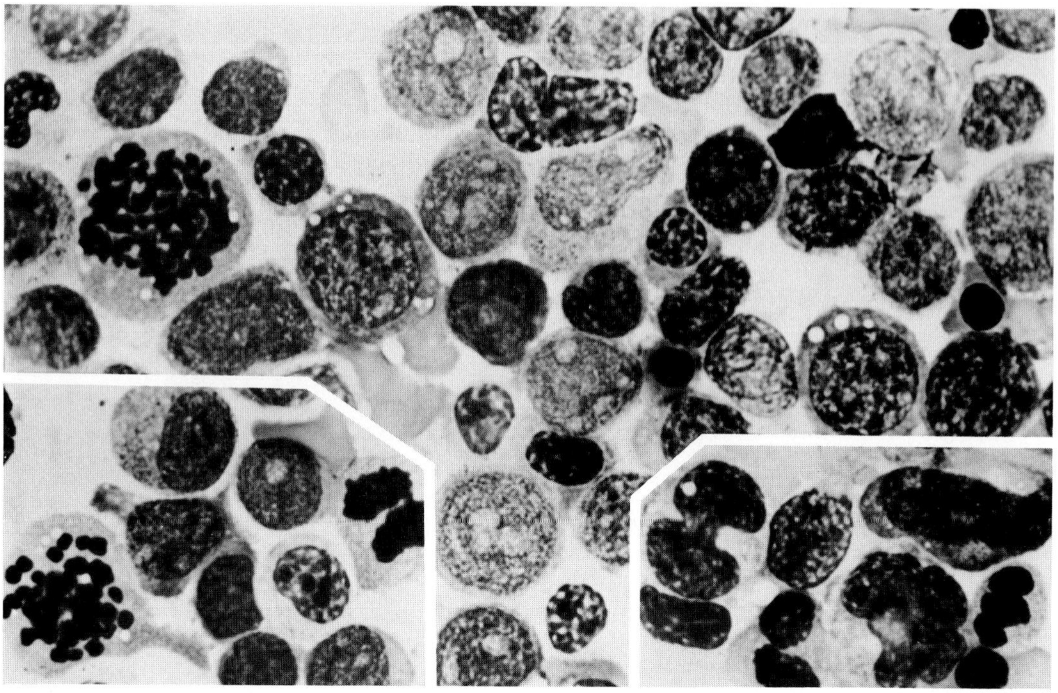

Figure 1-88. Bone marrow in smoldering acute leukemia, showing increased numbers of myeloblasts with prominent nucleoli as well as increased numbers of megaloblastoid intermediate macronormoblasts. Several erythroblasts contain multiple vacuoles. Additionally, some erythroblasts show marked karyorrhexis **(main photo and left inset)**. Inset **(right)** shows several monocytoid cells and an atypical band with dumbbell nuclear configuration.

Figure 1-89. Bone marrow in smoldering acute leukemia, showing increased numbers of eosinophils, myelocytes, myeloblasts, megaloblastoid intermediate macronormoblasts, and proerythroblasts. →

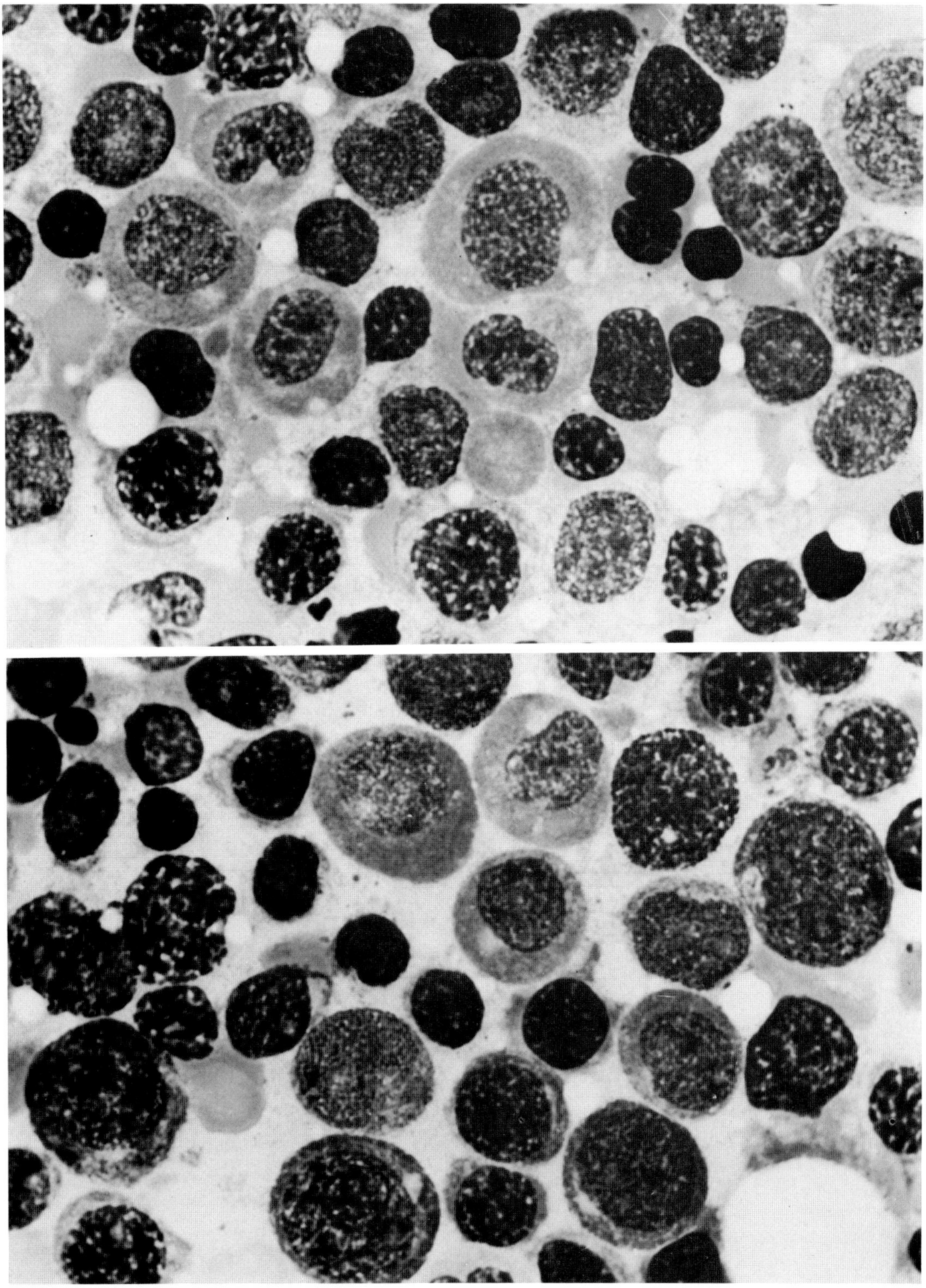

On H & E sections of biopsy of marrows of patients with smoldering acute leukemia, increased numbers of cells with pale vesicular nuclei and prominent nucleoli are seen (Figs. 1-90, 1-91). These represent proerythroblasts or myeloblasts. Increased numbers of eosinophils and plasma cells may be found.

Cytochemically, myeloblasts in smoldering acute leukemia contain scant activity of specific esterase and myeloperoxidase. Usually, erythroblasts do not contain detectable amounts of glycogen using the PAS reagent, and the occurrence of siderotic granules in erythroblasts is variable. In both myeloblasts and erythroblasts, activity of nonspecific esterase is weak and resistant to inhibition by sodium fluoride.

Subacute Myelomonocytic Leukemia

Subacute myelomonocytic leukemia (SAMML) is a disorder of neoplastic monocytes. It is characterized by macrocytic anemia, monocytosis, normoblastemia, and markedly hypercellular bone marrow. Clinically, patients with subacute myelomonocytic leukemia often have fever, hepatomegaly, and splenomegaly. Approximately 20% to 40% of the marrow cells are monocytes or monocytoid cells, and morphologic abnormalities can be detected in other marrow cells, indicating a panymyelopathy. Recognized under several names,[16,60,69,169,349,448,556,630,676,814,889,890,944,946,953,971,972,1067,1126–1128] subacute myelomonocytic leukemia is believed to be a monocytic variant of myeloblastic leukemia, in which proliferation of monocytes occurs in the bone marrow and visceral organs. Evolution into acute monocytic leukemia occurs in a variable number of patients. Others succumb to complications of pancytopenia or to infection or hemorrhage prior to the evolution of acute leukemia.

Peripheral blood findings include monocytosis (greater than 500/mm^3) of atypical monocytes having aberrant nuclear configurations, cytoplasmic "tail," and perinuclear clustering of granules, as well as macrocytosis of erythrocytes, thrombocytopenia, and normoblastemia. Bone marrows from patients with subacute myelomonocytic leukemia are hypercellular and contain 20% to 40% monocytes (Figs. 1-92 through 1-100, and Plate 2*A*). When the number of monocytes in the marrow exceeds 40%, it may be more appropriate to classify the specimen as acute monocytic leu-

(text continued on page 98)

Figure 1-90. H & E section of bone marrow biopsy in smoldering acute leukemia. Several mitotic figures are seen, along with pale vesicular nuclei that probably represent immature granulocytic cells such as promyelocytes. →

Figure 1-91. H & E section of bone marrow biopsy in smoldering acute leukemia. Darkly stained nuclei are erythroblasts. Large, pale vesicular nuclei with prominent nucleoli are probably proerythroblasts. →

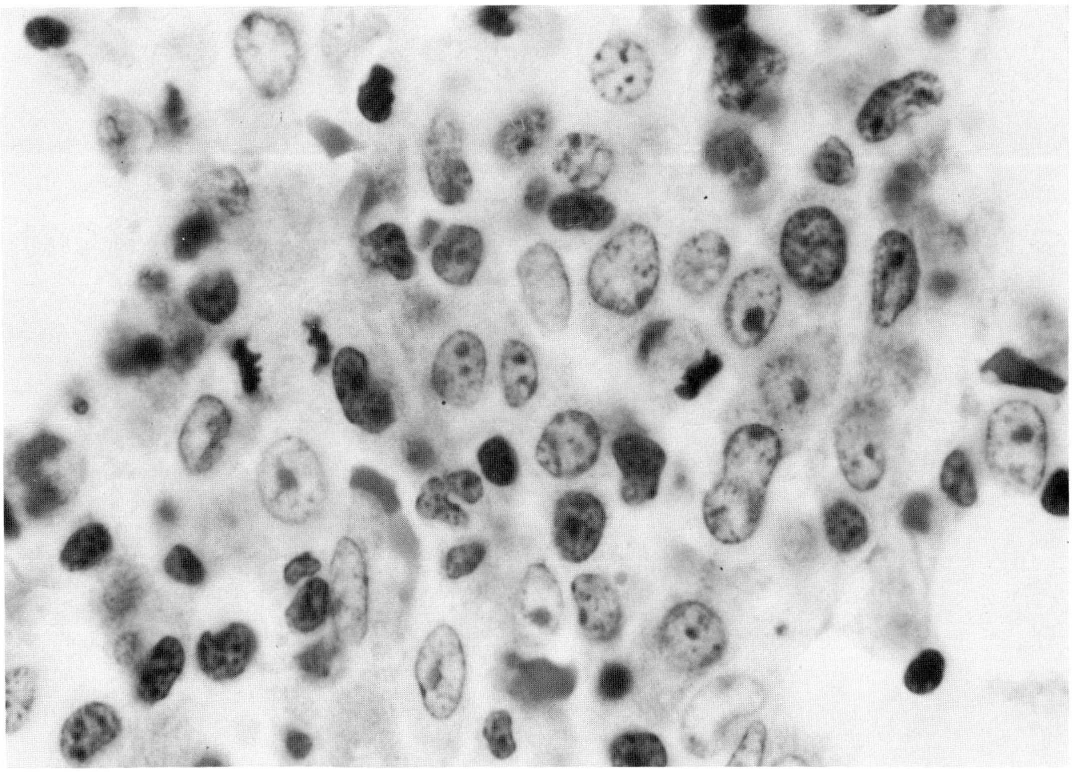

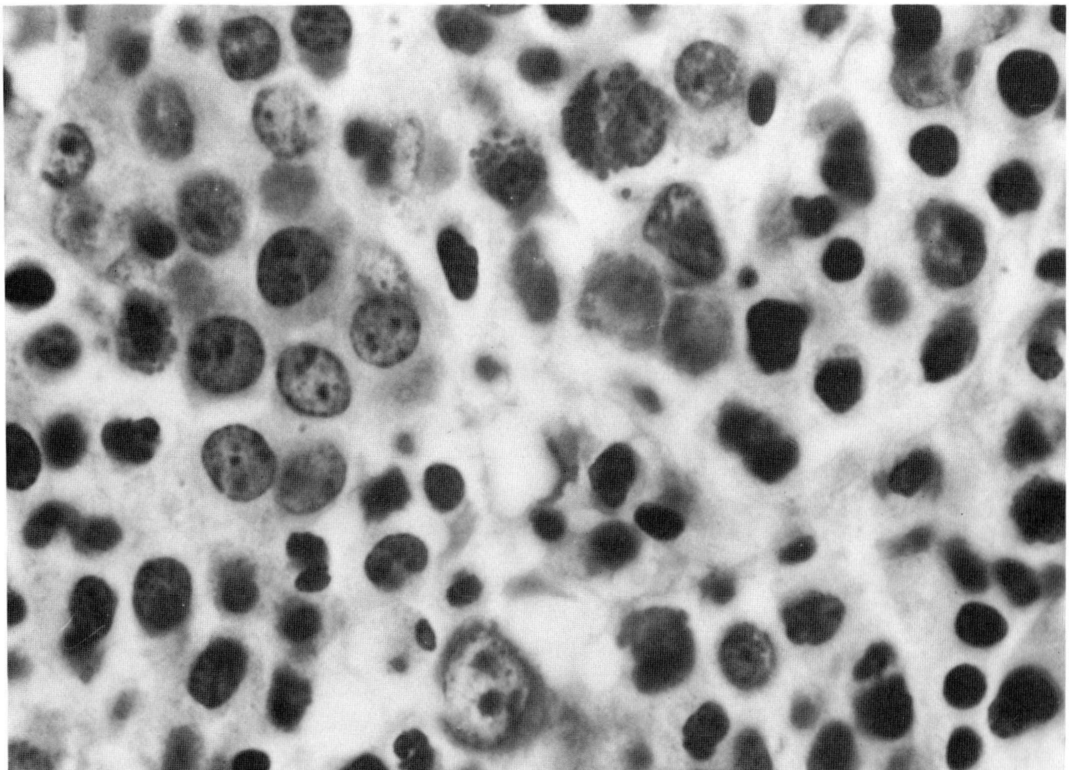

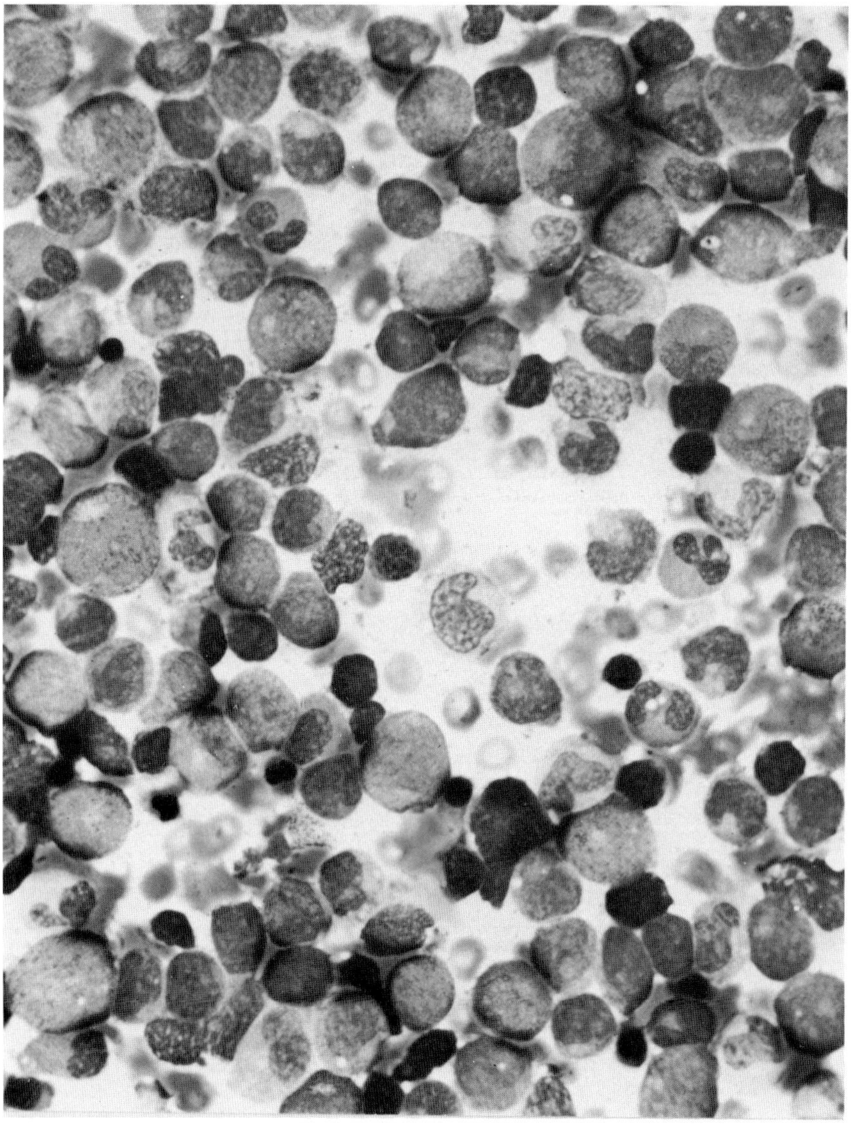

Figure 1-92. Low-power view of bone marrow in subacute myelomonocytic leukemia (SAMML), showing increased numbers of promyelocytes and cells with monocytoid-appearing nuclei.

Figure 1-93. Bone marrow in SAMML, showing increased number of cells with monocytoid-appearing convolutions and indentations of the nuclei, as well as late intermediate normoblasts and promyelocytes. ➡

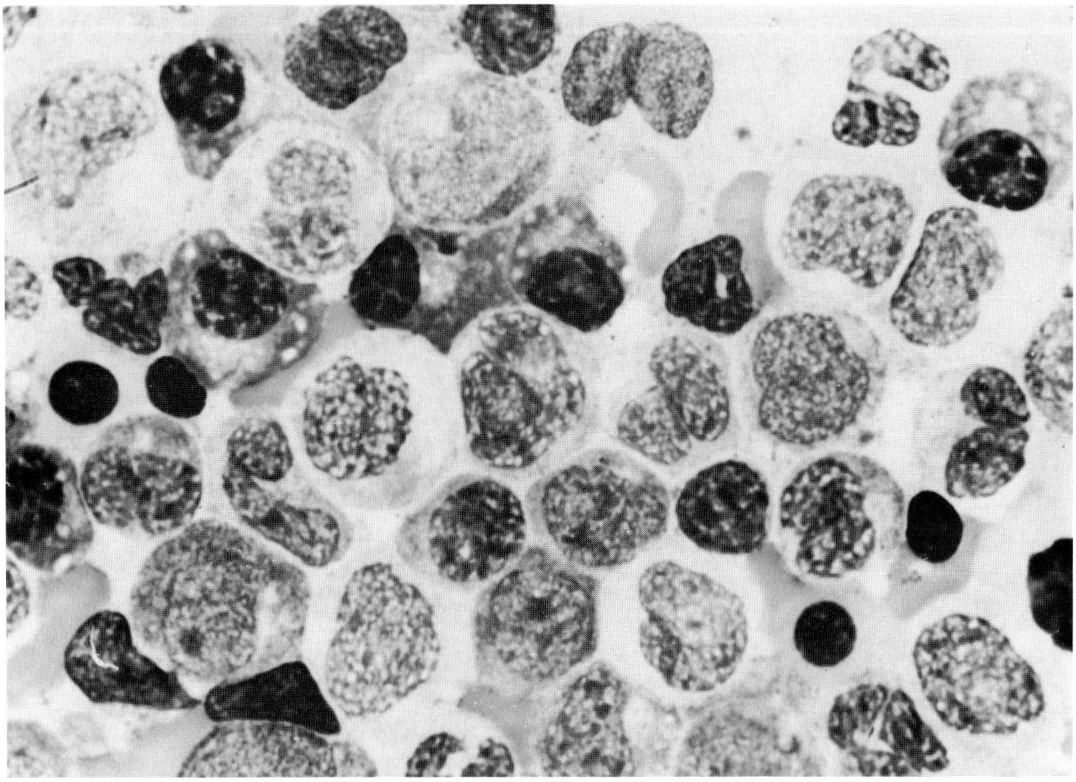

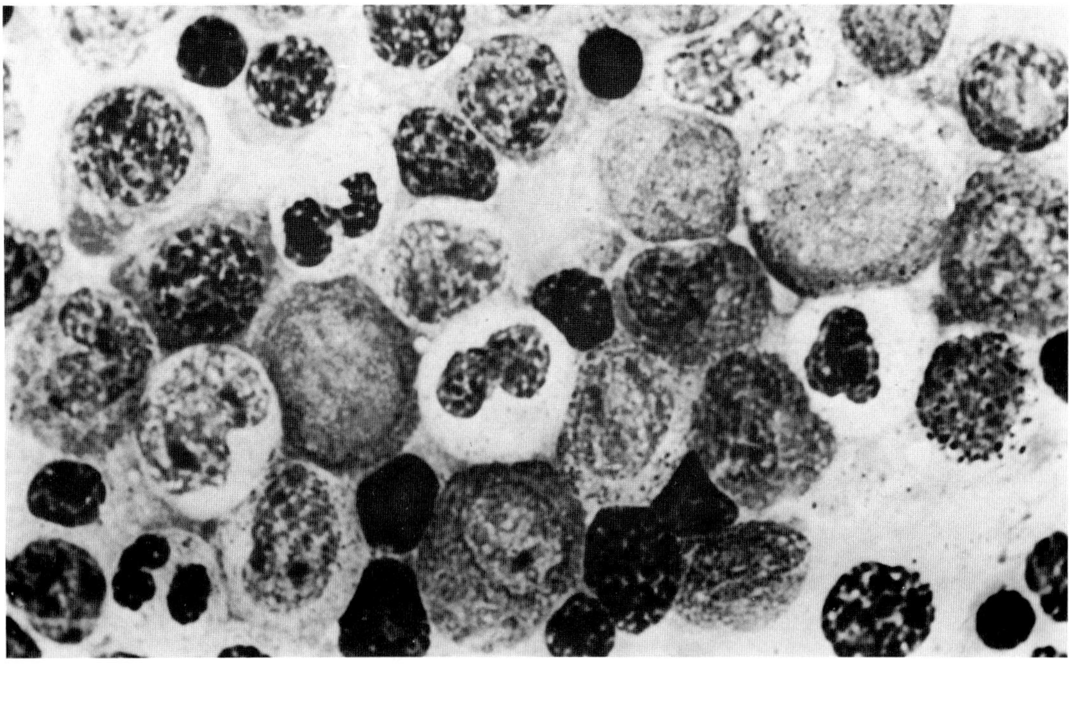

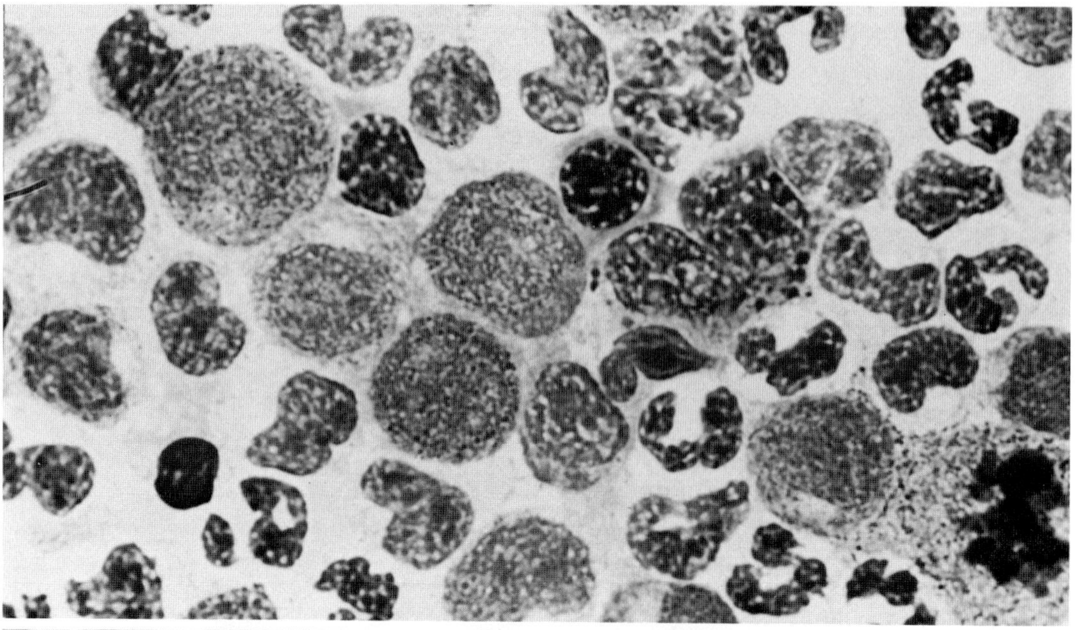

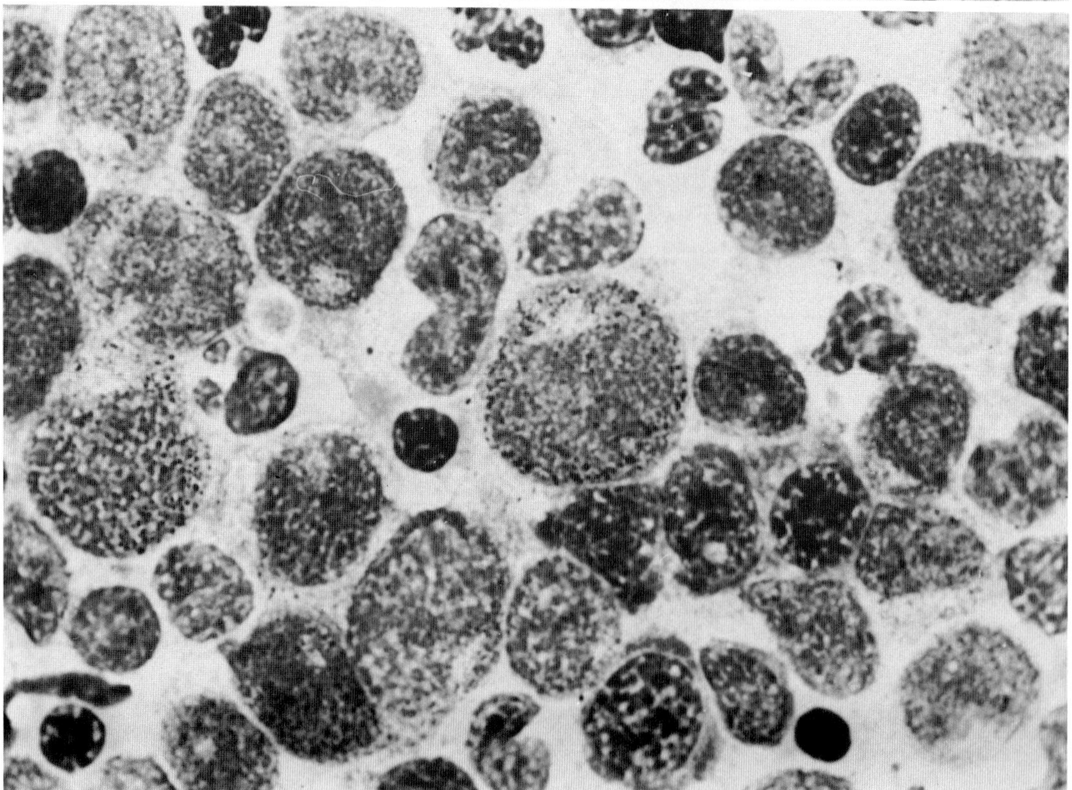

Figure 1-94. ← Bone marrow in SAMML, illustrating increased numbers of monocytoid cells, promyelocytes, metamyelocytes, bands with atypical nuclear configurations, and megaloblastoid intermediate macronormoblasts.

Figure 1-95. Bone marrow in SAMML, demonstrating increased numbers of promyelocytes, atypical mitotic figures in granulocytic cells, and increased numbers of bands.

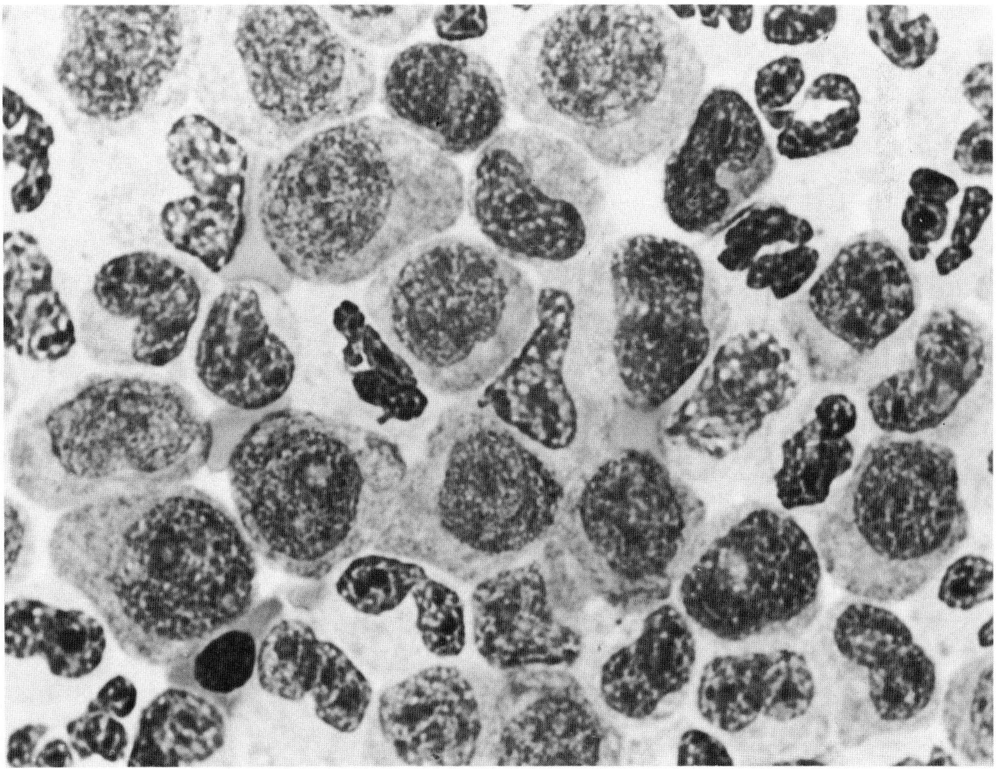

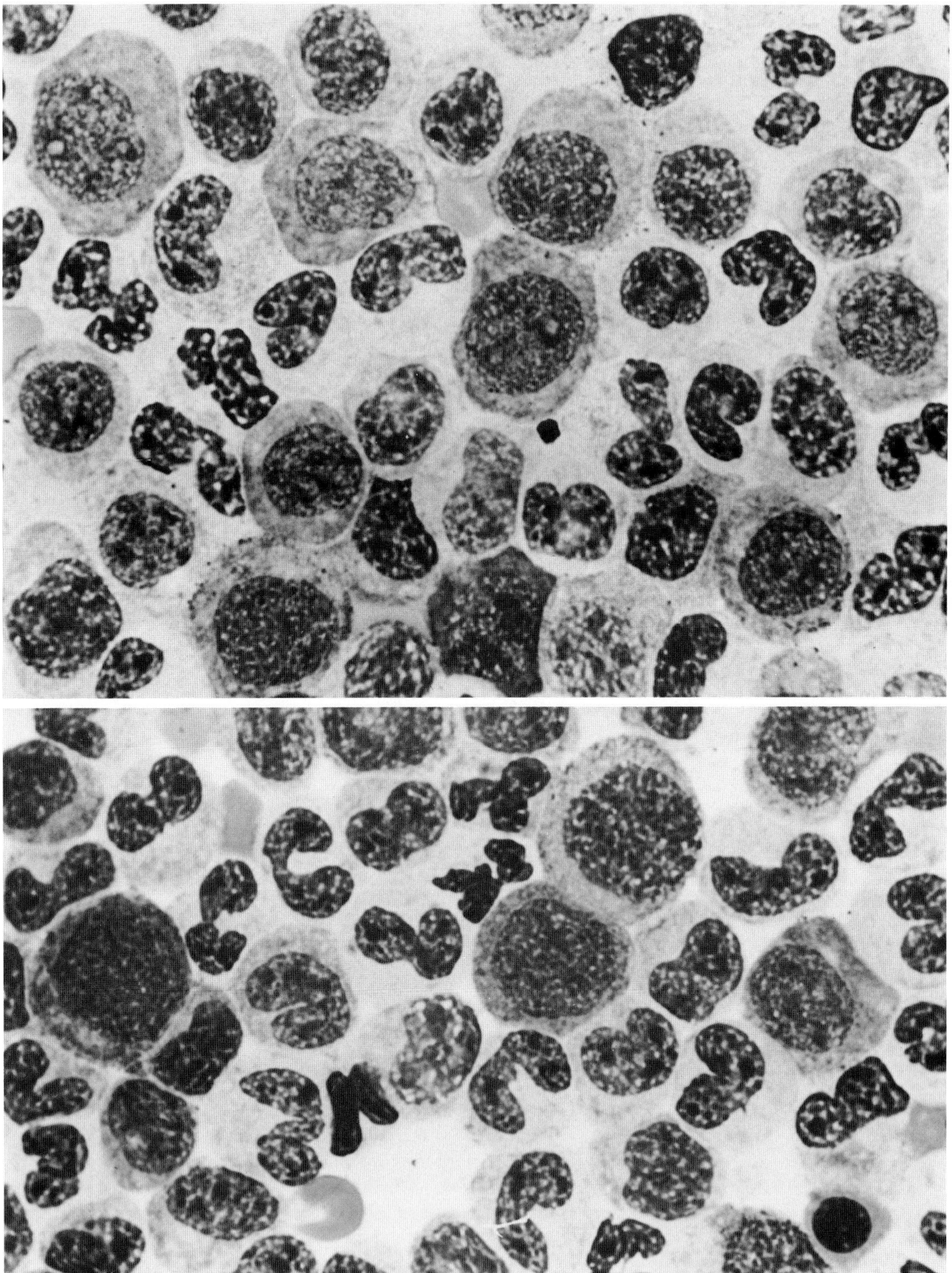

Figure 1-96. Bone marrow in SAMML, depicting increased numbers of bands showing aberrant nuclear configuration. Increased numbers of promyelocytes are shown.

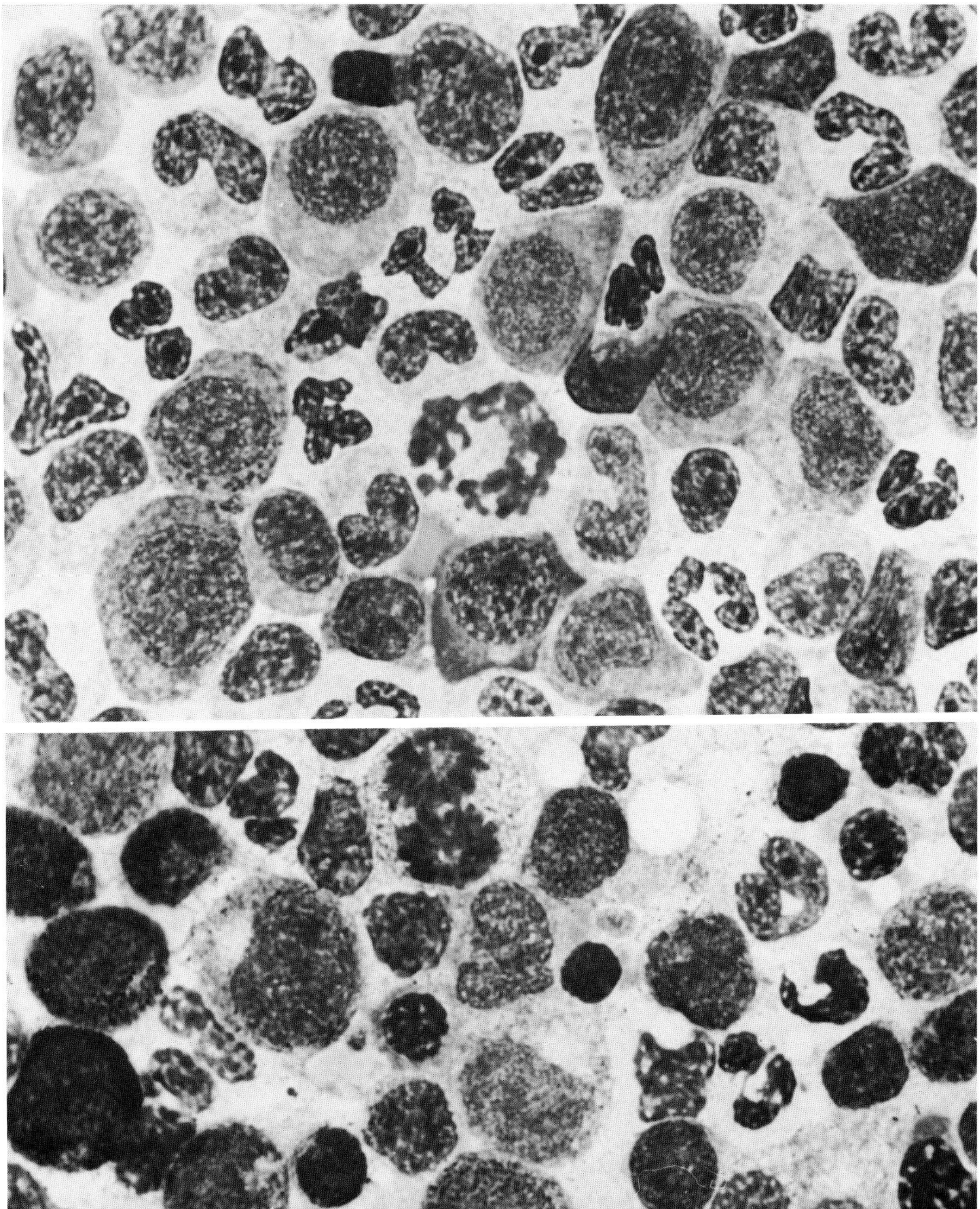

Figure 1-97. Bone marrow in SAMML, illustrating increased numbers of monocytoid cells and aberrant mitotic figures in granulocytic precursors, along with megaloblastoid intermediate macronormoblasts and bands with aberrant nuclear configurations.

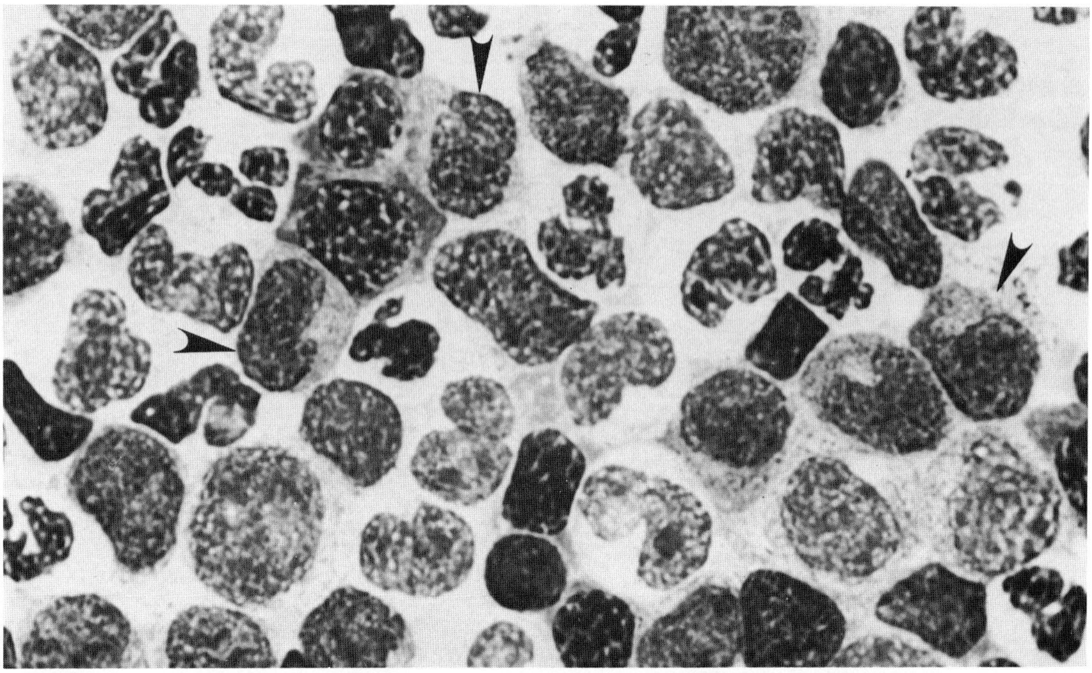

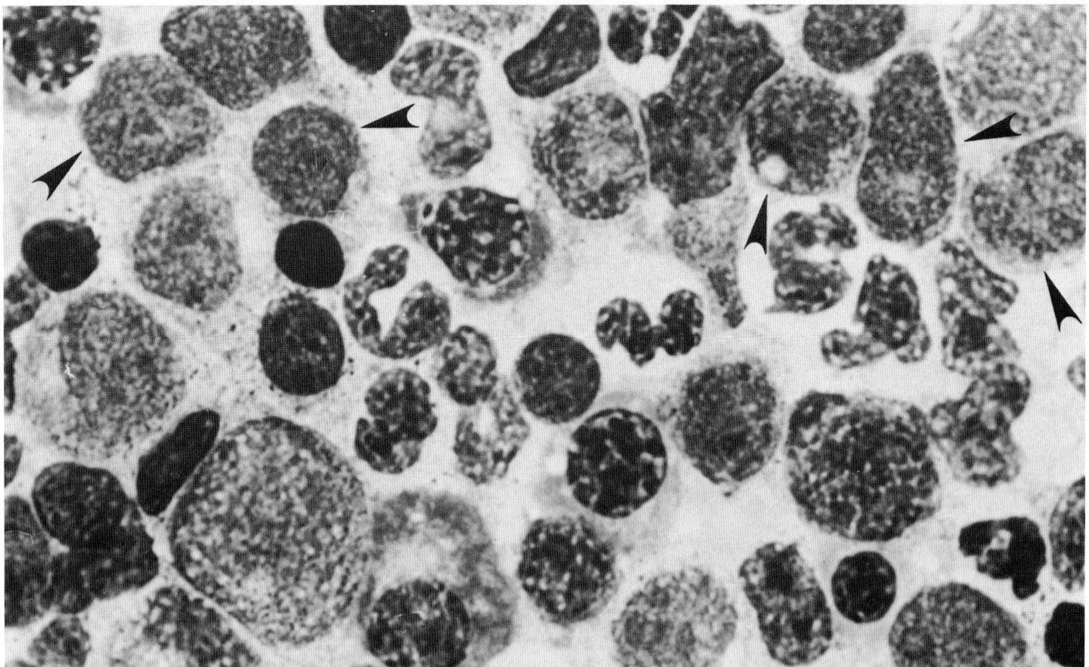

← Bone marrow in SAMML, illustrating increased numbers of monocytoid cells (**arrows**), promyelocytes, plasma cells, and bands with aberrant nuclear configuration. Several megaloblastoid intermediate macronormoblasts are shown. **Figure 1-98.**

Bone marrow in SAMML, showing increased numbers of promyelocytes with prominent granulation and large nucleoli, large metamyelocytes, monocytes, and megaloblastoid intermediate macronormoblasts. **Figure 1-99.**

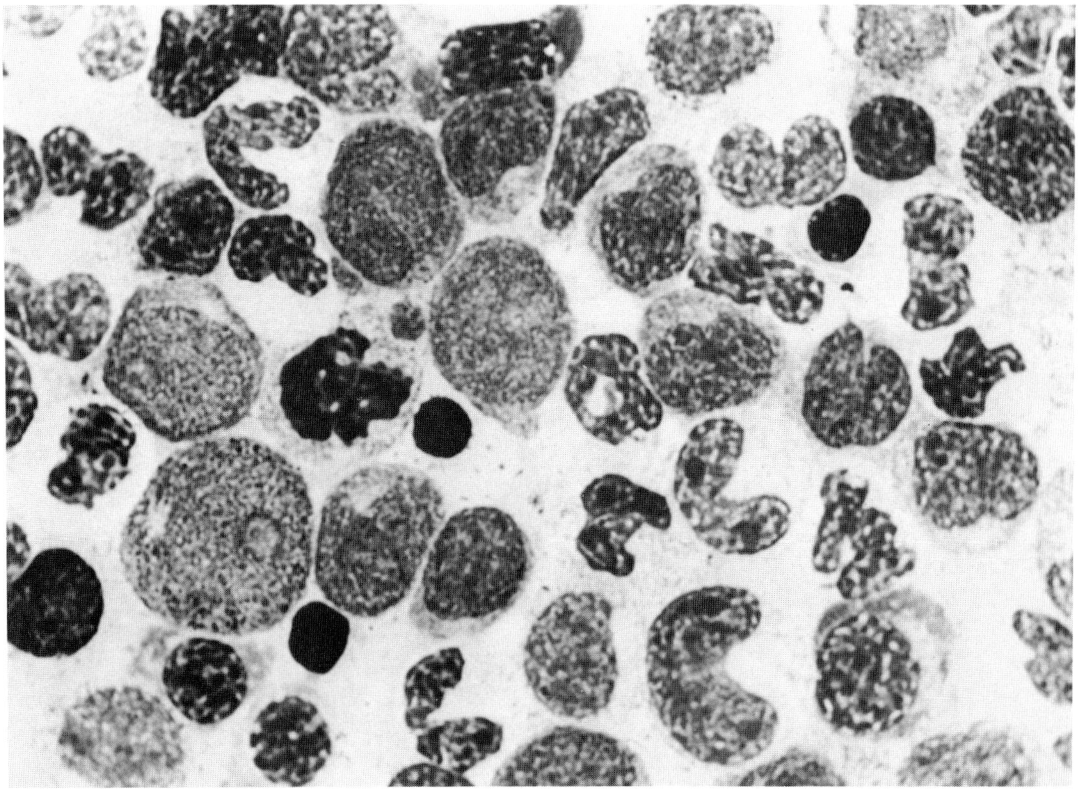

kemia. In subacute myelomonocytic leukemia, myeloblasts are seen rarely. Promyelocytes, myelocytes, and metamyelocytes are increased in number, and some of the metamyelocytes have atypical nuclear configurations (Figs. 1-92 through 1-100). Eosinophils and plasma cells are increased in number (Figs. 1-101 through 1-104).

Erythroblasts are usually reduced in number, but may be increased. In most instances, they demonstrate megaloblastoid nuclear chromatin patterns (Fig. 1-105). Some erythroblasts are multinuclear (Fig. 1-105). Proerythroblasts may be increased and may show multiple cytoplasmic pseudopodia (Fig. 1-106). In more mature erythroblasts, aberrant nuclear configurations are found. Some erythroblasts are gigantic and multinuclear. Megakaryocytes are decreased and vary in configuration. Some are unusually large with substantial cytoplasmic volume and increased lobulations of the nucleus; others are micromegakaryocytes.

On H & E sections of biopsies from patients with subacute myelomonocytic leukemia, marrows are hypercellular with increased numbers of granulocytic cells and increased numbers of cells with twisted, convoluted nuclei (Fig. 1-107). These probably represent leukemic monocytes. Erythroblasts are also increased in number, and some are multinuclear. Usually, megakaryocytes are decreased in number.

Cytochemically, in erythroblasts from patients with subacute myelomonocytic leukemia, the PAS reagent shows variable amounts of glycogen. In most cases, the staining pattern is diffuse. Rarely, erythroblasts in subacute myelomonocytic leukemia have blocklike aggregates of glycogen similar to those found in erythroblasts in erythroleukemia. Activities of specific esterase, acid phosphatase, and myeloperoxidase can be identified in granulocytic cells and, to a lesser extent, in monocytes.

Monocytes are identified easily with the stain for nonspecific esterase activity (Fig. 1-108). Using alpha naphthyl acetate or alpha naphthyl butyrate as the substrate along with hexazotized pararosaniline or new fuchsin as the dye coupler, nonspecific esterase activity is unusually intense in normal and leukemic monocytes and in phagocytes. Activity of monocytic type nonspecific esterase is sensitive to inhibition by sodium fluoride. Using Prussian blue, ringed sideroblasts occur occasionally in marrows from patients with subacute myelomonocytic leukemia.

(text continued on page 108)

Figure 1-100. Bone marrow in SAMML, depicting increased numbers of monocytes **(arrows)**, promyelocytes, atypical-appearing bands and neutrophils, and megaloblastoid intermediate macronormoblasts. ➔

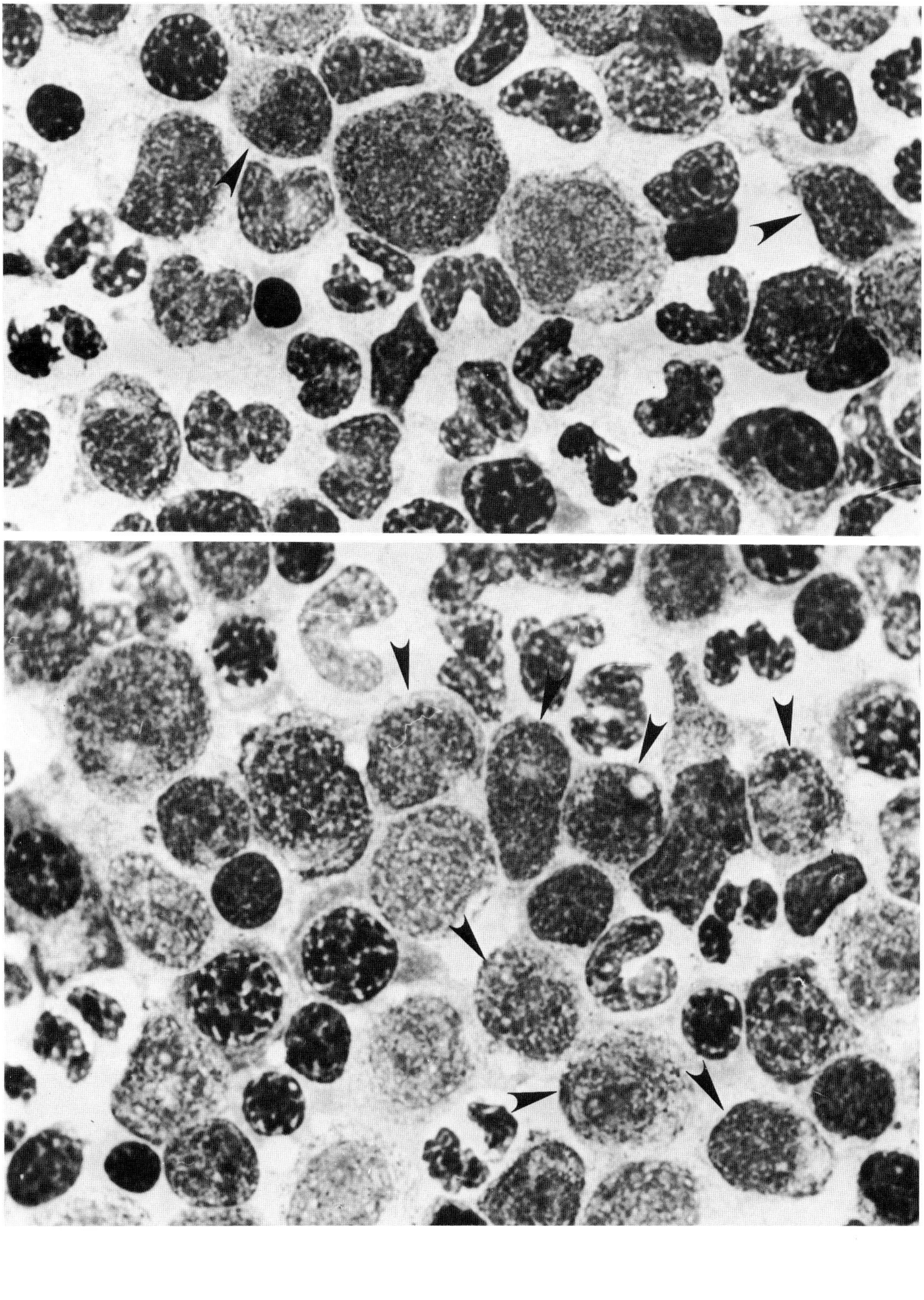

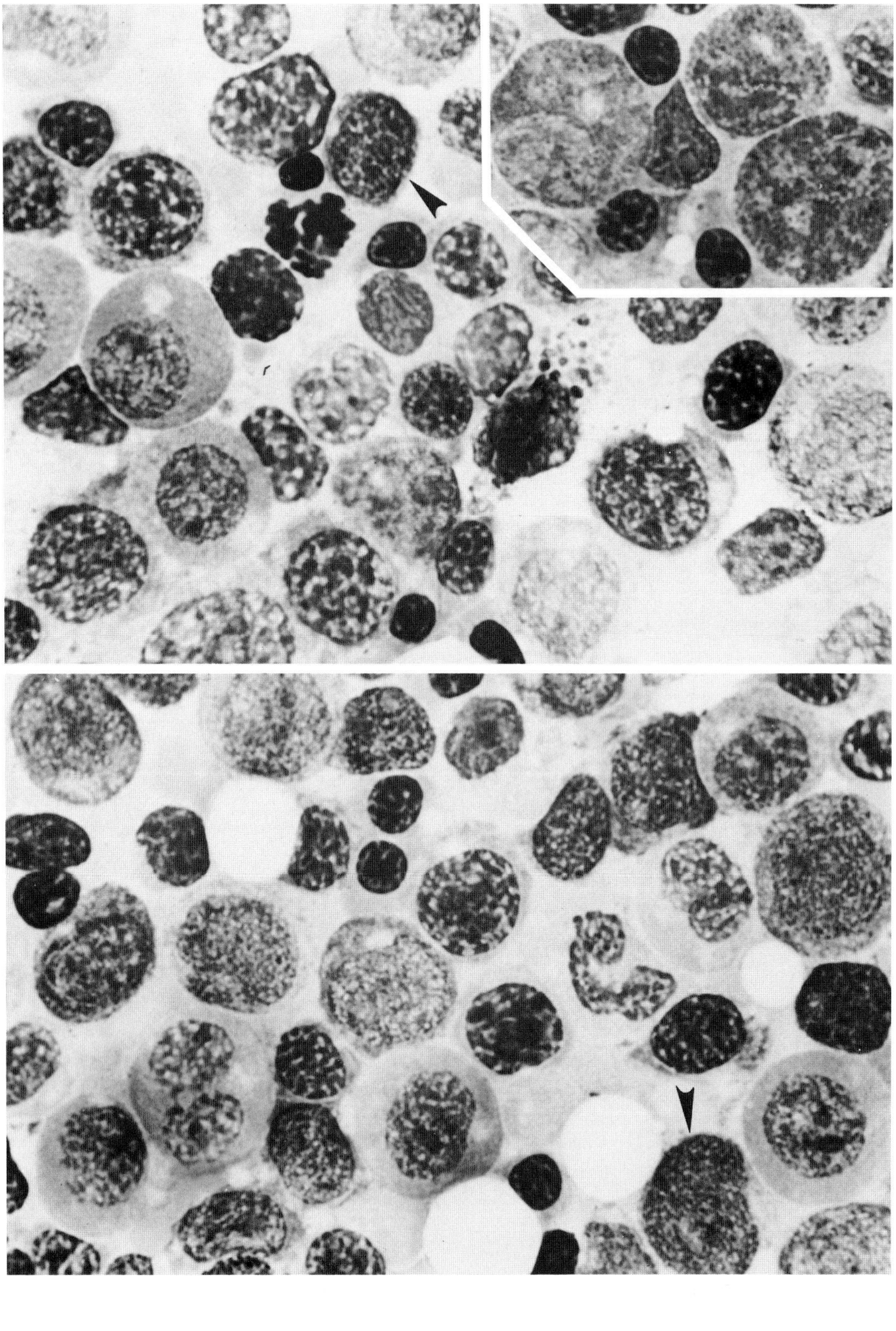

Figure 1-101. ← Bone marrow in SAMML, showing increased numbers of eosinophils along with monocytes and megaloblastoid intermediate macronormoblasts.

Figure 1-102. Bone marrow in SAMML, showing increased numbers of monocytes along with increased numbers of eosinophil myelocytes, plasma cells, and megaloblastoid intermediate macronormoblasts (**top**). H & E section of bone marrow biopsy, showing increased numbers of eosinophils and darkly staining nuclei of erythroblasts (**bottom**). Several plasma cells are seen (**right**).

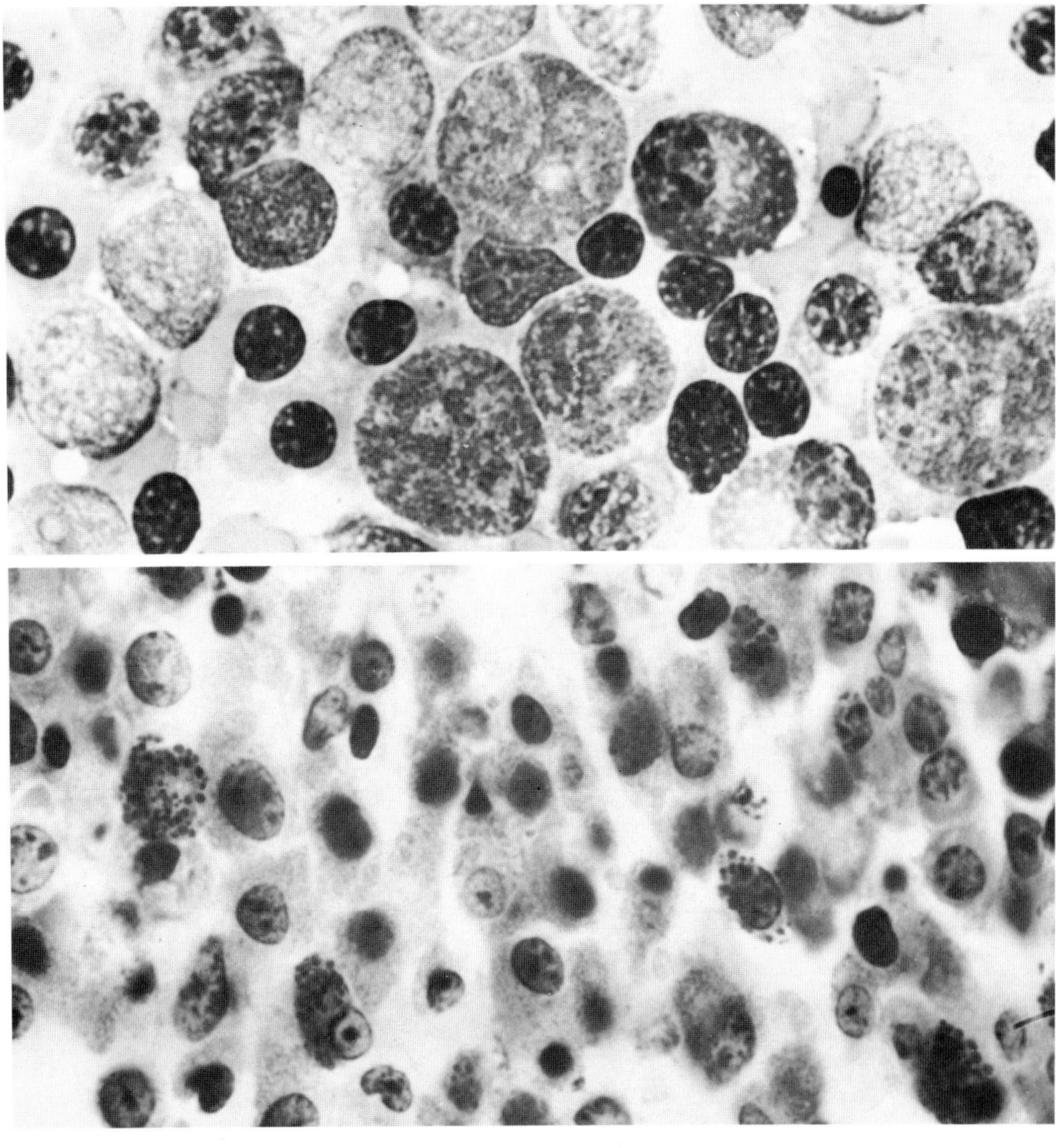

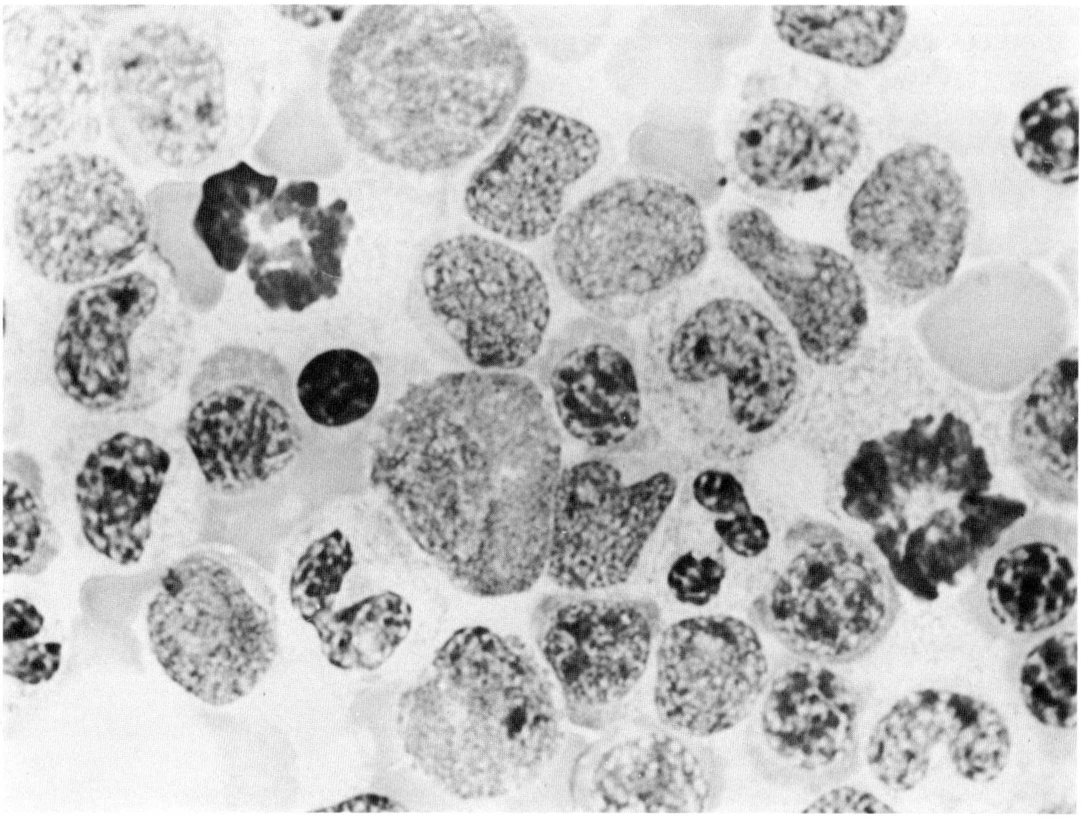

Figure 1-103. Bone marrow in SAMML, showing increased numbers of eosinophils, monocytes, and megaloblastoid intermediate macronormoblasts.

Figure 1-104. Bone marrow in SAMML, showing increased numbers of monocytes, promyelocytes, and mature-appearing plasma cells, some of which contain vacuoles. ➡

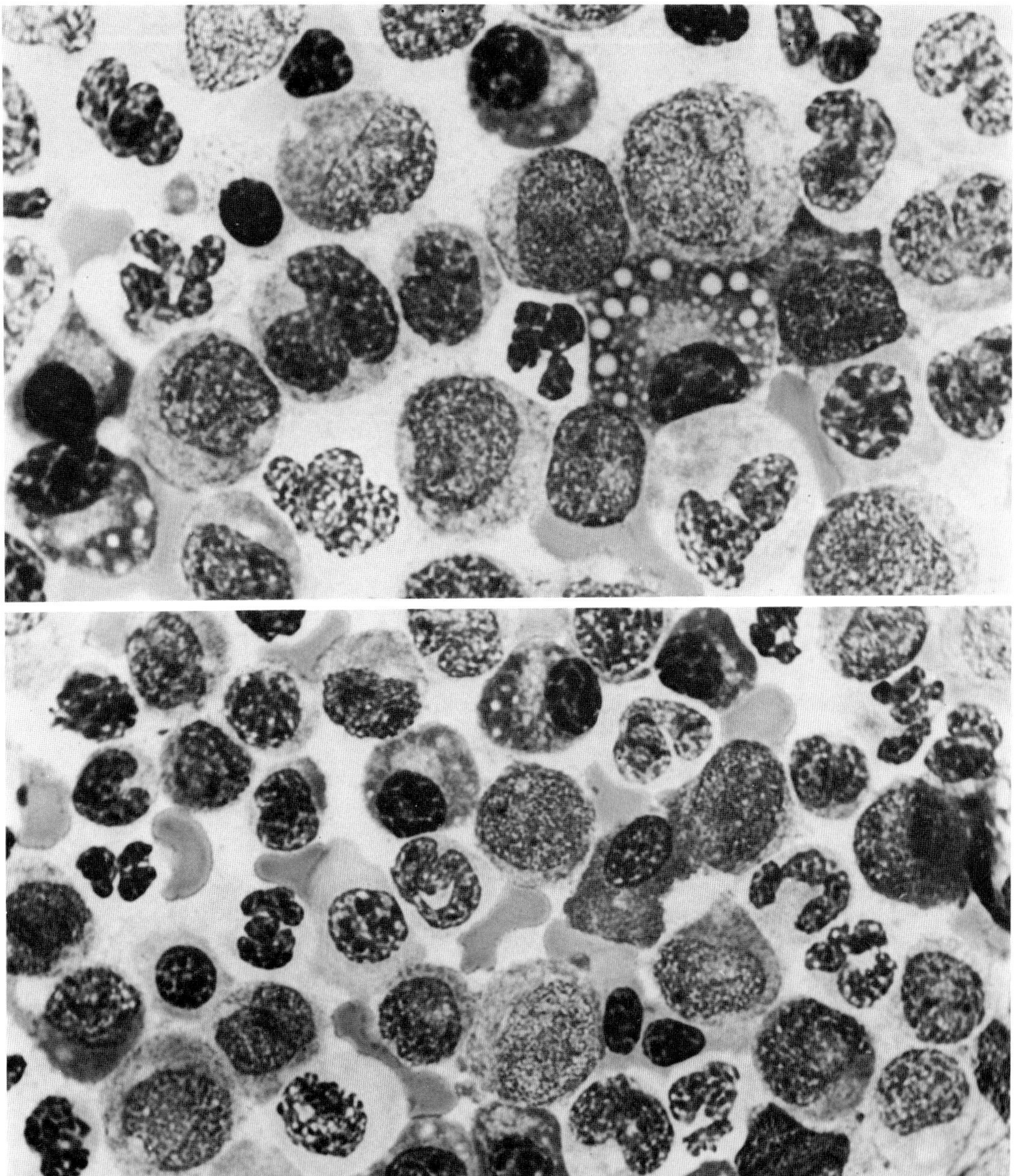

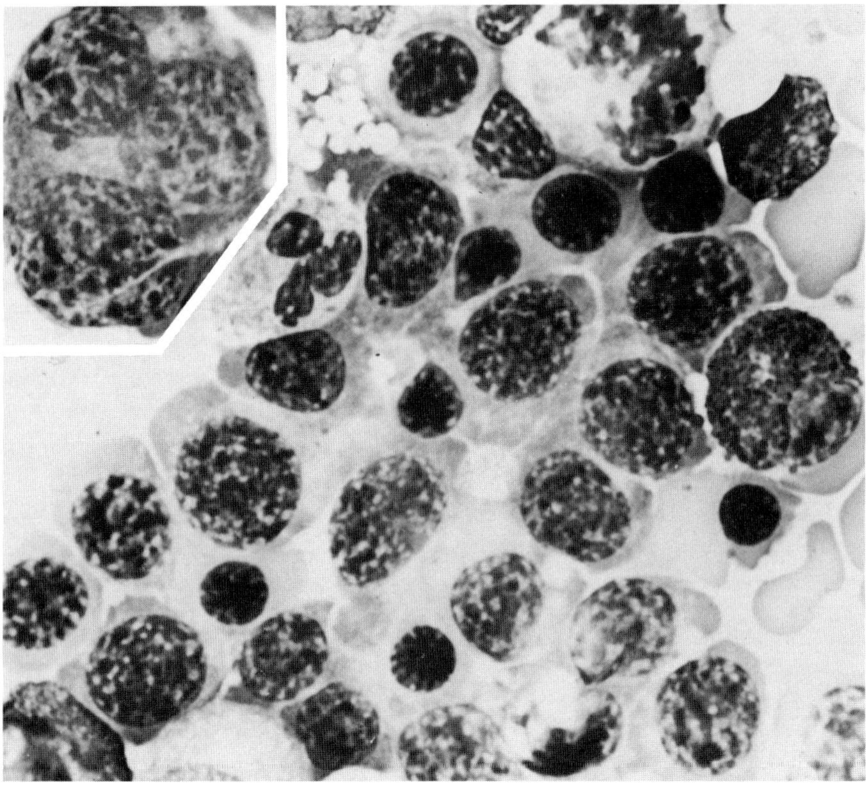

Figure 1-105. Bone marrow in SAMML, showing marked megaloblastoid erythroblastic hyperplasia **(top)**, multinuclear erythroblast with megaloblastoid nuclear chromatin pattern **(top, left)**, and increased numbers of erythroblasts.

Figure 1-106. Proerythroblasts in SAMML, showing cytoplasmic shedding and pseudopodia **(arrows)**.

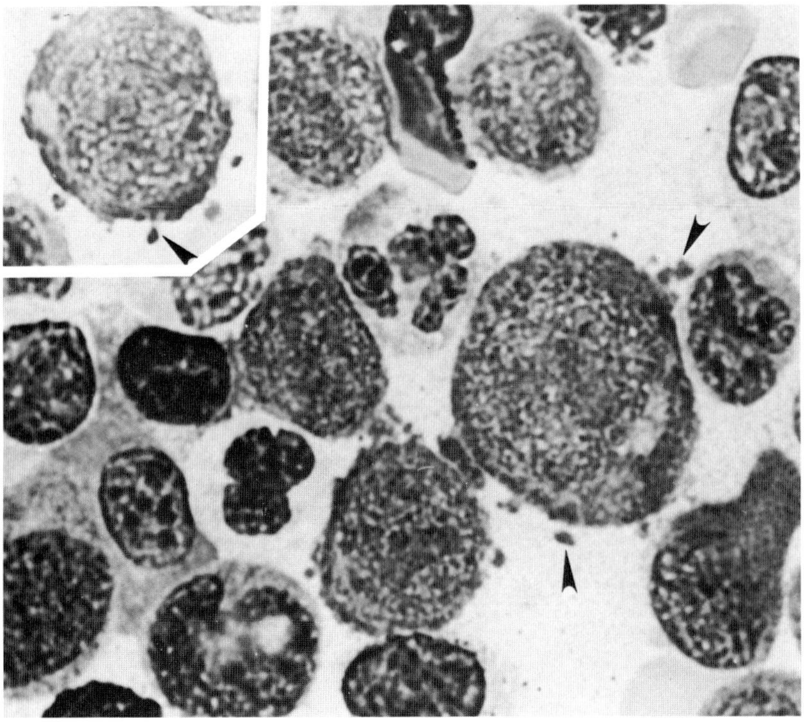

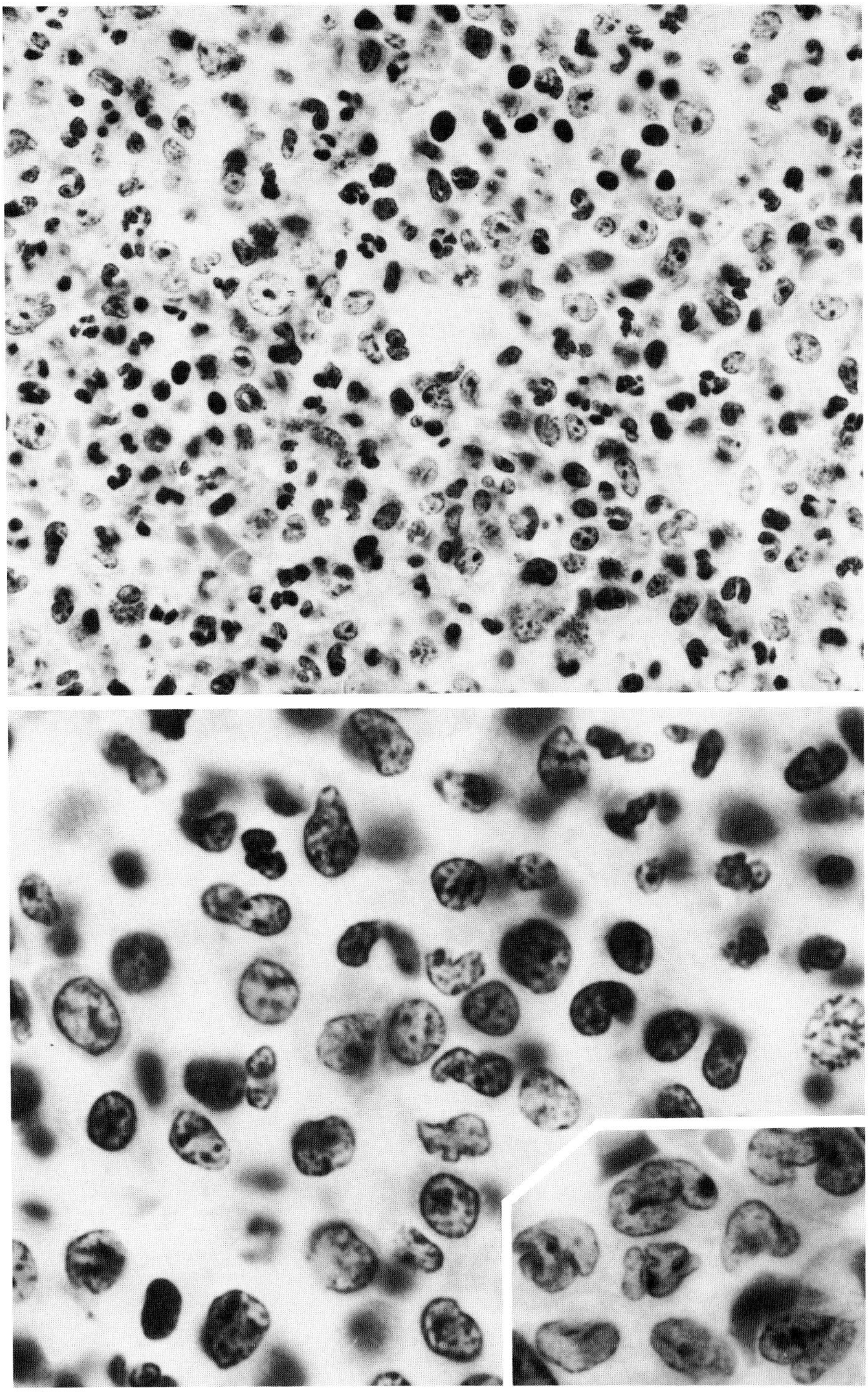

Bone marrow in SAMML. Low-power view of the H & E stained biopsy section (**top**) shows hypercellularity and increased numbers of cells with twisted, convoluted nuclei, along with darkly stained nuclei. Higher-power view of the marrow (**bottom**) demonstrates cells with irregular nuclear configurations that represent monocytes (**main photo and inset**).

Figure 1-107.

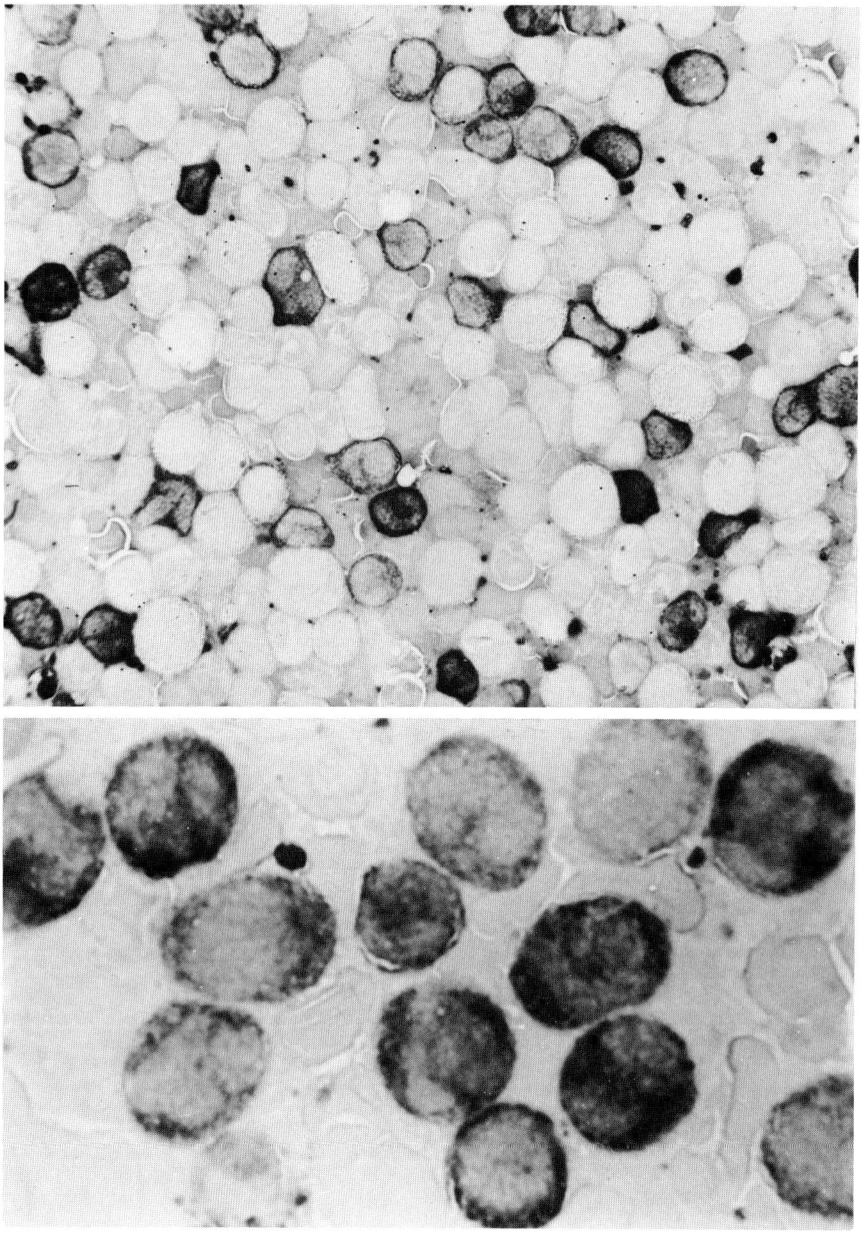

Figure 1-108. Bone marrow in SAMML, demonstrating nonspecific esterase stain using alpha naphthyl acetate as the substrate and hexazotized paraosaniline as dye coupler. Low-power view demonstrates monocytes that stain intensely with this stain **(top)**. Higher-power view shows intense staining of the cytoplasm of leukemic monocytes by the reagents **(bottom)**. In these leukemic monocytes as well as normal monocytes, nonspecific esterase activity is intense and diffuse, and sensitive to inhibition by sodium fluoride.

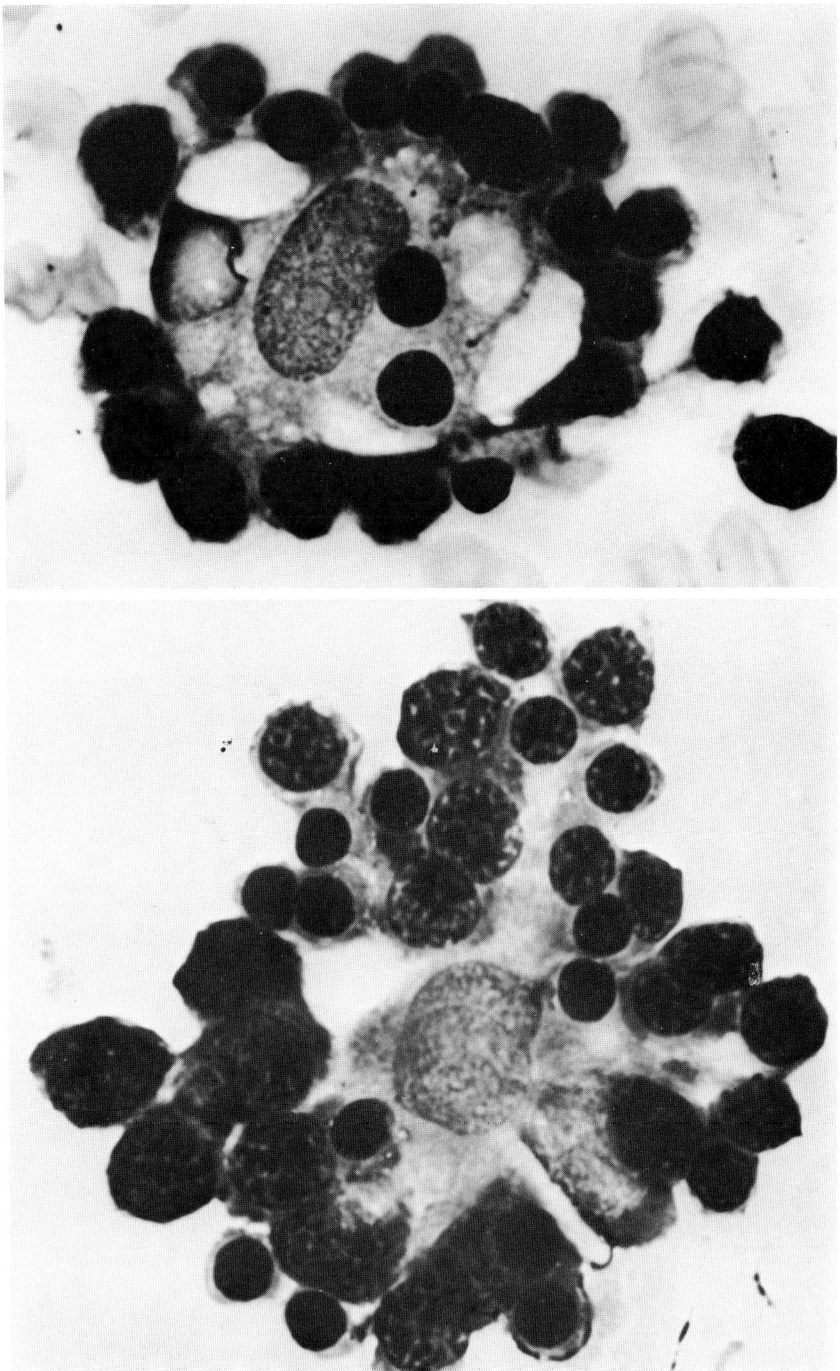

Erythroblastic islands in dilute bone marrow aspirate from patient with agnogenic myeloid metaplasia. Central macrophage contains nuclei of erythroblasts and cytoplasmic debris and is surrounded by developing erythroblasts.

Figure 1-109.

Agnogenic Myeloid Metaplasia

Agnogenic myeloid metaplasia is a disorder characterized clinically by splenomegaly and osteosclerosis, and hematologically by anemia, marked anisocytosis and poikilocytosis of erythrocytes, the occurrence of giant platelets and dwarf megakaryocytes in the peripheral blood, and fibrosis and megakaryocytosis of the bone marrow. Probably, this disorder has been recognized for many years under a variety of names.[115,426,461,1061,1077] Recently, agnogenic myeloid metaplasia has been included in the broad category of myeloproliferative disorders[560,1077] Some patients who have agnogenic myeloid metaplasia have had an antecedent diagnosis of polycythemia vera, chronic granulocytic leukemia, or essential thrombocythemia.[374,1077] In other patients, the disease does not have an apparent hematologic antecedent. Some patients with agnogenic myeloid metaplasia demonstrate evolution into acute leukemia as a terminal event.[104,272]

Typically, peripheral blood films from patients with agnogenic myeloid metaplasia show variability in numbers of white blood cells, as well as varying proportions of mature and immature granulocytes, increased numbers of eosinophils, basophils, and normoblasts, and poikilocytosis and basophilic stippling of erythrocytes. Giant hypergranular platelets, large hypogranular platelets, and dwarf megakaryocytes are seen frequently.

Characteristically, bone marrow aspirations in patients with agnogenic myeloid metaplasia result in a "dry tap." On touch preparations of the bone marrow core obtained on biopsy of the bone marrow, a few erythroblastic islands can be seen (Fig. 1-109). These represent clusters of erythroblasts that have been dislodged from their site of proliferation within marrow sinusoids during forceful aspiration of the marrow. Occasionally, several small fragments of marrow demonstrating strikingly increased numbers of micromegakaryocytes, megakaryocytes in mitosis, and variable numbers of erythroblasts and granulocytes can be found.

H & E sections of bone marrow biopsies show a disorganized pattern of fibrosis (Fig. 1-110). Numerous megakaryocytes are seen (Fig. 1-111); some are small, and others are large with multilobulation of the nucleus, as shown in Figure 1-112.[139] Mitotic figures are seen in some megakaryocytes. Fibroblasts are numerous (Figs. 1-110 through 1-113). Clusters of erythroblasts occur and represent proliferation of these cells within sinusoids (Figs. 1-114 through 1-116). Most of the erythroblasts are intermediate normoblasts. Proerythroblasts may be increased, and paratrabecular fibrosis is intense.

Pathogenetically, relationships between megakaryocytic and fibroblastic proliferation are thought to be intimate, and deposition of collagen usually occurs in close proximity to megakaryocytes.[999] In biopsies of marrow from some patients with agnogenic myeloid metaplasia, areas of dense fibrosis may alternate with areas of marked hypercellularity as described above. It is believed that hypercellularity and megakaryocytosis of

(text continued on page 116)

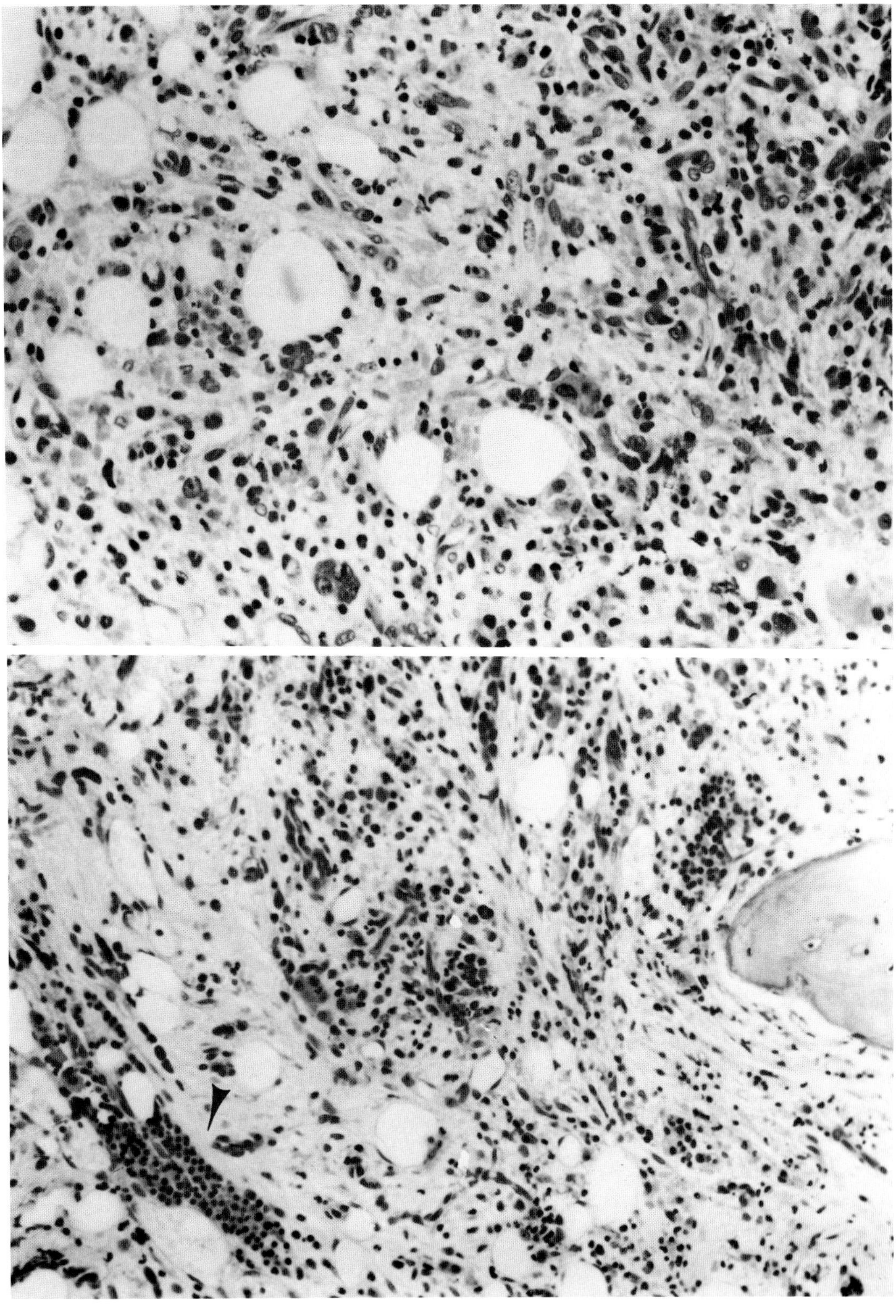

Figure 1-110. Bone marrow in agnogenic myeloid metaplasia. Dense fibrosis is seen along with apparent sinusoidal collections of developing erythroblasts (**arrow**).

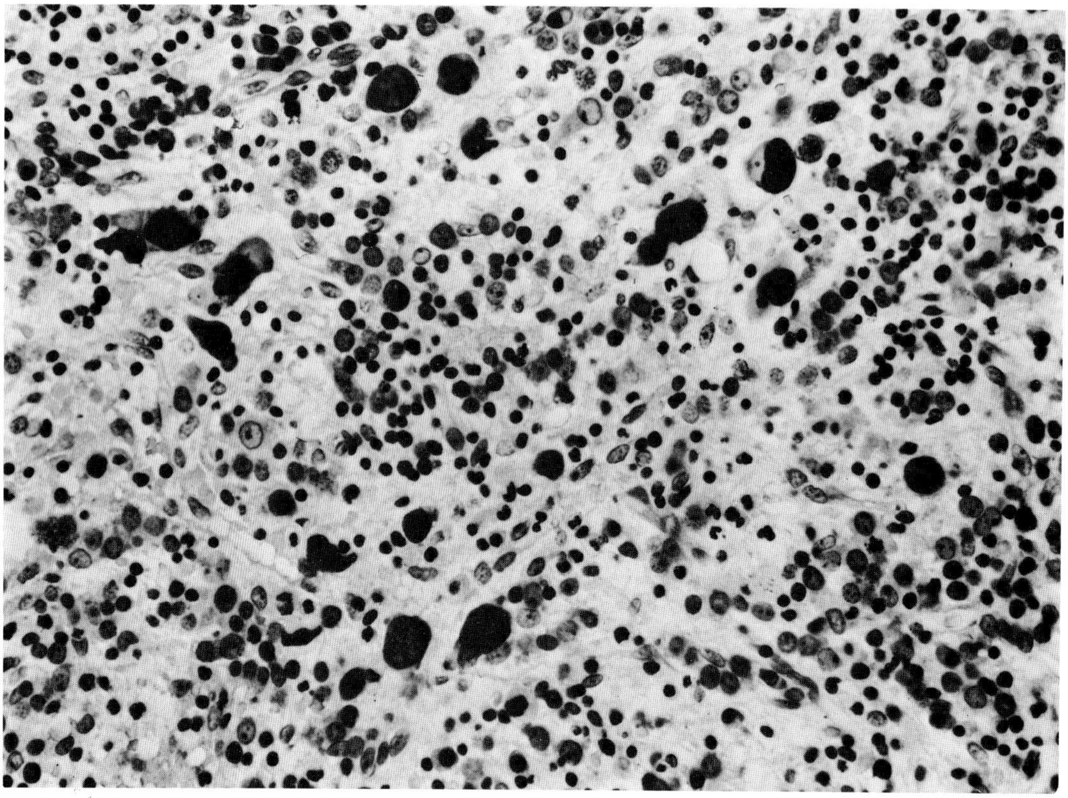

Figure 1-111. Bone marrow in agnogenic myeloid metaplasia. Many darkly staining nuclei of megakaryocytes as well as erythroblasts and fibroblasts are seen.

Figure 1-112. Bone marrow in agnogenic myeloid metaplasia, showing atypical ➡ appearing megakaryocytes in increased numbers.

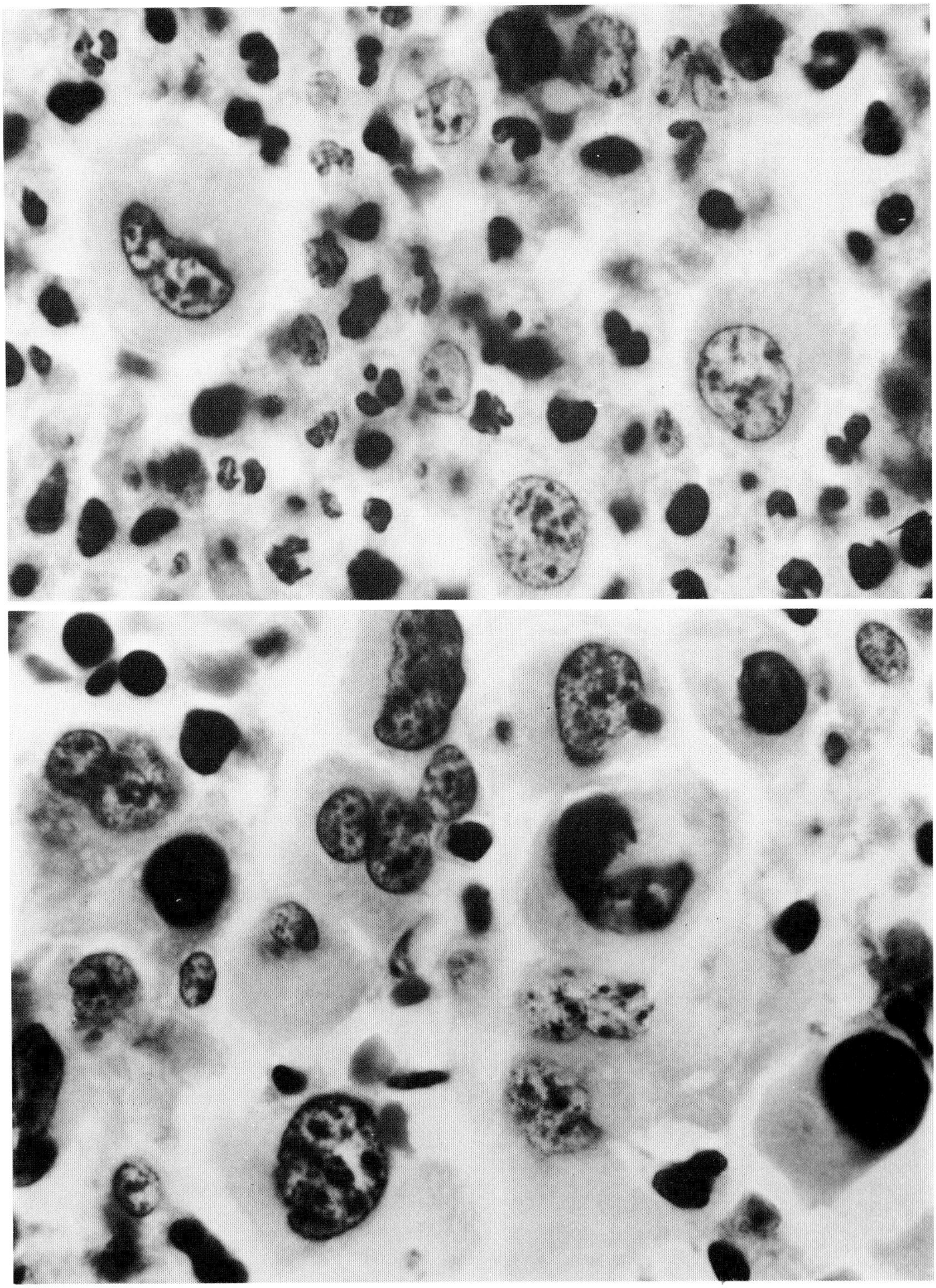

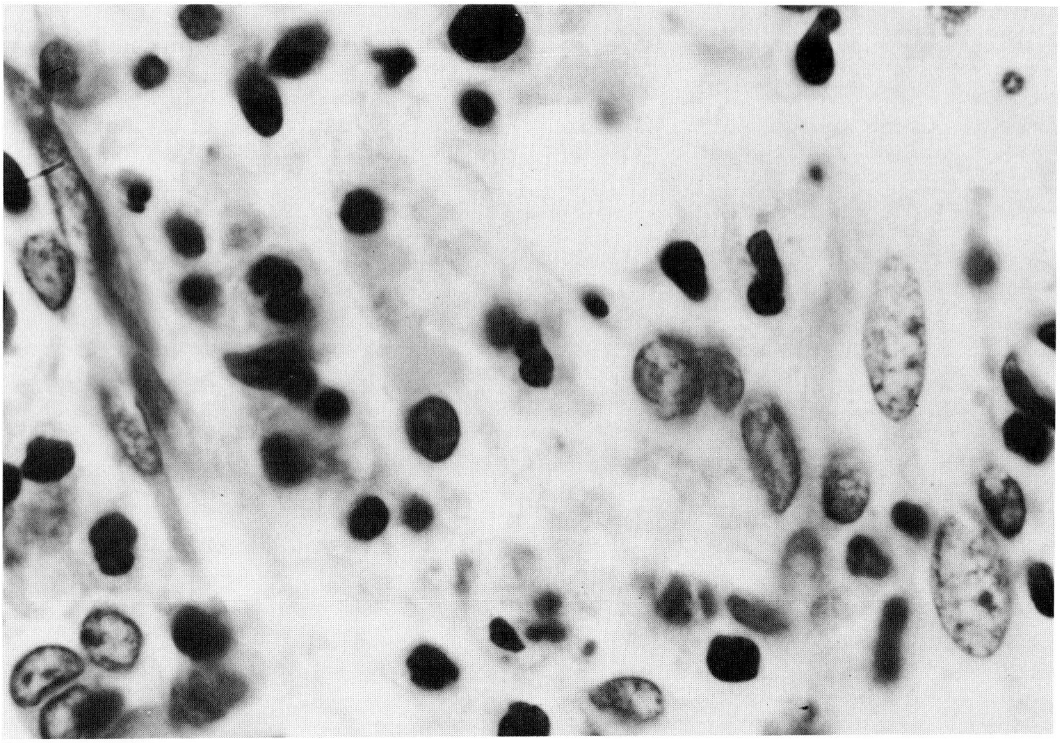

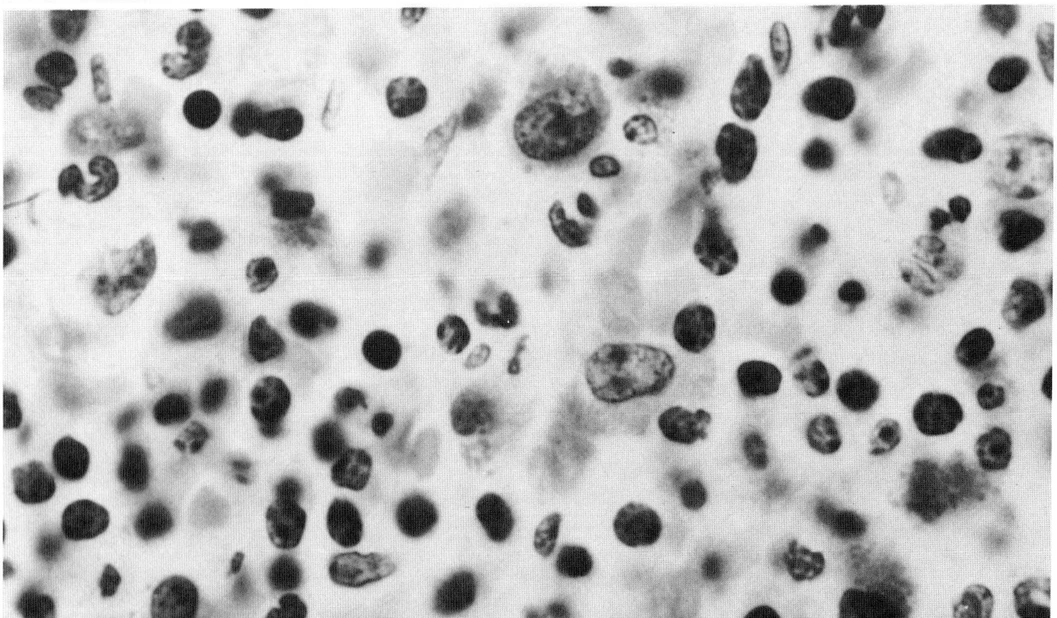

Figure 1-113. ← Bone marrow in agnogenic myeloid metaplasia, demonstrating increased numbers of fibroblasts along with increased numbers of band forms and erythroblasts with darkly staining nuclei.

Figure 1-114. Bone marrow in agnogenic myeloid metaplasia, illustrating disorganized patterns of fibrosis along with apparent sinusoidal collections of erythroblasts (**arrow**).

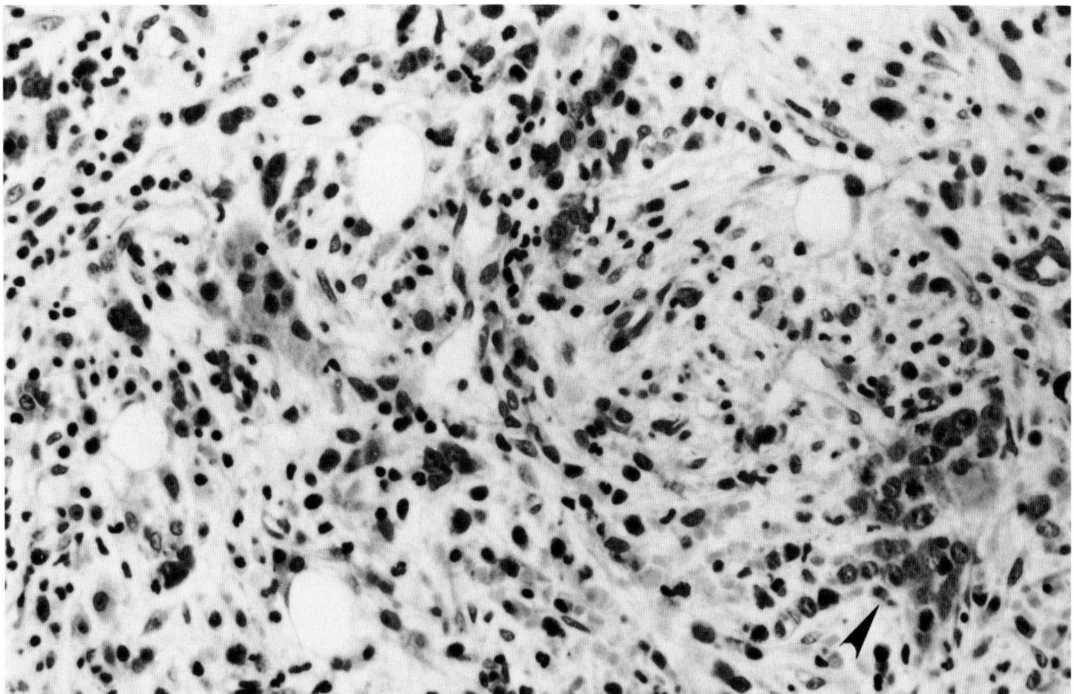

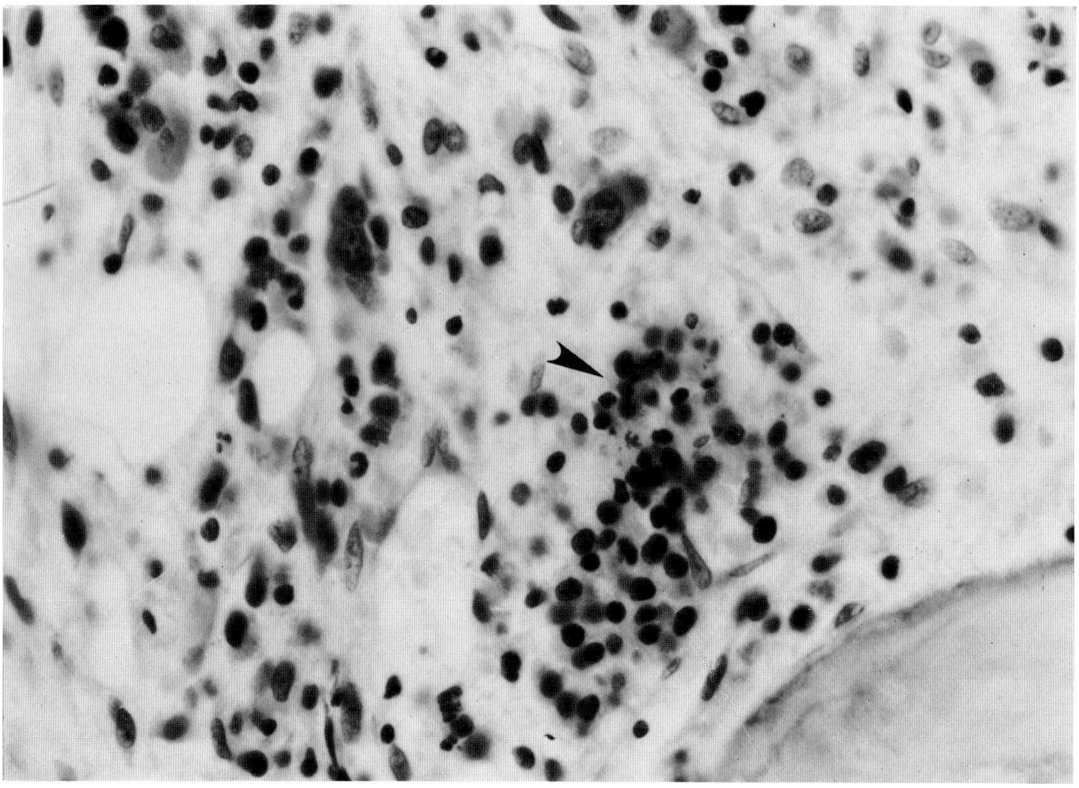

Figure 1-115. Bone marrow in agnogenic myeloid metaplasia, showing sinusoidal collections of erythroblasts with darkly stained nuclei (**arrow**).

Figure 1-116. Bone marrow in agnogenic myeloid metaplasia, showing increased numbers of erythroblasts with darkly staining nuclei, as well as cells with large, pale vesicular nuclei and prominent nuceloli, probably representing procrythroblasts. ➡

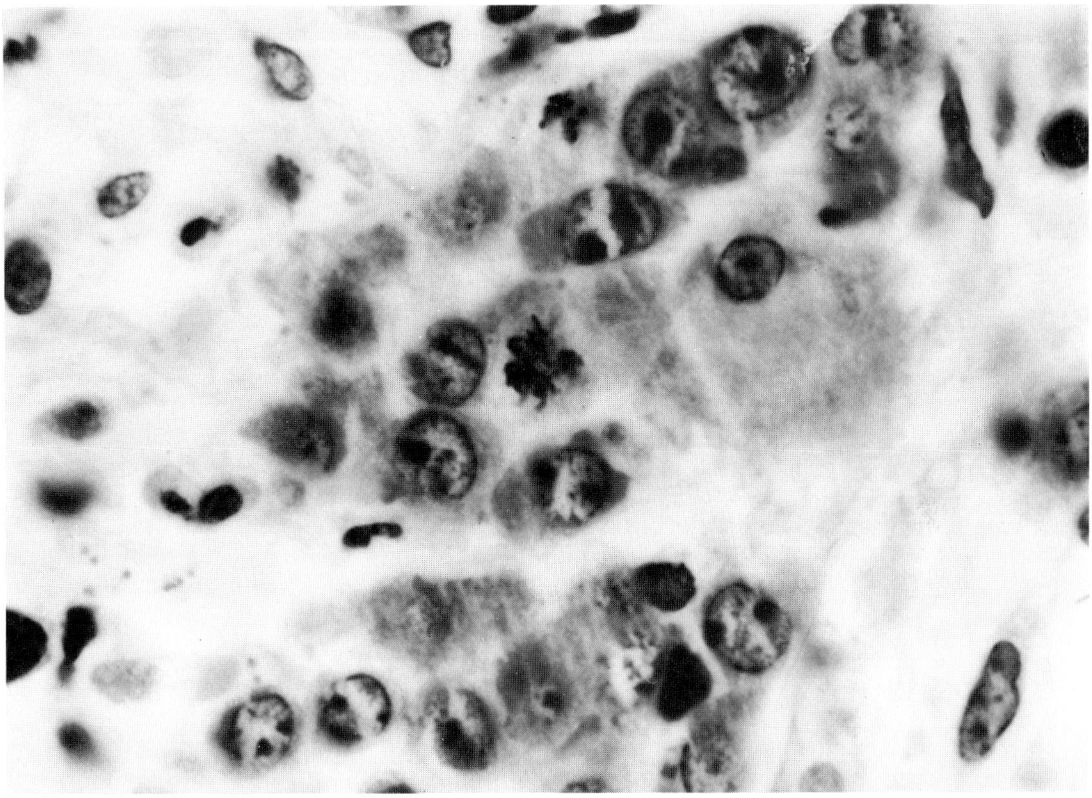

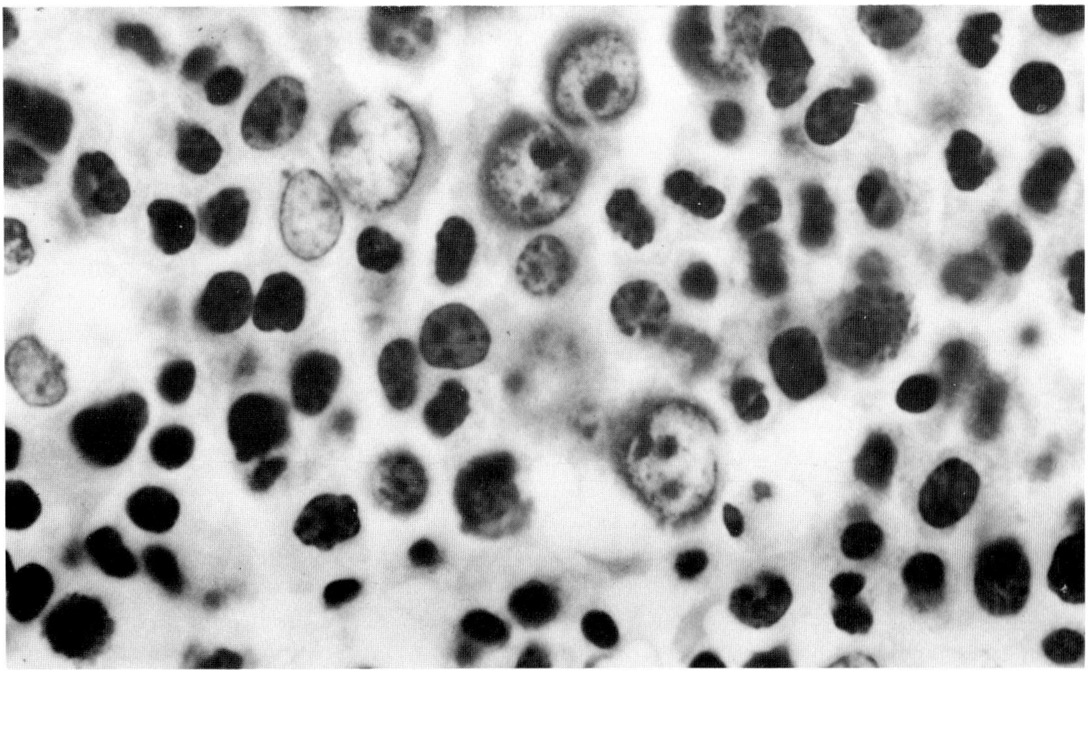

the bone marrow precede the development of extensive fibrosis in patients with agnogenic myeloid metaplasia.

Cytochemically, bone marrows from patients with agnogenic myeloid metaplasia contain markedly increased amounts of reticulin (Fig. 1-117). Using the PAS reagent, atypical megakaryocytes reveal abundant amounts of glycogen. They also contain substantial activity of nonspecific esterase. Using an immunoperoxidase technique, megakaryocytes can be identified on the basis of distinctive staining properties.[150,156,158,159,164,165]

Polycythemia Vera

Polycythemia vera is a disorder characterized clinically by splenomegaly and hepatomegaly, and hematologically by absolute erythrocytosis, leukocytosis, and thrombocytosis in typical cases. Now included in the myeloproliferative category of disorders, polycythemia vera may show evolution into agnogenic myeloid metaplasia, heralded by gradual enlargement of the spleen accompanied by anemia. Polycythemia vera can also show evolution into acute leukemia. Usually this is acute myeloblastic leukemia, but it may be erythroleukemia, monocytic leukemia, or even lymphoblastic leukemia.[58,623,674]

Peripheral blood specimens from patients with polycythemia vera show increased numbers of erythrocytes. In many cases, erythrocytes demonstrate hypochromia and microcytosis, reflecting iron deficiency due to increased utilization of iron or gastrointestinal bleeding. Increased numbers of neutrophils may occur, and occasionally myelocytes and metamyelocytes are present. Increased numbers of eosinophils and basophils are frequently found, and thrombocytosis is common. Sometimes, giant hypercellular platelets are observed.

Marrows from patients with polycythemia vera are hypercellular with increased numbers of megakaryocytes. Frequently, megakaryocytes are micromegakaryocytes. In other instances, megakaryocytes appear normal or may be unusually large and show multiple nuclear lobulations and hypergranularity of the cytoplasm. Erythroblastic proliferation is intense, and erythroblasts usually comprise up to 50% of the bone marrow cells in a differential cell count (Fig. 1-118). Erythropoiesis is normoblastic and proceeds in an orderly fashion from proerythroblasts to mature erythrocytes. Similarly, granulocytic development is normal (Fig. 1-118).

Cytochemically using the Prussian blue stain, decreased or absent amounts of iron are found, resulting from the accelerated erythropoiesis. Using other cytochemical stains readily available, granulocytic cells and megakaryocytes do not show distinctive abnormalities.

On H&E sections of biopsies of marrows from patients with polycythemia vera, specimens are hypercellular and show increased numbers of megakaryocytes and increased numbers of granulocytic cells and erythroblasts (Fig. 1-119). These are identified on the basis of darkly-staining nuclei with a checkerboard nuclear chromatin pattern. In uncomplicated cases, paratrabecular fibrosis is not found.

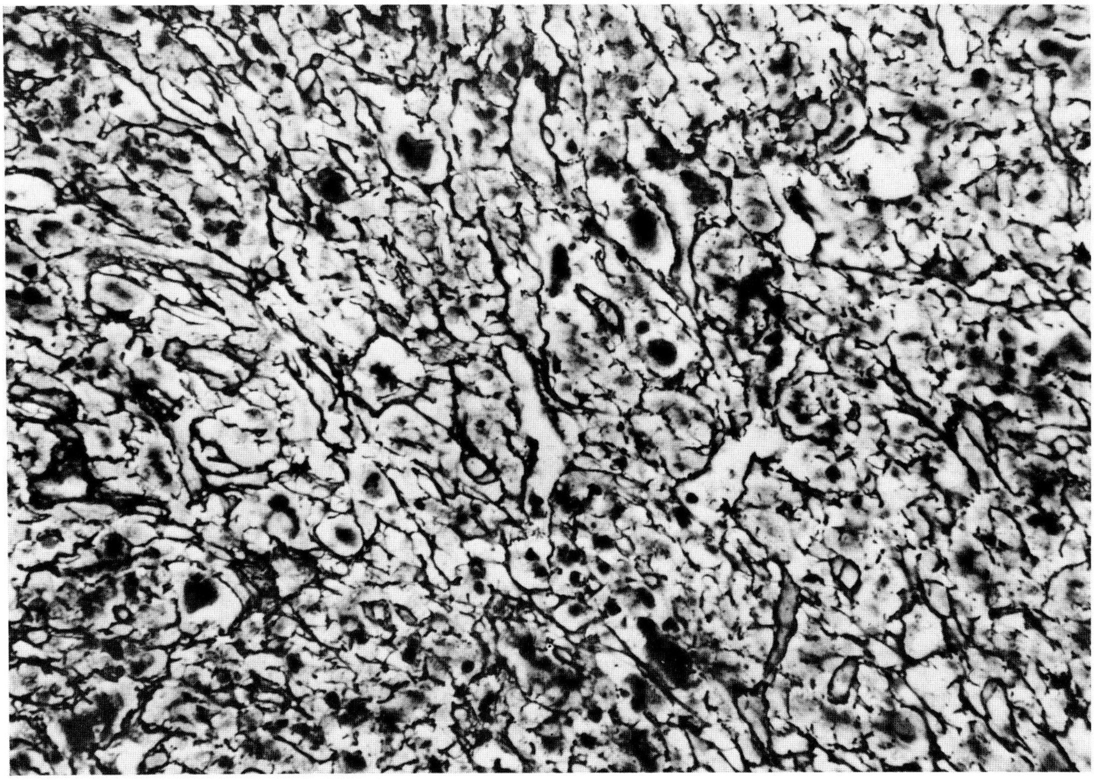

Bone marrow in agnogenic myeloid metaplasia, depicting increased amounts of reticulin using Snook's reticulin stain. **Figure 1-117.**

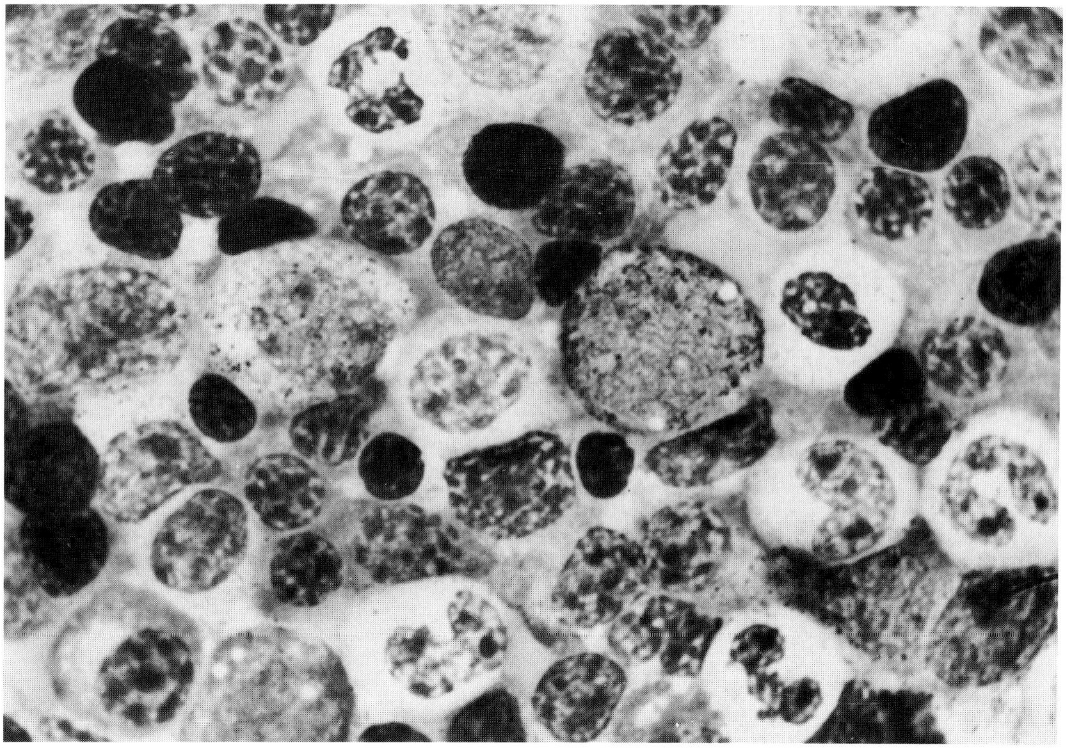

Figure 1-118. Bone marrow in polycythemia vera, showing hypercellularity along with increased numbers of granulocytic precursors and erythroblasts.

Figure 1-119. H & E section of bone marrow biopsy in polycythemia vera, illustrating increased numbers of erythroblasts with darkly staining nuclei and checkerboard nuclear chromatin pattern. ➔

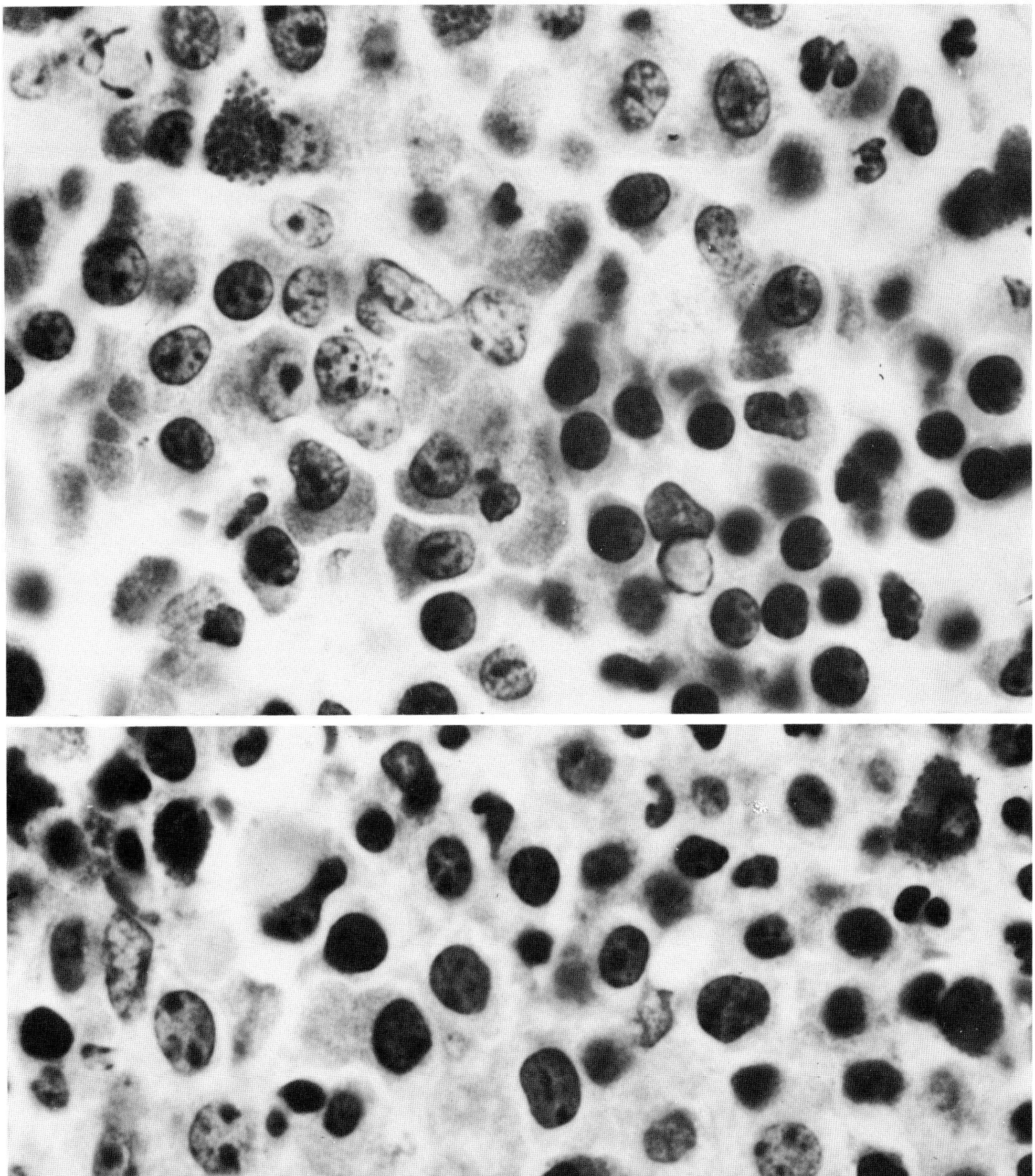

2

Myeloblastic Leukemia and Its Variants

Acute Myeloblastic Leukemia

Acute myeloblastic leukemia is a disorder of neoplastic myeloblasts that proliferate in the blood, bone marrow, and visceral organs. Hematologically, it is characterized by anemia, normoblastemia, thrombocytopenia, and variable numbers of myeloblasts in the blood. Clinically, it is often characterized by hepatomegaly, splenomegaly, and fever in some cases, and by consequences of pancytopenia, such as hemorrhage and infection. In acute myeloblastic leukemia, greater than 40% of the marrow cells are leukemic myeloblasts. When present, the Auer rod forms a distinctive and valuable marker for leukemic myeloblasts or monocytes. Detection of Auer rods rapidly distinguishes cells of granulocytic origin, such as myeloblasts, from leukemic lymphoblasts that do not have Auer rods.

Although acute myeloblastic leukemia is usually fatal in 3 to 4 months after diagnosis if untreated, over the past several decades it has been recognized that there are wide variations in the survival of patients with acute myeloblastic leukemia, probably reflecting differences in proliferative capabilities of leukemic myeloblasts. In some patients, acute myeloblastic leukemia is fulminating. In others, it is indolent and has as its primary clinical and hematologic manifestation a refractory anemia requiring frequent transfusions over a long period of time.

Although most cases of acute myeloblastic leukemia seem to arise *de novo*,[217,576] other cases have antecedent hematologic disorders, such as drug-induced aplastic anemia, chronic granulocytic leukemia, Fanconi's syndrome, erythroleukemia, chronic erythremic myelosis, smoldering acute leukemia, preleukemia, agnogenic myeloid metaplasia, polycythemia vera, and granulocytic sarcoma or chloroma.[66,122,140,209,334,352,980] In still other circumstances, acute myeloblastic leukemia seems to have one or more predisposing factors. Some of these may be genetic abnormalities,[120,134,335,399,415,624,678,681–683,905] whereas others bear a close if not etiologic relationship to ionizing radiation[116,218] or exposure to volatile solvents such as benzene,[11,240,296,1064] phenylbutazone,[475] or chloramphenicol.[704] Some other cases of myeloblastic leukemia have a familial incidence.[17,100,387,539,857]

More recently, acute myeloblastic leukemia has been identified as a terminal event in some patients with other types of hematologic malignancies and solid tumors treated with a variety of drugs.[65,872] Of these, alkylating agents and procarbazine seem to be implicated as leukemogens.[122] Myeloblastic leukemia has been recorded as a terminal event in patients who have multiple myeloma treated with alkylating agents,[555] as well as in patients with untreated multiple myeloma,[1047] lymphocytic lymphoma,[122,531,648,1060,1120] Hodgkin's disease, particularly when chemotherapy and radiotherapy are combined,[65,122,182,211] and in patients who have other types of solid tumor malignancies.[65,122] It is as yet uncertain whether the evolution of these disorders into acute myeloblastic leukemia is part of their natural history, or if myeloblastic leukemia is induced or provoked by exposure to various chemotherapeutic drugs or radiotherapy used to treat the primary malignancy.[122]

Peripheral blood samples from some patients with acute myeloblastic leukemia demonstrate pancytopenia. In others, normoblasts, basophilic stippling of erythrocytes, giant macrocytes, atypical monocytes, and metamyelocytes, bands, and neutrophils with aberrant nuclear configurations can be identified. In patients with acute myeloblastic leukemia, the peripheral blood may contain many, few, or virtually no myeloblasts. Usually, a few myeloblasts can be identified in buffy coat preparations from patients regarded as aleukemic on the basis of a peripheral blood film alone. Some of the myeloblasts contain vacuoles.

As a characteristic cytoplasmic inclusion, some leukemic myeloblasts contain one or more Auer rods.[3,34,156,336,337,365,606] Using panoptic stains, Auer rods appear sharply angular and stain bright red to red orange. Generally, Auer rods are seen more frequently in young granulocytic cells, such as promyelocytes and myeloblasts, than in mature granulocytes. Auer rods are believed to originate from the coalescence of membranes of granules.[660,727] They contain activities of acid phosphatase, specific esterase, and myeloperoxidase, suggesting their lysosomal derivation.[3,161,316,1050,1092] Auer rods are valuable markers for cells of granulocytic origin because they are found exclusively in leukemic myeloblasts. This traditional view has been challenged by a recent report of apparent Auer rods in leukemic lymphoblasts.[474] In all likelihood, these Auer rods were atypical azurophil inclusions that occur in some leukemic lymphoblasts. Unlike typical Auer rods,[506] cytoplasmic inclusions in leukemic lymphoblasts do not contain activity of myeloperoxidase.

According to the recent French-American-British (FAB) classification of the acute leukemias,[87] acute myeloblastic leukemia has three morphologic variants. In the first, designated as *M1*, leukemic blasts show some evidence of granulocytic differentiation consisting of several azurophil granules or a few granules that are myeloperoxidase positive (Plate 2*B*). In the second variant, called *M2*, there is maturation beyond the promyelocyte stage (Plate 2*C*). Auer rods are almost always single. In unusual cases, those metamyelocytes and myelocytes that are present are all eosinophils. The third morphologic variant is promyelocytic leukemia (*vide infra*) and has been designated as *M3* (Plate 2*D*).

Bone marrow aspirates from most patients with typical acute myeloblastic leukemia are hypercellular, as seen in Figures 2-1 through 2-9.[1,2,87–88,96,99,106,108,125,128,129,205,234,246,247,259,285,290,311,312,332,340,373,396,400,434,479,604,

(text continued on page 132)

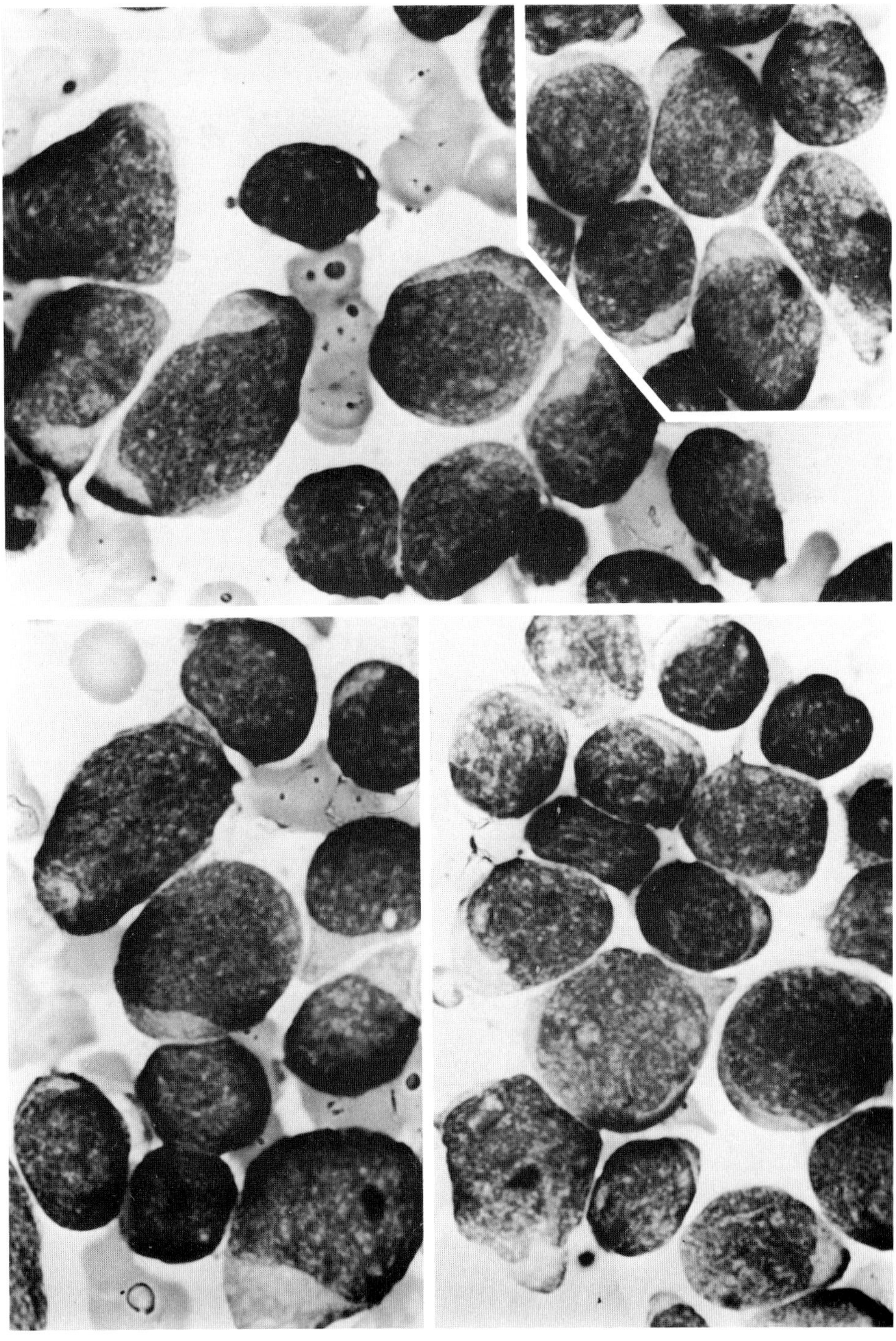

Large and small myeloblasts in acute myeloblastic leukemia (FAB M1). **Figure 2-1.**

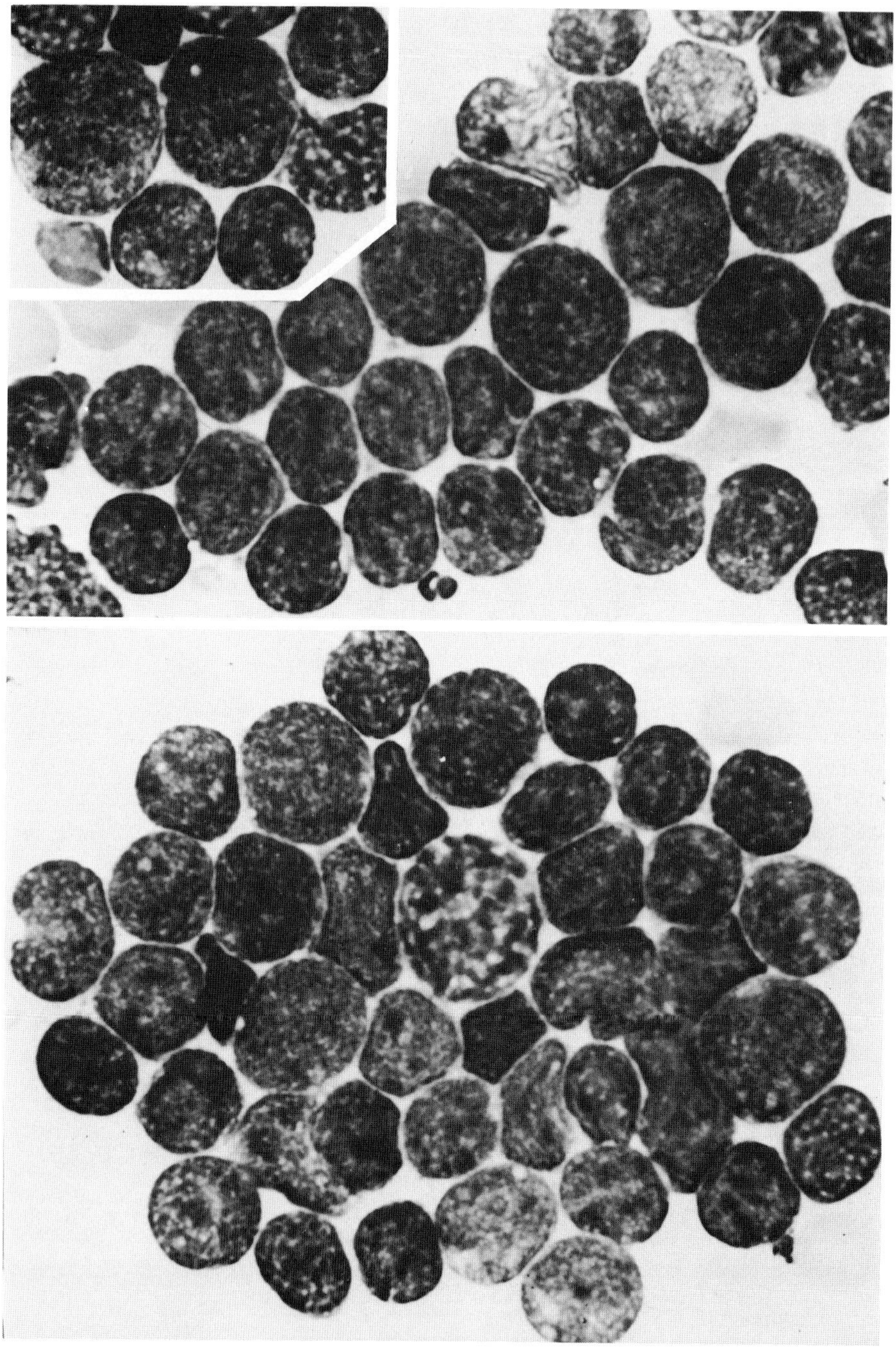

← Large and small myeloblasts in acute myeloblastic leukemia. **Figure 2-2.**

Large and small myeloblasts in acute myeloblastic leukemia. **Figure 2-3.**

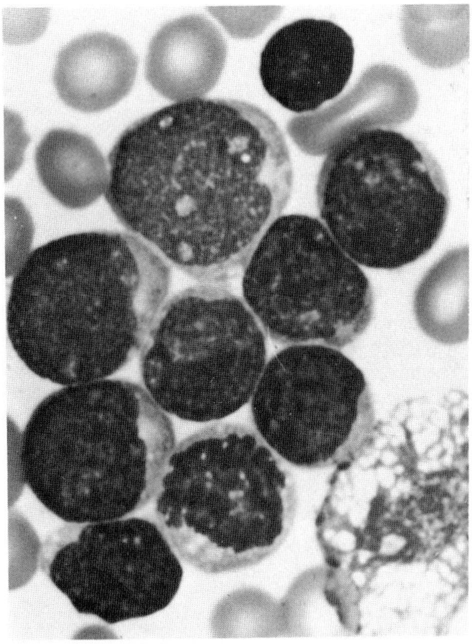

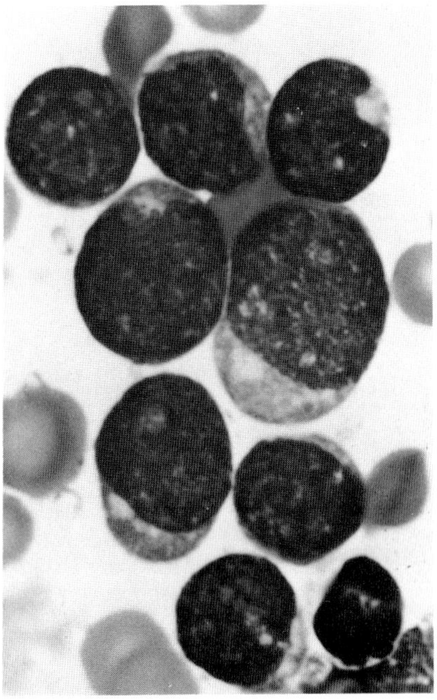

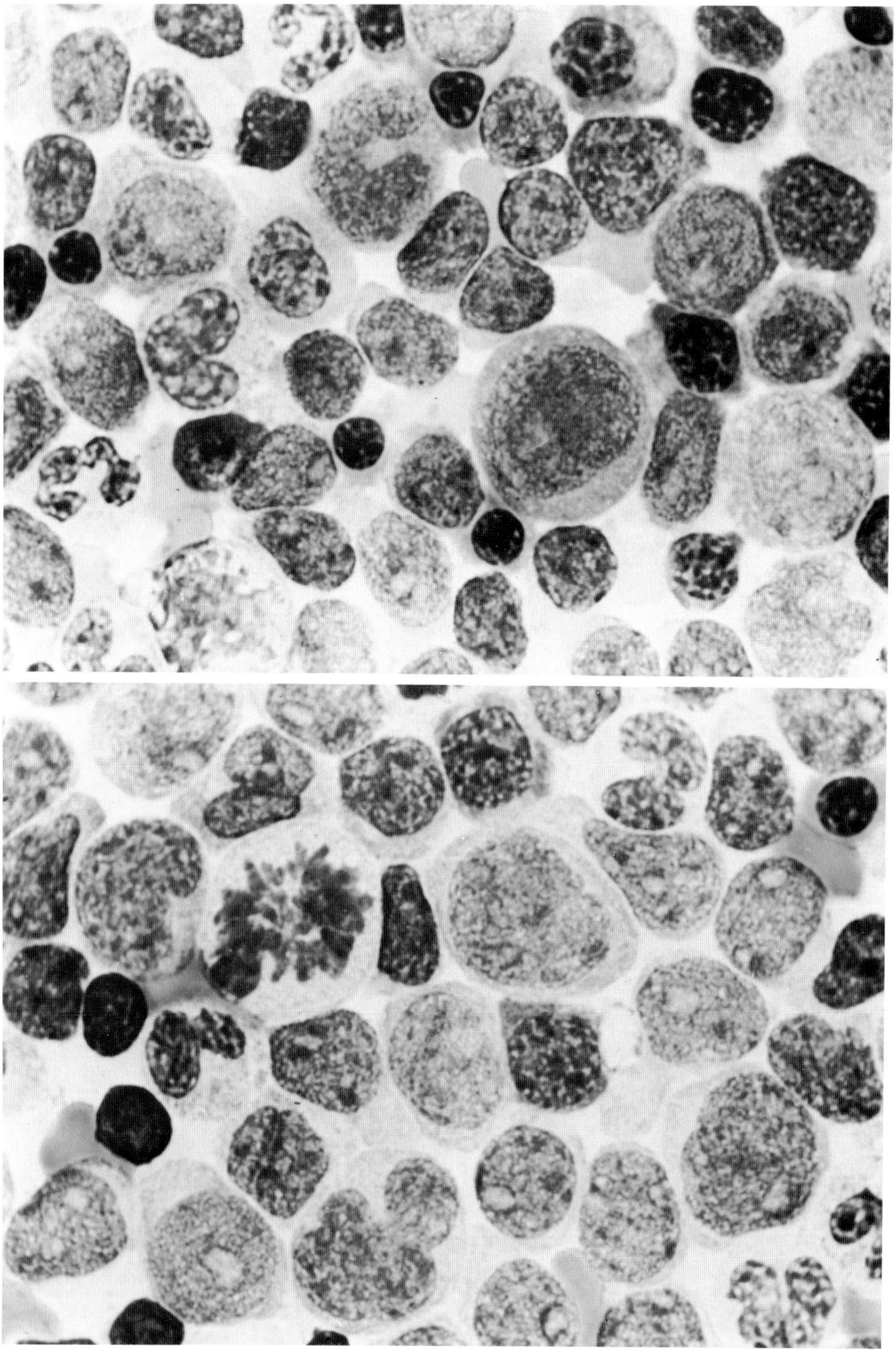

Figure 2-4. Bone marrow in acute myeloblastic leukemia, showing increased numbers of leukemic myeloblasts, with prominent nucleoli, atypical-appearing metamyelocytes, and few erythroblasts (FAB M2).

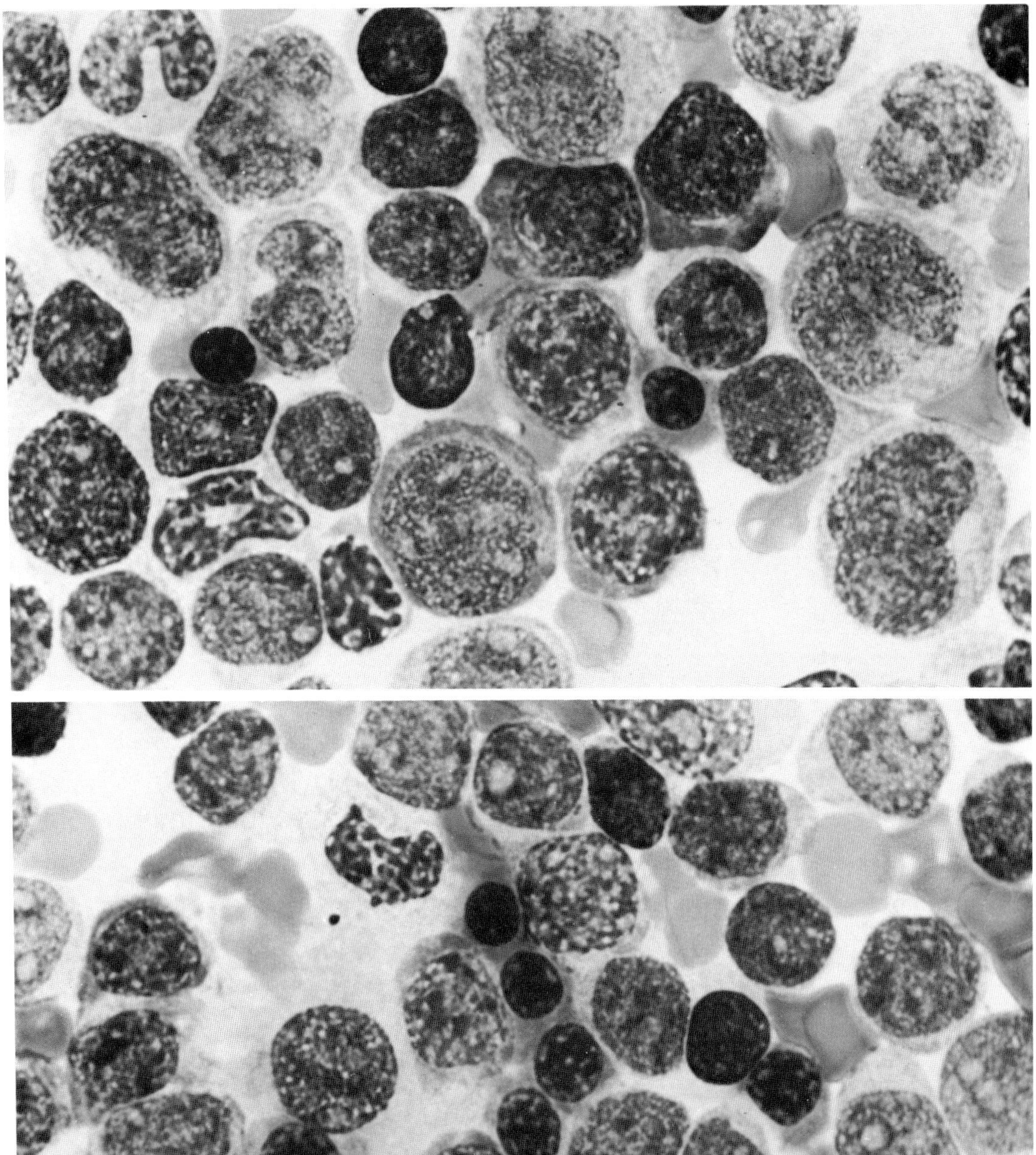

Bone marrow in acute myeloblastic leukemia, illustrating increased numbers of myeloblasts, atypical metamyelocytes, and megaloblastoid intermediate macronormoblasts.

Figure 2-5.

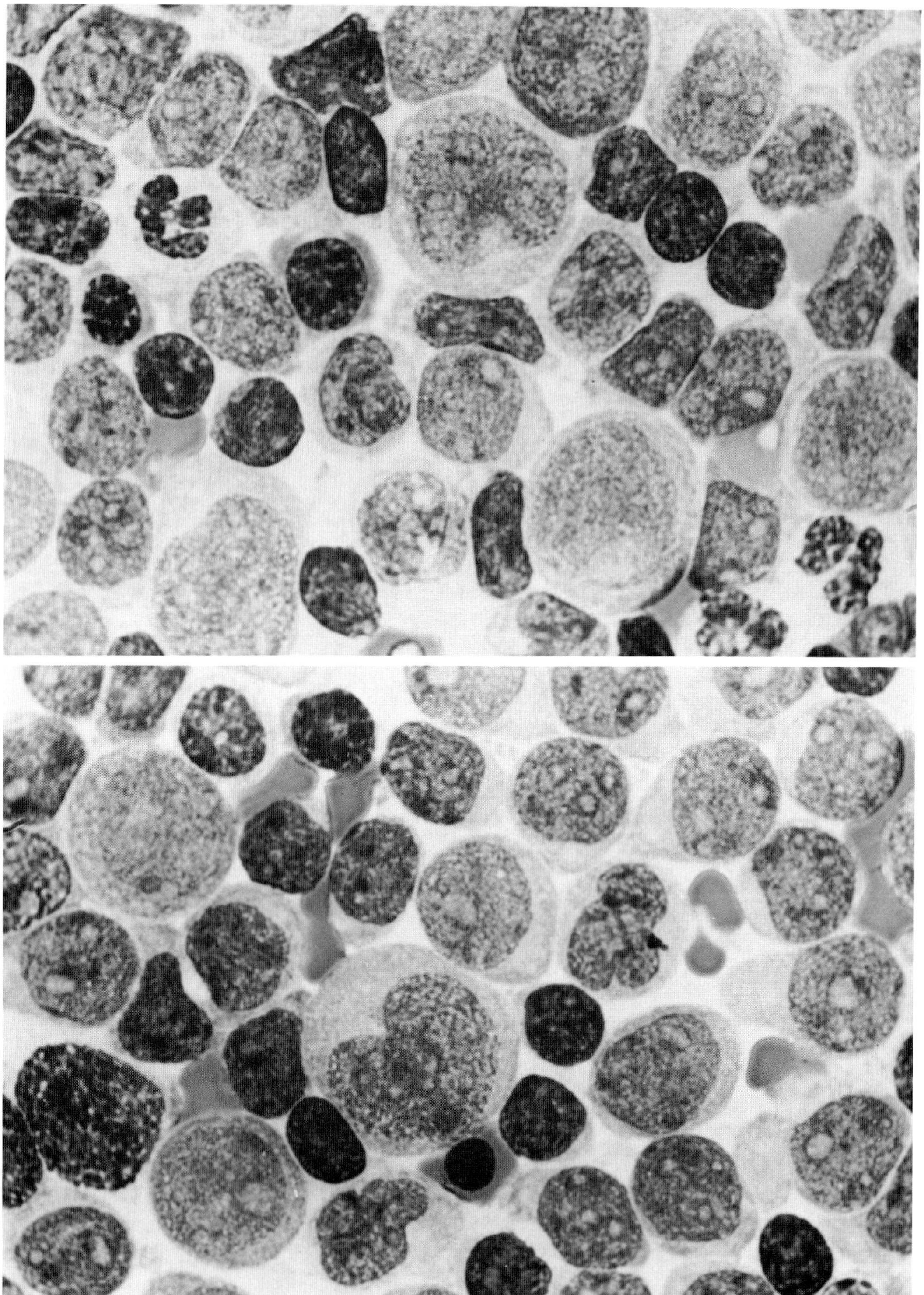

Figure 2-6. Bone marrow in acute myeloblastic leukemia, demonstrating increased numbers of myeloblasts having prominent nucleoli and delicate chromatin pattern.

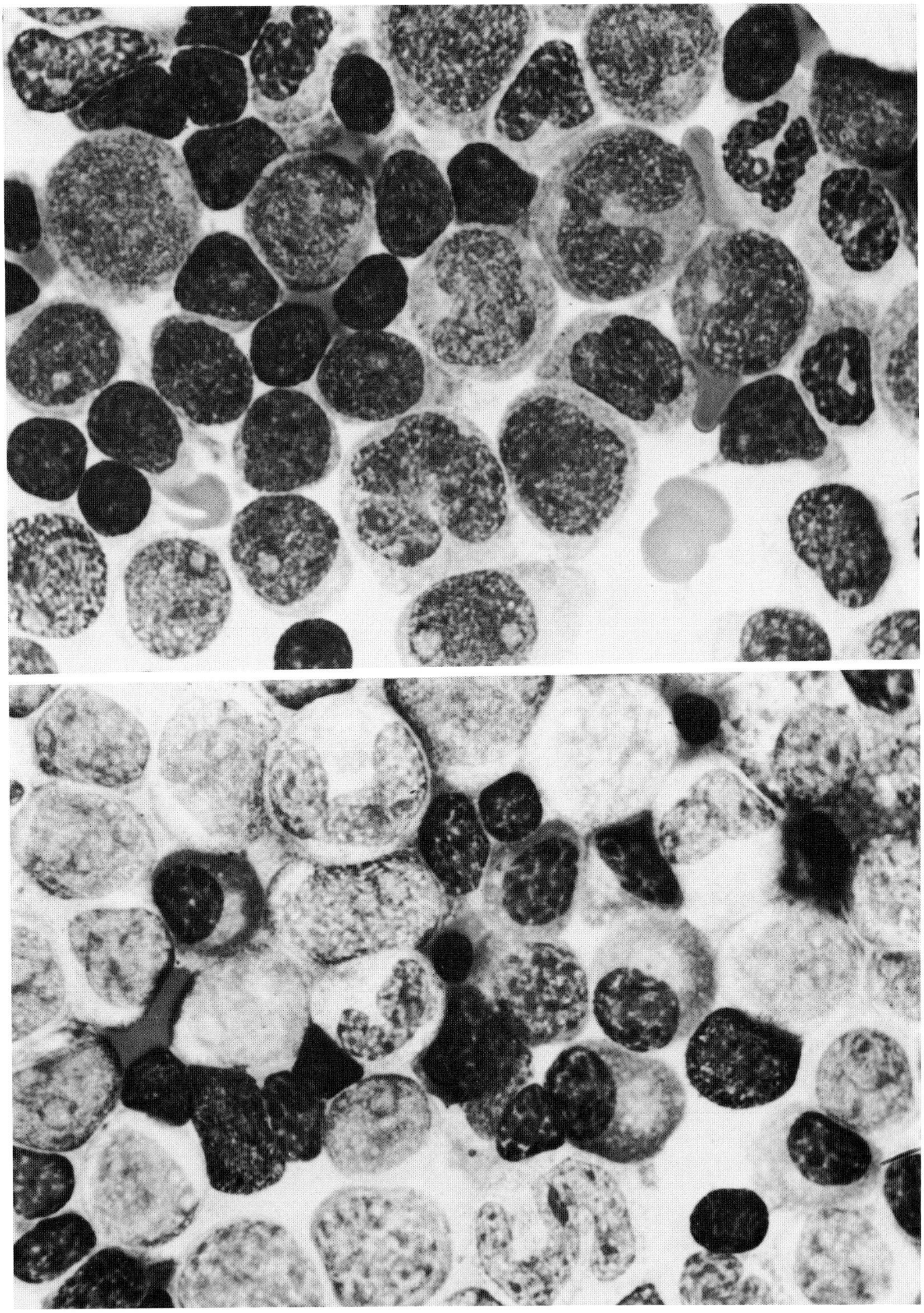

Figure 2-7. Bone marrow in acute myeloblastic leukemia, showing increased numbers of myeloblasts, giant promyelocytes, atypical metamyelocytes, and plasma cells (FAB M2).

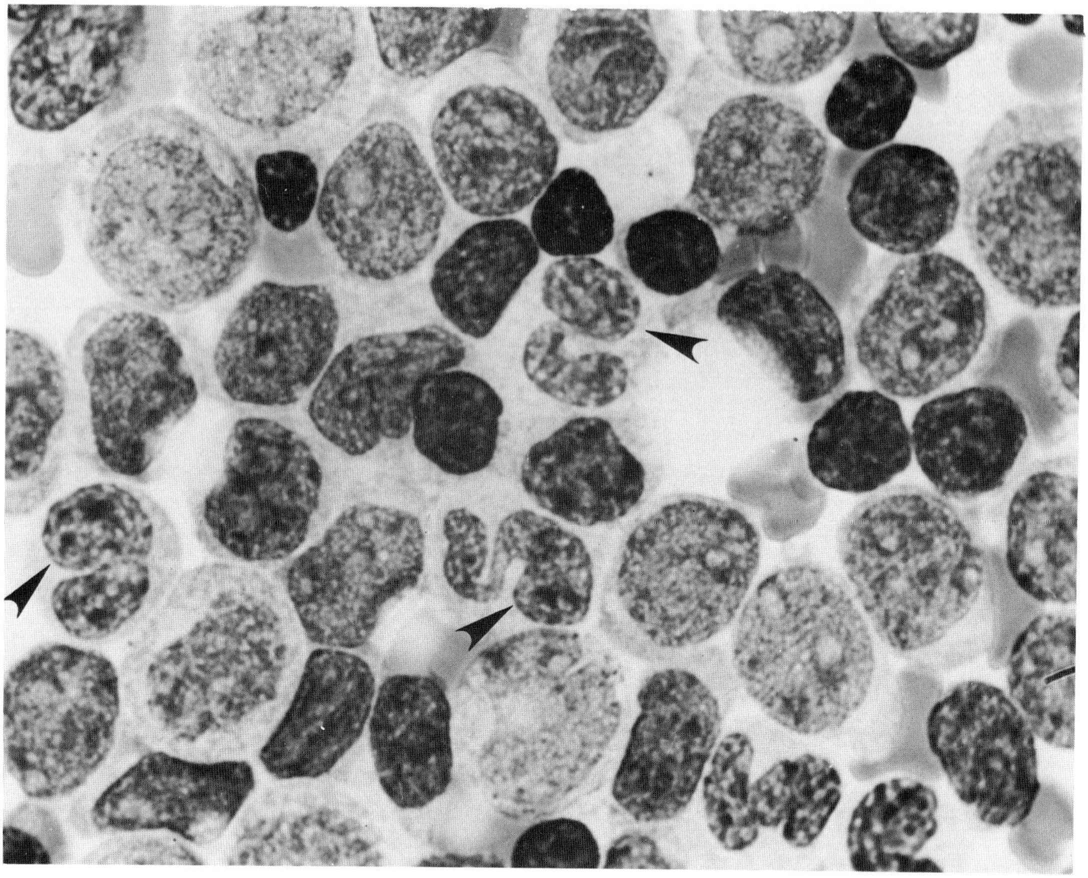

Figure 2-8. Bone marrow in acute myeloblastic leukemia, depicting increased numbers of myeloblasts, promyelocytes, and bands with atypical nuclear configurations in which one end of the nucleus appears as a bulbous protrusion (**arrows**).

Figure 2-9. Bone marrow in acute myeloblastic leukemia, demonstrating atypical megakaryocytes and increased numbers of myeloblasts. →

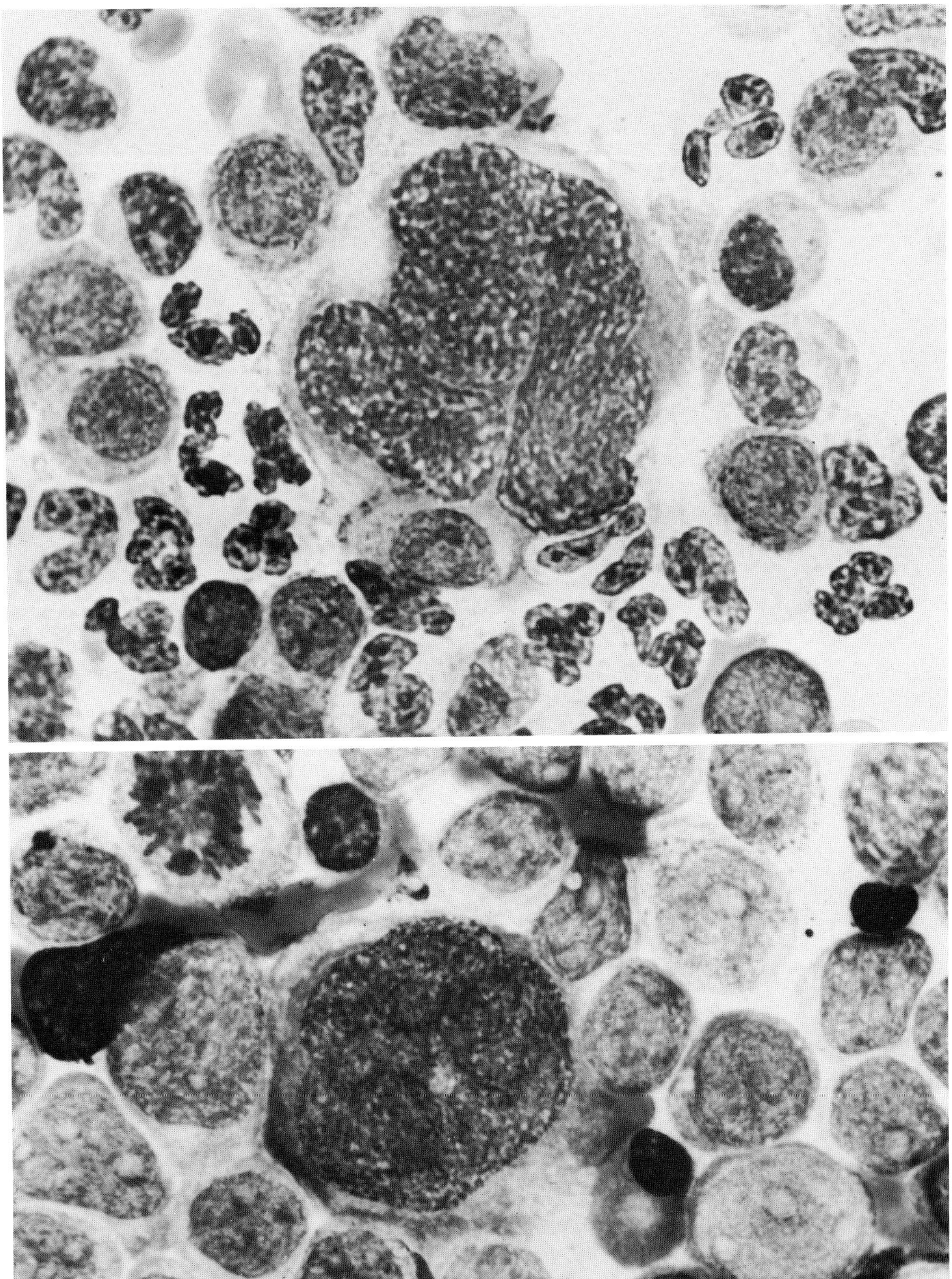

[610,612,649,650,694,699,707,723,724,758,881,882,934,1028] Myeloblasts are greatly increased in number. They constitute at least 40%, and in some instances they constitute up to 80% to 90% of the cells in a differential cell count of the bone marrow.[122] More mature granulocytic cells are usually reduced substantially in number and in some cases may be virtually absent. In the variant designated *M1*,[87] both large and small myeloblasts are found, as seen in Figures 2-1 through 2-3 and Plate *2B*.[700,712,713] Proportions of large and small myeloblasts vary from case to case. Large myeloblasts contain a nucleus with delicate-appearing nuclear chromatin strands that appear close together, as shown in Figures 2-4, 2-5, 2-10, and Table 2-1.[146,265,267,321] Aggregates of nuclear chromatin are few, and if present they are usually small (see Figs. 2-1 through 2-9). Nucleoli are small and may be multiple. There is usually perinucleolar condensation of chromatin. Using panoptic stains, cytoplasm of large myeloblasts is deeply basophilic, as in other types of malignant nonhematopoietic cells that can resemble them.[185] Auer rods may be present and can be multiple. They are usually rod-shaped but may be splinter-shaped or even coccoid-shaped.[34,565,852] Small myeloblasts contain a nucleus that shows numerous large aggregates of nuclear chromatin. These aggregates are connected to each other and to the nuclear

Table 2-1. MAJOR DISTINCTIONS BETWEEN TYPICAL LEUKEMIC LYMPHOBLASTS AND LEUKEMIC MYELOBLASTS

	Lymphoblast	Myeloblast
Configuration of nucleus	round to oval	round to oval
Contour of nuclear membrane	even	even
Caliber of chromatin strands	coarse	delicate
Proximity of chromatin strands	far apart	close
Number of chromatin aggregates	increased	decreased
Size of chromatin aggregates	large	small
Shape of chromatin aggregates	blocklike and stripe	dotlike
Nucleolus	yes	yes
Perinucleolar condensation of chromatin	prominent	inconspicuous
Cytoplasmic volume	scant	scant
Intensity of basophilia	intense	weak
Nonspecific azurophilic granules	rare	never
Auer rods	never	sometimes
PAS reaction for glycogen	strong, blocklike	weak, dustlike
Methyl green-pyronine stain for RNA	strong	weak
Specific esterase activity	absent	weak
Nonspecific esterase activity	weak in B+ null-cell, unipolar in T-cell	weak
Acid phosphatase activity	weak in B+ null-cell, unipolar in T-cell	weak
Terminal deoxynucleotidyl transferase	present	absent

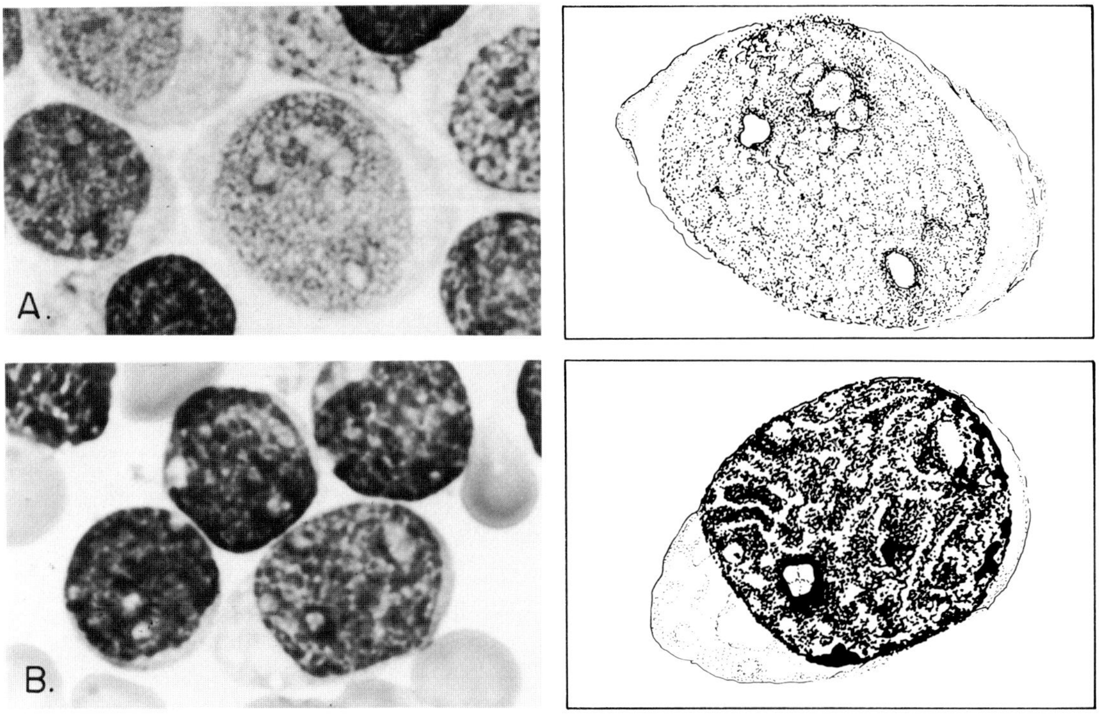

(**A**) Leukemic myeloblast, showing delicate nuclear chromatin strands that are closely approximated, few small aggregates of chromatin, and prominent nucleoli in photomicrograph (**left**) and artist's rendering (**right**). (**B**) Leukemic lymphoblast, demonstrating coarse-appearing strands of nuclear chromatin that appear widely separated, large, prominent aggregates of chromatin, and prominent nucleoli in photomicrograph (**left**) and artist's rendering (**right**). **Figure 2-10.**

Sudan black B positivity of promyelocytes and myelocytes in acute myeloblastic leukemia. **Figure 2-11.**

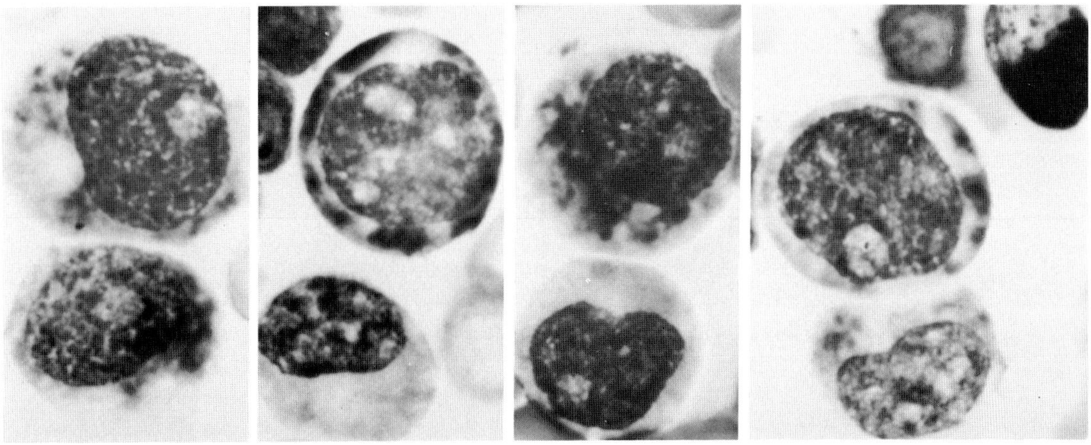

membrane by delicate strands of chromatin (see Figs. 2-1 through 2-4). Nucleoli are usually small and may be inconspicuous.

In the variant of acute myeloblastic leukemia designated *M2* in the FAB classification,[87] differentiation beyond the promyelocyte stage is seen (see Figs. 2-4 through 2-8, and Plate 2C). Promyelocytes may show splinter-shaped granules.[923] Metamyelocytes and bands often demonstrate aberrant nuclear configurations, such as bulbous nuclear ends, hyposegmentation, clubbing, and doughnut-hole (see Figs. 2-4 through 2-8). Some of these metamyelocytes are gigantic (see Figs. 2-4 through 2-6). Similar nuclear abnormalities are found in neutrophils,[926,1059] as shown in Figures 2-7 and 2-8. Plasma cells may be increased in number (see Fig. 2-8). Erythropoiesis is reduced and is usually megaloblastoid (see Figs. 2-4, 2-5). In contrast, erythropoiesis in acute lymphoblastic leukemia is usually normoblastic or, in some instances, megaloblastic. In acute myeloblastic leukemia, erythroblasts are sometimes gigantic with multiple nuclei and prominent nucleoli. More mature erythroblasts show accelerated pyknosis and abnormal karyorrhexis. Megakaryocytes are reduced in number and may be either large or micromegakaryocytes (see Fig. 2-9).

H & E sections of biopsies of bone marrow from patients with acute myeloblastic leukemia show hypercellularity and virtual replacement of the marrow with a proliferation of cells containing pale, vesicular nuclei and often one or more nucleoli. On the basis of H & E sections alone, it is not possible to confidently distinguish leukemic myeloblasts from leukemic lymphoblasts.

Cytochemically, the PAS reagent reveals a diffuse pattern of glycogen distribution in leukemic myeloblasts.[4,89,408,412,548,549,583,1002] In some myeloblasts, the PAS reagent demonstrates glycogen in a dustlike pattern surrounding the nucleus. Activity of acid phosphatase and nuclear aryl sulfatase are usually decreased in leukemic myeloblasts.[137] Sudan black B and Oil red O stains demonstrate lipid.[182,195,403–408,412,593,786,803,954,957,992,1001] Usually, these stains are strongly positive in the cytoplasm of the leukeumic myeloblasts and promyelocytes, and weaker in more mature granulocytic cells such as metamyeloctyes and neutrophils (Figs. 2-11, 2-12). Activity of myeloperoxidase is variable,[85,110,142,143,382,629,800,820,821,924,950,1110] and in typical cases is found in scant amounts in the cytoplasm (Fig. 2-12). Neutrophils may show little or no activity of myeloperoxidase.[191] Similarly, activity of specific esterase is weak in leukemic myeloblasts, as seen in Figure 2-12.[548,629] If Auer rods occur in the myeloblasts, they may show activities of myeloperoxidase, specific esterase, and acid phosphatase.[3] Methyl green-pyronine stain for RNA is strongly positive in some cases and weakly positive in others, and it does not constitute a diagnostic cytochemical test. Using either the direct enzyme assay for terminal deoxynucleotidyl transferase[449,656] or the newly developed direct immunofluorescent stain for this enzyme,[990] leukemic myeloblasts appear to have negligible activity. Activity of nonspecific esterase is usually weak and resistant to inhibition by sodium fluoride. Prussian blue stain shows occasional ringed sideroblasts.

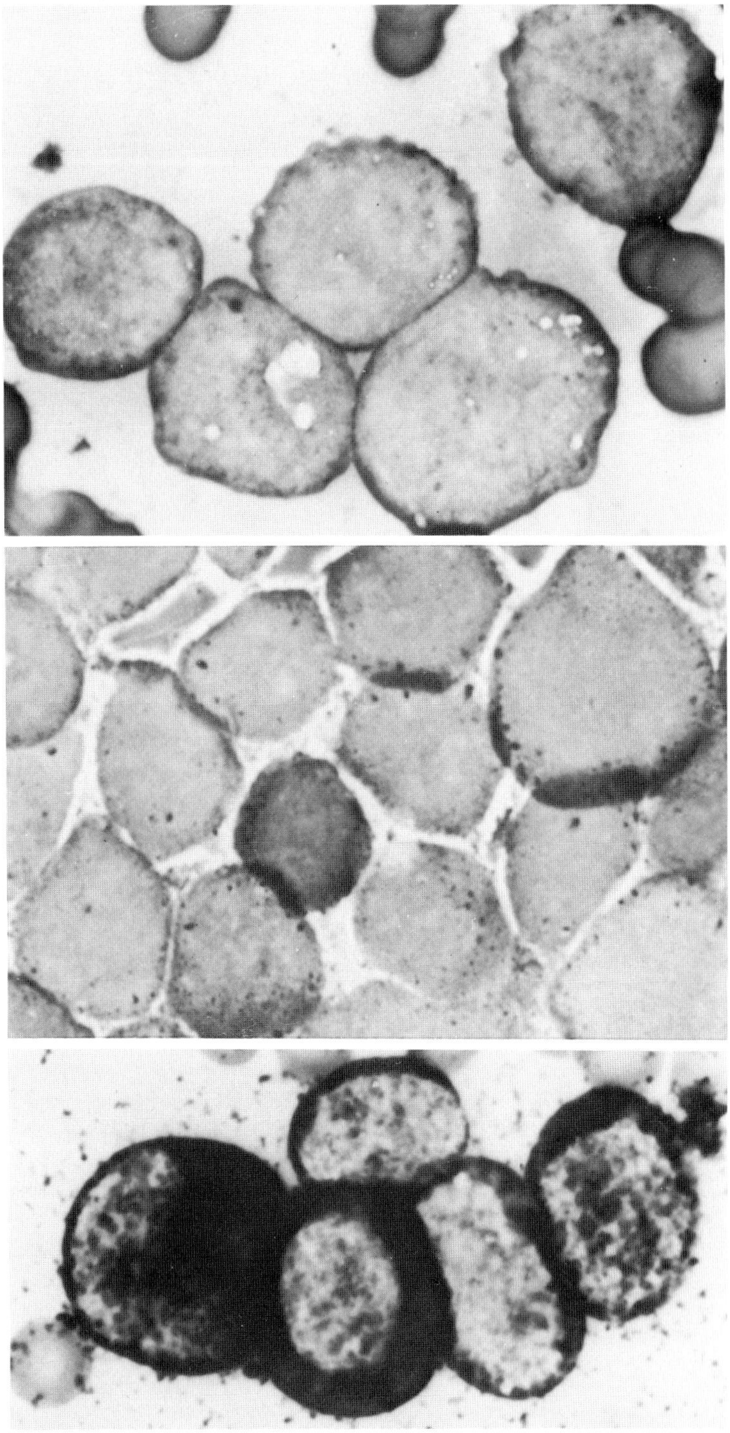

Figure 2-12. Myeloblasts in acute myeloblastic leukemia. Increased activity of myeloperoxidase, using o-tolidine (**top**). Increased activity of specific esterase, using naphthol AS-D chloroacetate as substrate and fast blue BBN as indicator (**center**). Increased staining of cytoplasm of leukemic myeloblasts by sudan black B (**bottom**).

Acute Promyelocytic Leukemia

Acute promyelocytic leukemia is a disorder of neoplastic promyelocytes that proliferate in the blood, bone marrow, and visceral organs. It is believed to be a morphologic variant of acute myeloblastic leukemia. Although leukemic promyelocytes can vary morphologically from case to case, in most instances they contain intense nonspecific granulation, and they often contain multiple Auer rods. According to the recent FAB classification,[87] this variant is called *hypergranular promyelocytic leukemia* or *M3*.

Acute promyelocytic leukemia may seem to arise *de novo*, or it may be a terminal event in patients with chronic granulocytic leukemia or other preleukemic erythroblastic and myeloblastic disorders.[86–88,101,108,109,124,340,356,361,375,376,431,480,1033] Clinically, promyelocytic leukemia is characterized by fever, splenomegaly, and an aggressive clinical course marked by a hemorrhagic diathesis. This bleeding tendency is due to disseminated intravascular coagulation.[376,564,1013] It is thought to arise from disruption of promyelocytes. In these cells, the large abnormal granules contain a high content of plasminogen. As the leukemic promyelocytes disintegrate, granules are disrupted, plasminogen is released, and a hemorrhagic disorder occurs.

Peripheral blood samples from patients with acute promyelocytic leukemia contain variable numbers of leukemic promyelocytes. These large mononuclear cells have a prominent nucleus with delicate-appearing nuclear chromatin, few aggregates of chromatin, and one or more nucleoli with prominent perinucleolar condensation of chromatin (Fig. 2-13). Cytoplasm is abundant and contains numerous darkly-staining granules of various sizes. In some leukemic promyelocytes, granules are large and splinter-shaped, unlike the giant globular-shaped granules of the Chédiak-Higashi syndrome.[31,237,367,686,768,1044] Typically, numerous leukemic promyelocytes in acute promyelocytic leukemia contain many Auer rods.[161,1033] Their multiplicity has been compared to a "stack of hay" or "bundle of twigs." Auer rods in these cells are usually slender and elongated, but they are sometimes short and angular. As in other myeloblasts, Auer rods in leukemic promyelocytes stain bright orange to red with conventional panoptic stains.

Figure 2-13. Leukemic promyelocytes, peripheral blood, in acute promyelocytic leukemia. These cells are gigantic and have prominent nucleoli and large darkly staining granules. ➡

Figure 2-14. Leukemic promyelocytes in acute promyelocytic leukemia, showing prominent granulation. ➡

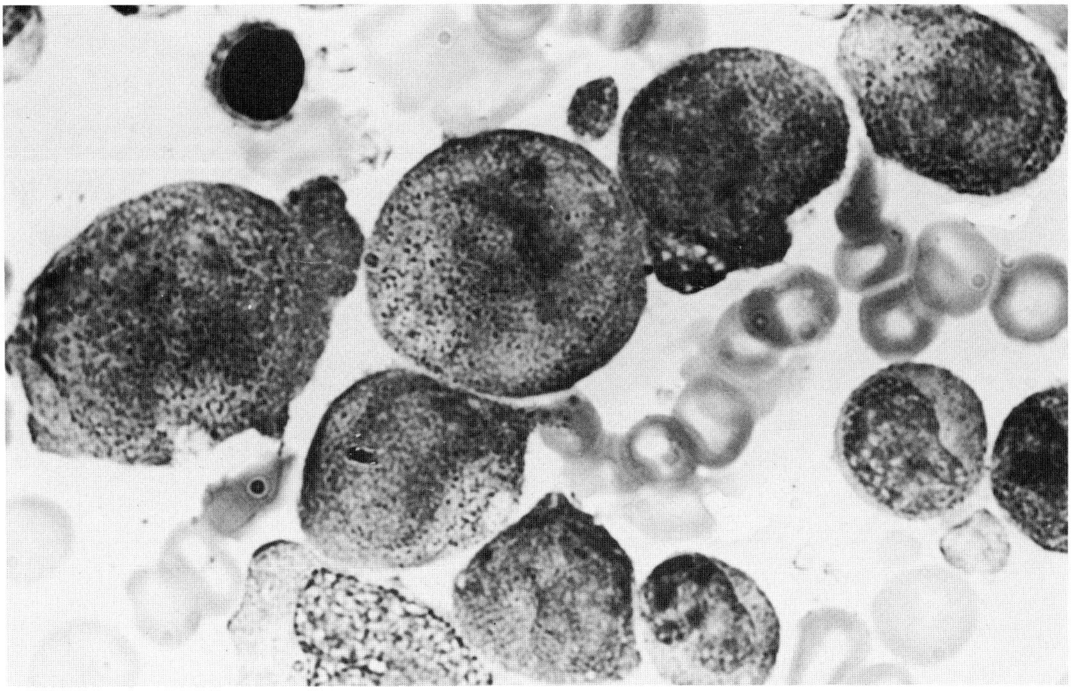

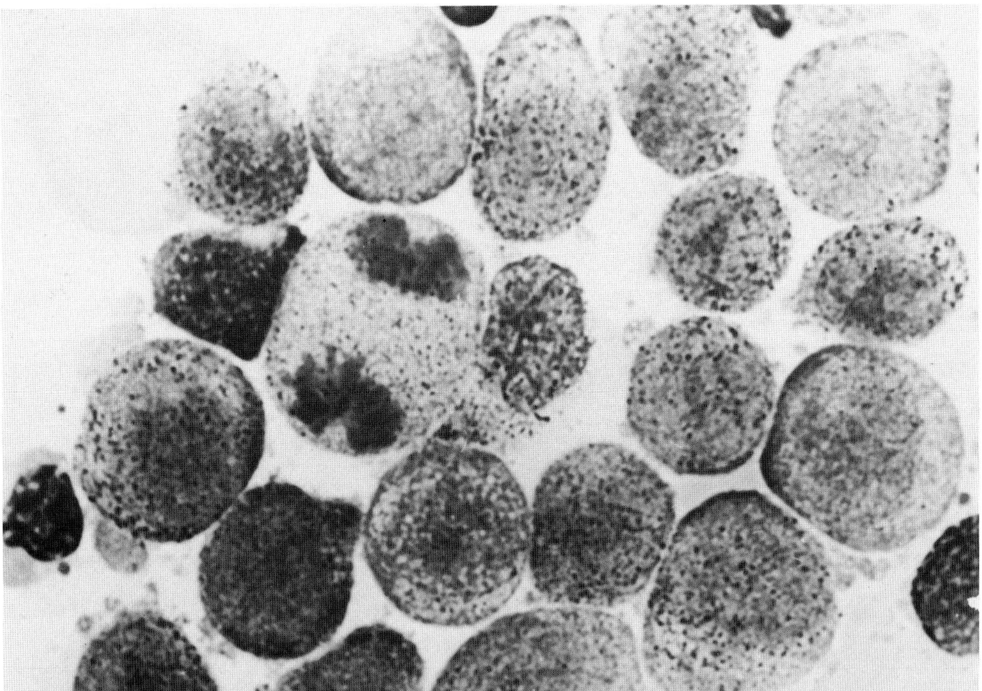

In the peripheral blood of patients with acute promyelocytic leukemia, normoblasts, aberrant-appearing monocytes, and neutrophils with nuclear abnormalities can also be found. Erythrocyte abnormalities include gigantism, macrocytosis, and basophilic stippling.

Aspirates of bone marrow from patients with acute promyelocytic leukemia show marked hypercellularity (Figs. 2-14 through 2-18, and Plate 2*D*). Megakaryocytes are greatly reduced in number in typical cases, and they may be micromegakaryocytes. Similarly, erythroblasts are reduced in number. Those that are present often show megaloblastoid nuclear chromatin pattern (see Fig. 2-18). Granulocytic cells at maturational ages beyond the promyelocyte stage are greatly reduced in number. For the most part, the marrow is composed of leukemic promyelocytes having large round to oval nuclei, delicate nuclear chromatin pattern with few aggregates of chromatin, and one or more small nucleoli with perinucleolar condensations of chromatin.[86–88,101,108,109,124,340,356,431,480,1033] Cytoplasm is abundant and usually basophilic to lightly acidophilic using conventional panoptic stains. Within the cytoplasm are numerous granules of various sizes and shapes that stain red to purple (see Figs. 2-13 through 2-17). Many leukemic promyelocytes contain multiple Auer rods resembling a "haystack" or a "bundle of twigs" (Figs. 2-19, 2-20, and Plate 2*D*). Sometimes, small myelocytes or micromyelocytes occur (Fig. 2-21).

H & E sections of bone marrow from patients with acute promyelocytic leukemia show hypercellularity (Fig. 2-22). Megakaryocytes, erythroblasts, and mature neutrophils are greatly reduced in number. Most of the cells are large and mononuclear, with pale vesicular nuclei, prominent nucleoli, and abundant cytoplasm. Sometimes, a few granules are seen on H & E section.

Cytochemically, the PAS stain in leukemic promyelocytes shows numerous dustlike particles of glycogen distributed evenly throughout the cytoplasm. Activities of specific esterase, acid phosphatase, and myeloperoxidase are intense in the cytoplasm,[379,601] reflecting the high content of granules and lysosomes (Figs. 2-23, 2-24). Sudan black B stain is strongly positive in these cells.[404–408,575] Auer rods show the expected cytochemical reactions consisting of intense activity of myeloperoxidase, specific esterase, and acid phosphatase. Methyl green-pyronine stain for RNA is normal to low in intensity, reflecting cellular content of RNA. Nonspecific esterase activity is weak and fluoride resistant.

(text continued on page 142)

Figure 2-15. Prominent granulation in leukemic promyelocytes in acute promyelocytic leukemia. →

Figure 2-16. Group of leukemic promyelocytes in acute promyelocytic leukemia. →

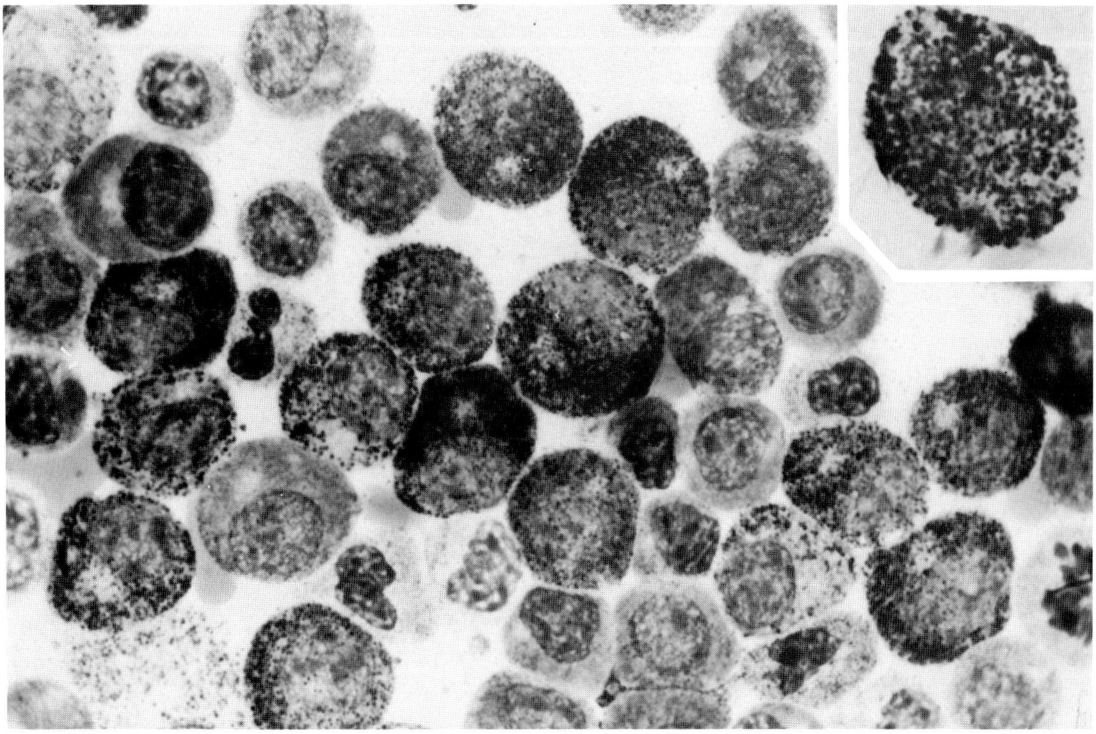

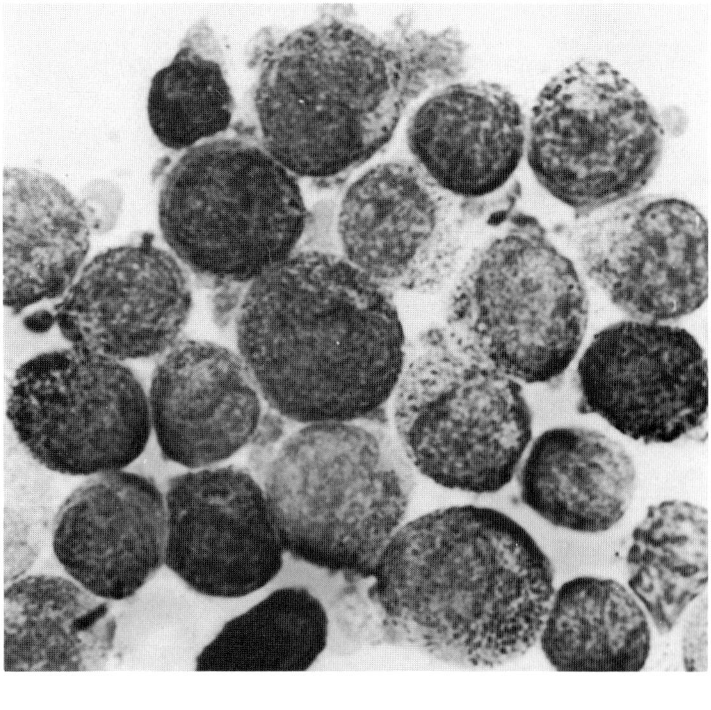

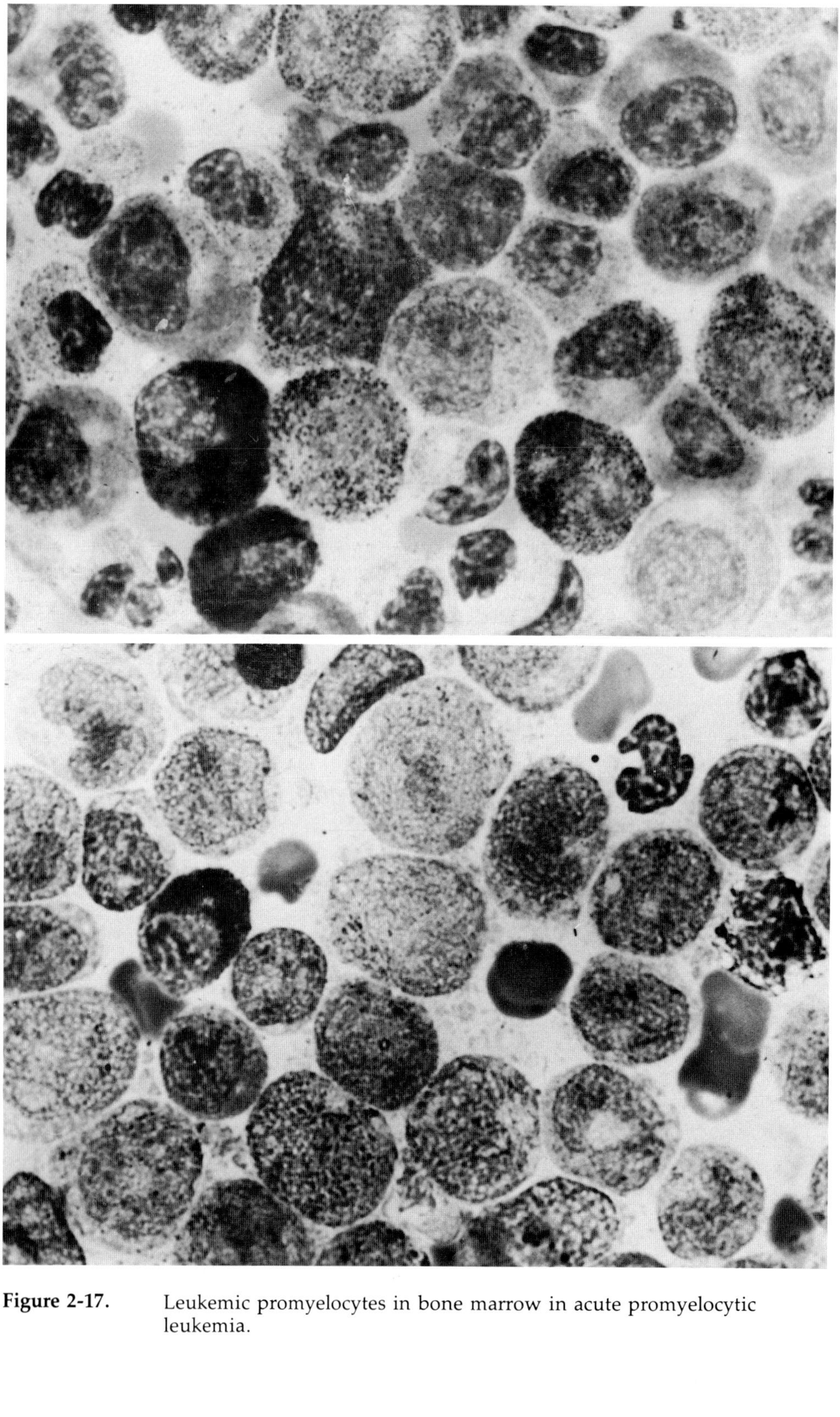

Figure 2-17. Leukemic promyelocytes in bone marrow in acute promyelocytic leukemia.

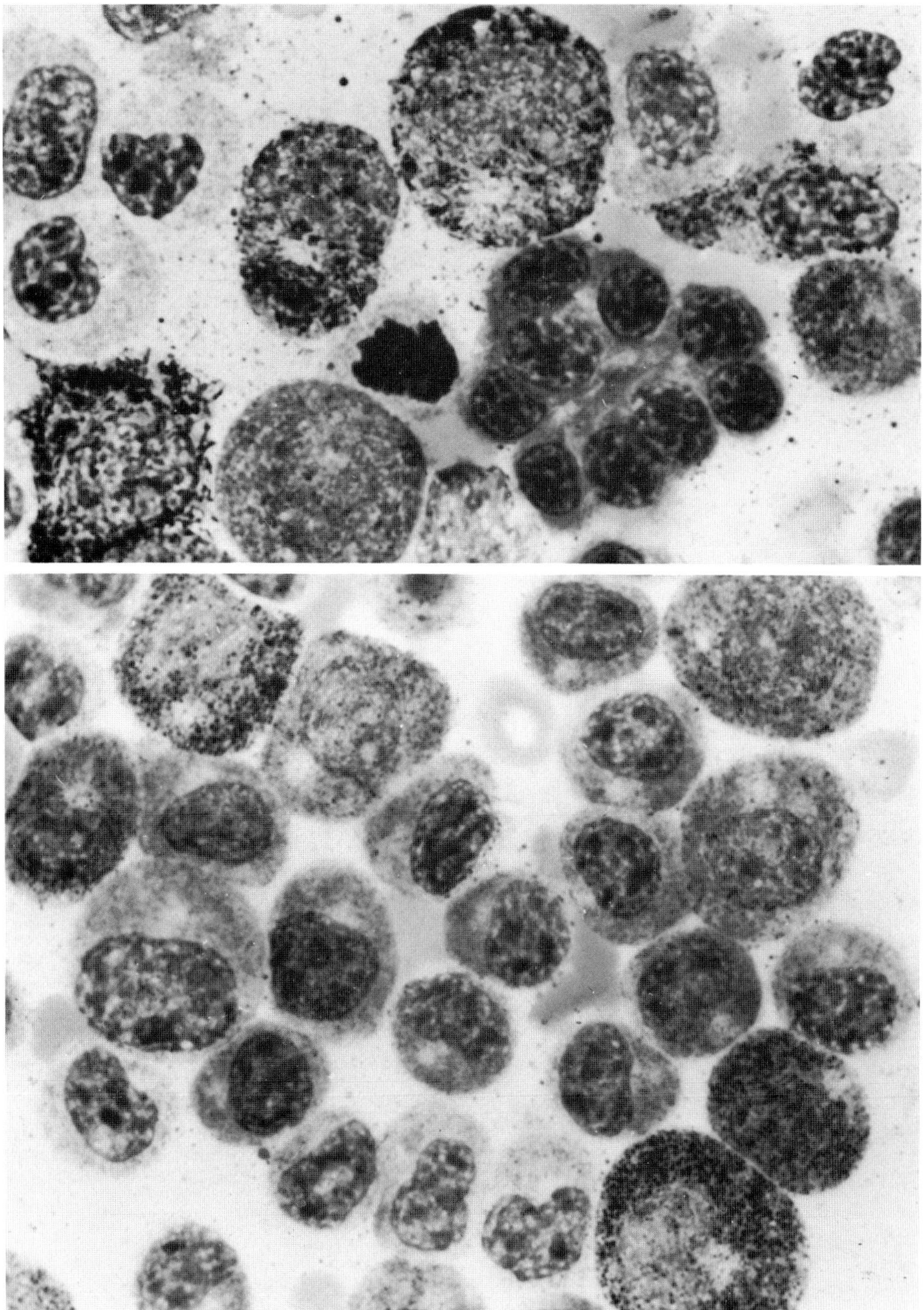

Acute promyelocytic leukemia, showing leukemic promyelocytes with splinter-shaped granules, increased numbers of erythroblasts (**top**), and micromyelocytes (**bottom**). **Figure 2-18.**

Recently, a "microgranular" variant of acute promyelocytic leukemia has been described.[88,901] In this variant, leukemic blasts have an unusual monocytoid configuration of the nuclei, resembling that found in cases of atypical monocytic leukemia (Figs. 3-13, 3-20). Some blasts contain many dustlike azurophilic granules. Others contain multiple Auer rods. Cytochemically, these leukemic blasts contain strong activity of myeloperoxidase and specific esterase. It may be difficult to distinguish cases like these from variants of acute monocytic leukemia, particularly *M4* and *M5*. Sometimes, leukemic blasts in these monocytic variants contain dustlike granules and multiple Auer rods like those described in the micro-granular type of acute promyelocytic leukemia. Demonstration of non-specific esterase activity may help to distinguish the promyelocytic variant from atypical cases of monocytic leukemia.[901] In the microgranular variant, nonspecific esterase activity is weak to absent[901] in contrast to the strong fluoride-sensitive nonspecific esterase activity found in leukemic monocytes, including those that contain Auer rods.

(text continued on page 147)

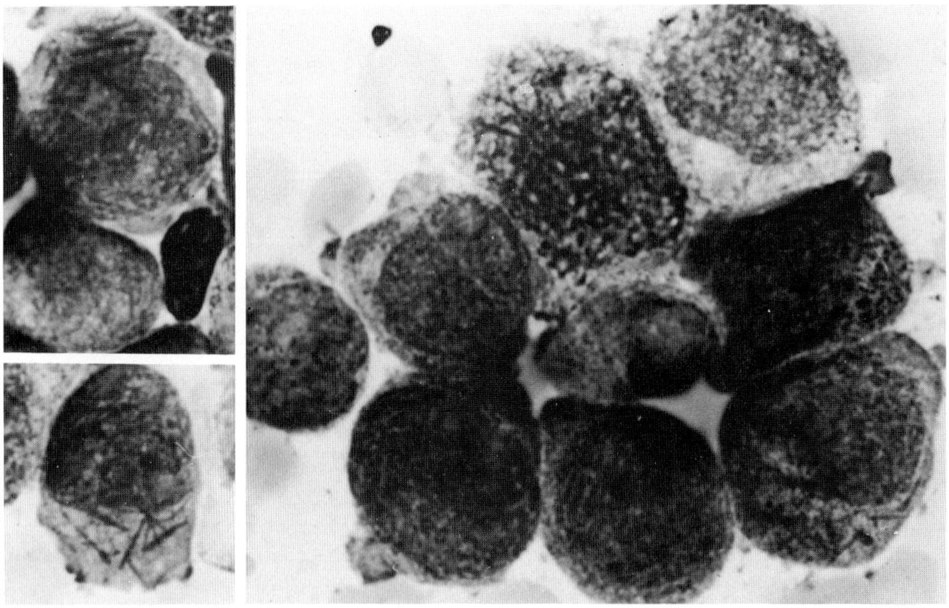

Figure 2-19. Leukemic promyelocytes in acute promyelocytic leukemia, showing multiple Auer rods in a "stack of twigs" configuration (FAB M3).

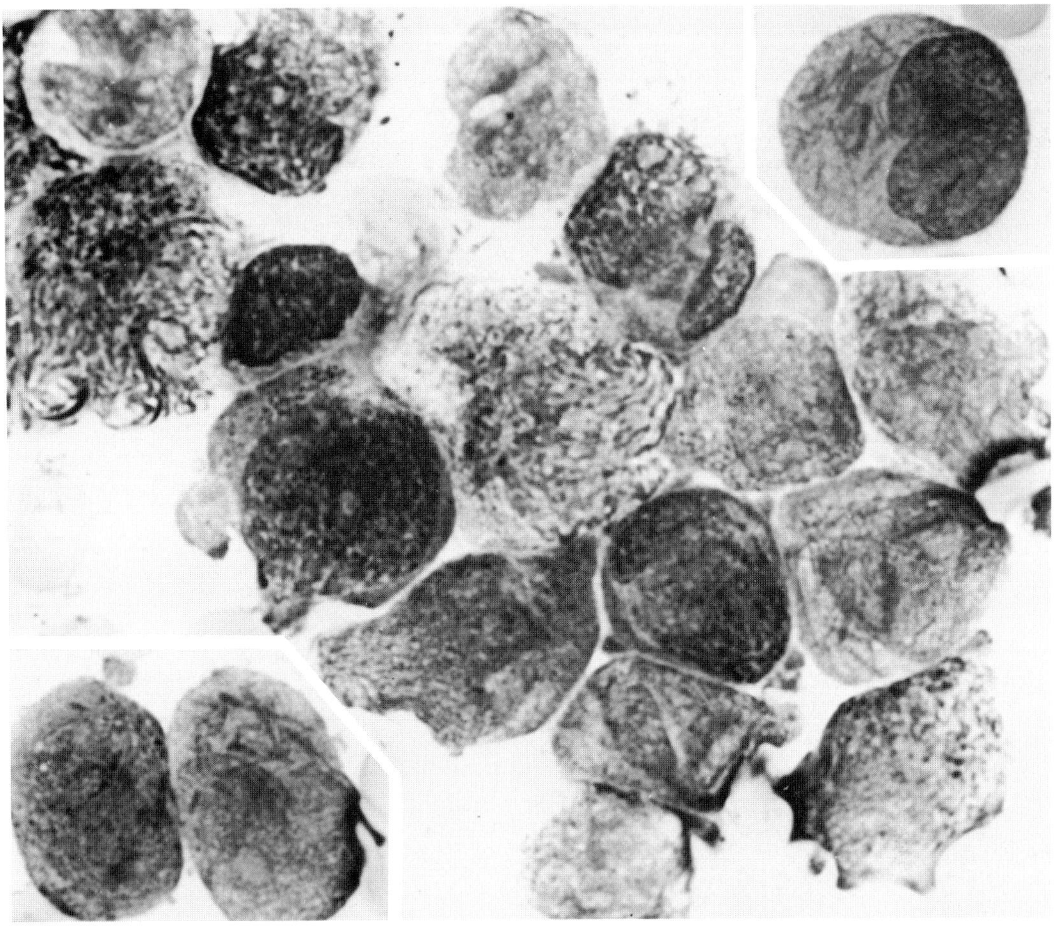

Multiple Auer rods in leukemic promyelocytes in acute promyelocytic leukemia (FAB M3). **Figure 2-20.**

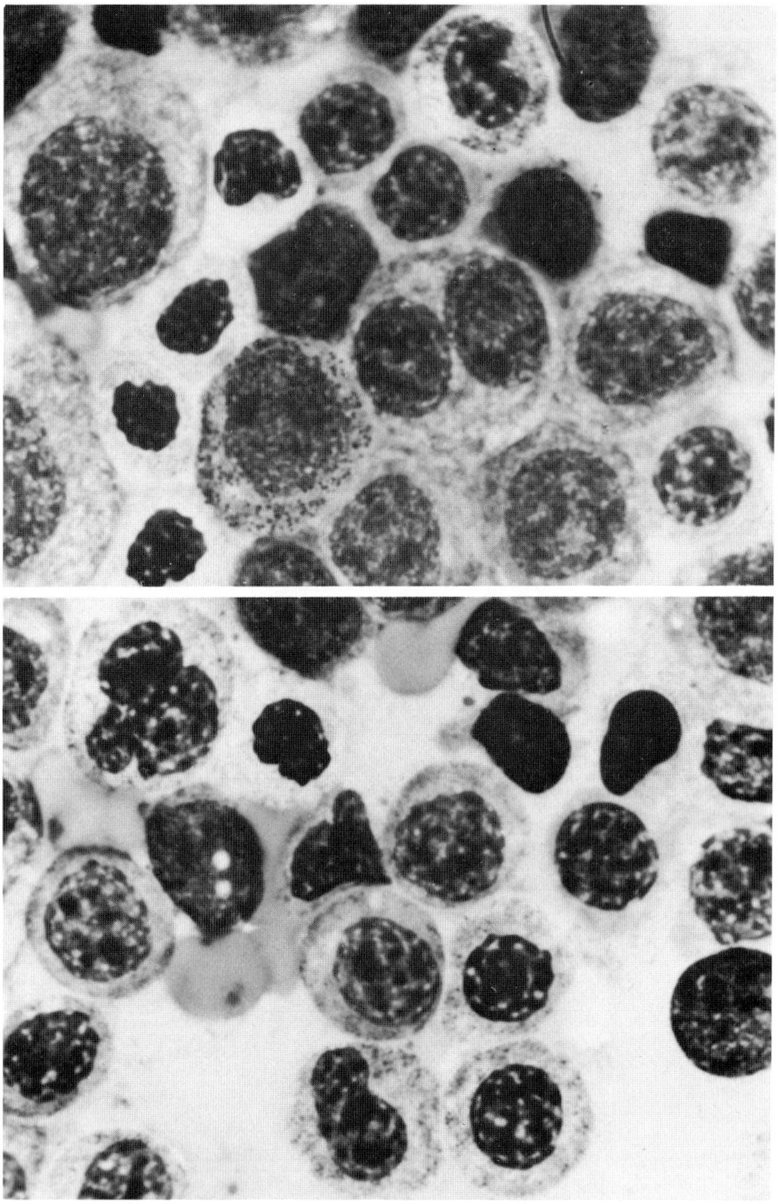

Figure 2-21. Micromyelocytes in acute promyelocytic leukemia.

Figure 2-22. H & E section of bone marrow biopsy, in acute promyelocytic leukemia, showing increased numbers of primitive-appearing cells with pale vesicular nuclei, representing leukemic promyelocytes. ➡

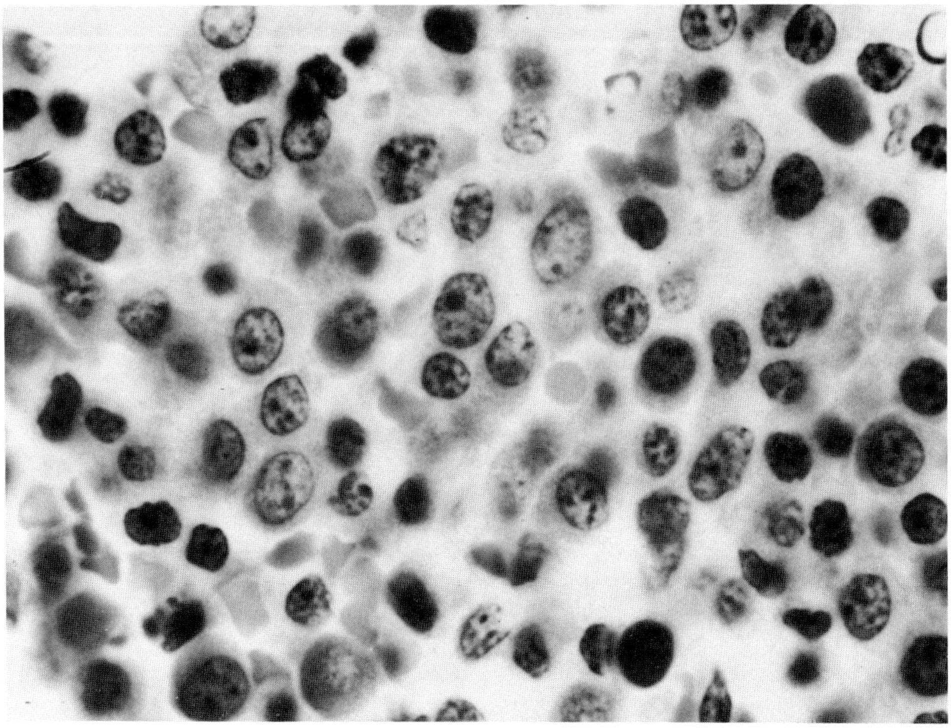

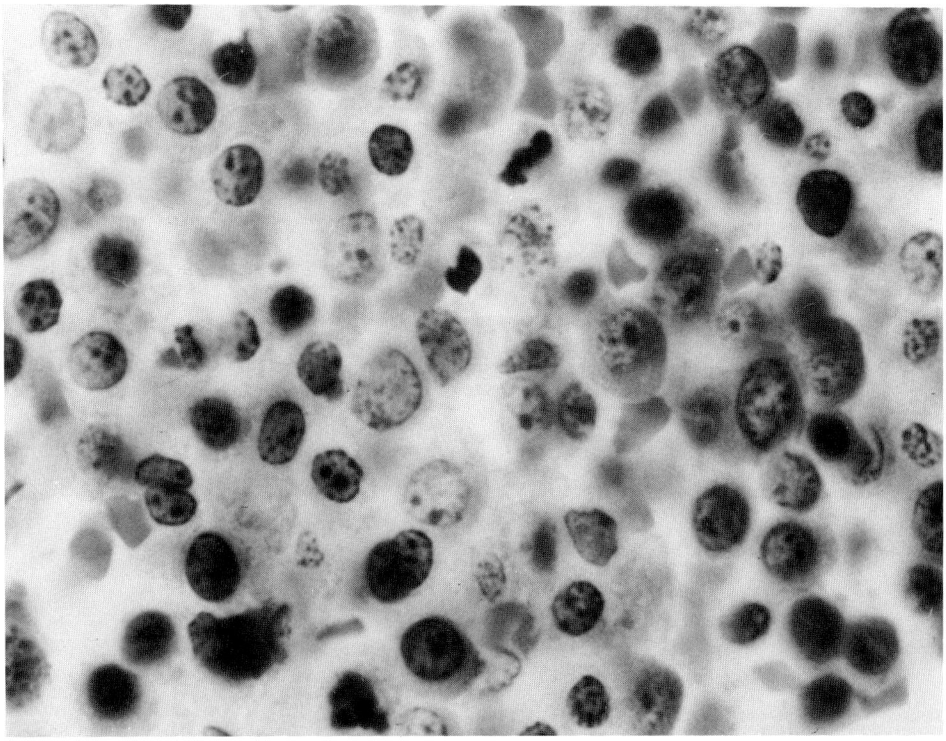

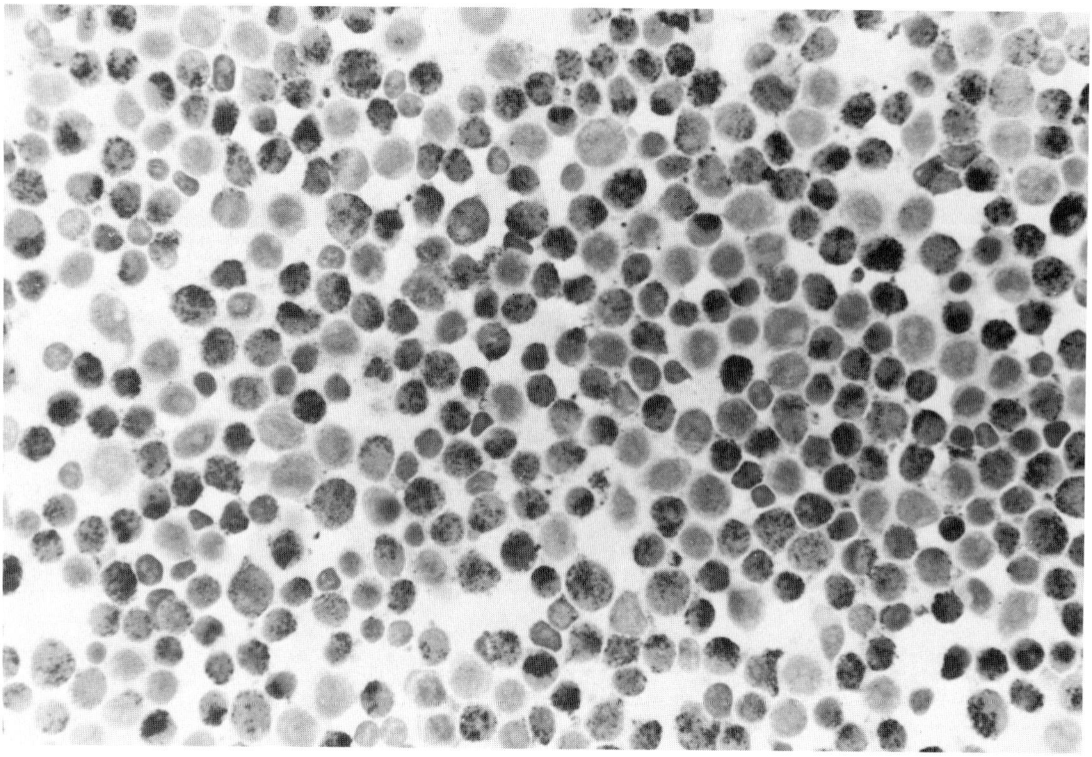

Figure 2-23. Low-power view of myeloperoxidase reaction in acute promyelocytic leukemia, demonstrating increased numbers of cells with strongly positive reaction for myeloperoxidase (benzidine stain).

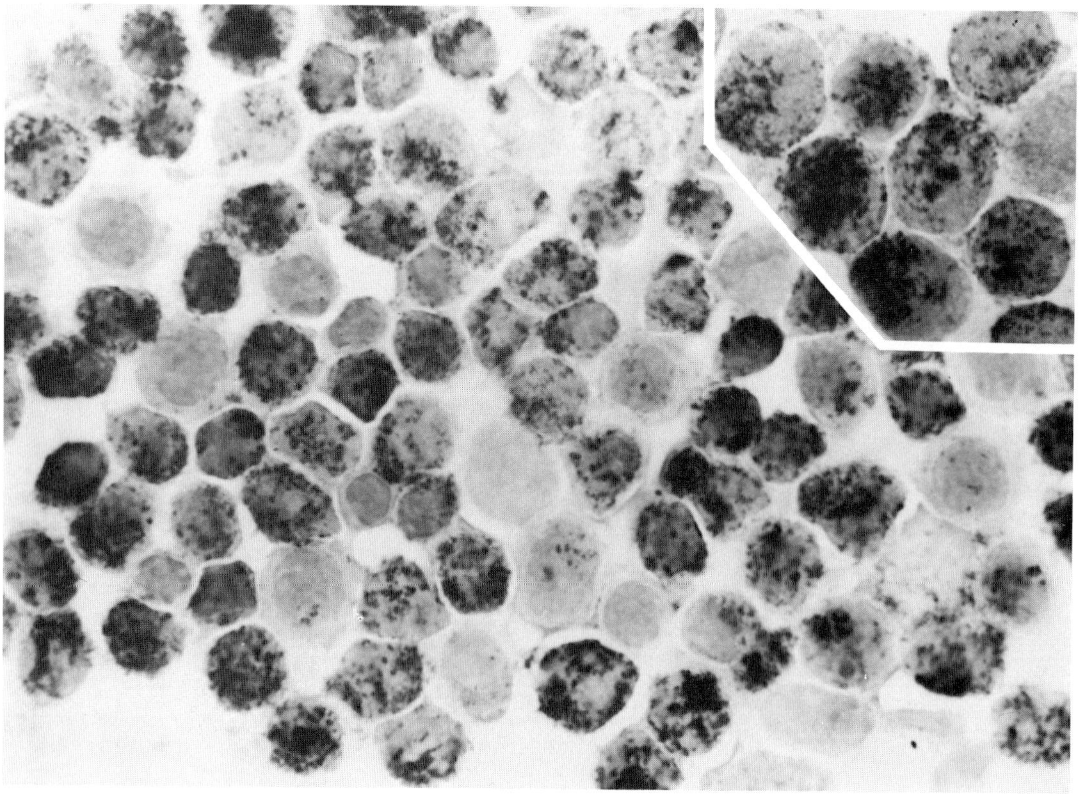

Chronic granulocytic leukemia is a disorder of neoplastic granulocytic cells that proliferate in the blood, bone marrow, and visceral organs. Hematologically, it is characterized by leukocytosis, the occurrence of immature white blood cells in the peripheral blood, and anemia and thrombocytopenia. Sometimes, thrombocytosis and dwarf megakaryocytes occur in the blood. Clinically, chronic granulocytic leukemia is typified by hepatomegaly, splenomegaly, and hyperuricemia. Cytogenetically, approximately 90% of cases of chronic granulocytic leukemia contain the Ph′ chromosome, a distinctive cytogenetic abnormality representing translocation of the long arm of the 22nd chromosome (G-group) to one of several other chromosomes.[732,873]

Currently, chronic granulocytic leukemia is regarded as a facet of the myeloproliferative syndrome, a group of closely related diseases that includes chronic granulocytic leukemia, agnogenic myeloid metaplasia, Di Guglielmo syndrome, acute myeloblastic leukemia, and paroxysmal nocturnal hemoglobinuria.[1004] Like other diseases within the myeloproliferative syndrome,[1004] chronic granulocytic leukemia may show evolution into other more aggressive hematologic disorders. For example, some cases diagnosed initially as chronic granulocytic leukemia have evolved into agnogenic myeloid metaplasia and subsequently into acute leukemia.[1007] In other cases, an aggressive phase is characterized by marked proliferation of eosinophils, basophils, or megakaryocytes in the bone marrow, with few manifestations of leukemia in the blood and little if any fibrosis of the bone marrow as found in agnogenic myeloid metaplasia.

In up to 80% of patients who have chronic granulocytic leukemia, a terminal acute leukemic phase develops.[33,104,187,274,397,464,494,543,584,613,757,778,790,871,894,900,952] Development of the blast phase is heralded clinically by fever, increasing splenomegaly, anemia, rising numbers of basophils or leukemic blasts in the blood, rising leukocyte alkaline phosphatase score, and abrupt cytogenetic changes.[538,1034] In most of these patients, the terminal blast phase is myeloblastic leukemia. In a smaller number, acute monocytic leukemia or promyelocytic leukemia develops.[33,871] In a few patients, erythroleukemia ensues,[870] and in rare individuals, acute megakaryocytic myelosis,[165] eosinophilic leukemia, and basophilic leukemia occur.[871]

Recently, cases have been described in which chronic granulocytic leukemia has evolved into acute lymphoblastic leukemia.[123,124,656,900,952,990] In these patients, leukemic blasts showed cytochemical findings that were typical for leukemic lymphoblasts, including coarse PAS positive material in the cytoplasm and absent activity of myeloperoxidase. Many of these cases showed substantial activity of terminal deoxynucleotidyl transferase, providing further evidence for lymphoid derivation of the leukemic blasts. In other cases, chronic granulocytic leukemia was preceded by chronic lymphocytic leukemia.[1090] These and other cases suggest that the leukemic myeloblast in chronic granulocytic leukemia has a wide potential for different pathways of morphologic expression and differentiation.[656,812]

Figure 2-24. ← Higher-power view showing intense peroxidase activity in leukemic promyelocytes (benzidine stain).

Peripheral blood samples from patients with typical chronic granulocytic leukemia contain granulocytic cells at all maturational ages.[278,968] Usually, myeloblasts comprise less than 5% of the differential cell count of the peripheral blood. In most cases, Auer rods are not observed. Promyelocytes, myelocytes, and metamyelocytes are increased in numbers. Sometimes, splinter-shaped granules are seen in promyelocytes. Normoblasts are found in the peripheral blood of most patients, and appear well hemoglobinized. Giant hypergranular platelets and large, pale hypogranular platelets are found frequently; as in agnogenic myeloid metaplasia, dwarf megakaryocytes may be found. Lymphocytes are reduced in number, and monocytes with aberrant-appearing nuclei are seen. Increased numbers of eosinophils and basophils are found frequently.

Bone marrow aspirates from patients with typical chronic granulocytic leukemia are markedly hypercellular (Fig. 2-25). Megakaryocytes are increased in number, and many of them are micromegakaryocytes with richly granular cytoplasm and a single, unlobulated nucleus. Erythroblasts are reduced in number and usually demonstrate normoblastic chromatin pattern. Granulocytic cells are increased markedly in number (Fig. 2-26, 2-27). Although maturation is normal, increased numbers of promyelocytes and myelocytes are found.[779] In some cases, granules in these cells stain intensely. In other cases, granules stain poorly and the immature cells appear hypogranular. Marrows from some patients with chronic granulocytic leukemia contain large macrophages containing green to light-blue tissue-paper-appearing material in the cytoplasm. Morphologically, these cells resemble Gaucher cells[118,437,486,529] and are thought to arise as a result of phagocytosis of degenerating granulocytes by macrophages (Figs. 2-28, 2-29). Cytochemically, these storage cells in chronic granulocytic leukemia differ from those found in sea-blue histiocytosis.[965,1023]

On H & E stained biopsies of bone marrow from patients with chronic granulocytic leukemia, specimens are markedly hypercellular (Fig. 2-25). Megakaryocytes are increased, sometimes strikingly. Usually, these are small megakaryocytes and probably represent micromegakaryocytes as seen on panoptically stained aspirates. Markedly increased numbers of granulocytic cells are seen. These cells have pale-appearing nuclei with dustlike nuclear chromatin scattered throughout the nucleus and adherent to the nuclear membrane. Rarely, nucleoli are seen in these cells. Increased numbers of metamyelocytes, band forms, and neutrophils are seen. In most cases, increased numbers of eosinophils and basophils are seen. Erythroblasts are reduced in number.

Cytochemically, the most useful stain in the diagnosis of chronic granulocytic leukemia is the reaction for activity of alkaline phosphatase.[52,201,377,381,488,489,491,492,780,783,984] This enzyme is found in secondary granules of granulocytes.[45–48,52,53,93] In mature and immature granulocytes from patients with chronic granulocytic leukemia, activity of the enzyme is low.[52] Frequently it is undetectable. From the standpoint of differential diagnosis, cases of leukemoid reaction and leukocytosis usually have normal to high activity of alkaline phosphatase in leukocytes.[28] Granulocytic cells in chronic granulocytic leukemia show high activity of specific esterase and myeloperoxidase, (Fig. 2-30) as well as increased acid phosphatase. Using the PAS reaction for glycogen, mature and immature granulocytes in chronic granulocytic leukemia reveal reduced amounts of glycogen. Normoblasts in chronic granulocytic leukemia usually do not contain glycogen detectable by the PAS reaction. The methyl green-pyronine stain for RNA shows normal amounts of this substance in the cytoplasm of granulocytes. Activity of nonspecific esterase is normal in granulocytes and fluoride resistant. In Gaucher-type cells, activity of acid phosphatase and nonspecific esterase are strong. Usually, the Prussian blue stain for iron does not reveal ringed sideroblasts in marrows from patients with chronic granulocytic leukemia.

Diagnostically, chronic granulocytic leukemia must be distinguished from disorders that can resemble it. Principal among these are leukemoid reactions and leukocytosis. In leukemoid reactions, immature granulocytes are even myeloblasts are found in the blood, and normoblasts are seen occasionally. Leukemoid reactions are found in association with or in response to other disorders, such as carcinoma or overwhelming infection. On the basis of peripheral blood and marrow examination alone, it may be difficult to distinguish chronic granulocytic leukemia from a leukemoid reaction. In cases of neutrophilic leukocytosis, peripheral blood contains many neutrophils, as well as a few band forms and metamyelocytes. Cases of this type may be associated with carcinoma and severe infections. In leukemoid reactions and leukocytosis, the Philadelphia chromosome is not present. Furthermore, activity of alkaline phosphatase in leukocytes from patients with leukocytosis or leukemoid reaction is normal or elevated. These findings contrast with those in chronic granulocytic leukemia, in which the Philadelphia chromosome is usually present and activity of alkaline phosphatase in leukocytes is reduced markedly.

In atypical cases that do not have the Philadelphia chromosome or in which alkaline phosphatase activity in leukocytes is lowered marginally, it may be difficult to distinguish chronic granulocytic leukemia from leukemoid reactions. In these situations, Philadelphia-chromosome-negative chronic granulocytic leukemia is thought to have a more aggressive course clinically than Ph′-chromosome-positive chronic granulocytic leukemia. In those indefinite cases, it may be necessary to observe the clinical course over a period of time, watching for the usual clinical manifestations of chronic granulocytic leukemia such as hepatomegaly, splenomegaly, and hyperuricemia. When the provoking condition is treated successfully, cases of leukemoid reaction or leukocytosis often improve and blood abnormalities disappear.

(text continued on page 156)

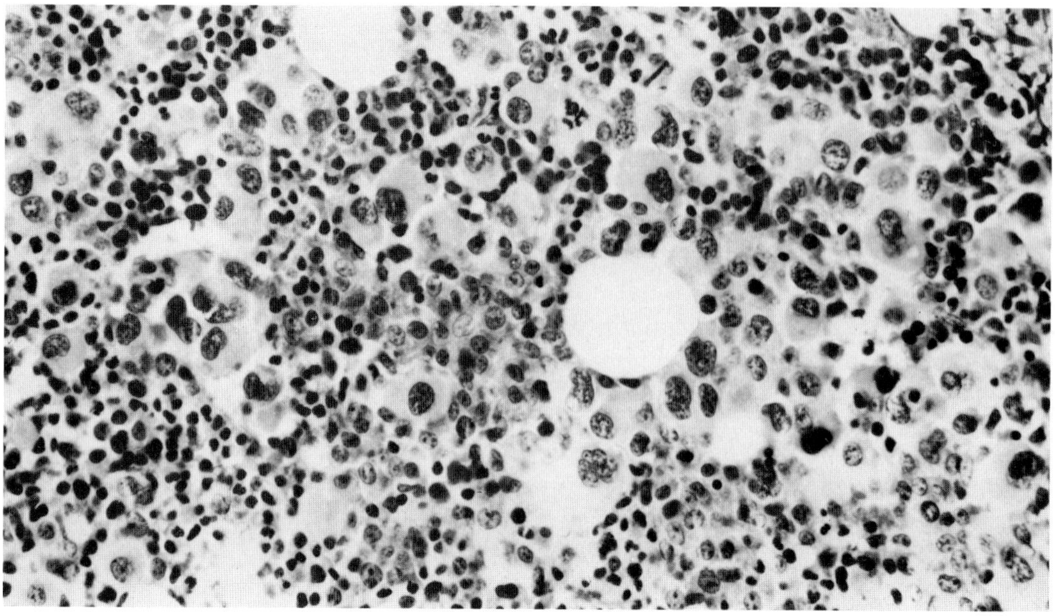

Figure 2-25. Bone marrow in chronic granulocytic leukemia, depicting increased numbers of megakaryocytes and granulocytic precursors.

Figure 2-26. Bone marrow in chronic granulocytic leukemia, showing increased numbers of promyelocytes, myelocytes, and erythroblasts. Several bands and neutrophils are shown. ➡

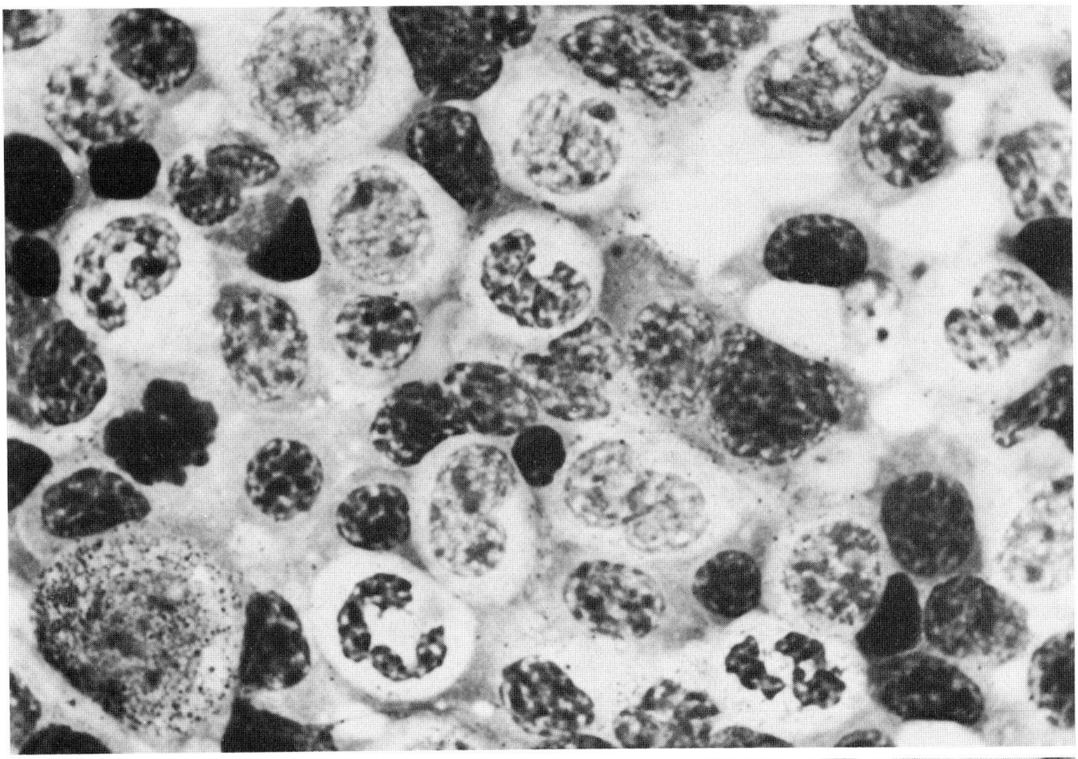

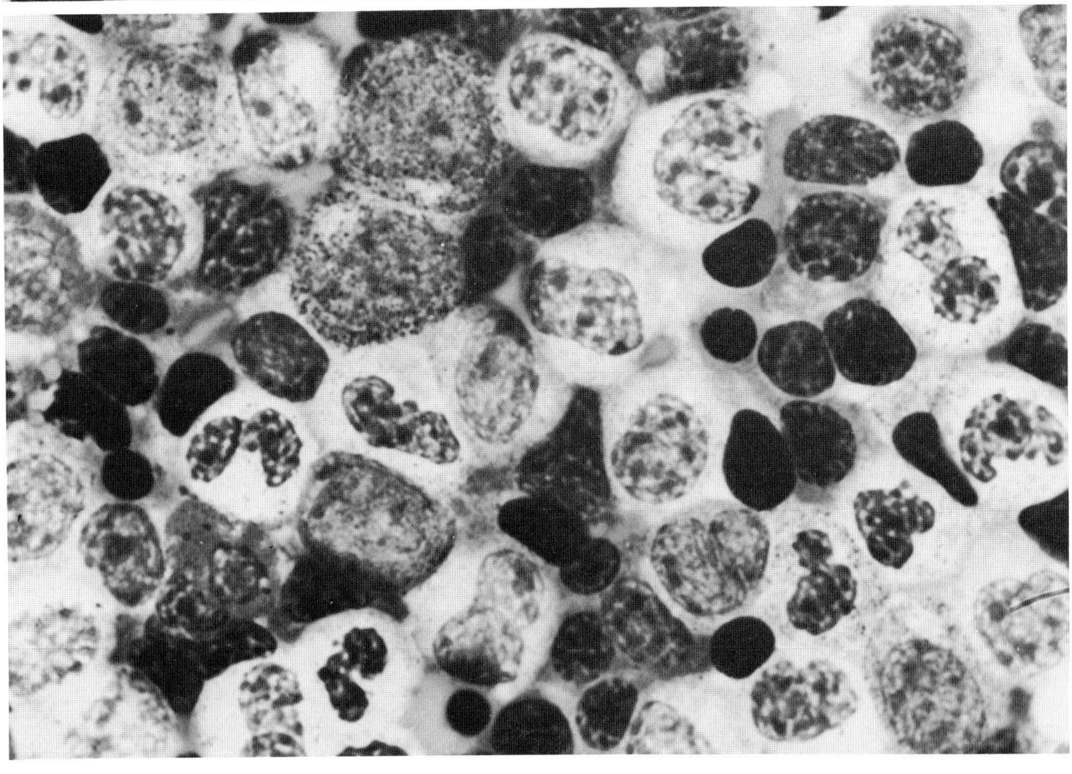

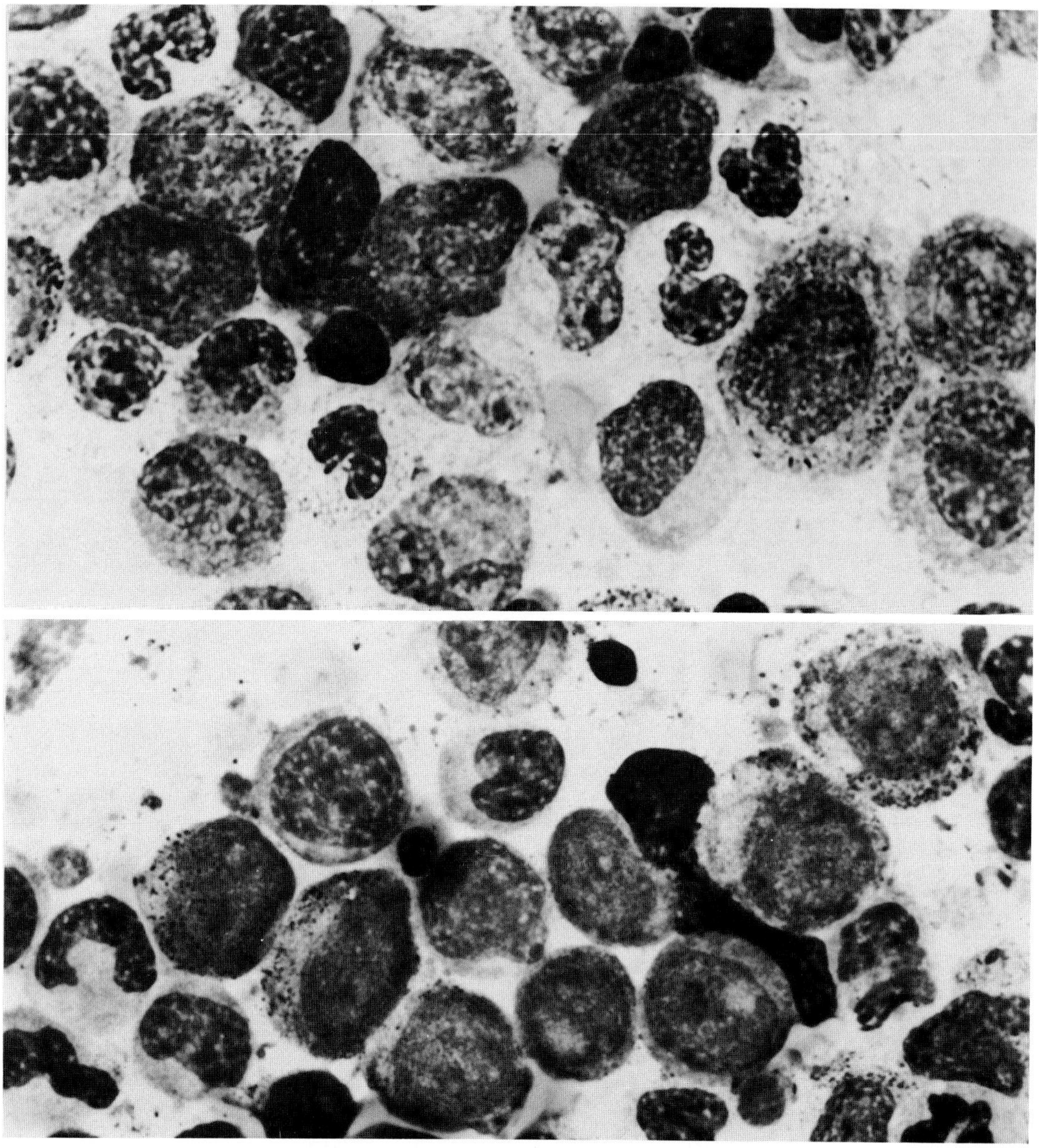

Figure 2-27. Bone marrow in chronic granulocytic leukemia, illustrating increased numbers of promyelocytes and myelocytes.

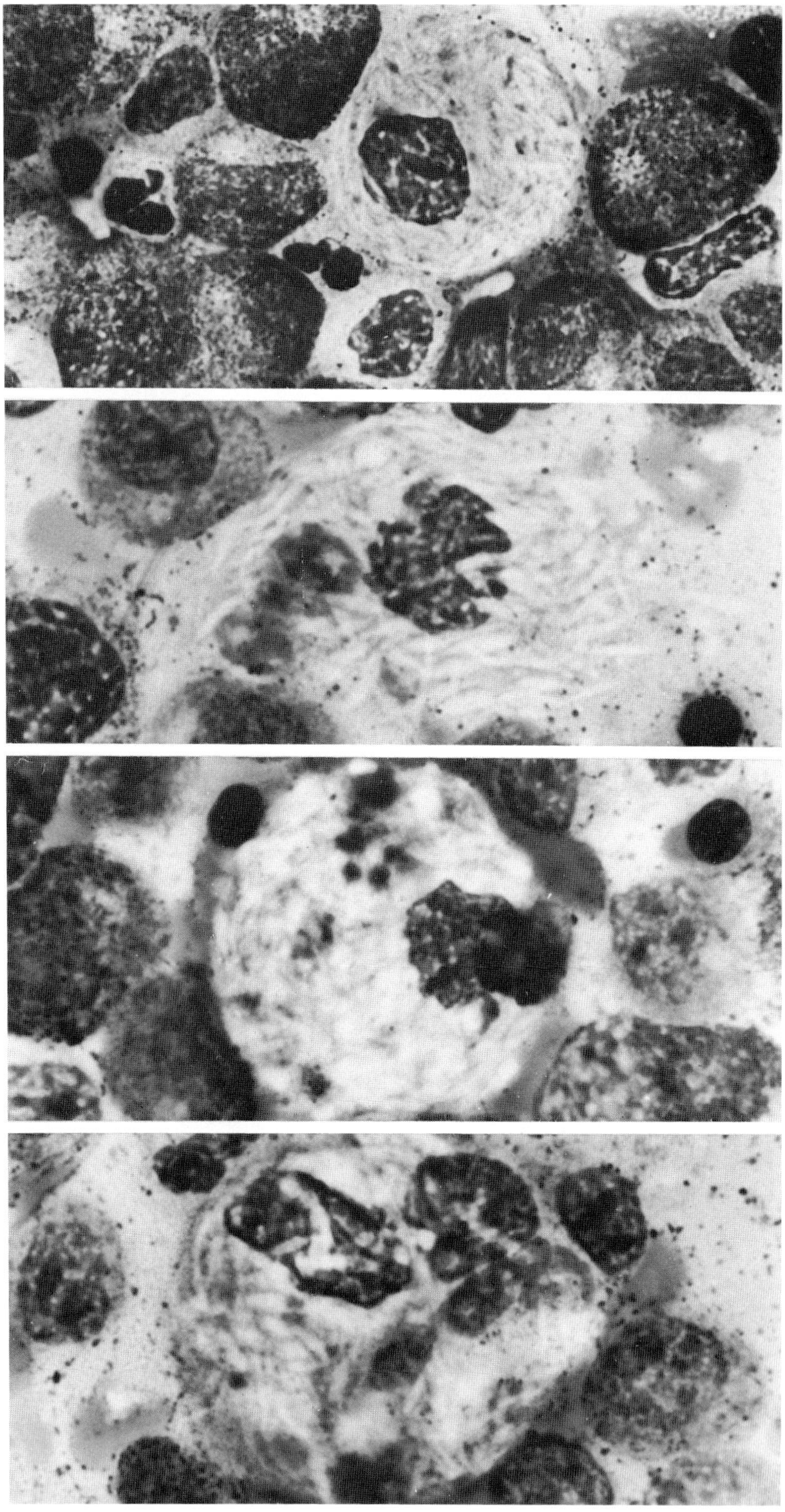

Figure 2-28. Bone marrow in chronic granulocytic leukemia, showing Gaucher-like cells.

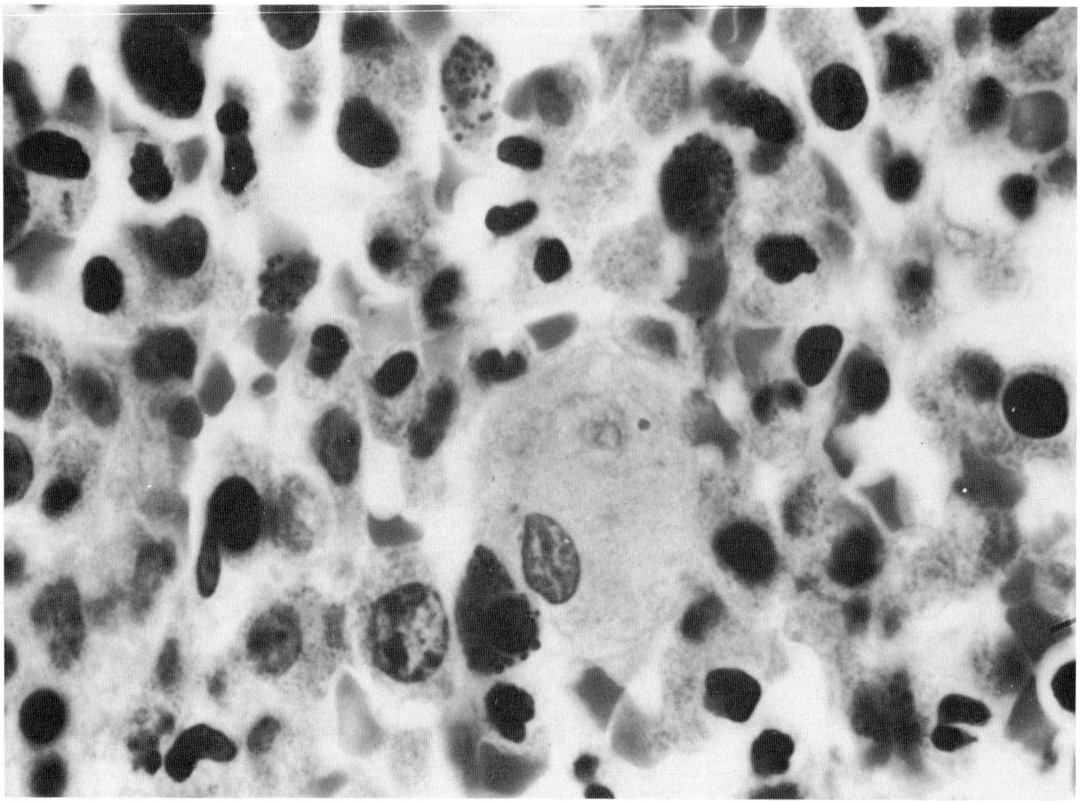

Figure 2-29. H & E section of bone marrow biopsy in chronic granulocytic leukemia, showing Gaucher-like (**center**) surrounded by immature granulocytes.

Figure 2-30. Bone marrow in chronic granulocytic leukemia, showing low- and high-power view of granulocytic cells with strong activity of myeloperoxidase (**top**). Intense activity of specific esterase in granulocytic cells (**bottom**). ➡

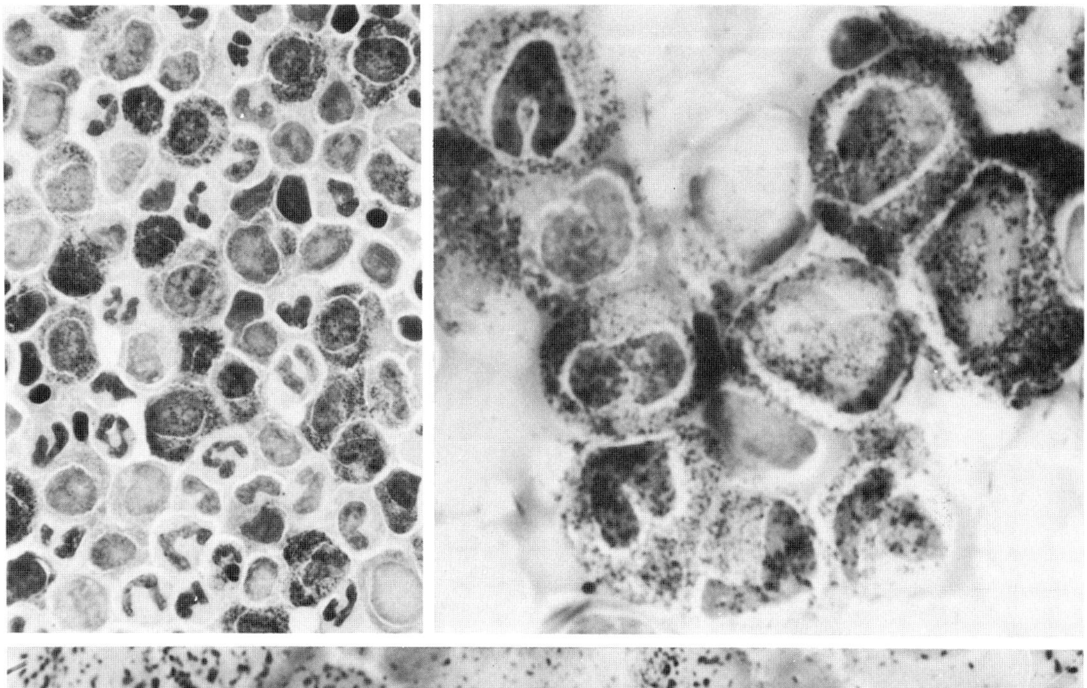

Basophilic Leukemia

This unusual morphologic variant of myeloblastic leukemia is characterized by striking proliferation of mature and immature basophils in the bone marrow and visceral organs.[230,478,908,932,963,1046] It can appear to arise *de novo*, or it may evolve in a patient who had chronic granulocytic leukemia. Peripheral blood specimens from patients with basophilic leukemia contain increased numbers of basophils (Fig. 2-31) as well as increased numbers of myeloblasts. Aspirates of bone marrow from these patients show a striking proliferation of mature and immature basophils and myeloblasts (Figs. 2-32, 2-33). In some patients, granules in these basophils are large and splinter-shaped; in others, they resemble the globular to ovoid granules found in normal basophils. Large numbers of basophilic myelocytes as well as promyelocytes with large nucleoli and abundant cytoplasm containing smaller numbers of basophilic granules are present. Increased numbers of typical-appearing myeloblasts are also found. Megakaryocytes, erythroblasts, and more mature neutrophils are greatly reduced in number. Large numbers of granules, presumably derived from disrupted basophils, are often seen on coverslip films of marrow aspirates.

On H & E sections of bone marrow biopsy, marrows from patients with basophilic leukemia show virtual replacement by mononuclear cells with pale vesicular nuclei, prominent nucleoli, and granules that stain metachromatically with toluidine blue under appropriate conditions of fixation. Cytochemically, mature and immature basophils in acute basophilic leukemia contain scant glycogen identifiable with the PAS reagent; strong activity of specific esterase, tyrosinase, and acid phosphatase; absent peroxidase activity; strong stainability of the cytoplasm by Sudan black B; and metachromatic staining of basophil granules by toluidine blue.[10,743]

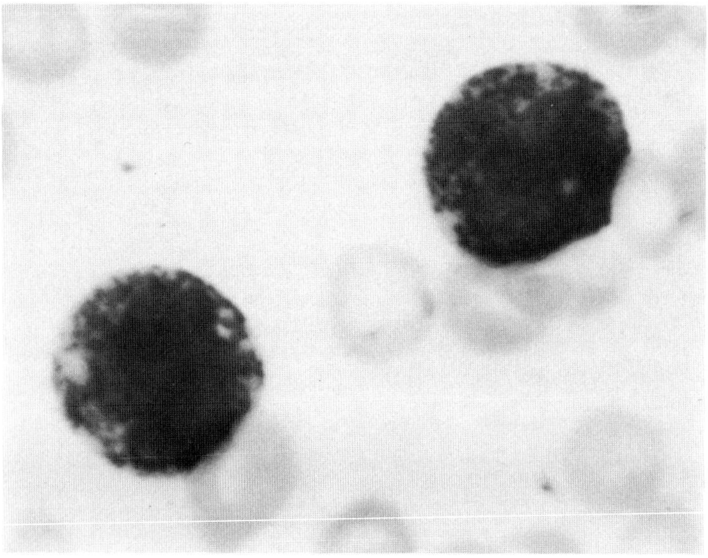

Figure 2-31. Peripheral blood in basophilic leukemia, showing two large basophils.

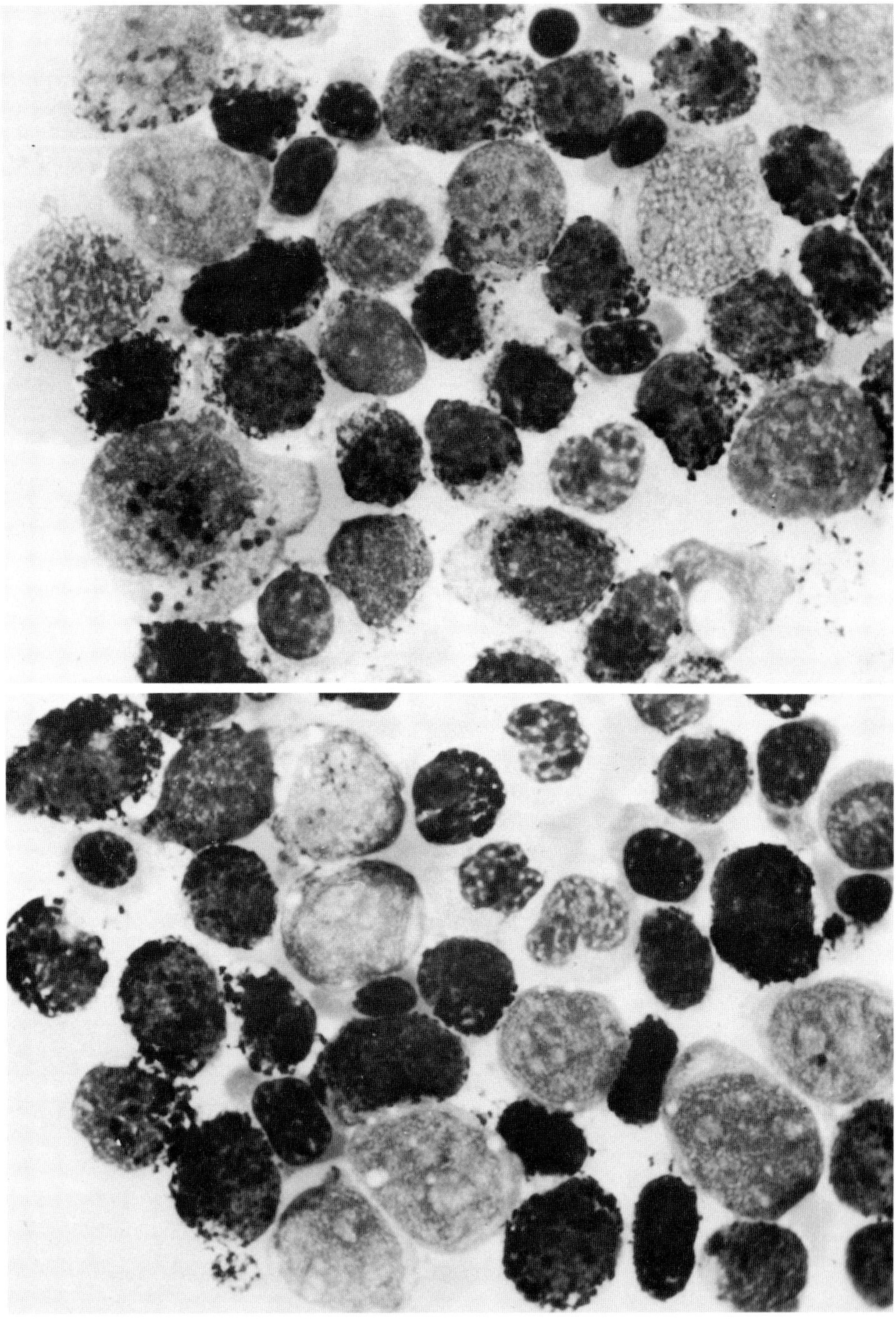

Bone marrow in basophilic leukemia, demonstrating increased numbers of mature and immature basophils and several myeloblasts with prominent nucleoli. **Figure 2-32.**

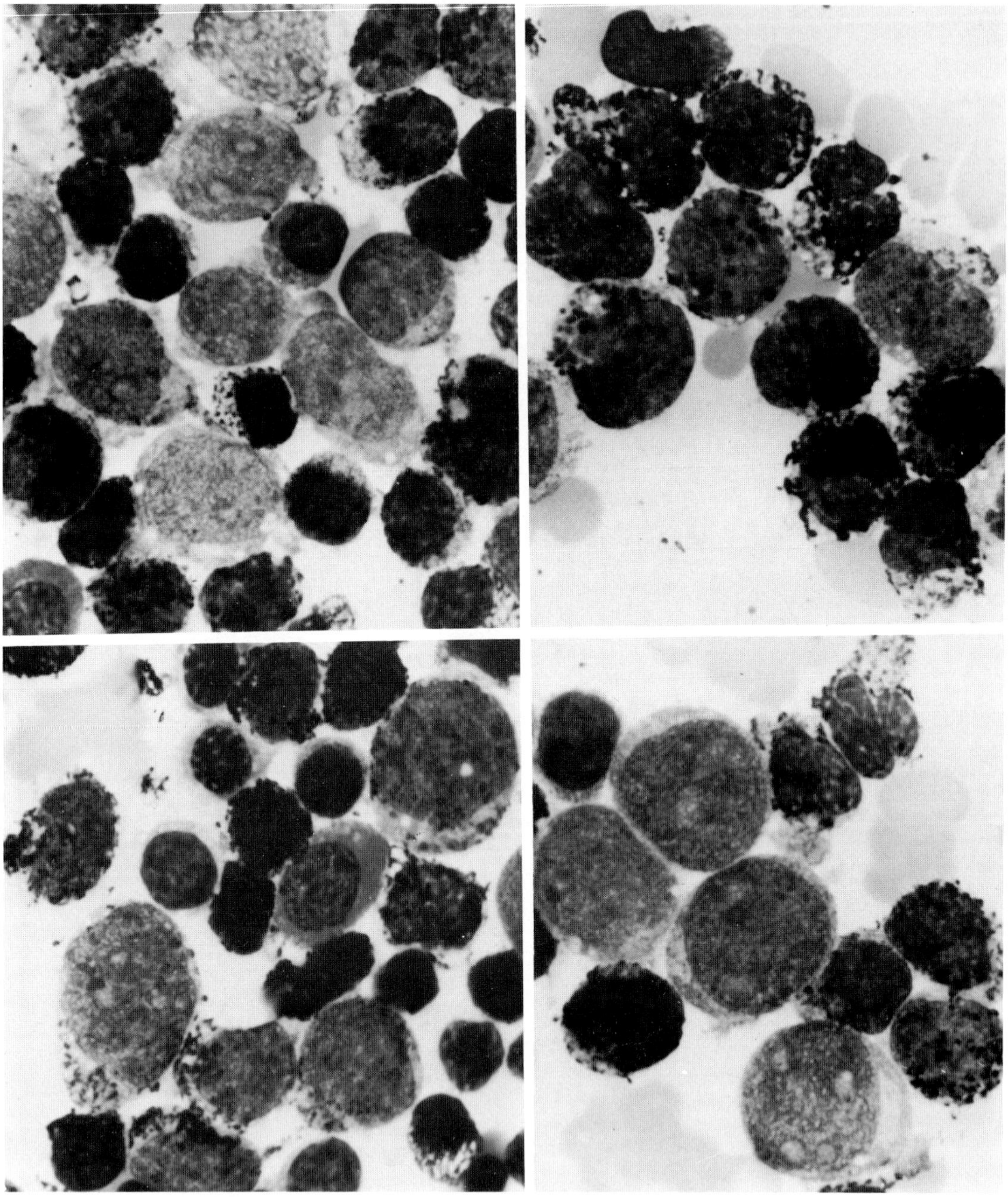

Figure 2-33. Bone marrow in basophilic leukemia, showing increased numbers of basophils as well as myeloblasts and immature granulocytes containing basophilic granules.

Eosinophilic Leukemia

In this rare morphologic variant of myeloblastic leukemia, peripheral blood and bone marrow samples reveal striking numbers of mature and immature eosinophils (Fig. 2-34). It is characterized by proliferation of eosinophils in the blood, bone marrow, and visceral organs.[19,327,813] Eosinophilic leukemia may occur *de novo* or as a terminal event in a patient who had chronic granulocytic leukemia. Some cases of eosinophilic leukemia demonstrate cytogenetic abnormalities and endocardial fibroelastosis.[327]

In the peripheral blood, mature eosinophils predominate. Occasionally, normoblasts and atypical monocytes are found. Rarely, giant platelets and dwarf megakaryocytes occur. Aspirates of bone marrow from patients with acute eosinophilic leukemia show markedly increased numbers of mature and immature eosinophils (Figs. 2-34, 2-35). Myeloblasts are usually less than 10% of the differential cell count of the bone marrow. Megakaryocytes are greatly reduced in number, as are erythroblasts and neutrophilic granulocytes. Sometimes, granules of leukemic eosinophils stain paler with routine panoptic stains than do normal eosinophils. In some leukemic eosinophils, granules appear small and sparse when compared with normal eosinophils. These hypogranular eosinophils occur in both peripheral blood and bone marrow of patients with eosinophilic leukemia. Some leukemic eosinophils are gigantic, with nuclear aberrations consisting of dumbbell-shaped nuclei and nuclear convolutions and lobulations. Gigantism is especially prominent in leukemic eosinophil metamyelocytes.

On H & E sections of bone marrow biopsies from patients with acute eosinophilic leukemia, marrows are virtually replaced by eosinophils (Fig. 2-35). Megakaryocytes, erythroblasts, and neutrophilic granulocytes are greatly reduced in number. Cytochemically, leukemic eosinophils show scant activity of specific esterase and strong activity of myeloperoxidase and acid phosphatase. In these eosinophils, myeloperoxidase is resistant to inhibition by sodium cyanide, as is the myeloperoxidase of normal eosinophils. Cytoplasmic staining by pyronine is faint.

(text continued on page 162)

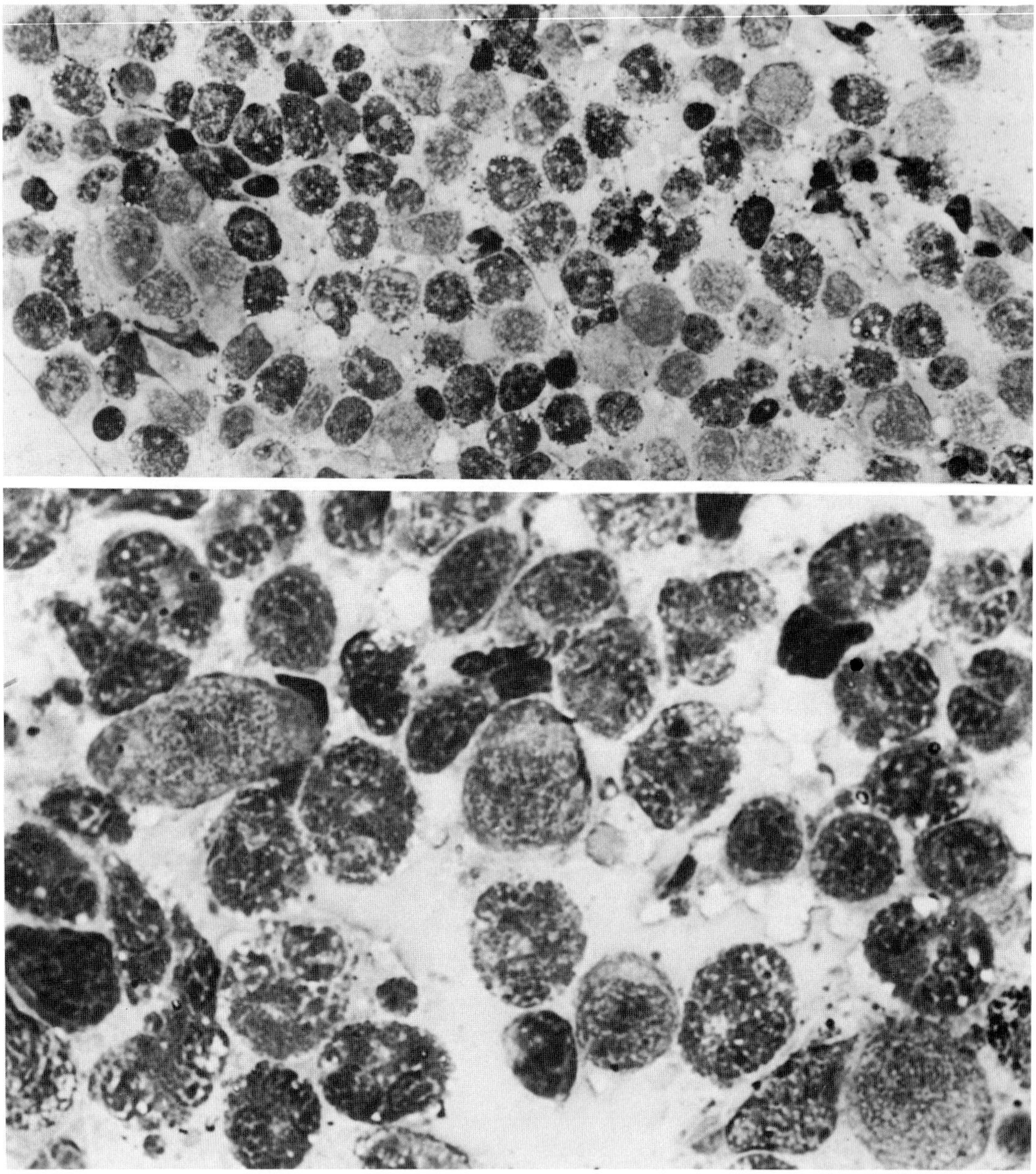

Figure 2-34. Bone marrow in eosinophilic leukemia. Low-power view (**top**) and higher-power view (**bottom**) show markedly increased numbers of eosinophils.

Figure 2-35. Bone marrow in eosinophilic leukemia, demonstrating increased ➔ numbers of eosinophils in Wright-stained film of bone marrow (**top**) and H & E section of bone marrow biopsy (**bottom**).

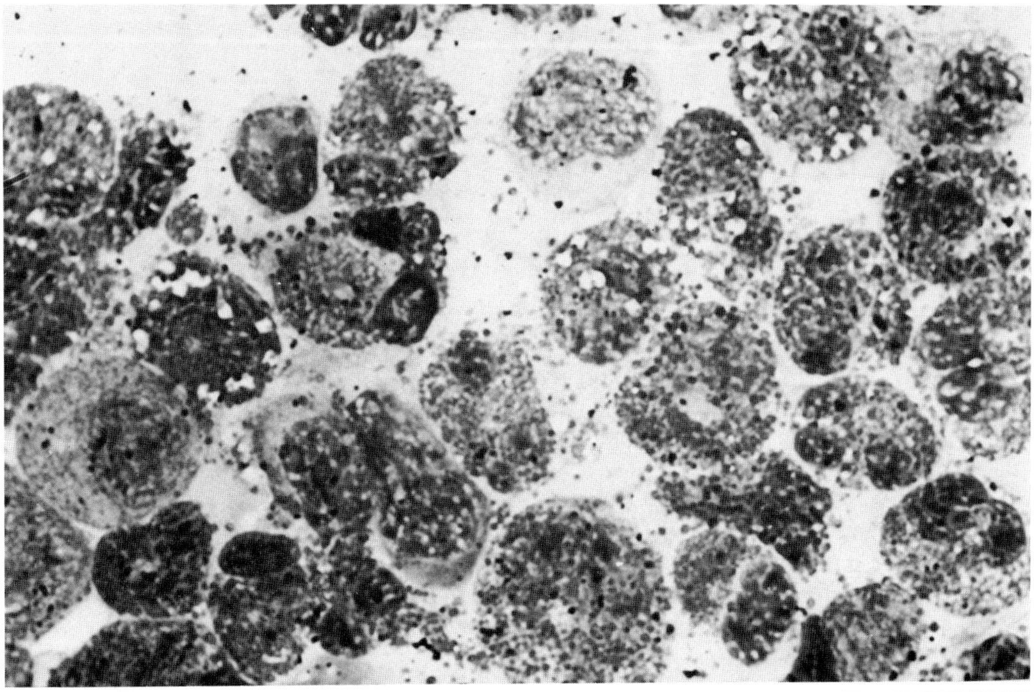

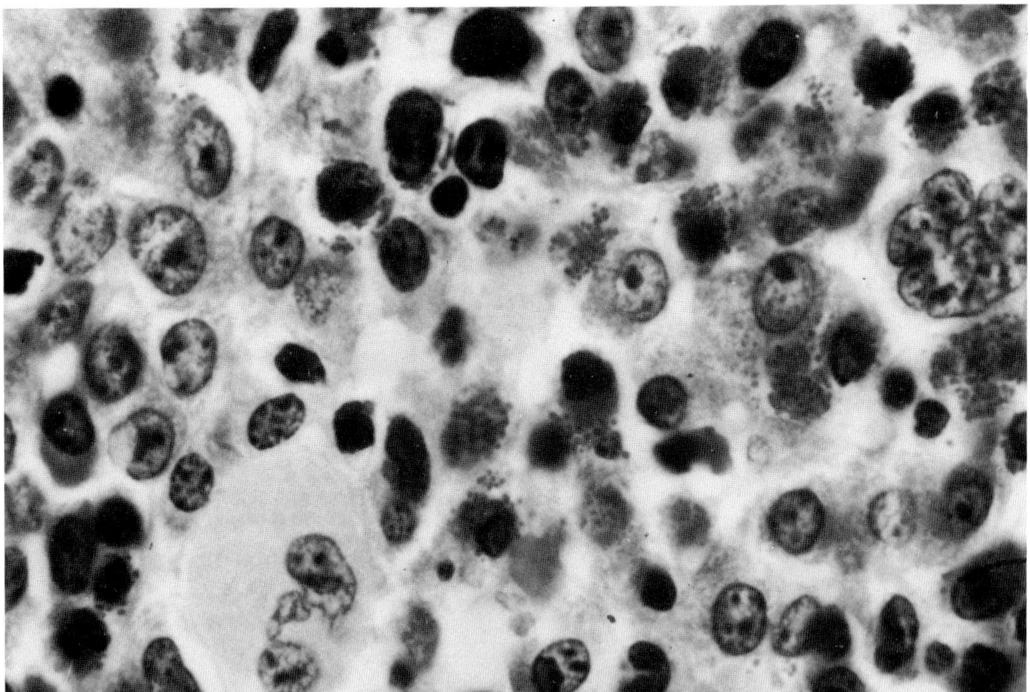

Acute Megakaryocytic Myelosis

This rare disorder is thought to represent an abnormal proliferation of neoplastic megakaryocytes in the bone marrow and visceral organs.[13,68,104,196,1084] It may seem to appear *de novo* in some cases, and in others it represents the terminal phase of chronic granulocytic leukemia,[165] polycythemia vera, or agnogenic myeloid metaplasia.[439]

Clinically, patients with acute megakaryocytic myelosis have fever and splenomegaly, and usually a fulminating course. Peripheral blood specimens are characterized by large numbers of dwarf megakaryocytes as well as large and small fragments of megakaryocytes, giant platelets, increased numbers of neutrophils and eosinophils, and anisocytosis and poikilocytosis of erythrocytes (Figs. 2-36, 2-37).

Aspirates of bone marrow from patients with acute megakaryocytic myelosis reveal a striking proliferation of atypical-appearing megakaryocytes. Most of these are normal-sized megakaryocytes and micromegakaryocytes. Cytoplasm is richly granular. Many of the megakaryocytes demonstrate mitotic division figures. Some of the megakaryocytes show indistinct nuclear and cytoplasmic details, and they may be degenerating. Other marrow cells, such as erythroblasts and granulocytic cells, are reduced in number.

H & E sections of bone marrow biopsies from patients with acute megakarycytic myelosis show virtual replacement of the marrow by atypical-appearing megakaryocytes (Fig. 2-38). These cells have abundant cytoplasm and vary in size and shape.

Cytochemically, megakaryocytes in acute megakaryocytic myelosis contain abundant glycogen demonstrable with the PAS reagent; no detectable activity of myeloperoxidase, acid phosphatase, or specific esterase; strong activity of nonspecific esterase; and strong staining of the cytoplasm by methyl-green pyronine, indicating a high cytoplasmic content of RNA. These megakaryocytes contain peroxidase activity demonstrable in a characteristic perinuclear distribution by light and electron microscopic techniques.[74,156,158,159,164,165,528] Although valuable as a stain for megakaryocytes in animals, cholinesterase activity is not readily demonstrable in human megakaryocytes.[460] Identification of serotonin storage organelles may represent a useful marker for megakaryocytes.[1093]

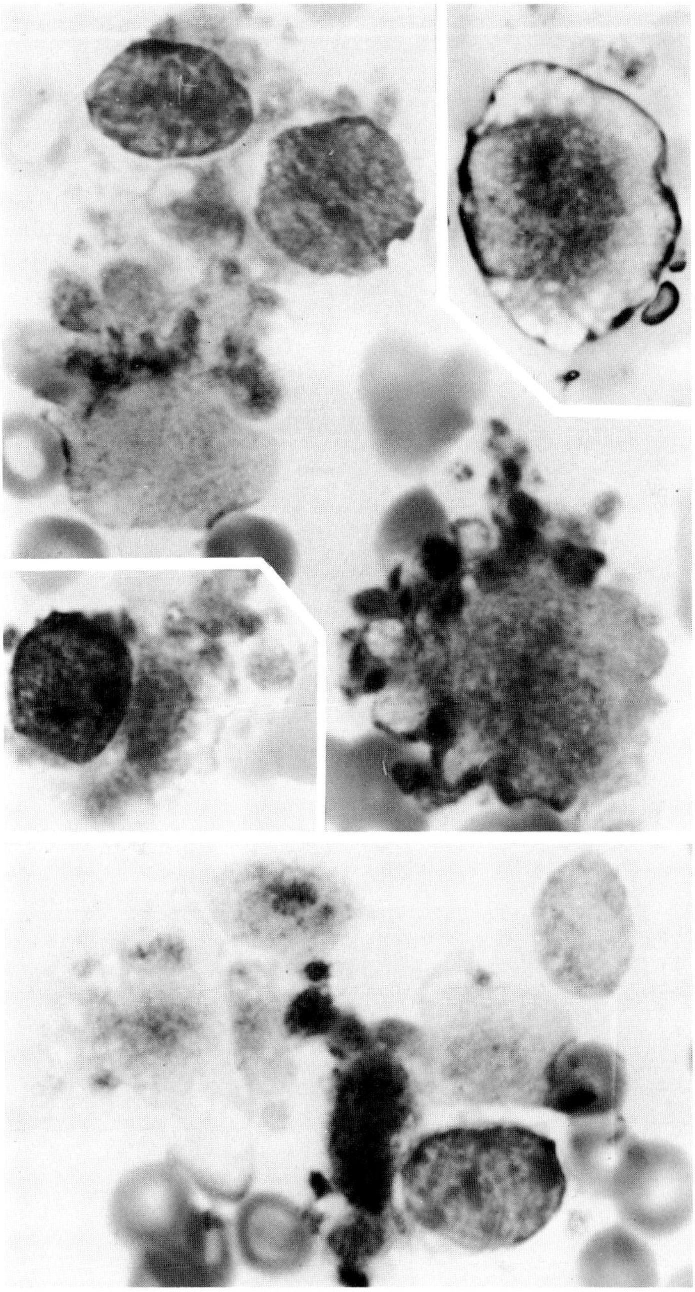

Peripheral blood in acute megakaryocytic myelosis, showing giant, atypical platelets and dwarf megakaryocytes. **Figure 2-36.**

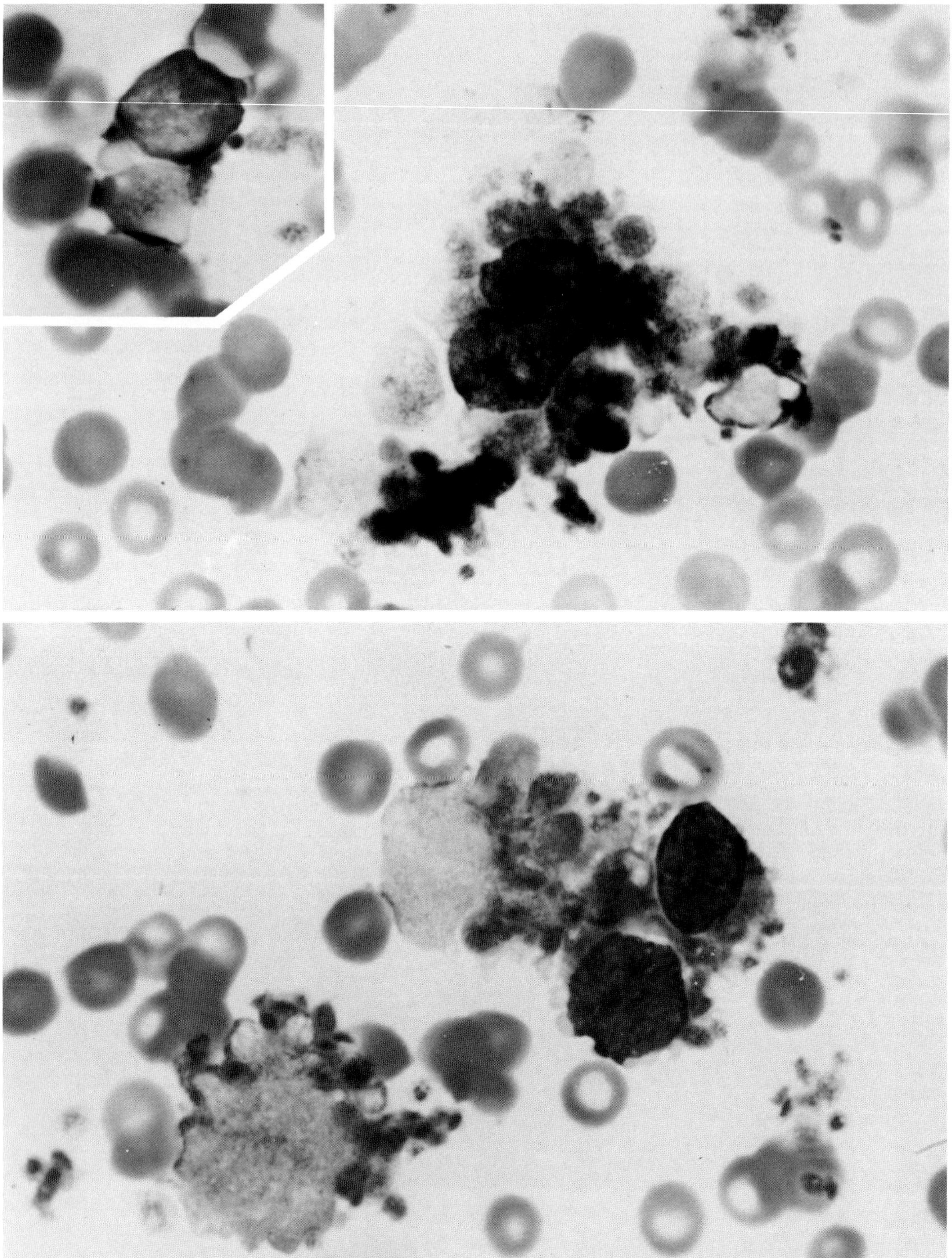

Figure 2-37. Peripheral blood in acute megakaryocytic myelosis, illustrating giant platelets and dwarf megakaryocytes.

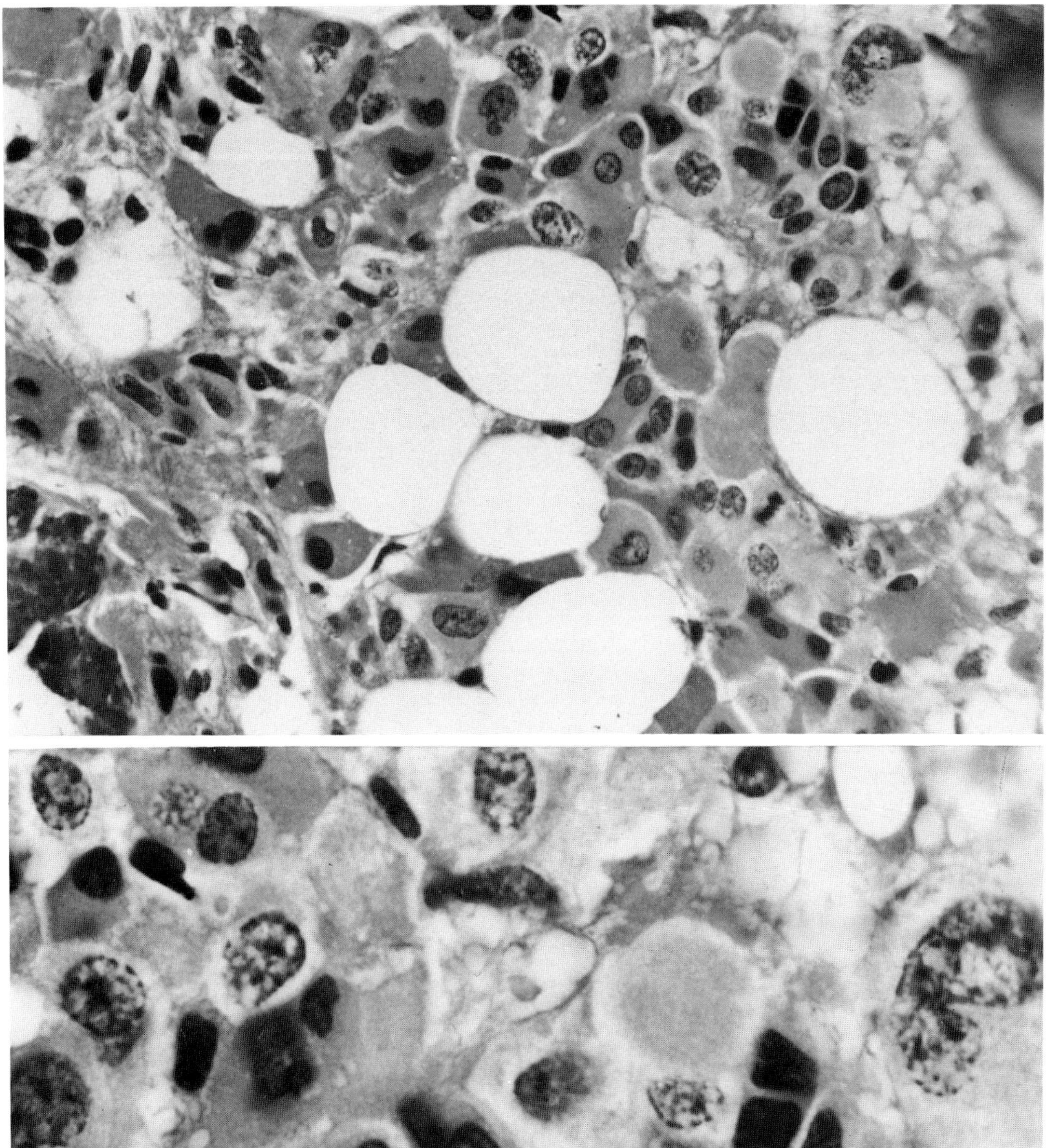

H & E section of bone marrow biopsy in acute megakaryocytic myelosis, showing increased numbers of megakaryocytes in low- and higher-power views.

Figure 2-38.

3

Monocytic Leukemia

Monocytic leukemia is a disorder of neoplastic monocytes that proliferate in blood, bone marrow, and visceral organs. Clinically, it is often characterized by fever, splenomegaly, and gingival hyperplasia. Hematologically, it is characterized by absolute monocytosis (greater than $500/mm^3$) of the blood and by greater than 40% monocytes in the bone marrow. Combined with monocytosis of the bone marrow, abnormalities are recognizable in other cell types, reflecting the panmyelotic disturbance.

Once divided sharply into myelomonocytic and histiomonocytic types,[261,266,463,518,846,916,1020] monocytic leukemia today is more appropriately conceptualized as a spectrum of variants of acute myeloblastic leukemia.[87,88,102,108,109,170,172,173,205,232,233,275,324,351,368,443,446,518,545,662,706,751–753,839,884,951,955,967,1014,1081,1083,1106] "Pure" monocytic leukemia probably does not exist as such, because cases of this type have been found to demonstrate specific esterase activity, a marker for granulocytic cells.[502,516]

Within the spectrum of monocytic leukemia, a cluster of cases at one end shows typical monocytes and monocytoid cells, many of which contain granules that resemble granules of neutrophils or eosinophils.[880] According to the FAB classification, this variant is labeled *M4*[87,88] and corresponds to myelomonocytic leukemia as conceptualized by Naegeli[713] and Downey.[266] At another point in the spectrum is a cluster of cases which show monocytes with multiple nuclear lobulations, abundant foamy-appearing cytoplasm, and an unusually aggressive clinical course. Using the FAB classification, this variant has been labeled *M5*[87,88] and corresponds to histiomonocytic or Schilling-type acute monocytic leukemia.[266,846,928] Monocytic leukemia can be a terminal event in virtually all instances in which myeloblastic leukemia has been described, suggesting that leukemic monocytes derive from leukemic myeloblasts in a manner analogous to the derivation of normal monocytes from promyelocytes.[470,726]

Peripheral blood samples from patients with acute monocytic leukemia contains few or many monocytes (Fig. 3-1). In some instances, white blood cell counts may be unusually high (greater than $500{,}000/mm^3$ and in some instances in excess of 1 million/mm^3). In cases showing extreme leukocytosis, some of the leukemic monocytes demonstrate adherence to one another, forming large aggregates of leukemic cells (Fig. 3-2). Presumably, this is due to interdigitation of cell membranes of these leukemic cells. Cellular aggregates of this type have been implicated in the pathogenesis of cerebral leukostasis and stroke in patients with acute monocytic leukemia.

A number of morphologic abnormalities are detectable in leukemic monocytes. These abnormalities occur alone or in combination, and they help to distinguish leukemic monocytes from normal monocytes that usually do not demonstrate them.[170,175,307,340,424,470,477,568,570,661,672,724,725,766,805,806,892,918,974,1009] First, leukemic monocytes may appear as so-called histiocytes, with round to oval nuclei, delicate-appearing nuclear chromatin pattern, nuclear folding, and abundant cytoplasm containing many vacuoles of varying sizes (Fig. 3-3). Second, leukemic monocytes can have multiple nuclear lobulations that seem to overlie one another, resembling fingerlike projections (Fig. 3-4). Third, leukemic monocytes have unusually delicate nuclear chromatin patterns, and chromatin strands appear closely approximated (Fig. 3-5). In some of these cells, nucleoli may be unusually large and show perinucleolar condensations of chromatin (Fig. 3-6).

Fourth, leukemic monocytes can be gigantic, with especially large nuclei and voluminous cytoplasm (Fig. 3-7). In some instances, the cytoplasm may contain large vacuoles in which crystalloids can be identified.[219] Fifth, leukemic monocytes may contain multiple cytoplasmic pseudopodia (Figs. 3-8, 3-9). These pseudopodia vary in size, and most of them appear clear and devoid of granules. Sixth, leukemic monocytes often demonstrate concentration or aggregation of granules around the nucleus (Fig. 3-10). These granules usually stain red to purple using conventional panoptic stains. Seventh, some leukemic monocytes contain orange to pink staining granules, closely resembling granules found in neutrophils and suggesting a close relationship between monocytes and cells of granulocytic origin (Figs. 3-11, 3-12). Some leukemic monocytes contain one or more Auer rods. Other leukemic monocytes show phagocytosis of mature erythrocytes and developing erythroblasts.

Eighth, some leukemic monocytes in a particular morphologic variant of acute monocytic leukemia have a prominent *U*-shape of the nucleus. Nuclear chromatin in some of these leukemic monocytes is unusually coarse, with large, blocklike and elongated aggregates connective to each other and to the nuclear membrane by smaller, more delicate-appearing strands of chromatin (Fig. 3-13). In addition to leukemic monocytes, peripheral blood of patients with monocytic leukemia may contain normoblasts, myelocytes and promyelocytes, neutrophils and metamyelocytes with aberrant nuclear configuration, giant platelets, dwarf megakaryocytes, and few lymphocytes. Ninth, leukemic monocytes sometimes have an eccentric nucleus and prominent cytoplasmic tail (Fig. 3-14). This tail may represent unusual fluidity of the cytoplasm of leukemic monocytes, or it may be an exaggeration of the cytoplasmic excrescences found in normal monocytes.

Wright-stained aspirates of bone marrows from patients with acute monocytic leukemia show several morphologic variants. In the myelomonocytic leukemia variant (*M4*), leukemic monocytes show multiple nuclear lobulations, delicate nuclear chromatin pattern with few if any aggregates, prominent nucleoli with perinucleolar condensations of chromatin, and abundant cytoplasm that frequently contains numerous orange-red

(text continued on page 178)

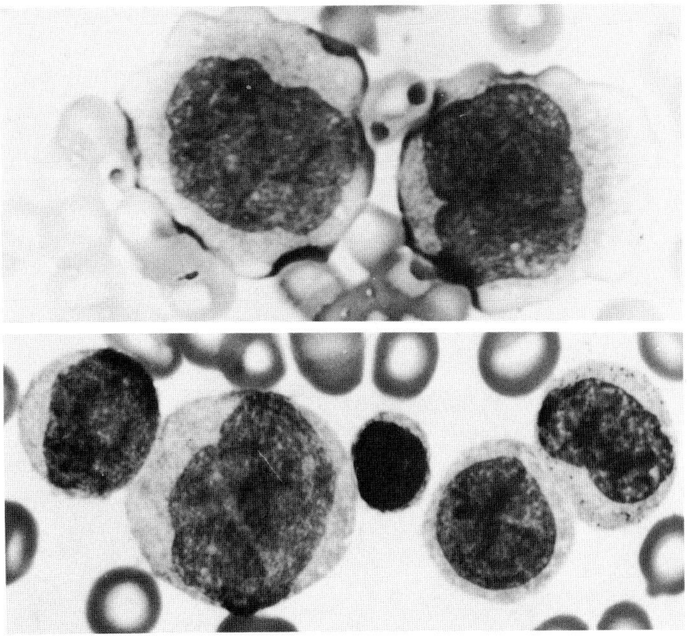

Figure 3-1. Monocytes in acute monocytic leukemia, peripheral blood film, showing increased numbers of nuclear lobulations and multiple pseudopodia in cytoplasm.

Figure 3-2. Leukemic monocytes in acute monocytic leukemia, depicting monocytoid nuclear configuration and prominent cytoplasmic granules and vacuoles.

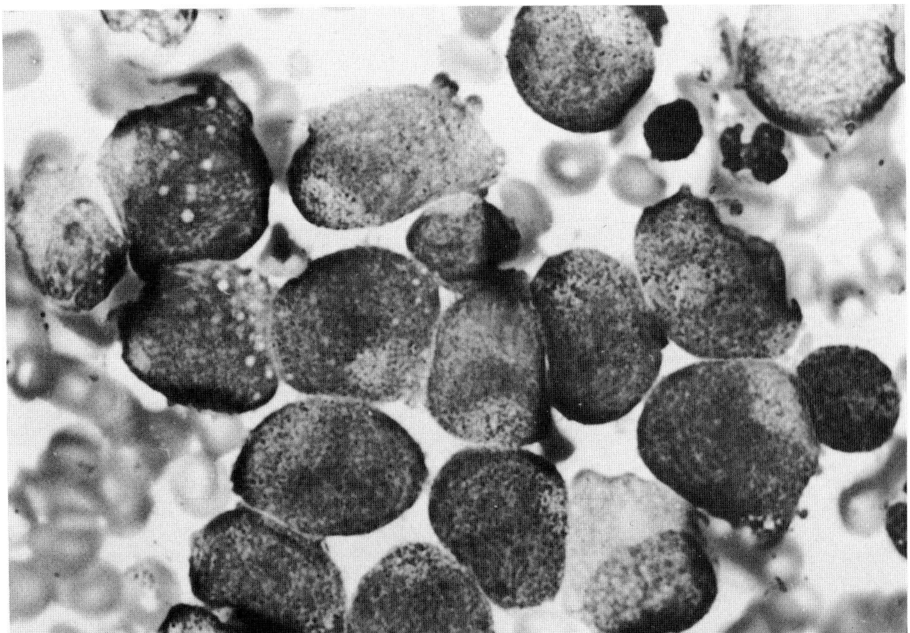

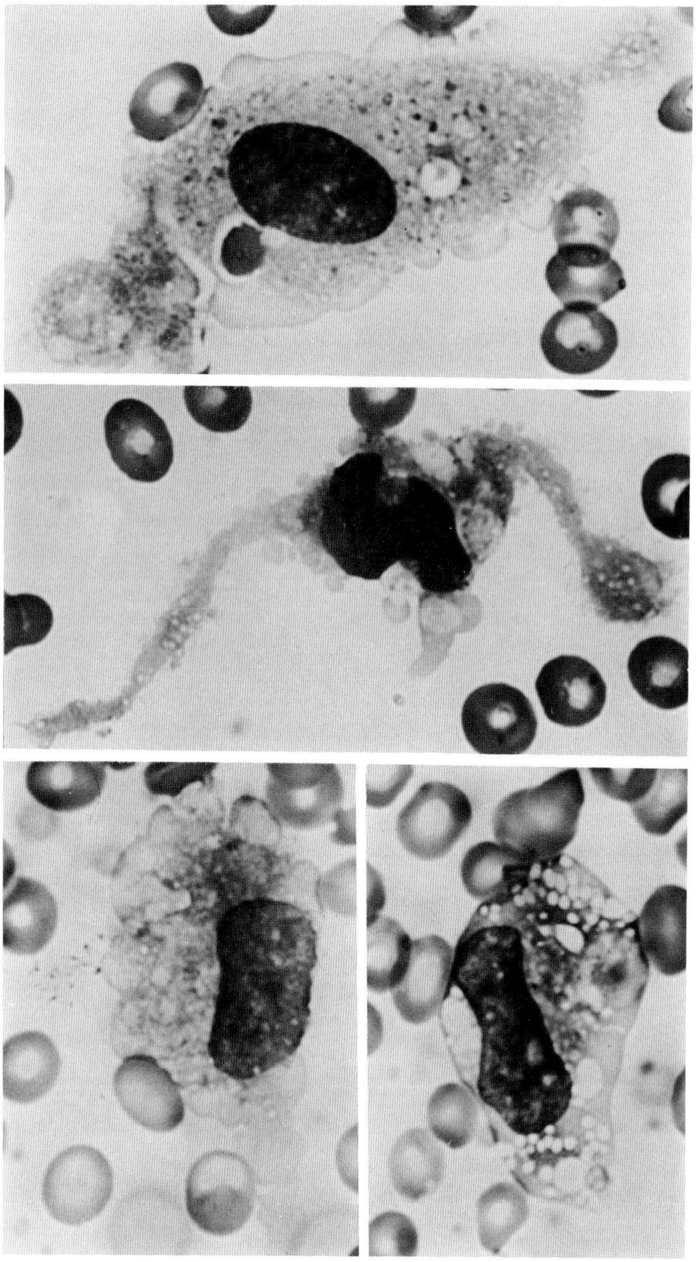

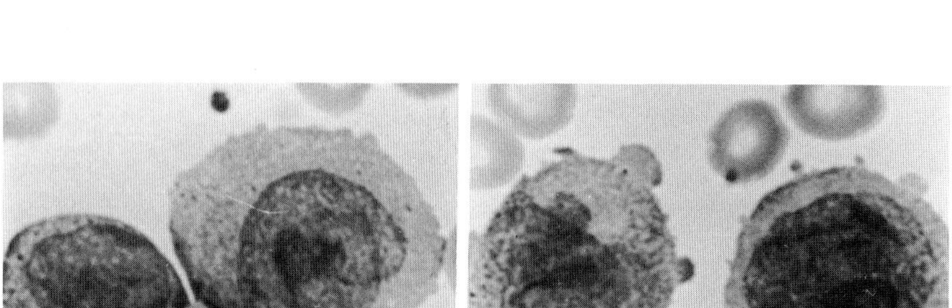

Figure 3-3. ← Monocytes in acute monocytic leukemia, illustrating erythrophagocytosis (**top**), prominent cytoplasmic processes, and vacuoles (**bottom, right**).

Figure 3-4. ← Leukemic monocytes in acute monocytic leukemia, demonstrating multiple nuclear lobulations overlying each other in fingerlike projections.

Figure 3-5. Leukemic monocytes in acute monocytic leukemia, showing delicate nuclear chromatin pattern.

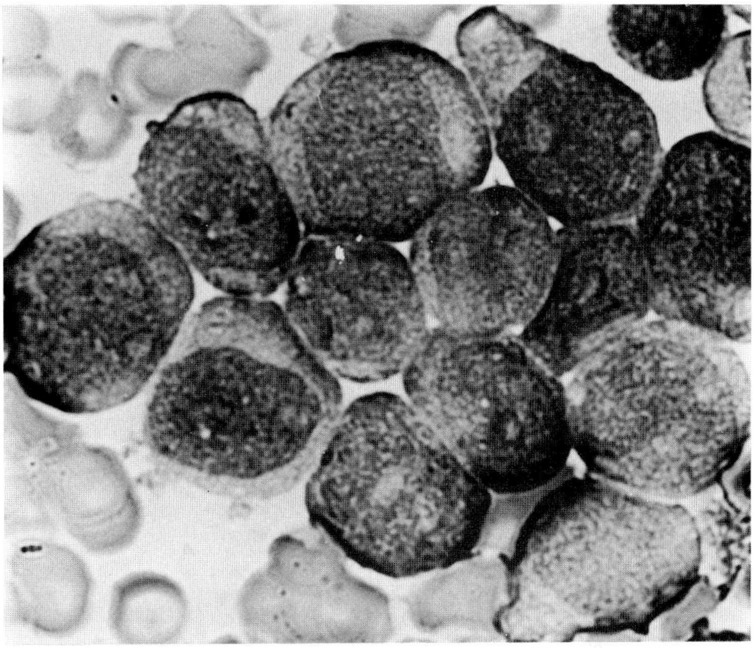

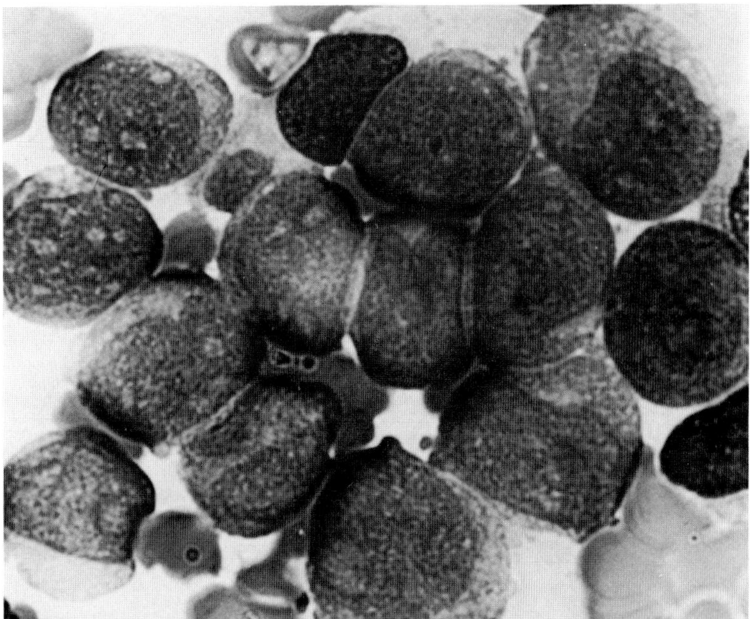

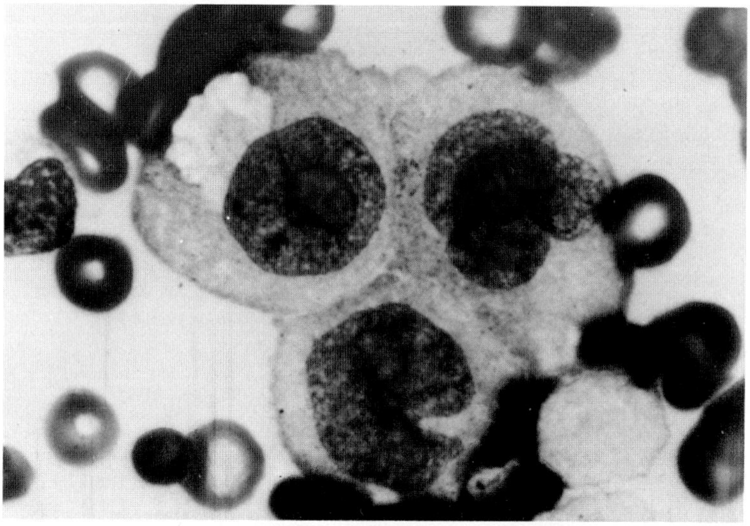

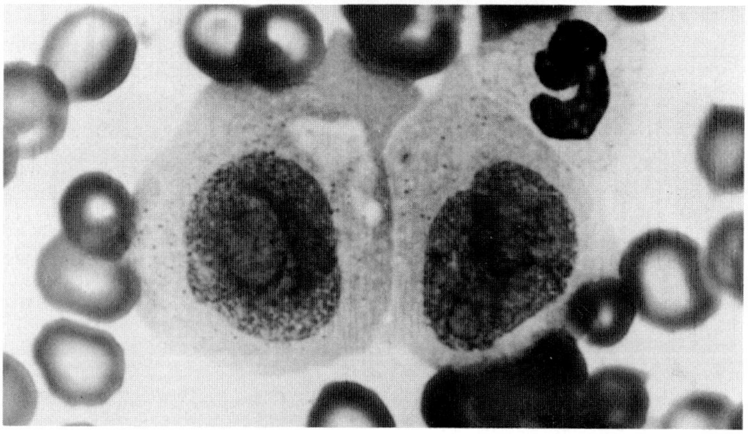

Figure 3-6. Leukemic monocytes in monocytic leukemia, showing giant nucleoli and abundant cytoplasm containing vacuoles.

Figure 3-7. Giant leukemic monocytes in acute monocytic leukemia. →

Figure 3-8. Leukemic monocytes in acute monocytic leukemia, depicting multiple cytoplasmic pseudopodia and erythrophagocytosis. →

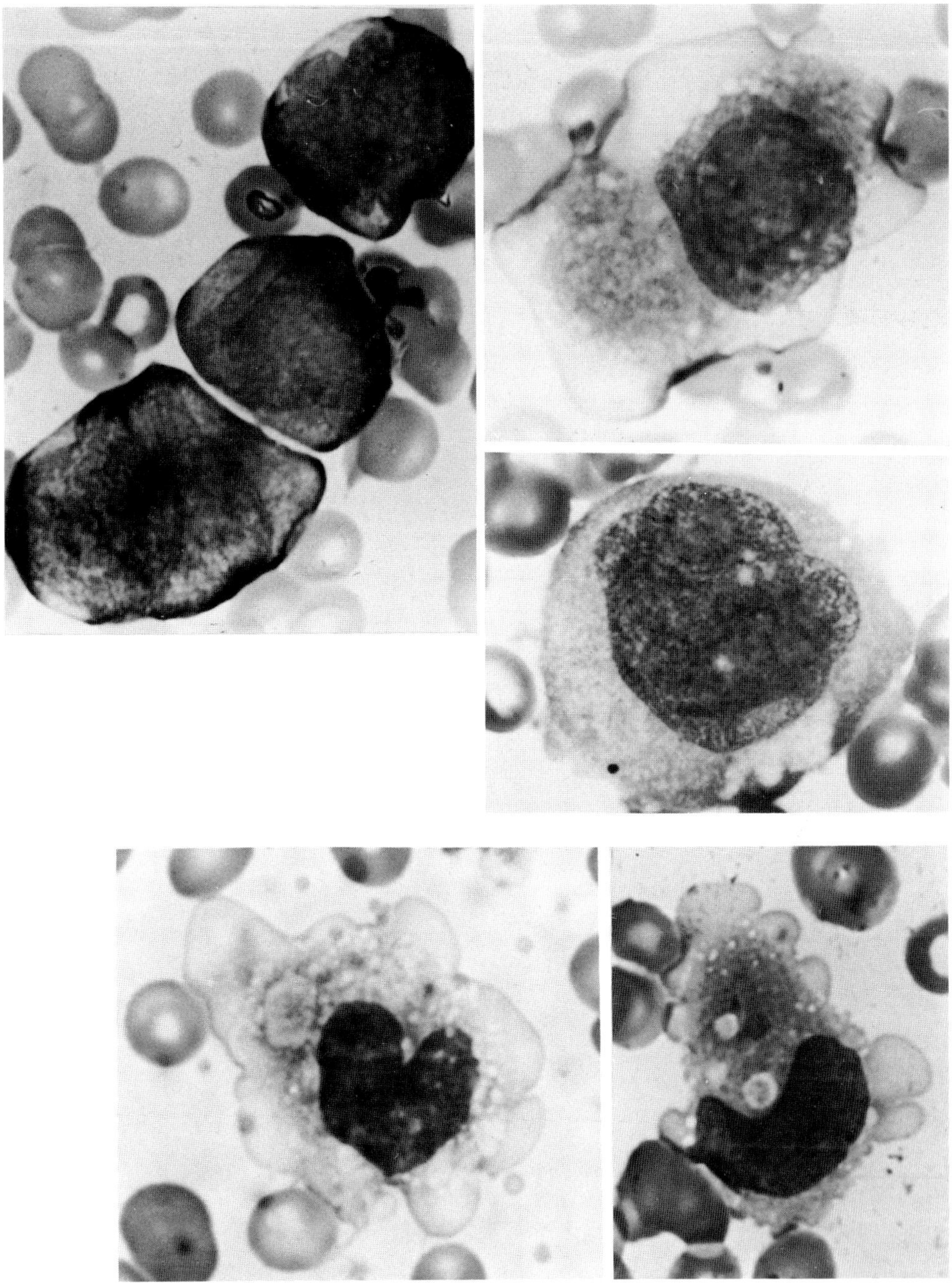

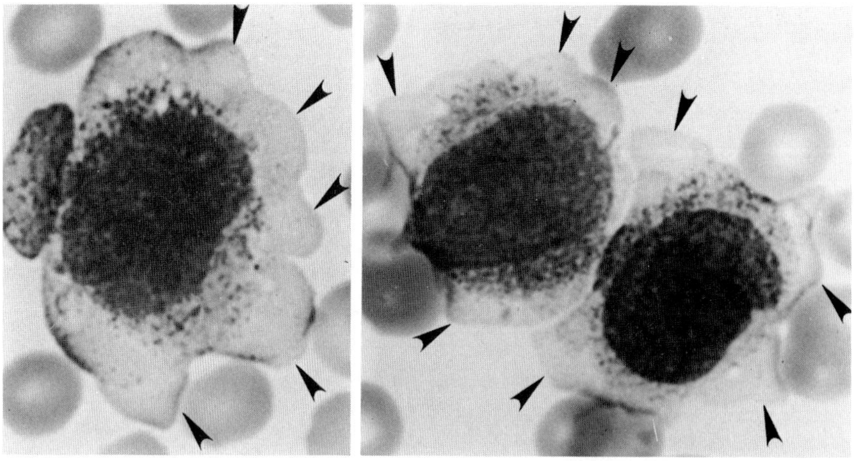

Figure 3-9. Leukemic monocytes in monocytic leukemia, demonstrating multiple clear cytoplasmic pseudopodia (**arrows**).

Figure 3-10. Leukemic monocytes in acute monocytic leukemia, illustrating increased numbers of granules clustered around the nucleus. Cells at the bottom are adherent.

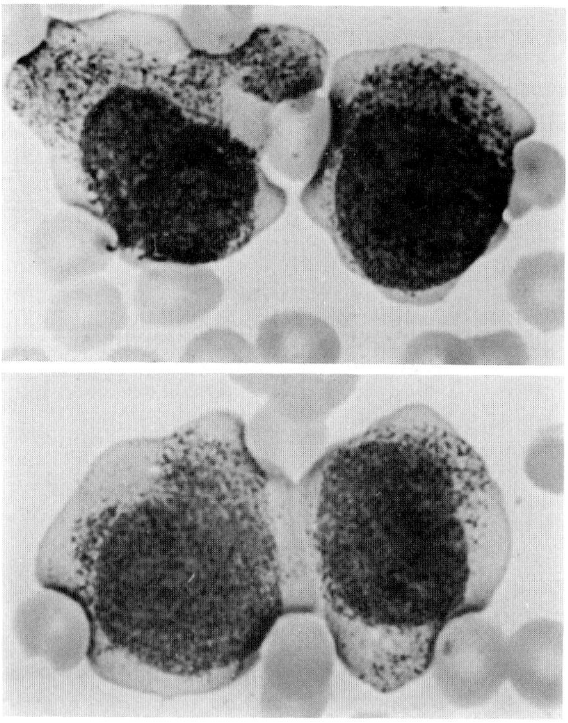

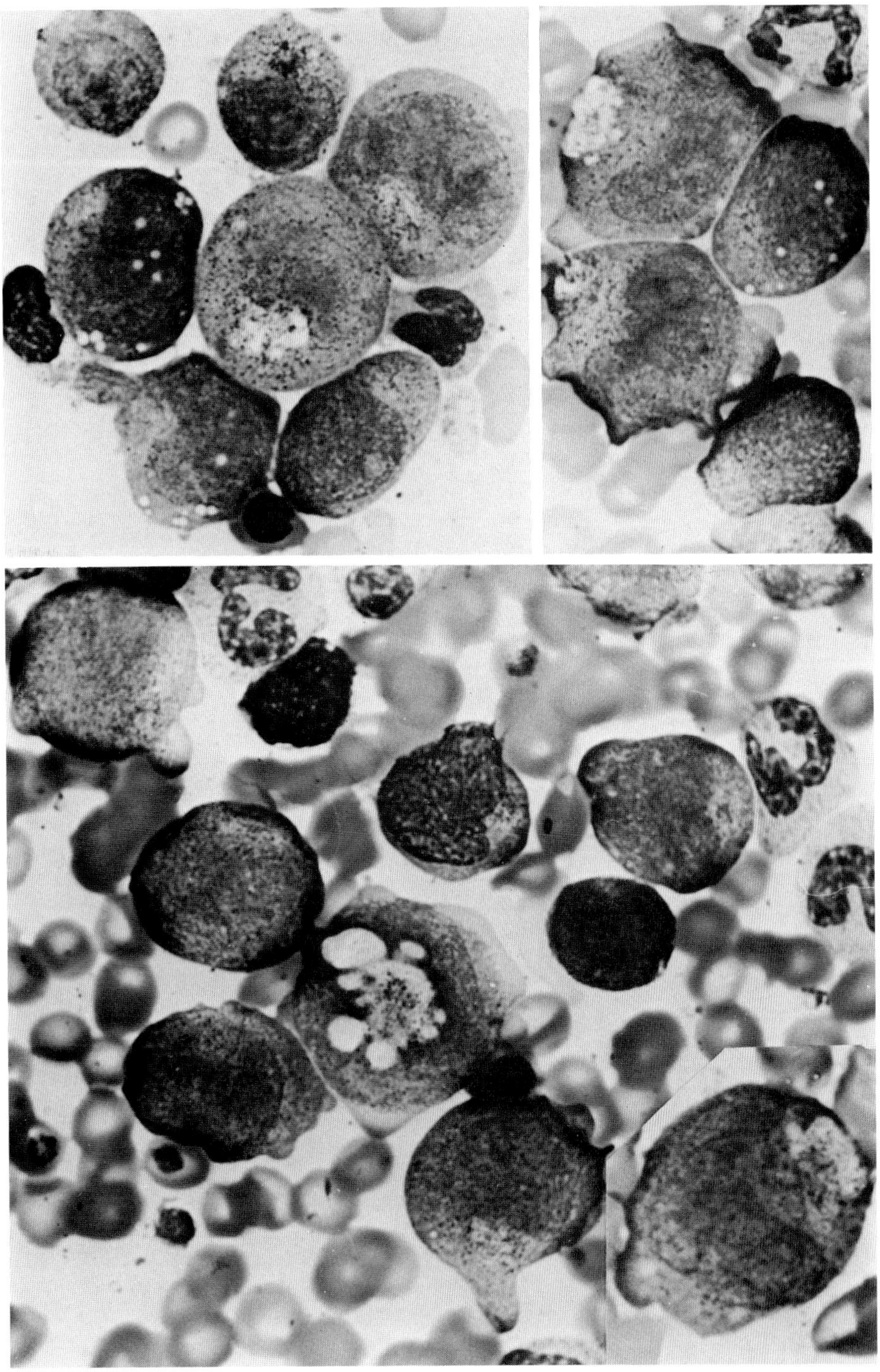

Monocytes in acute monocytic leukemia, myelomonocytic leukemia variant (FAB M4), showing increased numbers of cells with folded and convoluted nuclei, large nucleoli, prominent granules, and multiple pseudopodia. Some monocytes contain clusters of large vacuoles.

Figure 3-11.

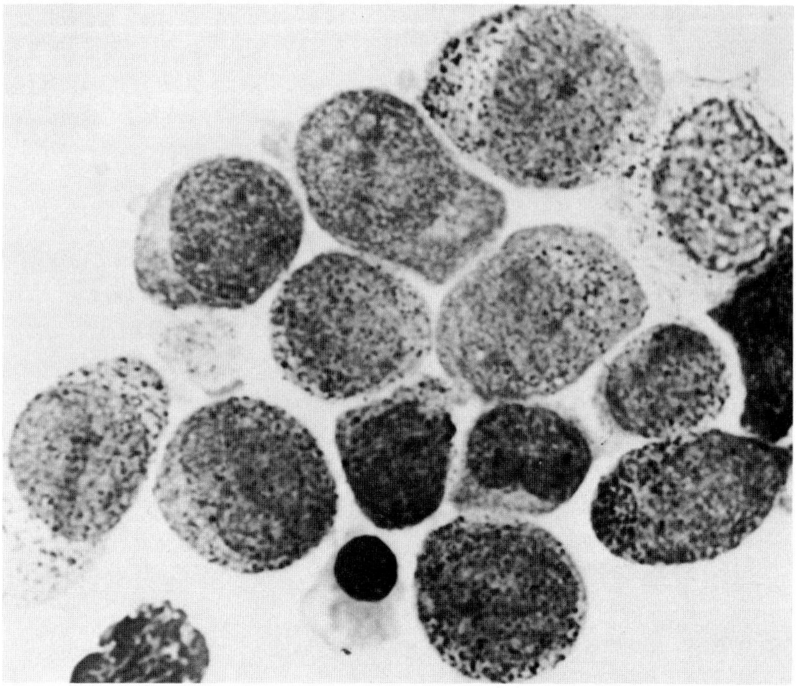

Figure 3-12. Monocytes in acute monocytic leukemia, demonstrating increased numbers of granules.

Figure 3-13. Leukemic monocytes in acute monocytic leukemia, illustrating prominent U-shaped nuclear configuration (**left**) and large, prominent aggregates of chromatin (**right**).

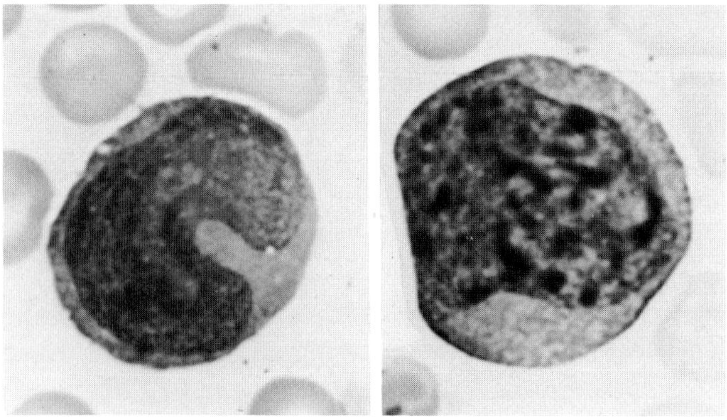

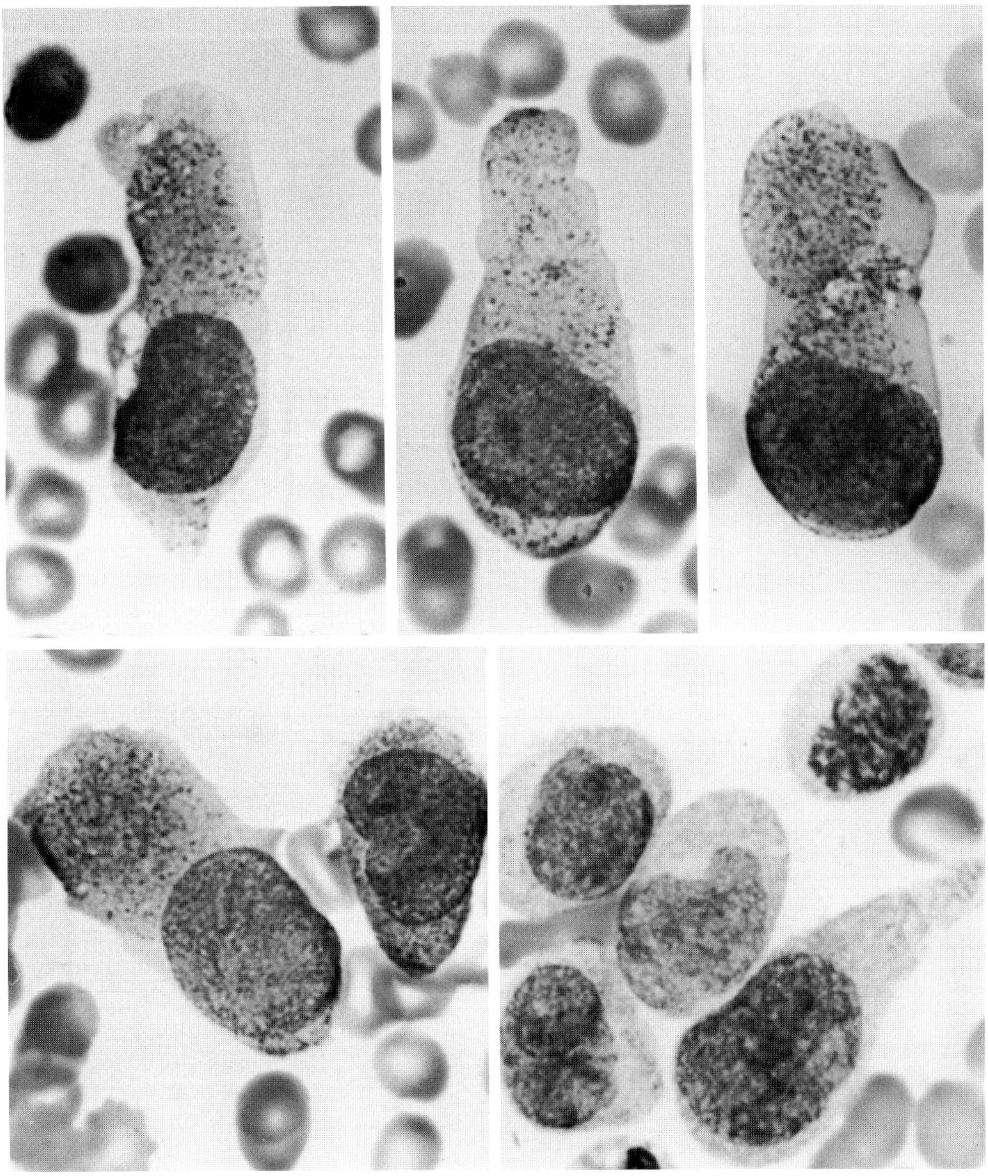

Figure 3-14. Monocytes in acute monocytic leukemia, showing prominent cytoplasmic "tails" containing increased numbers of nonspecific granules.

and purple staining granules using conventional panoptic stains (Figs. 3-15, 3-16, and Plate 3*A*). Some leukemic monocytes contain one or more Auer rods.[401] Other leukemic monocytes demonstrate multinuclearity.

Erythrophagocytosis of mature erythrocytes and erythroblasts can be found in some cells. Granulocytic cells are reduced in number. There may be increased numbers of promyelocytes, some of which contain splinter-shaped granules. Abnormalities of nuclear configuration in metamyelocytes and mature neutrophils can be found (Fig. 3-17). These consist of the pseudo-Pelger-Huët anomaly of nuclear hyposegmentation as well as hypersegmentation and twisted and convoluted nuclei of metamyelocytes.

Erythroblasts are reduced in number. Some show multinuclearity and megaloblastoid type nuclear chromatin pattern. Proerythroblasts are moderately increased in number, and some are binuclear. More mature erythroblasts show coarse basophilic stippling as well as accelerated pyknosis and abnormal karyorrhexis. Megakaryocytes are reduced in number, and some of them are micromegakaryocytes.

Recently, a morphologic variant of monocytic leukemia has been shown to contain striking numbers of mature and immature eosinophils,[994] as shown in Figures 3-18, *top*, and 3-19. In marrow specimens from these patients, megaloblastoid intermediate macronormoblasts are increased in number. Proerythroblasts may be multinuclear (Fig. 3-18, *bottom*). In macrophages from some patients who have large numbers of eosinophils in the marrow, Charcot-Leyden crystals can be found (Fig. 3-19, *top*). These crystals are believed to be derived from membranes of degenerating eosinophil granules.

In another morphologic variant of acute monocytic leukemia, monocytes have a *U*-shaped nucleus with unusually prominent aggregates of nuclear chromatin connected to each other and to the nuclear membrane by delicate strands of chromatin (see Figs. 3-13, 3-20, *top*). Cytoplasm is abundant and contains few dustlike granules. Some of these leukemic monocytes are gigantic. In marrows from patients with this particular morphologic variant of acute monocytic leukemia, increased numbers of histiocytes are found (Fig. 3-20, *bottom*). Some of these histiocytes show erythrophagocytosis of mature erythrocytes and developing erythroblasts.

In the *M5* variant of acute monocytic leukemia, monocytes show similar nuclear convolutions and foldings as well as delicate nuclear chromatin and nucleoli. Cytoplasm is usually abundant, and it may contain a few dustlike granules that stain pink to purple using conventional panoptic stains (Fig. 3-21 and Plate 3*B*). Erythroblasts are greatly reduced in number, as are megakaryocytes and granulocytes. Some megakaryocytes contain leukocytes that seem to be contained within cytoplasmic vacuoles.[559] For the most part, granulocytic cells show fewer nuclear aberrations in the traditional histiomonocytic type of monocytic leukemia than in the myelomonocytic variant. Occasional ringed sideroblasts are identified with the Prussian blue reagent.

H & E sections of biopsies of marrow from patients with monocytic leukemia show marked hypercellularity. Megakaryocytes, erythroblasts,

(text continued on page 184)

Leukemic monocytes containing nonspecific granulation in acute monocytic leukemia, myelomonocytic variant. Nucleoli are prominent in these cells. **Figure 3-15.**

Leukemic monocytes with prominent granulation in acute monocytic leukemia, myelomonocytic variant. **Figure 3-16.**

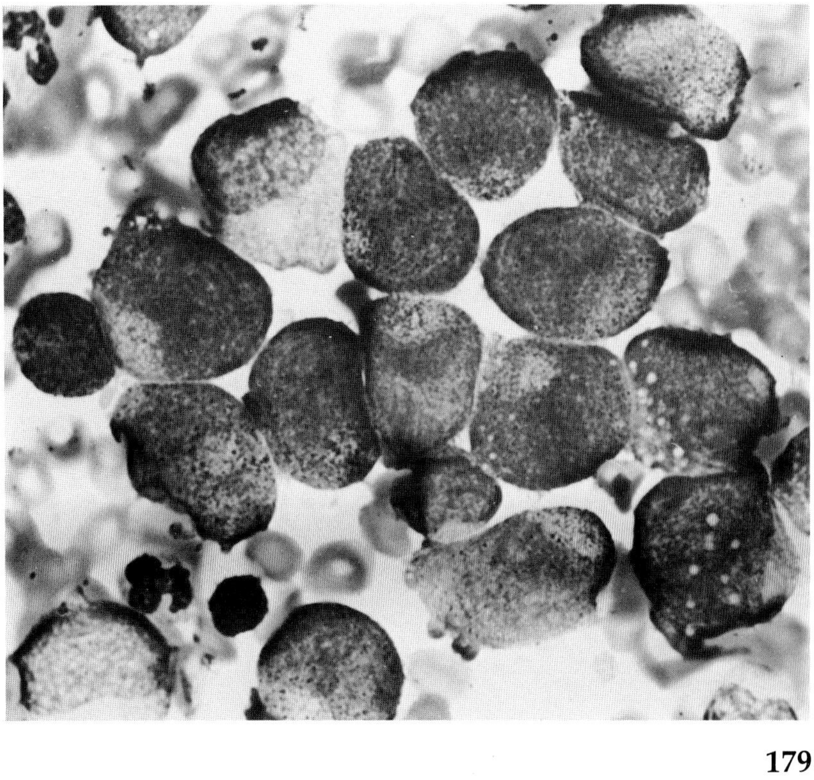

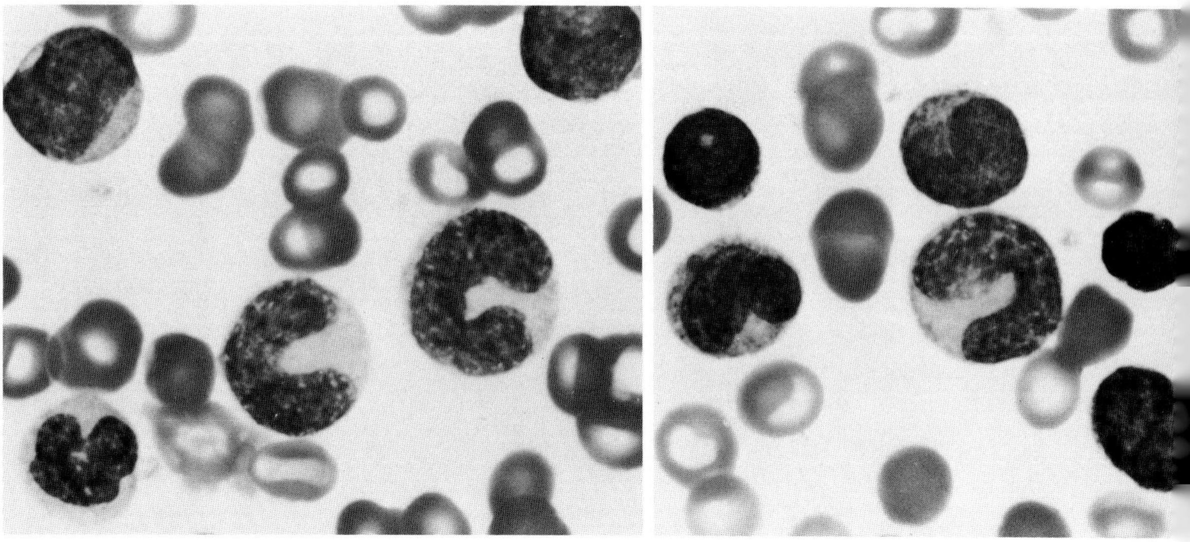

Figure 3-17. Atypical-appearing bands in monocytic leukemia, myelomonocytic variant.

Figure 3-18. Increased numbers of eosinophils and leukemic monocytes in acute monocytic leukemia, myelomonocytic variant (**top**). Giant, atypical proerythroblasts with delicate nuclear chromatin pattern in large nuclei, and condensed nuclear chromatin in smaller nuclei (**bottom left, arrows**). These may represent polar bodies. Giant multinuclear erythroblast with nuclear lobulations (**right**). ➡

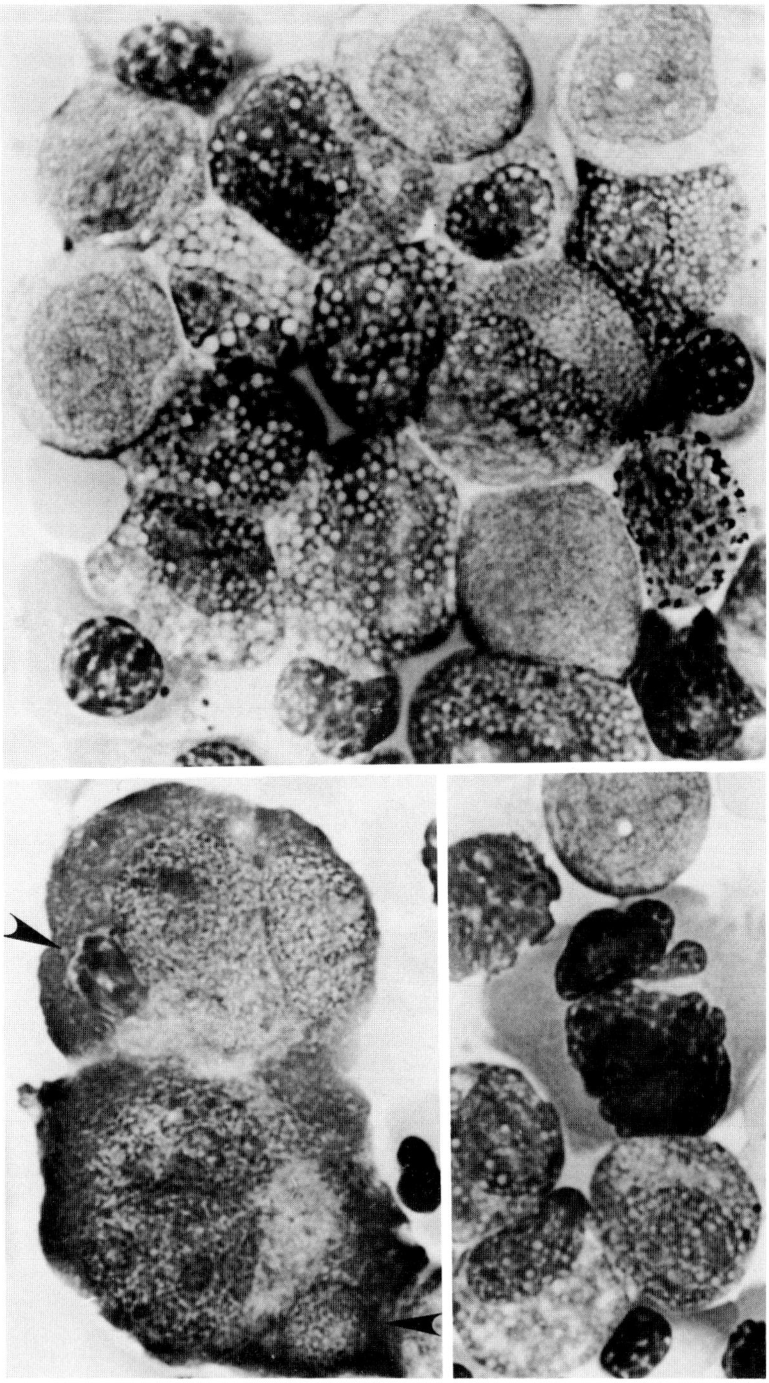

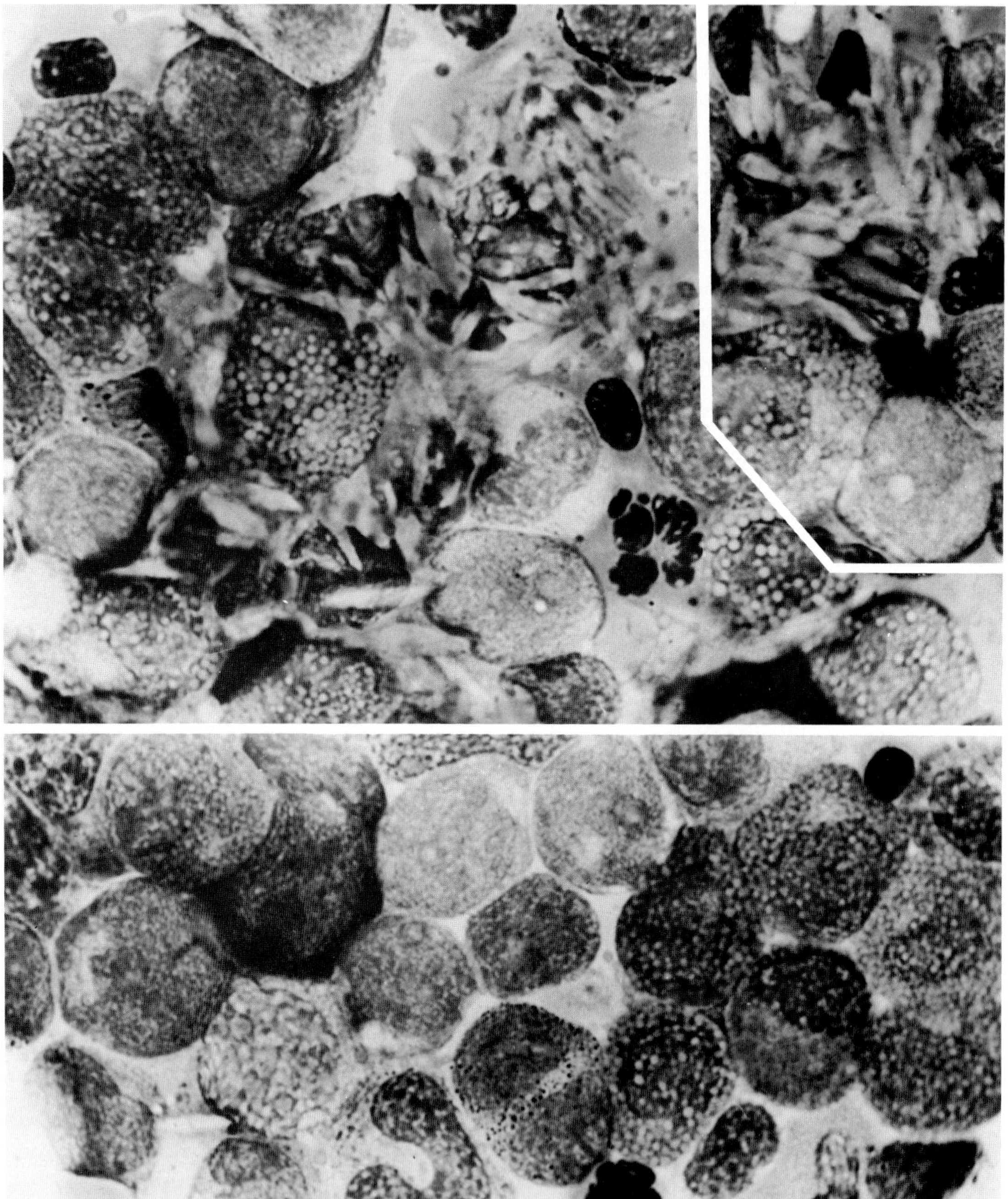

Figure 3-19. ← Increased numbers of eosinophils and monocytes in acute monocytic leukemia, myelomonocytic variant. Large macrophages contain needle-shaped Charcot-Leyden crystals (**top and inset**).

Figure 3-20. Leukemic monocytes with U-shaped nucleus and prominent aggregates of chromatin (**top**). Histiocytes showing erythrophagocytosis (**bottom, left**) and phagocytosis of normoblasts (**right**) along with leukemic monocytes having unusually prominent aggregates of nuclear chromatin, variant of monocytic leukemia.

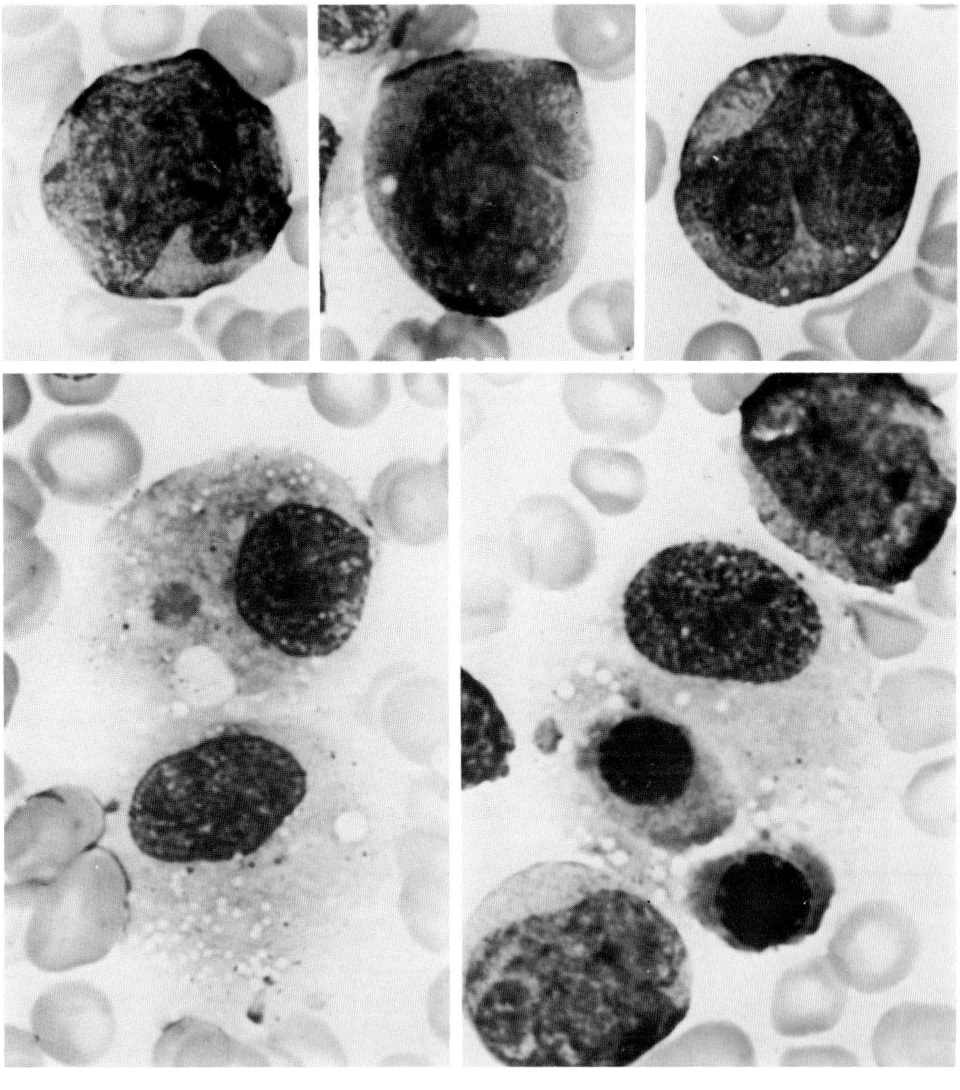

and granulocytes are greatly reduced. For the most part, the marrow is replaced by large mononuclear cells with prominent nuclei showing convolutions, infoldings, and lobulations as well as large acidophilic staining nucleoli (Fig. 3-22). Cytoplasm of the leukemic monocytes is abundant.

Cytochemically, the PAS stain in both morphologic variants of monocytic leukemia reveals scant amounts of glycogen. Usually, the pattern is diffuse, but occasionally leukemic monocytes show a dustlike pattern of glycogen distribution like leukemic myeloblasts. In most instances, weak activities of myeloperoxidase, acid phosphatase,[168] and specific esterase[92,1021] can be demonstrated in normal monocytes as well as in leukemic monocytes from both traditional variants of monocytic leukemia (Fig. 3-23), suggesting that they are of granulocytic origin. Activities of these enzymes are distributed evenly throughout the cytoplasm. With immunocytochemical[26,27,167,495,1021,1113] and direct staining[929] techniques, lysozyme (muramidase) can be demonstrated in leukemic monocytes as well as in normal monocytes and macrophages. Using methyl green-pyronine to demonstrate RNA, scant amounts of RNA are found in the cytoplasm of some leukemic monocytes, and markedly increased amounts in others.

At present, the most useful cytochemical stain in monocytic leukemia is the reaction for nonspecific esterase activity. Using alpha naphthyl acetate or alpha naphthyl butyrate as substrate and a variety of dye couplers as indicators, nonspecific esterase activity is intense in normal and in leukemic monocytes, as seen in Figure 3-24 and Plate 6*A*.[71,73,92,144,145,171,190,231,493,502,666,667,819,853,918–922,921,1091,1108,1110] As in normal monocytes, nonspecific esterase activity in leukemic monocytes is sensitive to inhibition by sodium fluoride, further enhancing the diagnostic specificity of this enzymatic test for monocytes.[137]

In the most immature leukemic monocytes as judged by size, cytoplasmic volume, and presence or absence of nucleoli, nonspecific esterase activity is weak. In more mature leukemic monocytes as identified by increased numbers of nuclear lobulations and the presence of granules, nonspecific esterase is more intense. Furthermore, nonspecific esterase activity appears to be more intense in the morphologic variant called *histiomonocytic* than in the myelomonocytic variant. In myelomonocytic leukemia, specific esterase, myeloperoxidase, and acid phosphatase activities are more intense than in leukemic monocytes from patients with histiomonocytic leukemia.[502,571] In a sense, the two variants differ partly in the proportions of nonspecific esterase, specific esterase, and myeloperoxidase activities, suggesting variability in differentiation along granulocytic lines, with preservation of both granulocytic and monocytic properties in both instances.

Recent studies using isoelectric focusing of nonspecific esterase in polyacrylamide gel have demonstrated differences in the isoenzymatic components of this enzyme in normal monocytes compared with leukemic monocytes.[513,514] Presumably, these isoenzymatic differences reflect fundamental biochemical distinctions between normal and leukemic monocytes, and suggest that elaboration of enzymatic components is related to cellular differentiation.[513,514]

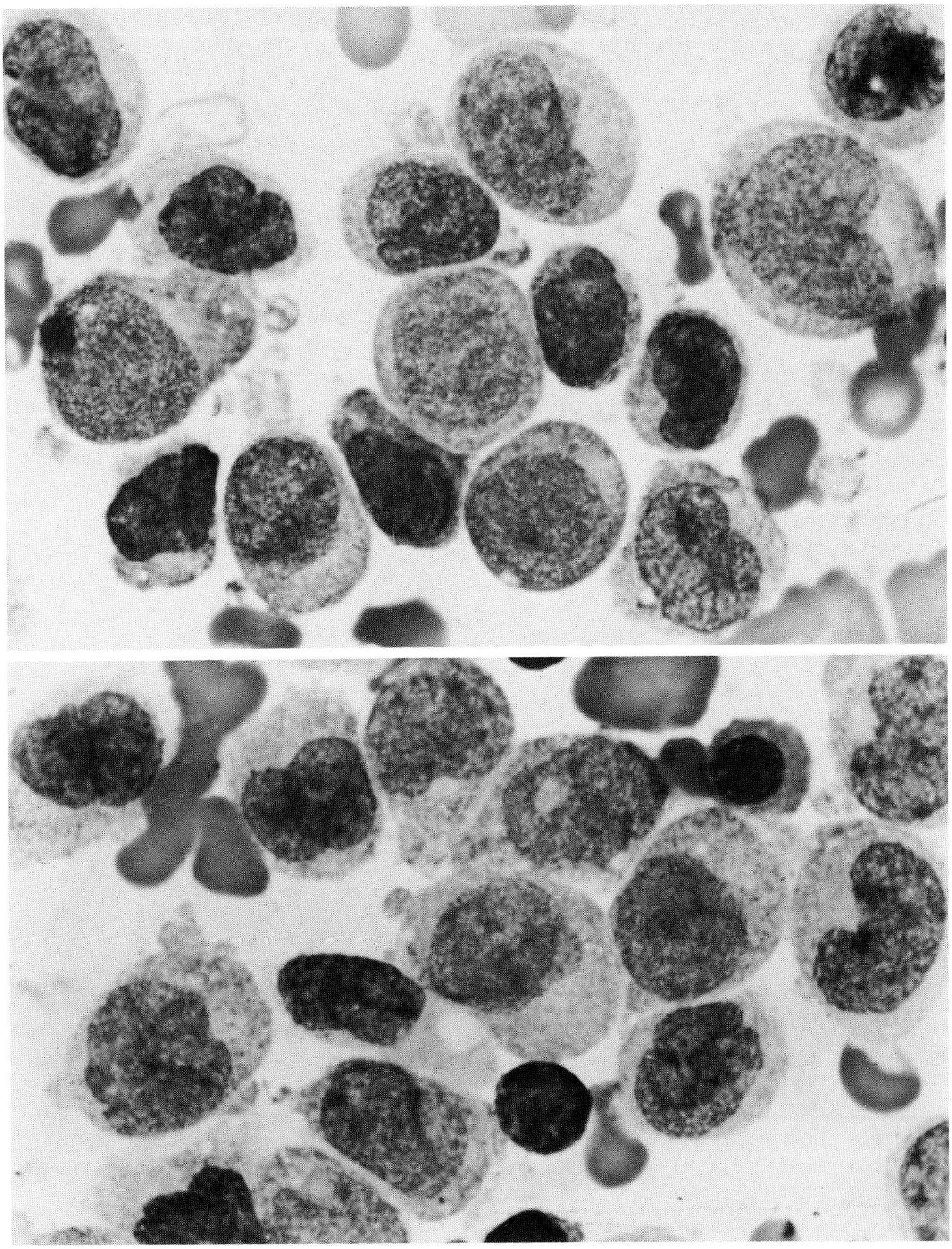

Leukemic monocytes in acute monocytic leukemia, histiomonocytic variant (FAB M5). **Figure 3-21.**

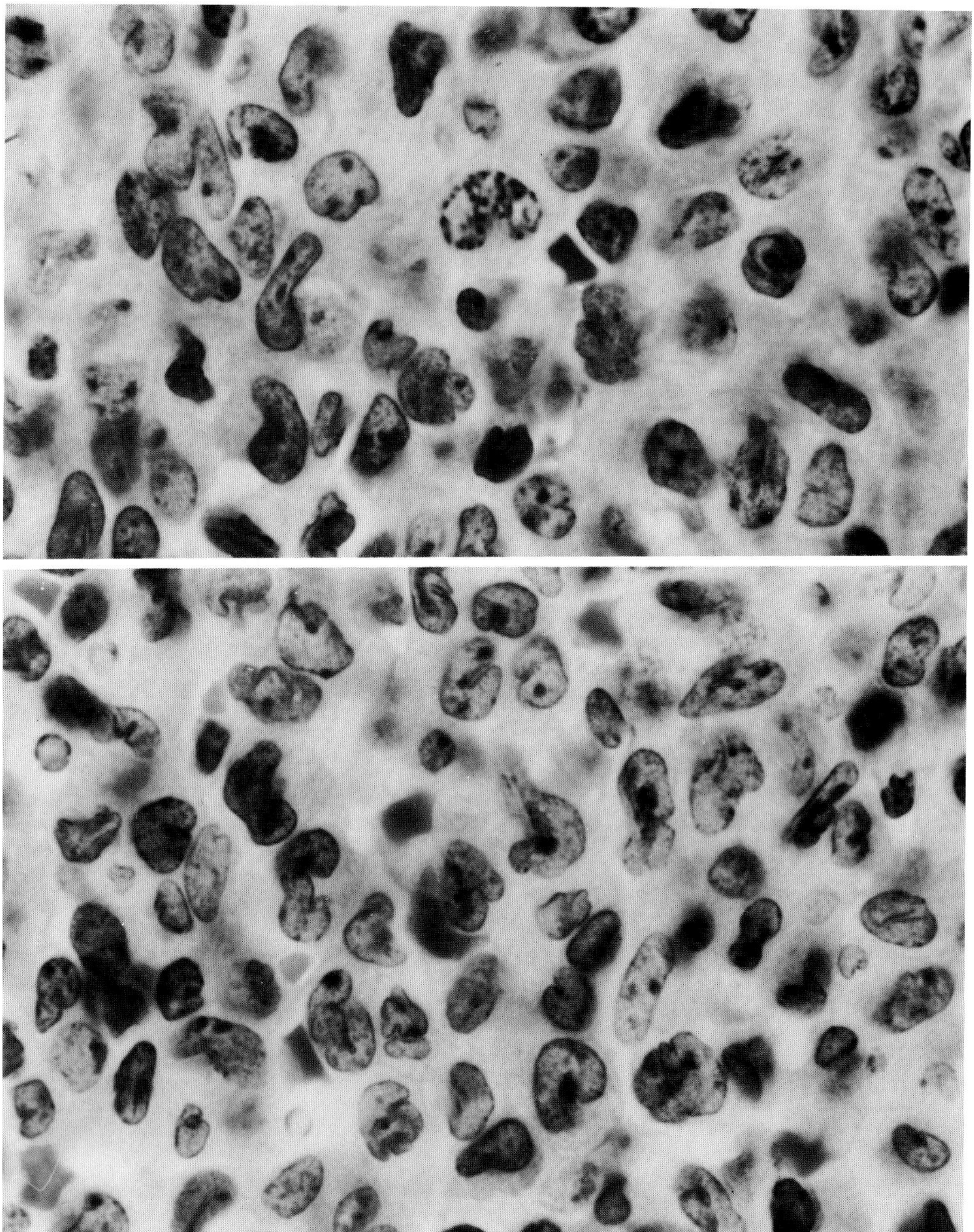

Figure 3-22. ← H & E section of bone marrow biopsy in acute monocytic leukemia. Many of the leukemic monocytes show pale vesicular nuclei and twisted, convoluted nuclei.

Figure 3-23. Strong activity of acid phosphatase in leukemic monocytes in acute monocytic leukemia.

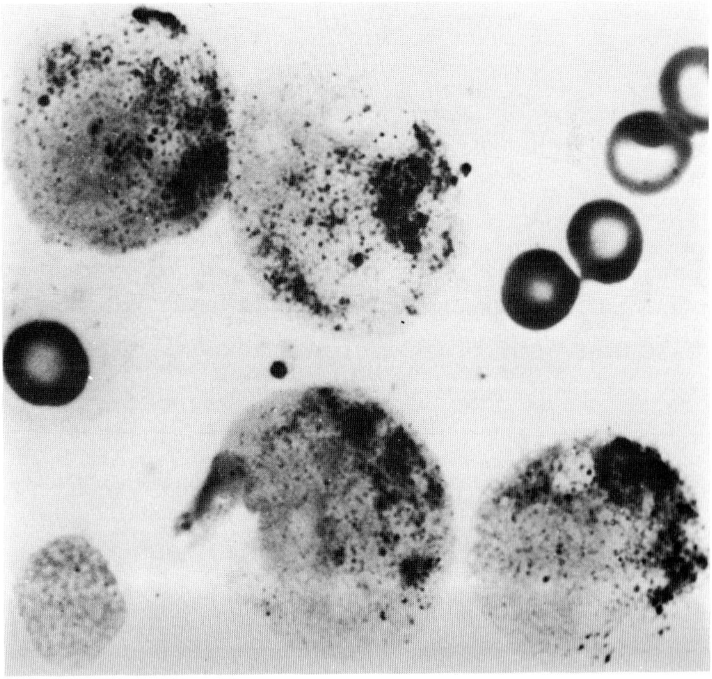

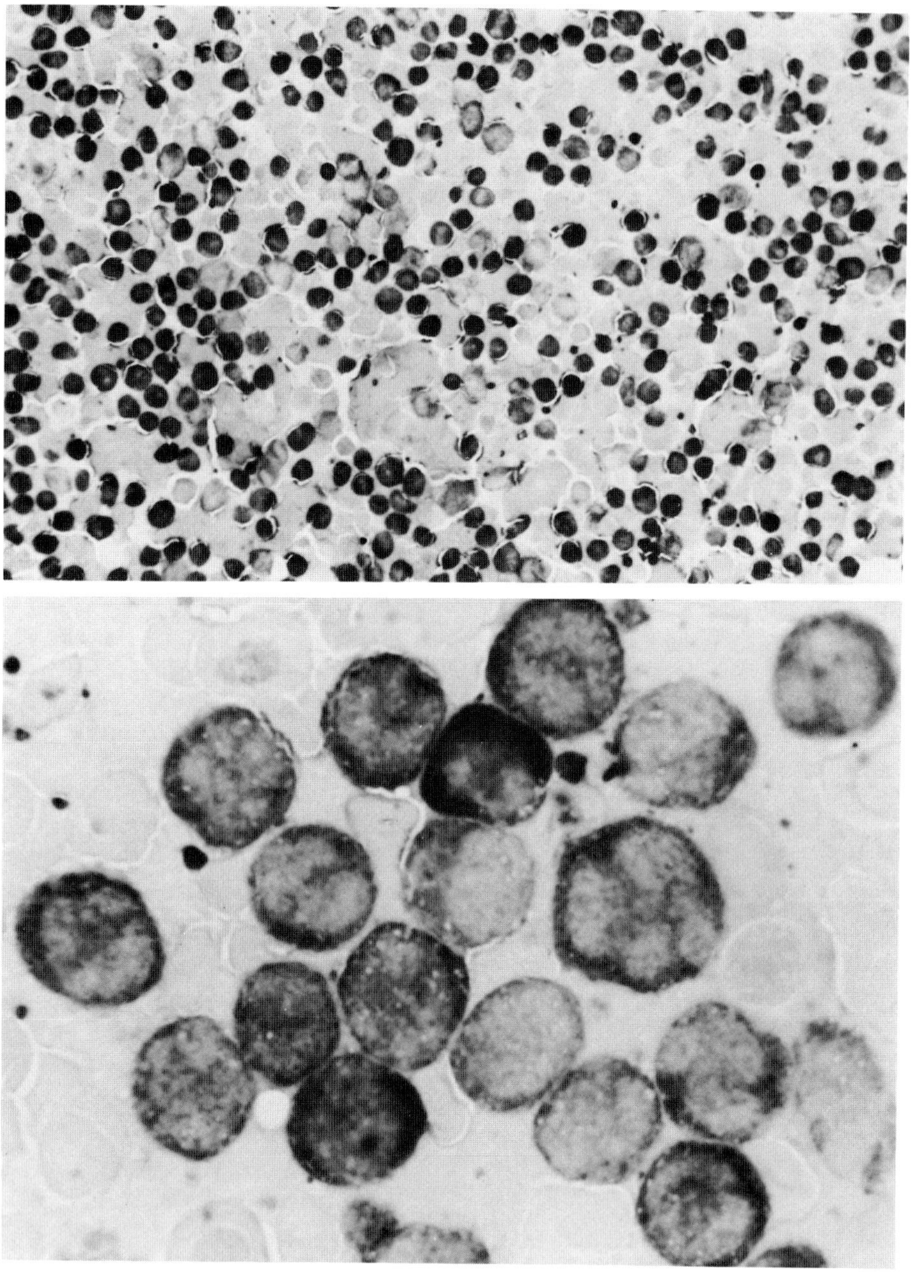

Figure 3-24. Nonspecific esterase reaction in acute monocytic leukemia, showing low-power view (**top**) and higher-power view (**bottom**). Alpha naphthyl acetate and hexazotized pararosaniline.

4

Lymphoid Leukemia

Acute lymphoblastic leukemia

Acute lymphoblastic leukemia is a neoplastic disorder characterized by proliferation of lymphoblasts in the blood, bone marrow, and visceral organs. Currently, acute lymphoblastic leukemia is believed to represent a clone of leukemic lymphoblasts engrafted upon the host rather than a generalized panmyelosis, as in acute myeloblastic or monocytic leukemia.

Generally regarded as a childhood disease, in recent years acute lymphoblastic leukemia has been identified with increasing frequency in young adults and older persons.[125,981] For reasons yet unknown, these cases in older persons respond less favorably to chemotherapy than do cases in children. Recently, acute lymphoblastic leukemia has been identified as a terminal event in some patients with chronic granulocytic leukemia,[123,124] suggesting that leukemic blasts in this disorder have the capacity to differentiate along multiple cell lines.[123,124] Similarly, acute lymphoblastic leukemia has evolved as a terminal event in patients with Hodgkin's disease[378] and Down's syndrome.[380]

Morphologic and cytochemical identification of leukemic lymphoblasts has been complemented by immunochemical identification of these cells.[24,72,73,76,77,168,200,235,413,430,441,452,490,536,544,585,636–645,653,658,664,675,677,703,709,719,721,782,805,806,809,831,832,835,951,985,1038,1042,1066,1071,1074,1079,1098,1102,1117] Thus far, T-cell, B-cell, and null-cell leukemic lymphoblasts have been identified with the use of immunologic techniques,[388,395] further improving diagnostic precision in evaluating these cells. Approximately 80% of cases of childhood acute lymphoblastic leukemia are of the null-cell type, 20% to 25% are T-cell type, and only 1% to 2% are B-cell type.[168,200,1042,1102]

In addition to the immunochemical distinctions, it has become apparent that immunologic subtypes of leukemic lymphoblasts are associated with distinctive pictures and responses to antileukemic therapy.[578] For example, in T-cell convoluted lymphoblastic leukemia, patients often have high white blood cell counts and high leukemic blast counts at the time of diagnosis. Furthermore, these patients may have a mediastinal mass due to enlarged mediastinal lymph nodes,[977] and a turbulent clinical course with poor response to currently available chemotherapy. Similarly, patients with the B-cell type of acute lymphoblastic leukemia respond poorly to chemotherapy now available. Although most cases of null-cell acute lymphoblastic leukemia respond favorably to chemotherapy with prolonged remissions, a substantial number do not respond to treatment. Recently, a "hand mirror" variant of leukemic lymphoblast has been described.[995] As yet, it is uncertain whether this "hand mirror" morphology is due in large part to contact with various substrates, or if it is due to an intrinsic or distinctive property of these lymphoblasts.

Peripheral blood samples from patients with acute lymphoblastic leukemia contain many lymphoblasts in some cases, and few if any in others. As in cases of acute myeloblastic leukemia that are thought to be aleukemic, buffy coat preparations from pancytopenic patients with acute lymphoblastic leukemia usually contain a few lymphoblasts. Some of the lymphoblasts contain cytoplasmic vacuoles. Cytoplasm of leukemic lymphoblasts is deeply basophilic and devoid of granularity.[724] Rarely, leukemic lymphoblasts contain angular or coccoid-shaped cytoplasmic inclusions. Nuclei of leukemic lymphoblasts contain chromatin strands that appear coarse and widely separated, in contrast to those found in leukemic myeloblasts (see Table 2-1 and Fig. 2-10). Aggregates of nuclear chromatin may be elongated and appear like stripes.[436] Nucleoli are usually small and may be multiple. In most instances, perinucleolar condensation of chromatin is present. Occasionally, increased numbers of mature-appearing lymphocytes are found. In some instances, these may represent small leukemic lymphoblasts rather than normal mature lymphocytes. Mature neutrophils may be hypersegmented. Occasionally, normoblasts are seen. Platelets are usually reduced in number.[792]

Wright-stained aspirates of bone marrow of patients with acute lymphoblastic leukemia are almost always hypercellular, as seen in Figures 4-1 through 4-4[29,147,326,340,360,386,451,465,551,577,579,636–645,763,838,840,969,970,1100] Megakaryocytes are greatly reduced in number, and those that are present appear normal. Erythroblasts are greatly reduced in number and show orderly maturation. In some early intermediate normoblasts, chromatin strands may be attenuated and aggregates of nuclear chromatin small and infrequent. These changes may be accentuated in some cells, and they may be frankly megaloblastic. Granulocytic cells are also reduced in number, and some neutrophils show hypersegmentation. In other marrows, numbers of mature-appearing plasma cells are increased (see Fig. 4-3). As yet, the pathogenetic implications and diagnostic significance of increased numbers of plasma cells in acute lymphoblastic leukemia are unknown.

According to the FAB classification (Table 4-1), there are three morphologic variants of acute lymphoblastic leukemia.[87] In marrows of patients who have the L1 variant of acute lymphoblastic leukemia,[976] nuclei of leukemic lymphoblasts contain coarse-appearing nuclear chromatin and many blocklike or stripe-shaped aggregates of nuclear chromatin, as shown in Figure 4-1 and Plate 4*A*, *top* (see Table 2-1).[91,108,109] Nucleoli are small and may be inconspicuous. In the L1 variant, lymphoblasts are generally of the same size and shape, and they are predominantly small lymphoblasts. Cytoplasm is deeply basophilic and does not contain granules.[30,59,82,83,87,88,757,838] In rare instances, some leukemic lymphoblasts contain azurophil inclusions that may represent gigantic mitochondria.

In the L2 variant of acute lymphoblastic leukemia, both large and small lymphoblasts are found, as shown in Figures 4-2, 4-3, and Plate 4*A*, *bottom*.[636–645] Nuclear chromatin appears coarser in small lymphoblasts than in large lymphoblasts, and aggregates of nuclear chromatin are larger and more numerous. Proportions of large and small lymphoblasts vary

(text continued on p. 194)

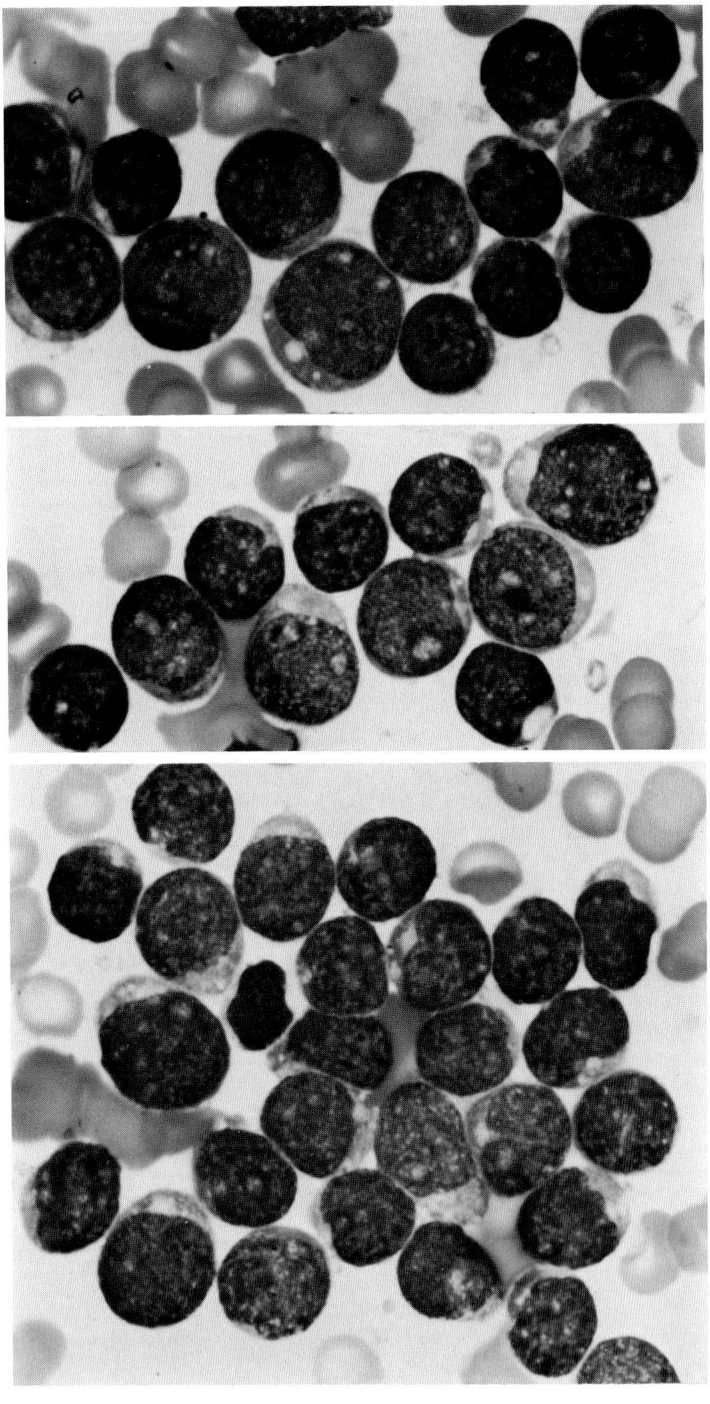

Figure 4-1. Lymphoblasts in acute lymphoblastic leukemia (L1 variant). Nucleoli are small, and there are many aggregates of nuclear chromatin.

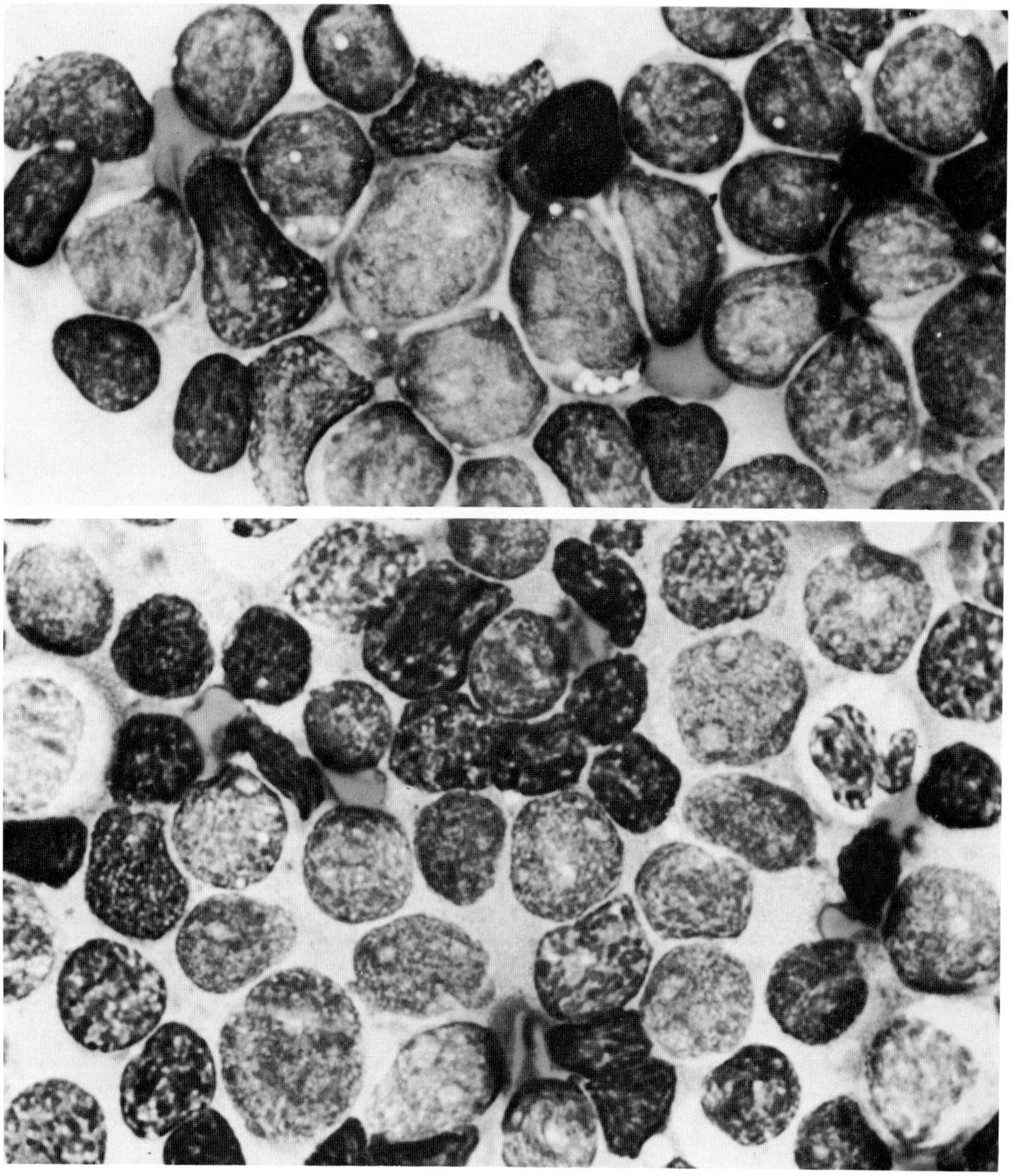

Figure 4-2. Lymphoblasts in acute lymphoblastic leukemia (L2).

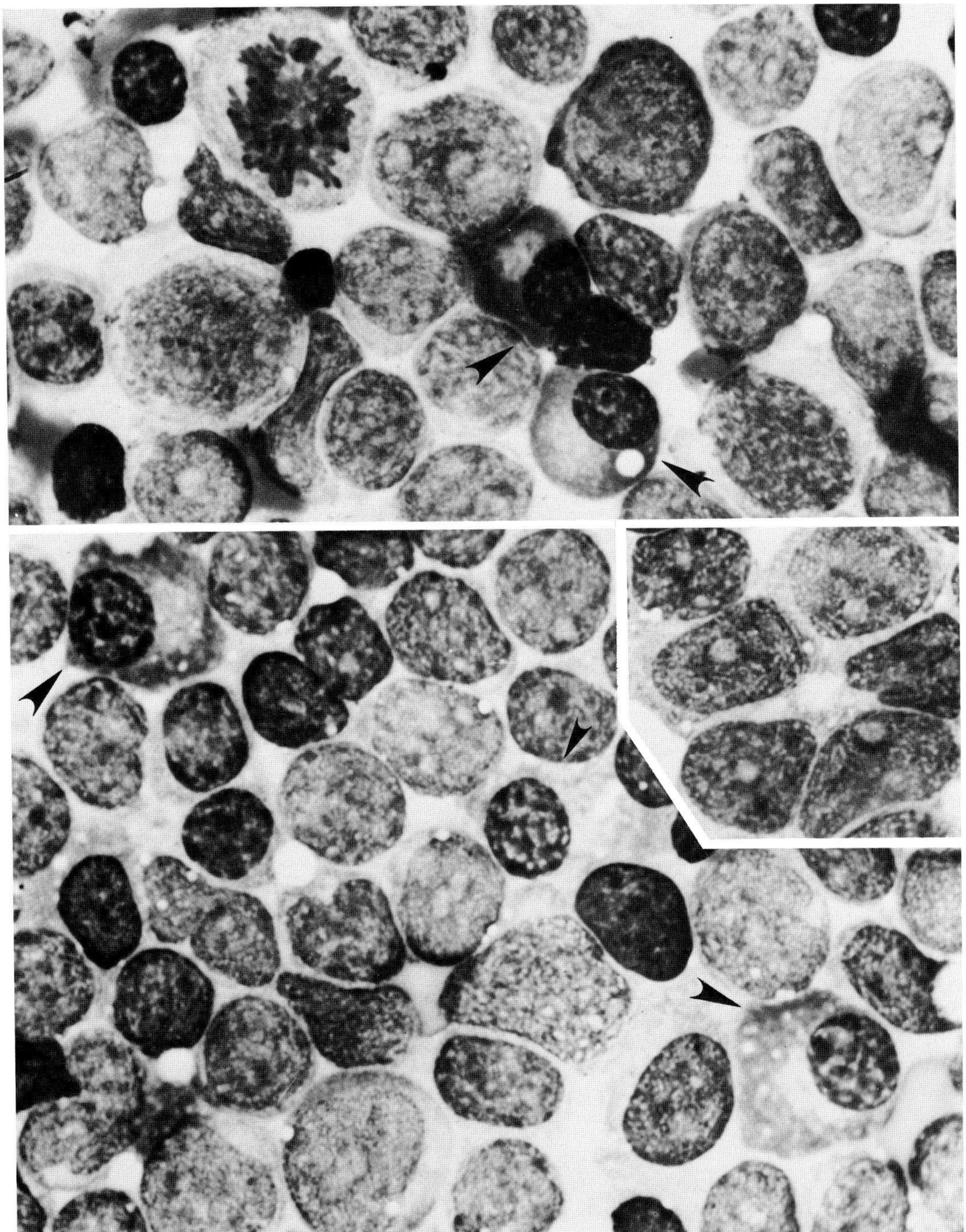

Figure 4-3. Lymphoblasts in acute lymphoblastic leukemia (L2). Increased numbers of plasma cells are seen (**arrows**). Leukemic lymphoblasts with prominent nucleoli and perinucleolar condensations of chromatin (**inset**).

from case to case. In the L2 variant, nucleoli are usually large and may be multiple. In most instances, perinucleolar condensations of chromatin are found (Figs. 4-2, 4-3). Nuclear clefting is seen often.

In bone marrows from patients with T-cell convoluted type of acute lymphoblastic leukemia, leukemic lymphoblasts have nuclei with coarse-appearing nuclear chromatin and numerous blocklike and stripe-shaped aggregates.[663,761] In many of these T-cell lymphoblasts, nuclear foldings and convolutions are present, as seen in Figure 4-4 and Plate 4*B*.[761] Nucleoli are usually small and inconspicuous and may be multiple. Perinucleolar condensations of nuclear chromatin are seen frequently. The L3 variant of acute lymphoblastic leukemia is the leukemic phase of Burkitt's lymphoma (*vide infra*).

Cytochemically, the PAS reagent has been of diagnostic value in distinguishing leukemic lymphoblasts from leukemic myeloblasts.[18,32,72,73,75–79,89,90,92,309,447,548,549,820,821,923,1032,1043,1110] This is especially so in the case of B-cell and null-cell acute lymphoblastic leukemia, and less so in the case of T-cell convoluted acute lymphoblastic leukemia. In B-cell and null-cell leukemic lymphoblasts, the PAS reagent often demonstrates glycogen in a blocklike pattern of coarse particles surrounding the nucleus (Fig. 4-5 and Plate 5C). This pattern is seen in both large and small leukemic lympho-

Table 4-1. FAB CLASSIFICATION OF ACUTE LYMPHOBLASTIC LEUKEMIA*

Cytologic Features	L1	L2	L3
Cell size	small cells predominate	large, heterogeneous in size	large and homogeneous
Nuclear chromatin	homogeneous in any one case	variable, heterogeneous in any one case	finely stippled and homogeneous
Nuclear shape	regular, occasional clefting or indentation	irregular, clefting and indentation common	oval to regular
Nucleoli	not visible, or small and inconspicuous	one or more present, often large	prominent, one or more vesicular
Amount of cytoplasm	scanty	variable, often moderately abundant	moderately abundant
Basophilia of cytoplasm	slight or moderate, rarely intense	variable, deep in some	very deep
Cytoplasmic vacuolation	variable	variable	often prominent

* Modified from Bennett JM, Catovsky D, Daniel MT et al: Proposals of classification of the acute leukaemias. Brit J Haematol 33:451, 1976

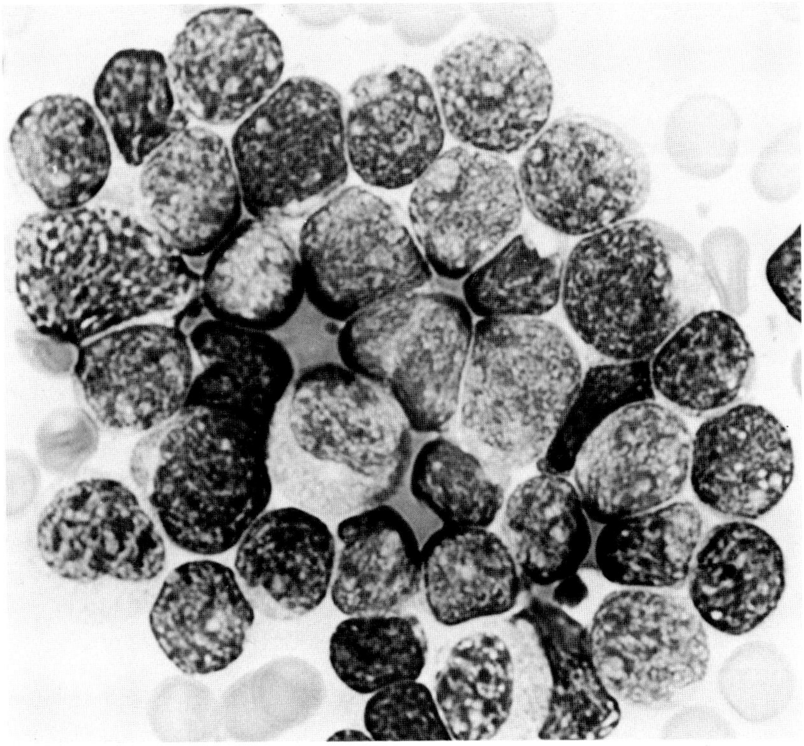

Figure 4-4. Lymphoblasts in T-cell lymphoblastic leukemia, showing irregular nuclear contours and prominent nucleoli.

Figure 4-5. Lymphoblasts in acute lymphoblastic leukemia, depicting prominent aggregates of glycogen in the cytoplasm demonstrable with the PAS reagent.

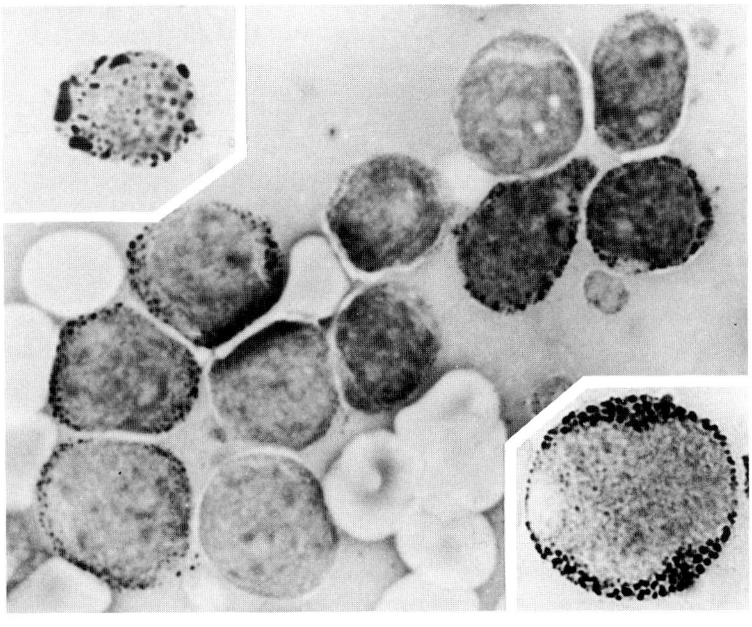

blasts. In some cases of T-cell leukemic lymphoblasts, glycogen may not be detectable with the PAS reagent, or it may be weak and diffuse rather than coarse and blocklike as in the B-cell and null-cell types.

Using stains that demonstrate granulocytic properties—such as myeloperoxidase, specific esterase, and Sudan black B—lymphoblasts from patients with acute lymphoblastic leukemia show a negative reaction.[92,305,408,412,802,803] Using the pyronine stain for RNA, leukemic lymphoblasts demonstrate a strongly positive reaction, indicating a high content of RNA. In most instances, this reaction is not as intense as that found in malignant lymphoid cells of Burkitt's lymphoma. Using the Prussian blue stain, ringed sideroblasts are usually not detectable. With further experience, demonstration of dihydrofolate reductase[1048] in leukemic lymphoblasts may provide insight into abnormalities of folate metabolism, as well as proving useful in monitoring results of chemotherapy.

In B-cell and null-cell leukemic lymphoblasts, activity of nonspecific esterase is usually weak and diffuse[189,976] and moderately resistant to inhibition by sodium fluoride. In T-cell convoluted lymphoblasts, activity of nonspecific esterase can be intense and localized in a unipolar distribution.[189] Similarly, activity of acid phosphatase is usually absent or weakly positive in B-cell and null-cell leukemic lymphoblasts.[188,189,192] In T-cell convoluted leukemic lymphoblasts, activity of acid phosphatase may be intense,[188,189,192,211,212,670] and, like nonspecific esterase activity, is localized in a unipolar area (Fig. 4-6 and Plate 6*E*). Activity of nuclear arylsulfatase is high in leukemic lymphoblasts, in contrast to leukemic myeloblasts, in which enzyme activity is weak or absent.[137]

By both direct enzyme assay[449,656] and immunofluorescent detection using fixed cells,[990] activity of terminal deoxynucleotidyl transferase is positive in all subtypes of leukemic lymphoblasts. Since activity of this enzyme is limited primarily to cells of lymphoid origin, including thymic lymphocytes and leukemic lymphoblasts,[449,656,657] its detection is valuable for the accurate designation of a leukemic blast as lymphoblastic or myeloblastic.[657,900] For the most part, leukemic lymphoblasts have high activity of terminal deoxynucleotidyl transferase, whereas most leukemic myeloblasts have weak or negligible activity.

Measurement of terminal deoxynucleotidyl transferase activity is sometimes useful in patients with chronic granulocytic leukemia who have entered a terminal blast phase.[656,657,900,990] In most cases, this blast phase is a myeloblastic or monocytic leukemia. In other instances, it is erythroleukemia, acute megakaryocytic myelosis, or promyelocytic leukemia. In still other cases, it is acute lymphoblastic leukemia, with the same cytochemical, immunologic, and enzymatic properties of lymphoblasts found in patients with acute lymphoblastic leukemia arising *de novo*. These properties include blocklike PAS positivity, immunologic specificity, and elevated levels of terminal deoxynucleotidyl transferase.

Recent reports indicate that the demonstration of terminal deoxynucleotidyl transferase activity may not be as specific for lymphoid cells as thought originally. As illustrated in Figure 4-7, approximately 15% of all cases of acute myeloblastic leukemia in which the diagnosis was estab-

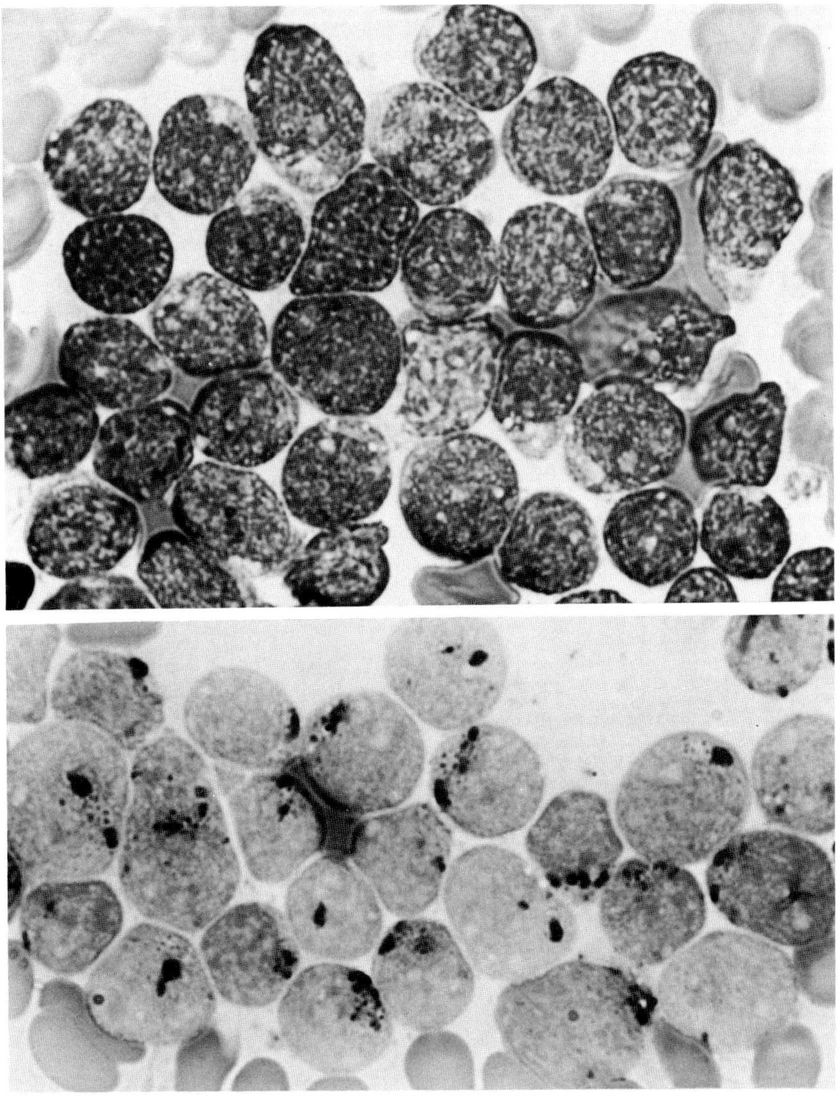

Figure 4-6. Lymphoblasts in T-cell convoluted type of acute lymphoblastic leukemia (**top**), showing unipolar localization of acid phosphatase (**bottom**).

lished by conventional morphologic and cytochemical criteria have high activities of terminal deoxynucleotidyl transferase.[987,988] Among these cases of acute myeloblastic leukemia that showed terminal transferase activity were cases that contained typical Auer rods, a reliable marker for cells of granulocytic origin. To complicate the picture further, F_c and C_3 receptors, once thought to be distinctive for lymphoid cells, are also detectable on the surface of granulocytic cells.[61] Studies of this type do not detract from the value of terminal transferase activity when combined with conventional morphologic and cytochemical criteria. They emphasize the protean nature of many cases of acute leukemia and suggest that in leukemic blasts from these atypical cases, properties of both lymphoblastic cells and myeloblastic cells can be demonstrated, indicating multiple pathways of cellular differentiation.[987,988]

H & E sections of biopsies of marrow from patients with acute lymphoblastic leukemia are almost always hypercellular (Fig. 4-8). Megakaryocytes are greatly reduced in number, as are erythroblasts and granulocytic cells. The marrow is composed largely of leukemic blasts with pale vesicular nuclei and scant cytoplasm. In some instances, the nuclei contain many aggregates of nuclear chromatin, some of which adhere to the nuclear membrane. However, it is not possible to reliably distinguish leukemic lymphoblasts from leukemic myeloblasts on the basis of H & E sections of bone marrow alone.

Abnormalities in the appearance of erythroblasts and granulocytic cells are found in marrows of patients with both acute myeloblastic leukemia and acute lymphoblastic leukemia. These abnormalities include nuclear hypersegmentation, gigantism of metamyelocytes, and chromatin aberrations of erythroblasts. Although the morphologic abnormalities sometimes appear similar in acute myeloblastic leukemia and acute lymphoblastic leukemia, it is believed that they have different mechanisms of pathogenesis. In acute myeloblastic leukemia and in preleukemic myeloblastic disorders, all cell lines are believed to be affected, and the morphologic abnormalities of these cells in the bone marrow are believed to be a part of a generalized panmyelotic disturbance. Probably, all morphologic cell types in myeloblastic leukemia and its variants are derived from a common myeloblastic stem cell, as suggested by cytogenetic abnormalities common to both erythroblasts and myeloblasts.

In acute lymphoblastic leukemia, the leukemic cells are believed to represent an abnormal clone that has been engrafted onto the bone marrow, displacing normal cells. When megaloblastic changes in erythroblasts and hypersegmentation of neutrophils are found in acute lymphoblastic leukemia, they are believed to represent morphologic expressions of folate deficiency due to increased utilization of the vitamin by leukemic lymphoblasts, rather than being viewed as part of a panmyelotic disturbance as in acute myeloblastic leukemia. Accordingly, in acute lymphoblastic leukemia, those erythroblasts and granulocytic cells that are found in the bone marrow are thought to be normal erythroblasts and granulocytic cells that have remained in the marrow.

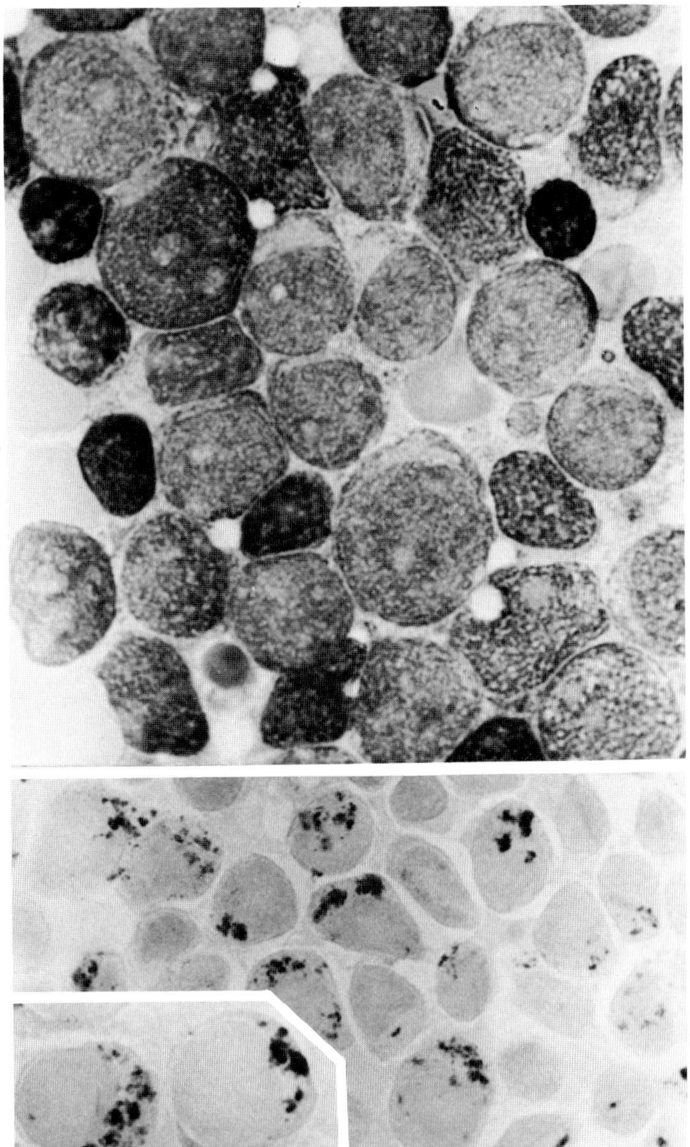

Figure 4-7. Bone marrow in atypical acute myeloblastic leukemia, with levels of terminal deoxynucleotidyl transferase as high as that found in typical acute lymphoblastic leukemia (25 $\mu/10^8$ cells). Wright-stained film of bone marrow illustrates pleomorphic population of leukemic blasts, some of which have unusually large nucleoli with perinucleolar condensations of chromatin (**top**). There is substantial activity of myeloperoxidase in most of the leukemic blasts (**bottom, inset**). Unusual cases of this type suggest that derepression of the genome for terminal deoxynucleotidyl transferase can occur in these cells, leading to unusually high activity of the enzyme. Such cases cast doubt as to the specificity for lymphoid cells claimed for terminal deoxynucleotidyl transferase.

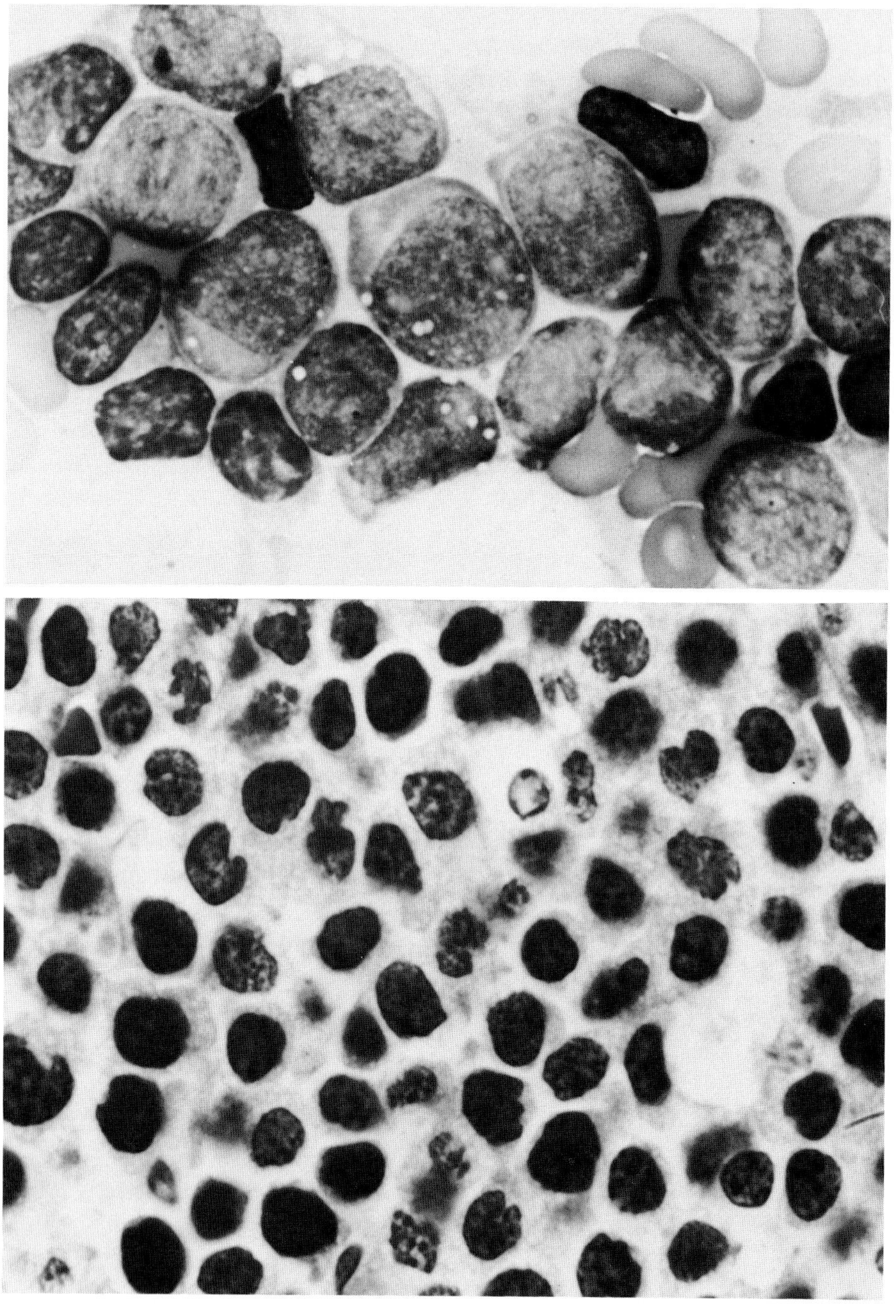

Figure 4-8. Lymphoblasts in acute lymphoblastic leukemia, showing prominent blocklike and stripe-shaped aggregates of chromatin (**top**). On H & E section of bone marrow biopsy (**bottom**), lymphoblasts have irregular nuclear outlines and prominent aggregates of nuclear chromatin.

Chronic lymphocytic leukemia

Chronic lymphocytic leukemia is a disorder of neoplastic lymphocytes that proliferate in the blood, the bone marrow, and the visceral organs. Clinically, the disease is characterized by varying degrees of lymphadenopathy, hepatomegaly, and splenomegaly. Hematologically, the disorder is typified by absolute lymphocytosis of the blood and bone marrow and, in some instances, by anemia, thrombocytopenia, or dysproteinemia. Currently, most cases of chronic lymphocytic leukemia are regarded as disorders of B lymphocytes and as the leukemic phase of diffuse lymphocytic lymphoma.[75,206,586,618,620,646,647,659,759,760,837,847,864,911,931,976,1019,1057]

Morphologically, chronic lymphocytic leukemia includes a spectrum of variants. From the clinical standpoint, it is important to identify these variants at the time of diagnosis and throughout the patient's clinical course because the tempo of the clinical course often seems to be related to the degree of differentiation of the lymphocyte. In some instances, other hematologic disorders or manifestations are seen in association with chronic lymphocytic leukemia. These include autoimmune hemolytic anemia, autoimmune thrombocytopenia, hypergammaglobulinemia, and hypogammaglobulinemia. Sometimes, acute lymphoblastic leukemia occurs as part of the evolutionary pattern in patients with chronic lymphocytic leukemia. Rarely, cases of chronic lymphocytic leukemia evolve into acute myeloblastic leukemia as a terminal event. In still other cases, a chronic lymphocytic leukemia-like picture has been described in association with a dysproteinemia, such as IgD myeloma.

Several different types of morphologic abnormalities are seen in lymphocytes from patients with chronic lymphocytic leukemia.[370,371,595,619,696,971,1114] Although one morphologic type often predominates in a single patient, in many patients a wide variety of different morphologic abnormalities can be found in a single peripheral blood specimen. Interestingly, most of the morphologic abnormalities of lymphocytes in chronic lymphocytic leukemia are abnormalities of nuclear size, shape, and configuration. These morphologic abnormalities are of several types. First, some leukemic lymphocytes have a delicate nuclear chromatin pattern and seem to lack the prominent aggregates found in normal lymphocytes (Fig. 4-9). Second, other leukemic lymphocytes may have a prominent nucleolus with perinucleolar condensations of chromatin, as seen in Figure 4-10.[484] Third, as shown in Figure 4-11, additional lymphocytes reveal abundant cytoplasm with radial striations of basophilic staining material. These lymphocytes resemble the atypical lymphocytes found in infectious mononucleosis and other disorders.[114,268,333] Fourth, leukemic lymphocytes may show unusually prominent blocklike aggregates of nuclear chromatin (Fig. 4-12). Fifth, some leukemic lymphocytes have a deep cleft extending at least halfway through the nucleus (Fig. 4-13). Other leukemic lymphocytes are binuclear (Fig. 4-13). Sixth, occasional leukemic lympho-

(text continued on p. 204)

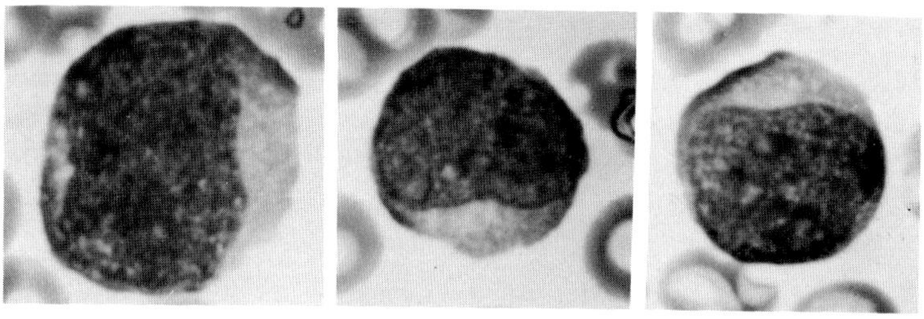

Figure 4-9. Leukemic lymphocytes, showing delicate nuclear chromatin and few, small aggregates.

Figure 4-10. Leukemic lymphocyte with prominent nucleolus, perinucleolar condensation of chromatin, and multiple, elongated aggregates of chromatin.

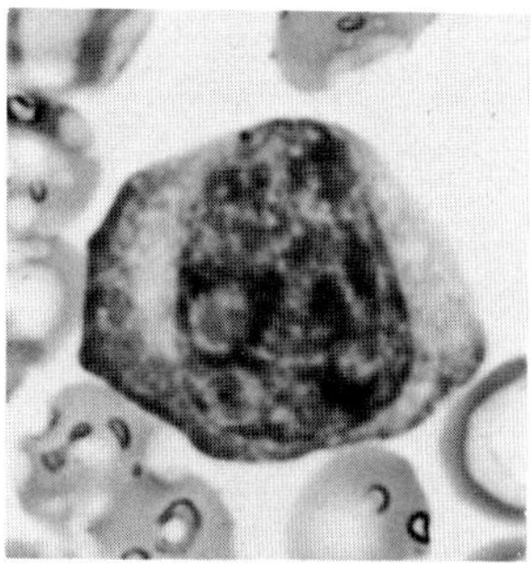

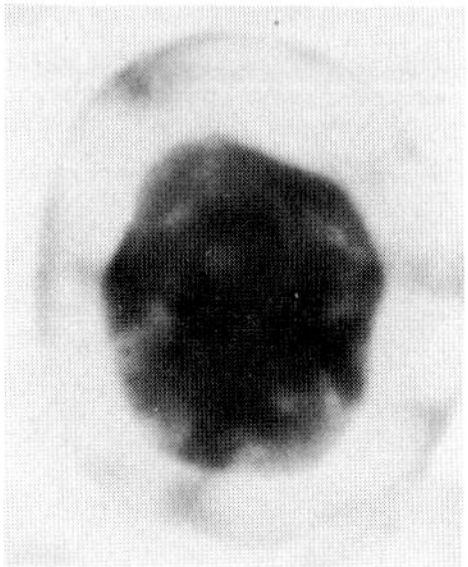

Figure 4-11. Leukemic lymphocyte, showing streaks of basophilic cytoplasm radiating from the periphery to the nucleus, resembling that seen in atypical lymphocytes in infectious mononucleosis or other viral illnesses.

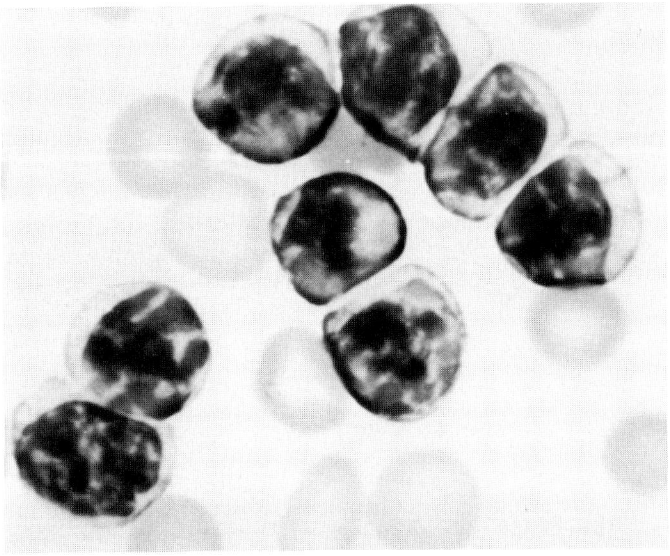

Figure 4-12. Leukemic lymphocytes, demonstrating large, prominent aggregates of nuclear chromatin.

Figure 4-13. Leukemic lymphoytes, depicting nuclear clefting and binuclearity of lymphocytes.

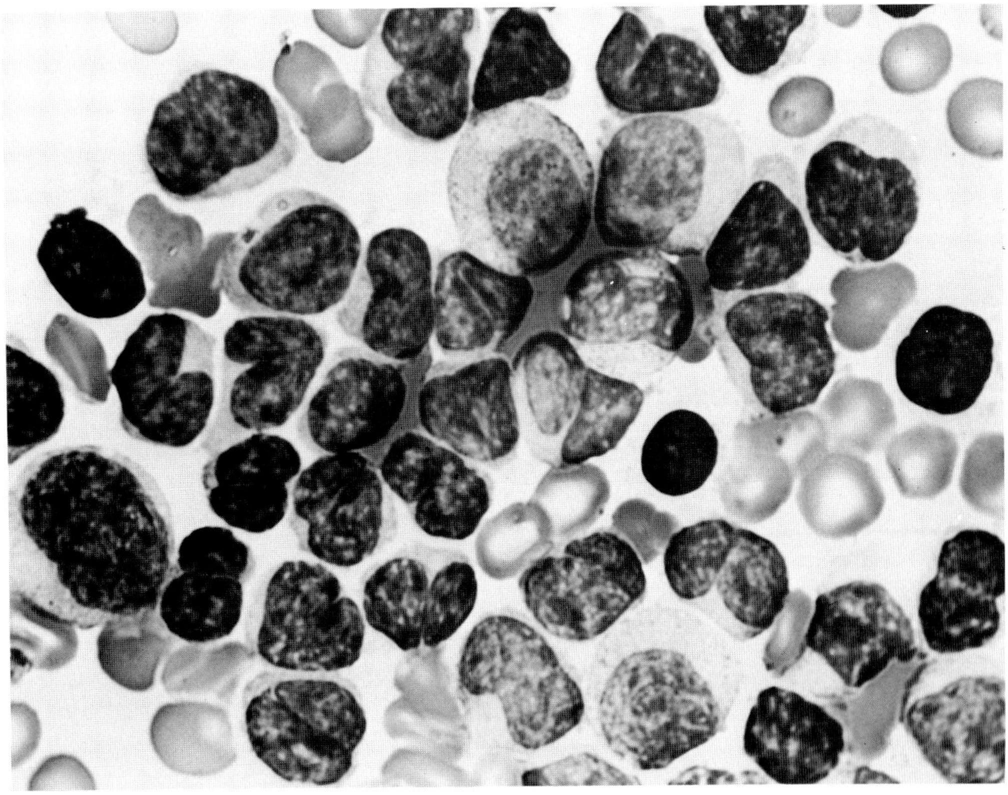

cytes exhibit a polar body, resulting from an asynchronous nuclear division (Fig. 4-14).

Seventh, rare leukemic lymphocytes are of the "lymphosarcoma cell" type, in which the nucleus has coarse nuclear chromatin strands, a large prominent nucleolus with perinucleolar condensations of chromatin, and abundant deeply basophilic cytoplasm containing many vacuoles of varying sizes (Fig. 4-15). Cells of this type have also been called *prolymphocytes*,[67,113,345,453,530,851] and are found in what has been called *prolymphocytic leukemia*.[67,453,851] Eighth, some poorly differentiated lymphocytes are unusually small, with dense nuclear chromatin and multiple clefts in the nucleus (Fig. 4-16). These clefts extend for various distances into the nucleus, imparting a mulberry or morula appearance to it. Cytoplasm of these poorly differentiated lymphocytes is scant. Lymphocytes with these abnormalities have been described in the aggressive, rare T-cell variant of chronic lymphocytic leukemia.[1049]

Clinically, patients who have higher proportions of cells with morphologic aberrations often have more aggressive clinical courses than most patients who have chronic lymphocytic leukemia with normal-appearing lymphocytes.[535,791] Clinical aggressiveness is expressed as decreased survival, increased tumefaction seen in marked lymphadenopathy and splenomegaly, and increased incidence of other hematologic complications, such as autoimmune thrombocytopenia and autoimmune hemolytic anemia.

Figure 4-14. Polar body in leukemic lymphocyte (**arrow**).

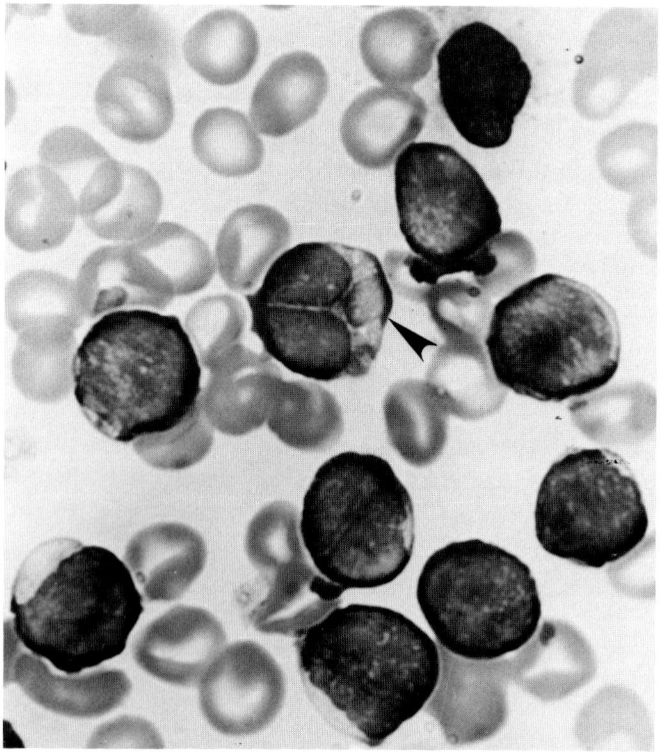

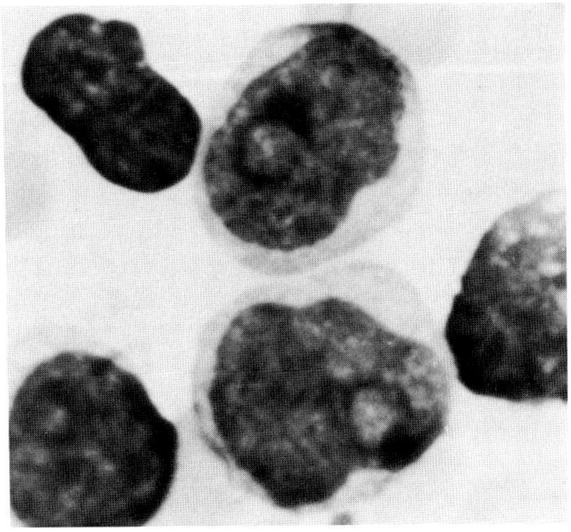

Figure 4-15. Lymphosarcoma cells, or prolymphocytes, illustrating large nucleolus, perinucleolar condensation of chromatin, and coarse chromatin network.

Figure 4-16. Leukemic lymphocytes, demonstrating multiple nuclear clefts and scant cytoplasm.

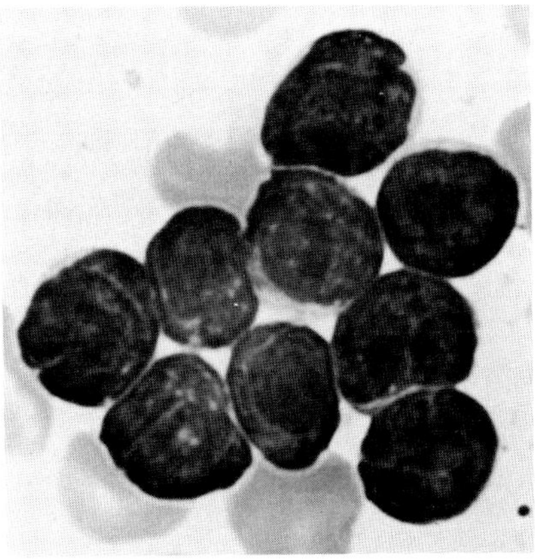

Chronic lymphocytic leukemia, well-differentiated type

In this variant, peripheral blood lymphocytes appear normal.[370,371,595,653] Granulocytes are usually markedly reduced in number, and there may be anemia and thrombocytopenia in severe cases. In other cases, anemia may be absent or minimal, and thrombocytopenia is variable.

In aspirates of bone marrow from patients with the well-differentiated variety of chronic lymphocytic leukemia, specimens are hypercellular. Megakaryocytes are markedly reduced in number; in uncomplicated cases, erythroblasts and granulocytic cells are also reduced. Usually, marrows are replaced by intermediate-sized lymphocytes that appear relatively normal (Fig. 4-17 and Plate 4C). In some instances, aggregates of nuclear chromatin appear more prominent and larger than in normal lymphocytes. Cytoplasm is scant, and cytoplasmic shedding occurs rarely.

In H & E sections of biopsies of marrow from patients with the well-differentiated type of chronic lymphocytic leukemia, hypercellularity of the marrow occurs (Fig. 4-17). Megakaryocytes, erythroblasts, and granulocytic cells are markedly reduced in number. Most of the cells are lymphocytes with round to oval nuclei, little if any irregularity of the contour of nuclear membrane, and prominent aggregates of nuclear chromatin dispersed throughout the nucleus and adherent to the nuclear membrane. Cytoplasm is scant.

Chronic lymphocytic leukemia, moderately poorly differentiated type

Peripheral blood lymphocytes in this morphologic variant show pleomorphism. Many exhibit clefting of the nucleus and binuclearity. Some of the cells are prolymphocytes with deeply basophilic cytoplasm and a nucleus with coarse nuclear chromatin and a prominent nucleolus. Variation occurs in size and shape of erythrocytes, thrombocytopenia, and neutropenia.

Aspirates of marrow from patients with the moderately poorly differentiated type of chronic lymphocytic leukemia are hypercellular (Fig. 4-18). Megakaryocytes are usually reduced in number, and most of them appear normal. Erythroblasts and granulocytic cells are decreased. Most of the cells in the marrow are lymphocytes having nuclear abnormalities (Fig. 4-18 and Plate 4*D*, *top*). These abnormalities include clefting, binuclearity, and occasional polar bodies. Prolymphocytes may also be present with coarse nuclear chromatin, a prominent nucleus with perinucleolar condensations of chromatin, and deeply basophilic cytoplasm containing many vacuoles.

On H & E sections of biopsies of marrows from patients with the moderately poorly differentiated type of chronic lymphocytic leukemia, specimens are hypercellular (Fig. 4-18). Megakaryocytes are markedly reduced in number, as are erythroblasts and granulocytic cells. Most of the marrow cells are intermediate-sized lymphocytes. These cells have nuclei that demonstrate irregular contours with wrinkling and indentations. Cytoplasm is scant.

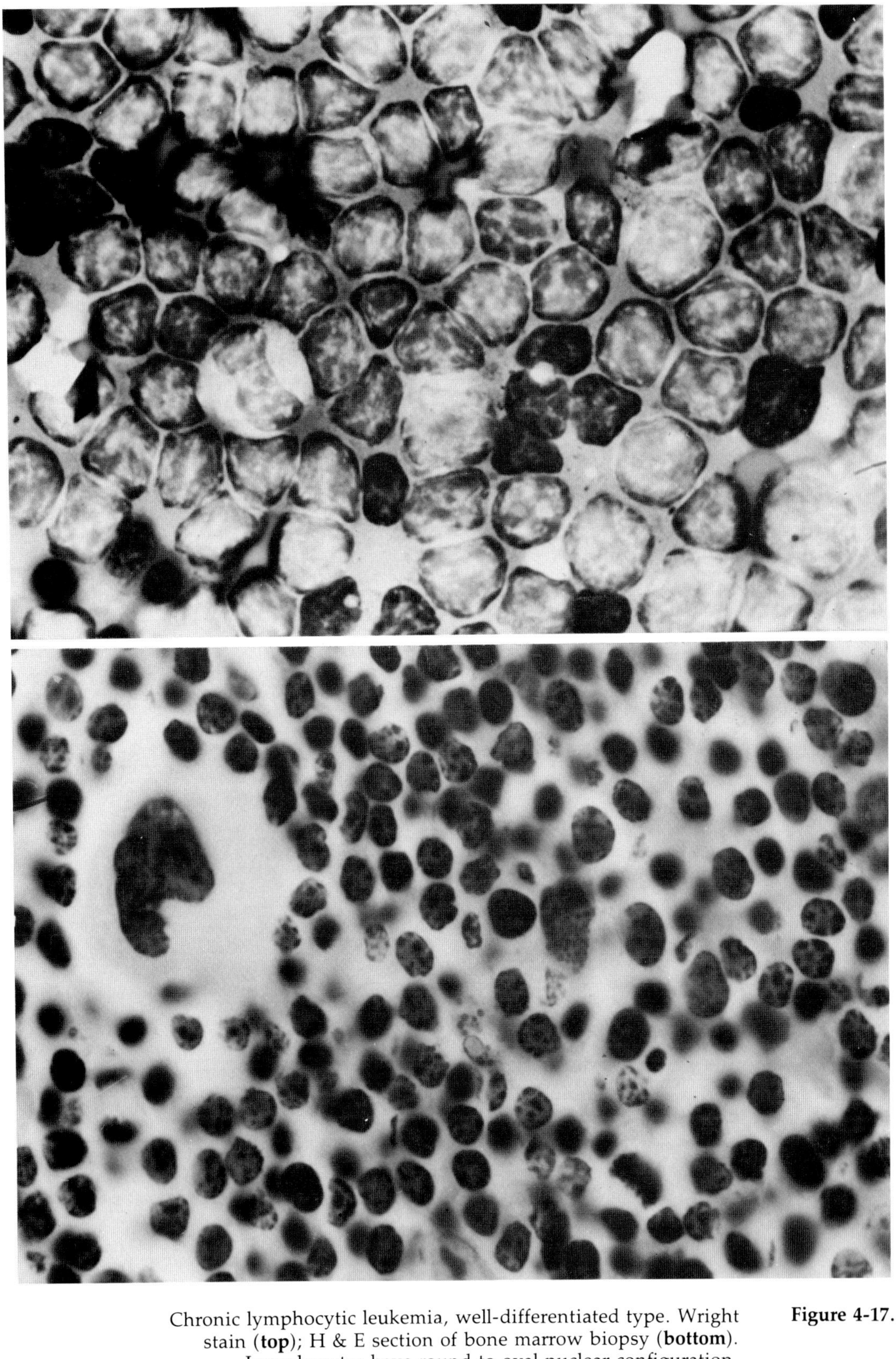

Chronic lymphocytic leukemia, well-differentiated type. Wright stain (**top**); H & E section of bone marrow biopsy (**bottom**). Lymphocytes have round to oval nuclear configuration.

Figure 4-17.

Chronic lymphocytic leukemia, poorly differentiated type

Clinically, this morphologic variant is associated with an unusually aggressive clinical course, marked tumefaction, and anemia and thrombocytopenia. In the past, such cases were identified as leukemic lymphosarcoma or the leukemic phase of lymphosarcoma or poorly differentiated lymphocytic lymphoma. Peripheral blood specimens from these patients contain many lymphocytes that have multiple nuclear clefts and infoldings. Nuclear chromatin appears dense with many aggregates, and areas of heterochromatin stain darkly with conventional panoptic stains. Cytoplasm of these cells is scant.

Megakaryocytes, erythroblasts, and granulocytic cells are markedly reduced in number in aspirates of marrow obtained from patients with the poorly differentiated type of chronic lymphocytic leukemia. Most of the marrow cells are poorly differentiated lymphocytes showing dense nuclear chromatin, many aggregates of chromatin, and multiple clefts and indentations (Fig. 4-19 and Plate 4*D*, *bottom*). Cytoplasm in these lymphocytes is scant.

On H & E sections of biopsies of bone marrow obtained from patients with this disorder, specimens are hypercellular. Megakaryocytes, erythroblasts, and granulocytic cells are reduced markedly in number. Most of the marrow cells are small lymphocytes with marked wrinkling and indentations of the nuclear membrane. Cytoplasm is scant.

Any of the three morphologic variants of chronic lymphocytic leukemia can be associated with other hematologic disorders. Among these are autoimmune hemolytic anemia, dysproteinemia, eosinophilia, and, as a terminal event, acute lymphoblastic or acute myeloblastic leukemia. In patients with chronic lymphocytic leukemia and coexistent autoimmune hemolytic anemia, peripheral blood samples demonstrate lymphocytosis along with increased numbers of reticulocytes, normoblasts, and variations in size and shape of erythrocytes. Bone marrow aspirates from these patients show increased numbers of lymphocytes and normoblasts (Figs. 4-20, 4-21). Most of these normoblasts exhibit the nuclear chromatin pattern found in normal erythroblasts. In patients who exhibit unusually aggressive hemolytic disease, folate utilization can be increased and megaloblastic changes can be detected in nuclei of some of the erythroblasts. Most of the erythroblasts are at the intermediate normoblast stage of development and show beginnings of hemoglobinization manifested by polychromasia. In some patients, proerythroblasts are increased.

On biopsies of bone marrow from these patients, H & E sections show increased numbers of lymphocytes and erythroblasts (Figs. 4-20, 4-21). *Lymphocytes* are identified as cells with round to oval nuclei. Occasionally, some lymphocytes show indentations of the nucleus (Fig. 4-21). Nuclear chromatin appears as large aggregates and may be adherent to the nuclear membrane. *Normoblasts* are identified as cells with round to oval nuclei that contain prominent aggregates of nuclear chromatin in a checkerboard pattern and stain darker than nuclei of lymphocytes.

(*text continued on p. 212*)

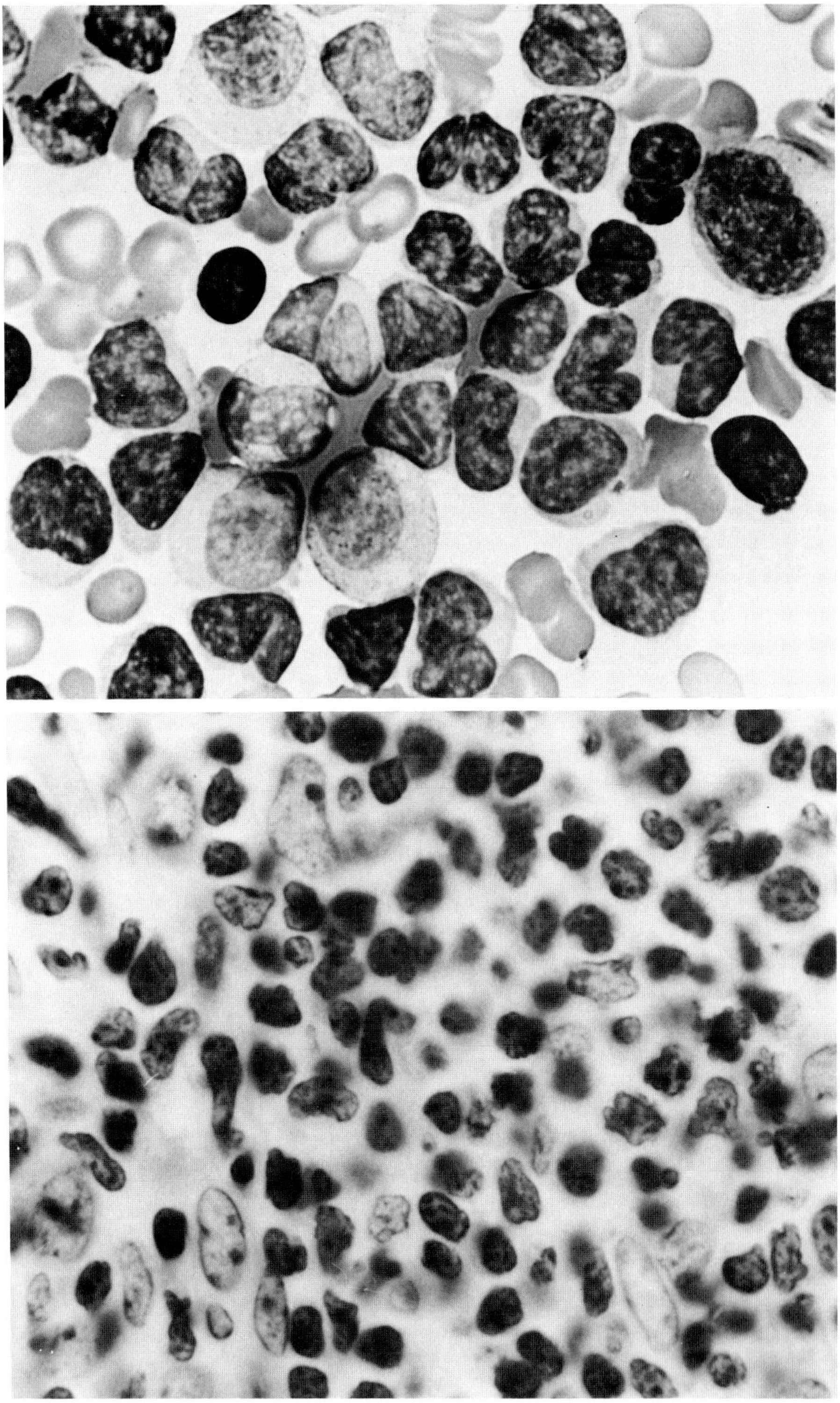

Figure 4-18. Chronic lymphocytic leukemia, moderately poorly-differentiated variety, showing nuclear clefting and binuclearity (**top**). On H & E section of bone marrow biopsy, cells demonstrate irregular nuclear configurations (**bottom**).

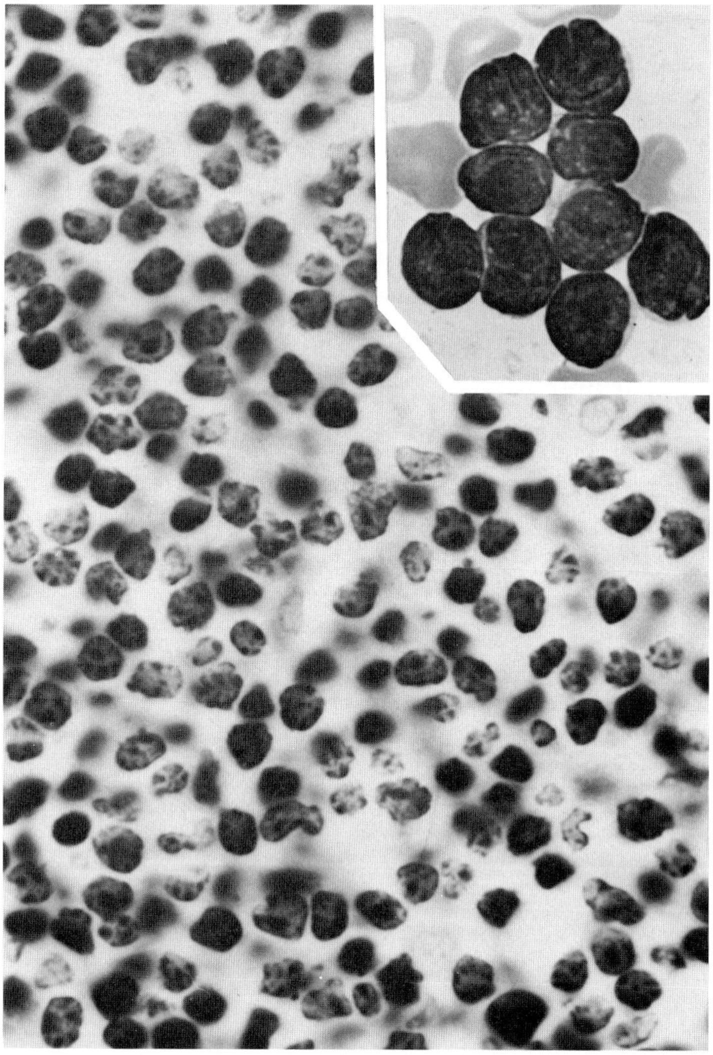

Figure 4-19. Chronic lymphocytic leukemia, poorly-differentiated type, showing small cells with multiple nuclear clefts (**inset**). H & E section of bone marrow biopsy illustrates cells with irregular nuclear configurations.

Figure 4-20. Chronic lymphocytic leukemia with autoimmune hemolytic anemia. Wright-stained aspirate (**top**) shows increased numbers of lymphocytes and normoblasts. On H & E section of bone marrow biopsy (**bottom**), lymphocytes and cells with darkly staining nuclei (erythroblasts) are seen. →

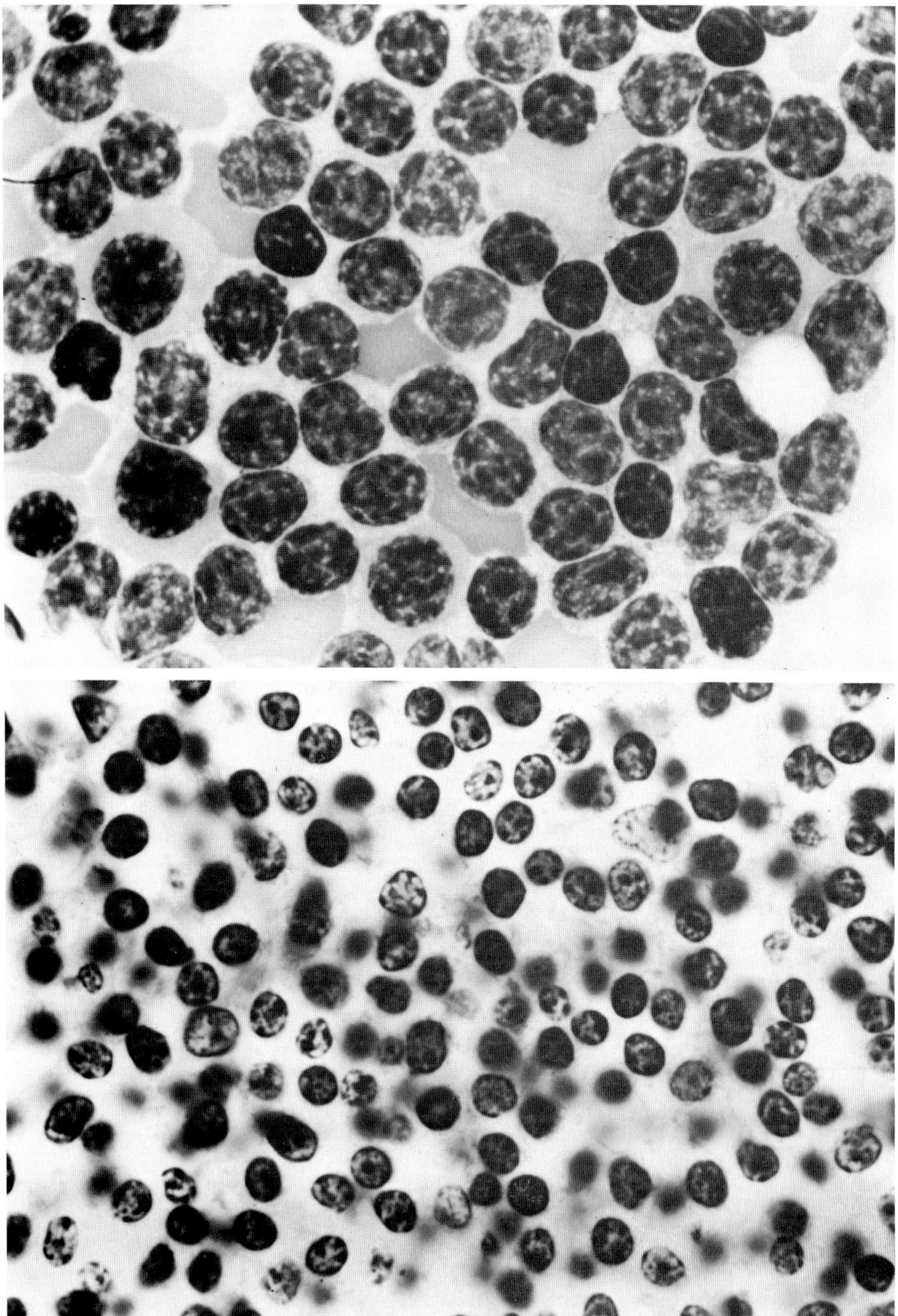

In patients who have chronic lymphocytic leukemia and coexistent dysproteinemia, peripheral blood samples demonstrate lymphocytosis and often rouleau formation of erythrocytes. Monocytes may be slightly increased in number, and platelets are normal or reduced. Aspirates of bone marrow from patients with chronic lymphocytic leukemia and dysproteinemia show increased numbers of lymphocytes and increased numbers (up to 20%–30%) of mature-appearing plasma cells (Fig. 4-22). In some instances, the coexistence of lymphocytes and plasma cells resembles that seen in Waldenström's macroglobulinemia. Some plasma cells may contain vacuoles or Russell's bodies. On H & E sections of bone marrow biopsies from patients with this disorder, increased numbers of lymphocytes and plasma cells are found. Megakaryocytes, erythroblasts, and granulocytic cells are reduced in number.

In patients who have chronic lymphocytic leukemia and coexistent eosinophilia, increased numbers of lymphocytes and increased numbers of eosinophils are seen in the peripheral blood. On aspirates of bone marrow from these patients, increased numbers of eosinophils and lymphocytes are found (Fig. 4-23). Most of the eosinophils are mature cells. On H & E sections of bone marrow biopsies from these patients, increased numbers of lymphocytes are identified by their round to oval nuclei and darkly stained nuclear chromatin. Eosinophils are identified easily by large, orange-staining cytoplasmic granules. Eosinophils are markedly increased in number (Fig. 4-24).

Rarely, an acute leukemic phase develops as a terminal event in patients with chronic lymphocytic leukemia. Although this is usually lymphoblastic leukemia, in some cases it is myeloblastic leukemia (Fig. 4-25). Peripheral blood films show a mixture of lymphocytes and myeloblasts. Similarly, aspirates of bone marrow from these patients show increased numbers of lymphocytes and cells with large, pale vesicular nuclei and prominent nucleoli, probably representing myeloblasts (Fig. 4-25). Megakaryocytes, erythroblasts, and granulocytic cells are reduced in number.

Cytochemically, lymphocytes from patients with chronic lymphocytic leukemia often show increased amounts of glycogen when stained with the PAS reagent, as seen in Figure 4-26.[688,824] Compared to normal lymphocytes,[761] amounts of glycogen and size of aggregates of glycogen are increased in patients with poorly differentiated types of lymphocytes. Similarly, activities of beta-glucoronidase and acid phosphatase are decreased in these lymphocytes,[264] in contrast to normal lymphocytes where these activities are higher.[402,1027] In lymphocytes from patients with chronic lymphocytic leukemia, activities of myeloperoxidase and specific esterase are absent and activity of nonspecific esterase is weak or absent.[20,264] Pyronine stain for RNA is usually weakly positive in lymphocytes from patients with chronic lymphocytic leukemia.

(text continued on p. 219)

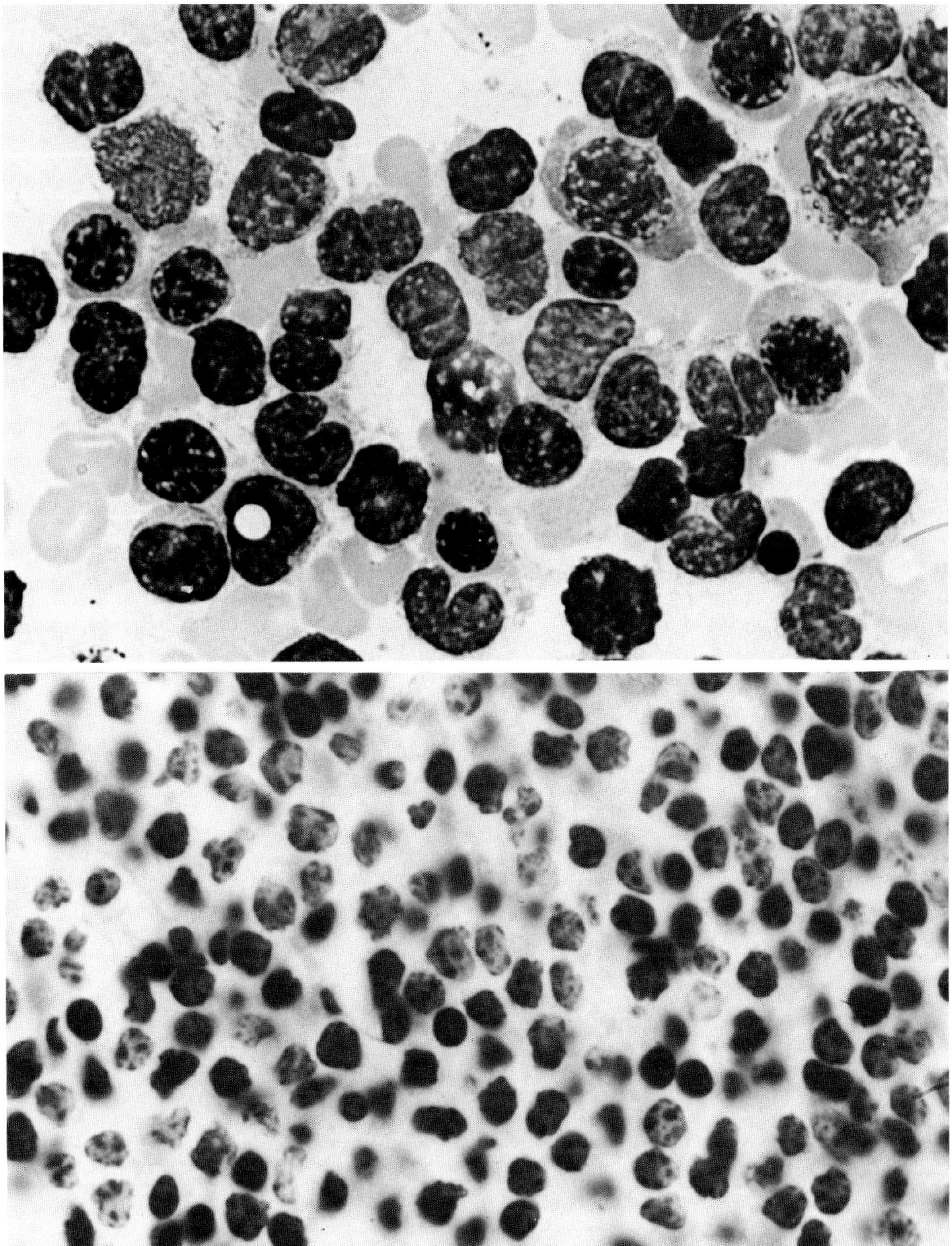

Figure 4-21. Chronic lymphocytic leukemia with autoimmune hemolytic anemia. On Wright-stained aspirate (**top**), poorly-differentiated lymphocytes with nuclear clefts are seen along with normoblasts. H & E section of bone marrow biopsy shows cells with irregular nuclear outline (lymphocytes) and darkly stained nuclei representing erythroblasts (**bottom**).

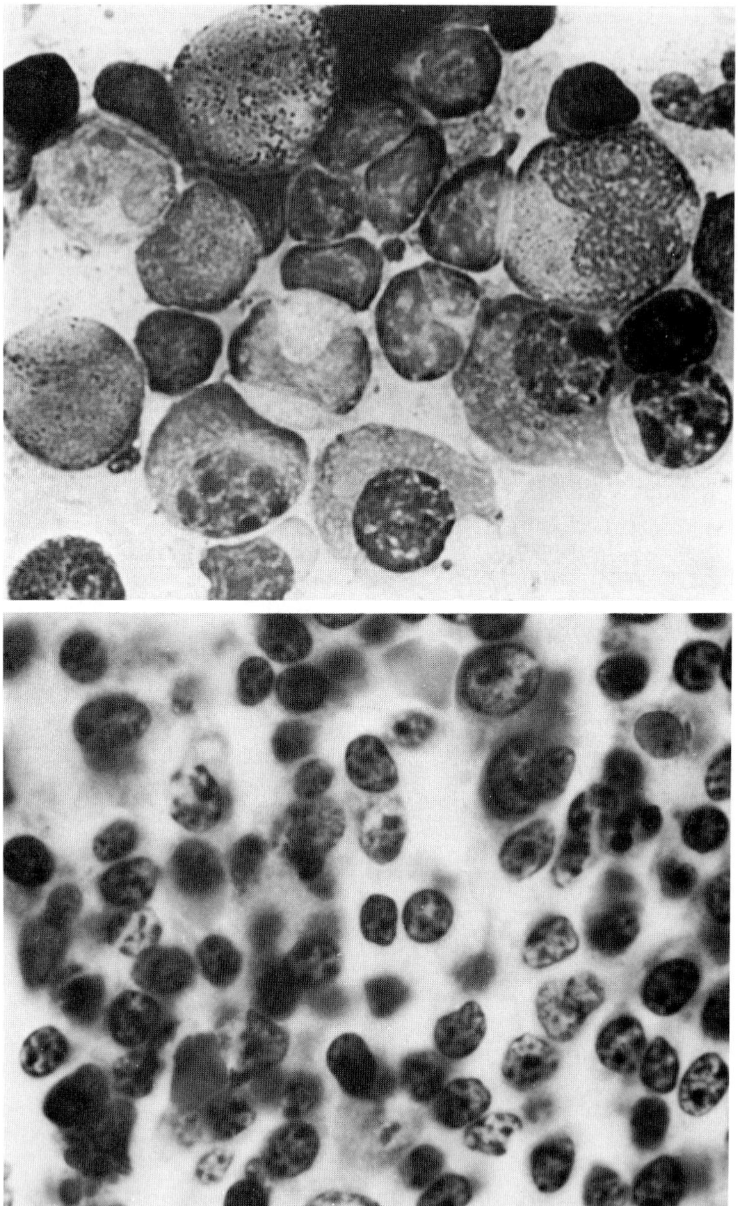

Figure 4-22. Chronic lymphocytic leukemia with dysproteinemia. On Wright-stained aspirate (**top**), lymphocytes show clefts and binuclearity, and increased numbers of plasma cells are found. H & E section of bone marrow biopsy (**bottom**), increased numbers of plasma cells and lymphocytes can be seen.

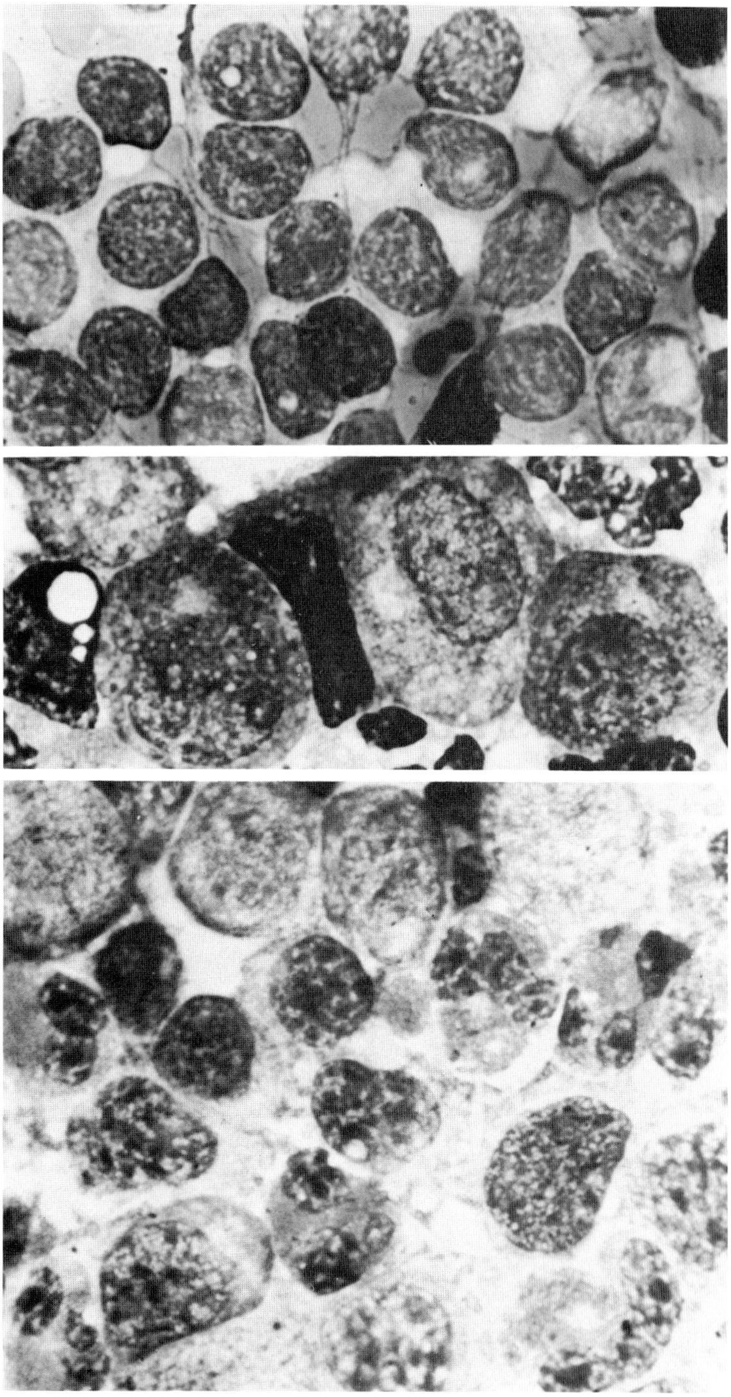

Figure 4-23. Chronic lymphocytic leukemia with eosinophilia. Lymphocytes (**top**); eosinophils (**center**); lymphocytes and eosinophils (**bottom**).

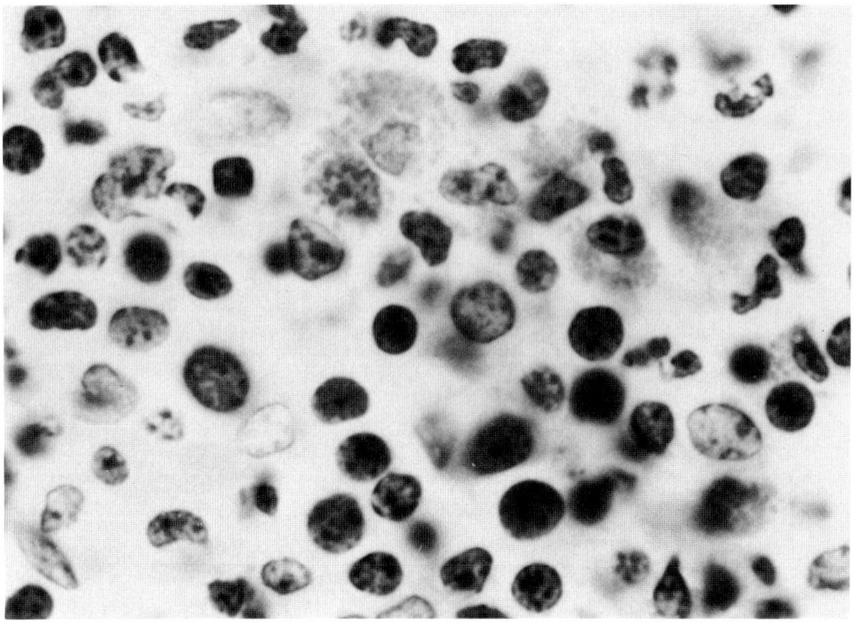

Figure 4-24. H & E section of bone marrow biopsy, showing increased numbers of eosinophils and lymphocytes.

Figure 4-25. Increased numbers of leukemic myeloblasts in chronic lymphocytic leukemia treated previously with alkylating agents. ➜

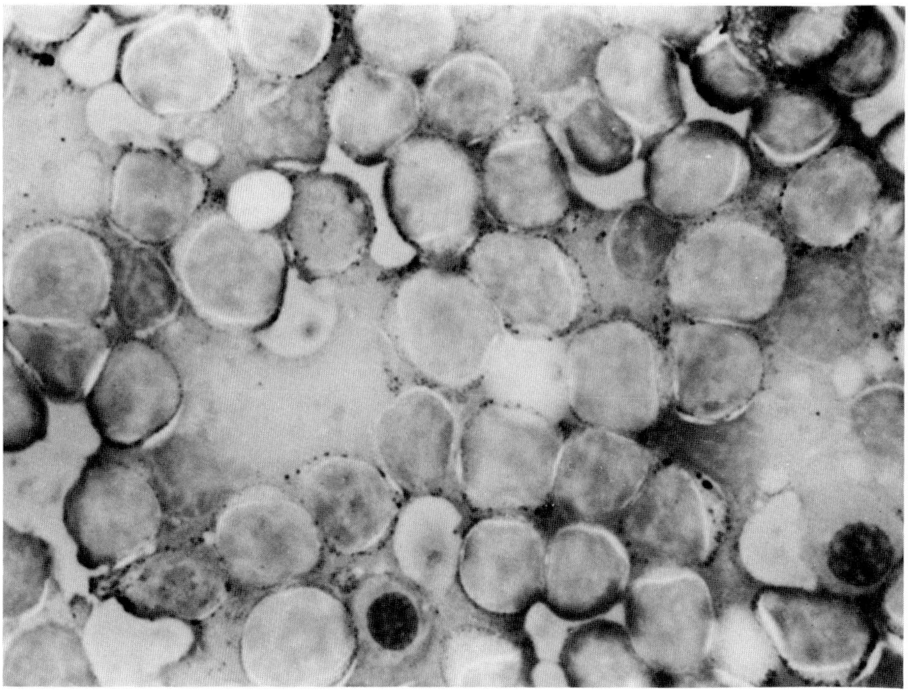

Figure 4-26. PAS positivity of leukemic lymphocytes in chronic lymphocytic leukemia.

Leukemic phase of histiocytic lymphoma

Leukemic phase of histiocytic lymphoma is a disorder characterized by proliferation of atypical mononuclear cells with giant nucleoli in the blood, the bone marrow, and the visceral organs. In approximately 3% to 5% of patients with an antecedent diagnosis of histiocytic lymphoma, a leukemic phase develops as a terminal event. In other patients, cells resembling those found in leukemic histiocytic lymphoma occur in the bone marrow, but there is no obvious antecedent history of histiocytic lymphoma documented on a biopsy of a lymph node or other tissue. At present, this disorder has a dire prognosis and is associated with rapid clinical deterioration characterized by recurrent severe infections, hepatomegaly, splenomegaly, and hypercalcemia. Although the origin of leukemic histiocytic cells is not well understood, most of the immunologic evidence to date favors a lymphoid origin rather than a monocytic or so-called histiocytic derivation. Histiocytic lymphoma is probably a disorder of B-lymphocytes.[868]

Peripheral blood samples from patients with leukemic histiocytic lymphoma contain many or few typical histiocytic lymphoma cells (Fig. 4-27). These are large mononuclear cells with round to irregular nuclear configurations and nuclear chromatin that appears coarse, with numerous blocklike and stripelike aggregates in some cells and few if any aggregates in others.[266,518,927] Characteristically, nucleoli are gigantic and have prominent perinucleolar condensations of chromatin. Using panoptic stains such as Wright's or Giemsa's, nucleoli in leukemic histiocytic lymphoma cells stain diffusely pale blue or gray. Cytoplasm is abundant and usually deeply basophilic. In most instances, these cells contain many cytoplasmic vacuoles of varying sizes. Some histiocytic lymphoma cells are giant and multinuclear. Gigantic erythrocytes, normoblasts, and aberrant-appearing monocytes are found in the peripheral blood of some patients. Thrombocytopenia and neutropenia are seen frequently.

Aspirates of bone marrow from patients with leukemic phase of histiocytic lymphoma are usually hypercellular. In some cases, marrow specimens are normocellular or even hypocellular, particularly if the patients have received chemotherapy or radiation therapy previously as a part of the treatment of diffuse histiocytic lymphoma. Megakaryocytes are decreased in number, as are erythroblasts and granulocytic cells. Distribution of leukemic histiocytic lymphoma cells in the bone marrow is variable. In some cases, these cells are distributed randomly throughout the marrow. In other cases, there may be nodular-appearing aggregates of leukemic histiocytic lymphoma cells. In large cells, nuclei show round to oval outlines and are often giant and multinuclear (Fig. 4-28 and Plate 3*D*). Chromatin strands are coarse and frequently show small aggregates. In typical cells, nucleoli are gigantic[266,518,617,927] and show prominent perinucleolar condensations of chromatin. In most cells, nucleoli are single, but some leukemic histiocytic lymphoma cells contain multiple nucleoli. Many of these nucleoli have an irregular, convoluted outline. In some instances, nucleoli in these cells appear lobulated. Cytoplasm is deeply basophilic and typically contains multiple vacuoles of varying sizes. Some vacuoles appear to coalesce.

In most instances, H & E sections of biopsies of bone marrow from patients with leukemic histiocytic lymphoma are hypercellular. In those cases that are normocellular or hypercellular, a few primitive-appearing cells with vesicular, prominent nuclei are found. These probably represent histiocytic lymphoma cells. In typical cases, the marrow contains large numbers of, or may be replaced by, cells with pale vesicular nuclei having an irregular nuclear configuration and prominent acidophilic staining nucleoli (Fig. 4-29). Cytoplasm is usually abundant, and there may be many cytoplasmic vacuoles. In other cases, involvement of the bone marrow by the lymphoma may seem to be nodular, in which nodules of lymphoma appear distinctive and separated from areas of more normal-appearing marrow. Megakaryocytes, erythroblasts, and granulocytic cells are greatly reduced in number.

Cytochemically, leukemic histiocytic lymphoma cells contain scant amounts of glycogen demonstrable by the PAS reagent. Activities of myeloperoxidase and specific esterase are not detectable in these cells. Activity of acid phosphatase is scant. In many histiocytic lymphoma cells, activity of nonspecific esterase is intense and often localized in a unipolar pattern,[507] as seen in Figure 4-30 and Plate *6B*. Nonspecific esterase activity in these cells is usually sensitive to inhibition by sodium flouride. With the pyronine stain for RNA, histiocytic lymphoma cells stain intensely, reflecting a high cytoplasmic content of RNA.

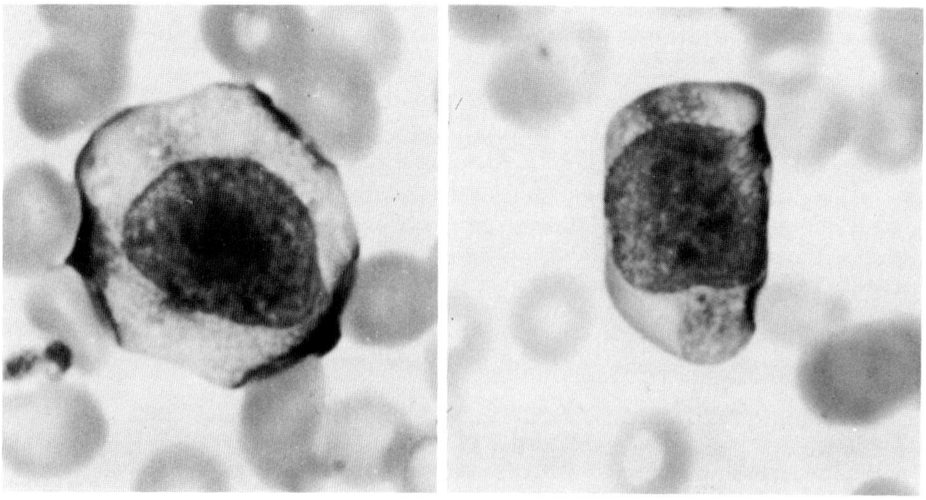

Figure 4-27. Leukemic histiocytic lymphoma cells, peripheral blood. Note the large nucleoli with perinucleolar condensations of chromatin.

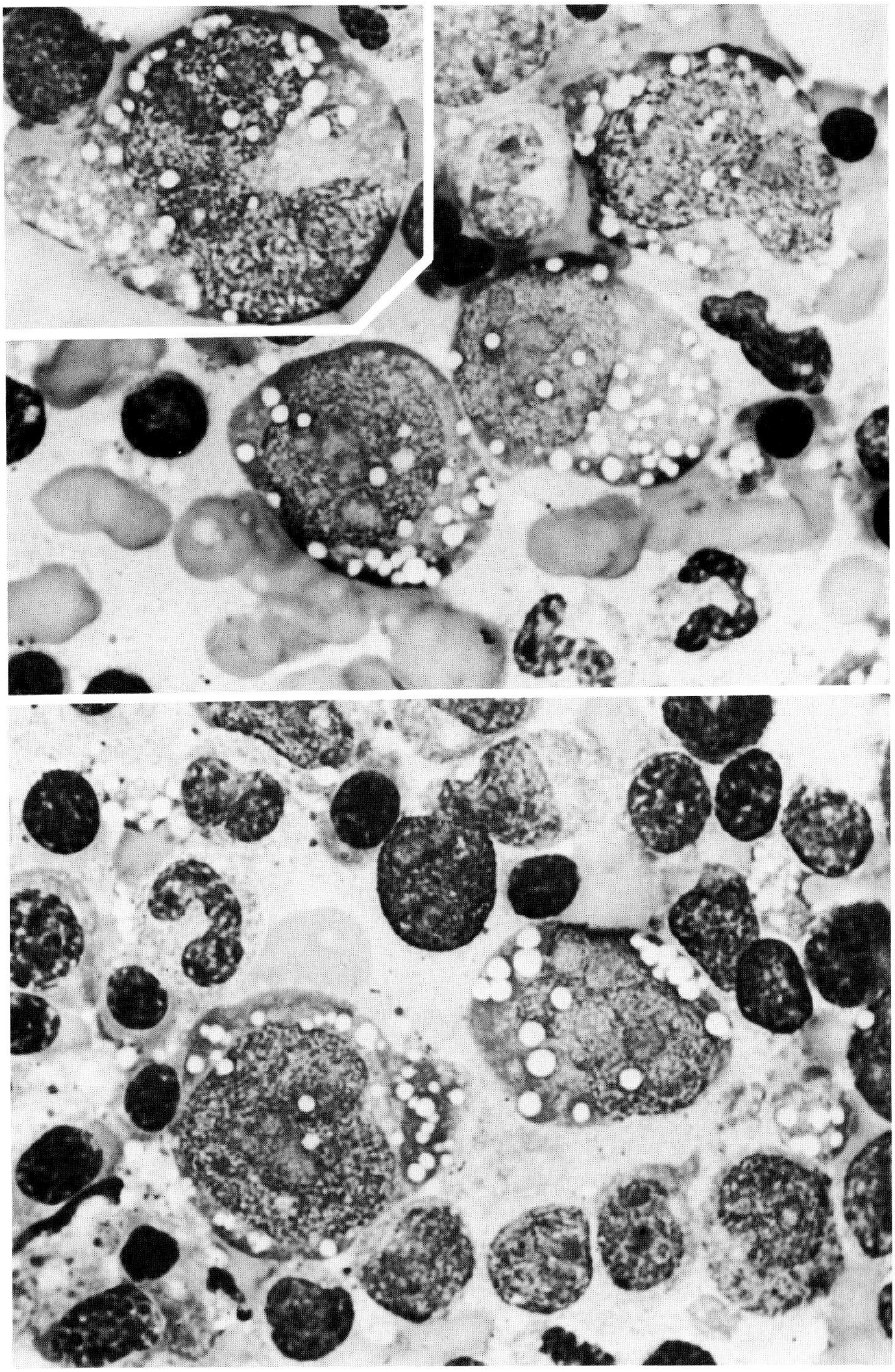

Bone marrow in leukemic histiocytic lymphoma. Malignant cells have gigantic nucleoli with perinucleolar condensations of chromatin, deeply basophilic cytoplasm with multiple cytoplasmic vacuoles, and multinuclearity (**inset**).

Figure 4-28.

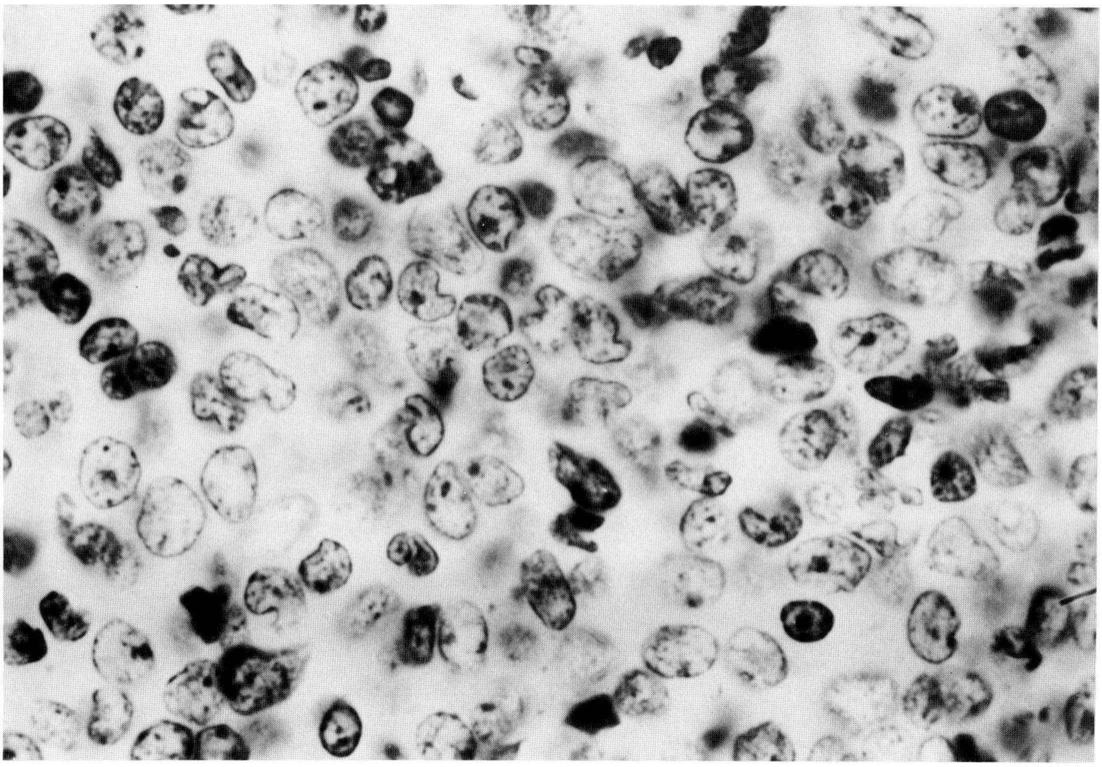

Figure 4-29. H & E section of bone marrow biopsy in leukemic histiocytic lymphoma, showing large cells with pale vesicular nuclei and prominent nucleoli.

Figure 4-30. Nonspecific esterase activity in unipolar distribution in leukemic histiocytic lymphoma cells, using alpha naphthyl acetate as substrate and hexazotized pararosaniline as dye coupler.

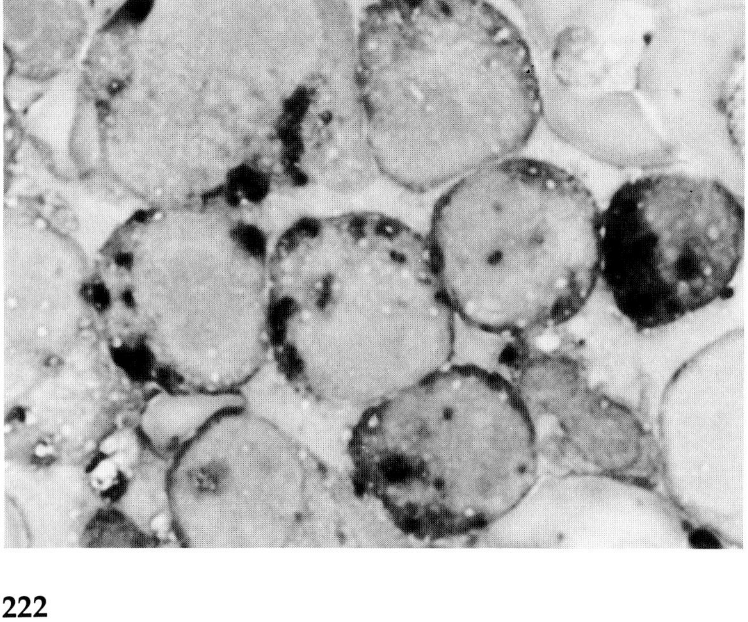

Hairy cell leukemia

Once identified under various names—including leukemic reticuloendotheliosis,[136,174,177,178,199,266,299,389,524,616,691,874,912,1040,1045] neoplastic leukemic reticulum cells in the blood,[689] chronic reticulolymphocytic leukemia,[880] and reticulum cell leukemia[579]—hairy cell leukemia[930] is recognized and described as a distinct morphologic and clinical entity. Clinically, it is typified by pancytopenia and splenomegaly. In some instances, the spleen may be enlarged massively. Many patients with hairy cell leukemia have infections due to unusual organisms such as *Torulopsis glabrata*, herpetic organisms, and *Listeria monocytogenes*. Consequences of pancytopenia, such as hemorrhages and soft tissue infections, are seen often. At present, the most effective remedy for the pancytopenia of hairy cell leukemia is splenectomy. Chemotherapy has not proved to be beneficial in most cases.

Peripheral blood samples from patients with hairy cell leukemia show many or few hairy cells. In most instances, patients with hairy cell leukemia have pancytopenia with few hairy cells in the blood. These cells are mononuclear and have a round to oval nucleus containing delicate nuclear chromatin, few if any aggregates of chromatin, and often one or more small light blue staining nucleoli with little perinucleolar aggregation of chromatin. Cytoplasm is abundant and stains pale gray to lilac with conventional panoptic reagents. Typically, hairy cells contain multiple cytoplasmic excrescences resembling hairs, befitting their graphic name (Fig. 4-31). These excrescences vary in size and length. In some patients, a few normoblasts, giant erythrocytes, and erythrocytes with basophilic stippling are found in the peripheral blood.

In typical cases, routine aspiration of bone marrow is met with a "dry tap," in which few if any marrow particles are obtained and primarily sinusoidal blood is aspirated from the marrow.[112,177,229,239,292,547] In those unusual instances in which a particle of marrow is removed and stained with Wright's stain as a conventional marrow coverslip film, the marrow is composed mainly of cells with large nuclei having an irregular configuration (Fig. 4-32). Some show folding and monocytoid features (Figs. 4-32, 4-33). Nuclear chromatin is often indistinct and shows few small aggregates. Cytoplasm is voluminous, causing the nuclei to appear widely separated. In these areas of dense accumulations of hairy cells, the characteristic hairy projections seen on the blood film are not apparent (Figs. 4-32, 4-33). However, in more dilute areas of the marrow preparation, individual hairy cells with typical cytoplasmic projections are found. Presumably, the dry tap occurs as the result of interdigitation of the hairy processes of adjacent cells, making it difficult to dislodge a particle of marrow during aspiration. Megakaryocytes are usually reduced in number, as are erythroblasts and granulocytic precursors. In some instances, erythropoiesis is megaloblastoid, and giant, bizarre-appearing erythroblasts can be identified.[97]

(text continued on p. 226)

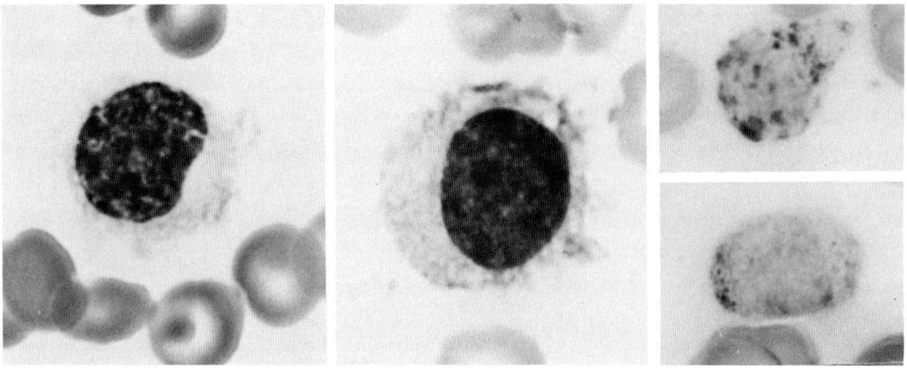

Figure 4-31. Hairy cells, peripheral blood, in hairy cell leukemia, showing activity of nonspecific esterase in hairy cells, using alpha naphthyl butyrate as substrate (**insets, right**).

Figure 4-32. Bone marrow aspirate in hairy cell leukemia, showing hairy cells with delicate nuclear chromatin, monocytoid nuclear configuration in some instances, and abundant cytoplasm. Several normoblasts are seen.

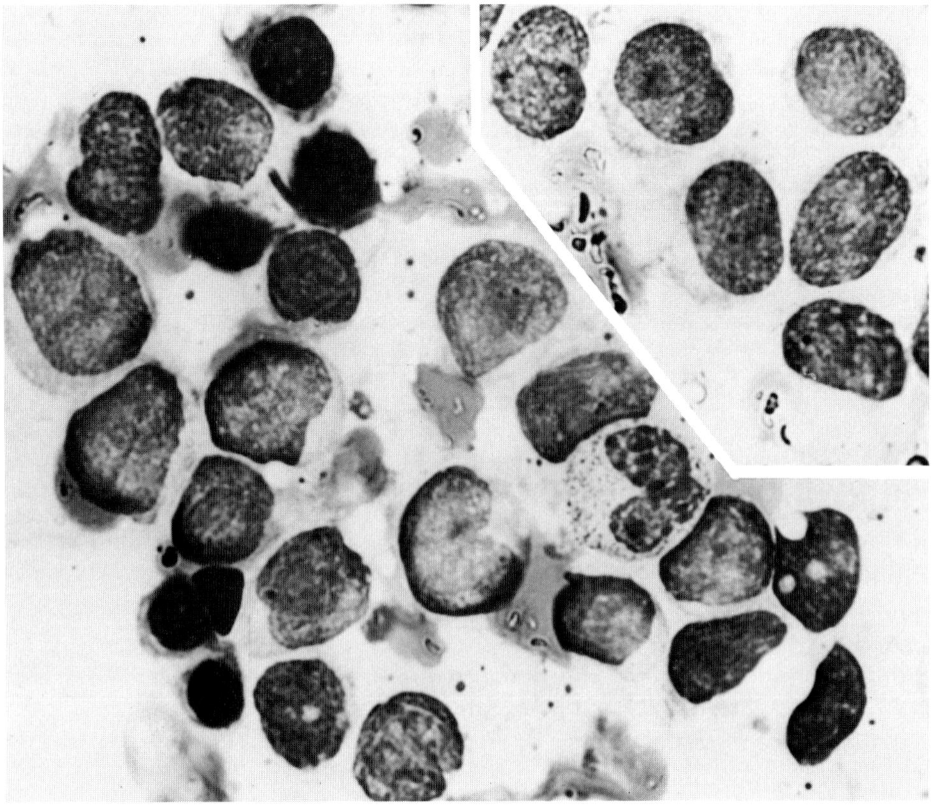

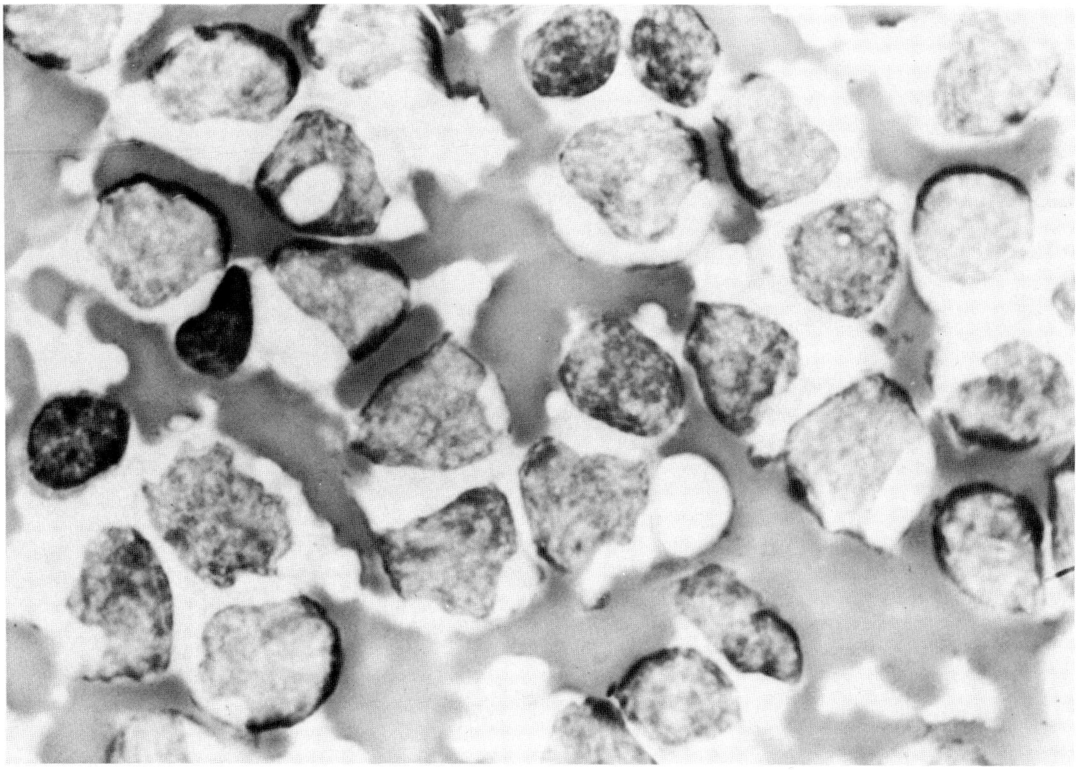

Bone marrow in hairy cell leukemia, illustrating hairy cells with irregular nuclear configuration, small nucleoli, and abundant cytoplasm lacking the cytoplasmic projections seen in peripheral blood films.

Figure 4-33.

H & E sections of bone marrow biopsies from patients with typical hairy cell leukemia show a widespread proliferation of cells with round to oval nuclei and irregular nuclear configurations. In some of these cells, small nucleoli are found. Because of the abundant cytoplasm, cells seem to be separated widely (Figs. 4-34, 4-35, 4-36). Interspersed between areas of hairy cell accumulations are small but relatively normal-appearing islands of marrow containing megakaryocytes, erythroblasts, and granulocytic cells. In other patients, these normal marrow islands are not visualized on the marrow biopsy, and the marrow seems to be composed almost entirely of hairy cells. Since dry taps occur in most cases of hairy cell leukemia, the confident diagnosis of this disorder rests on the findings of the bone marrow biopsy as well as on the cytochemical properties of hairy cells. These cells can be obtained from the peripheral blood film or, preferably, from dilute films of marrow or touch preparations of the marrow biopsy core.

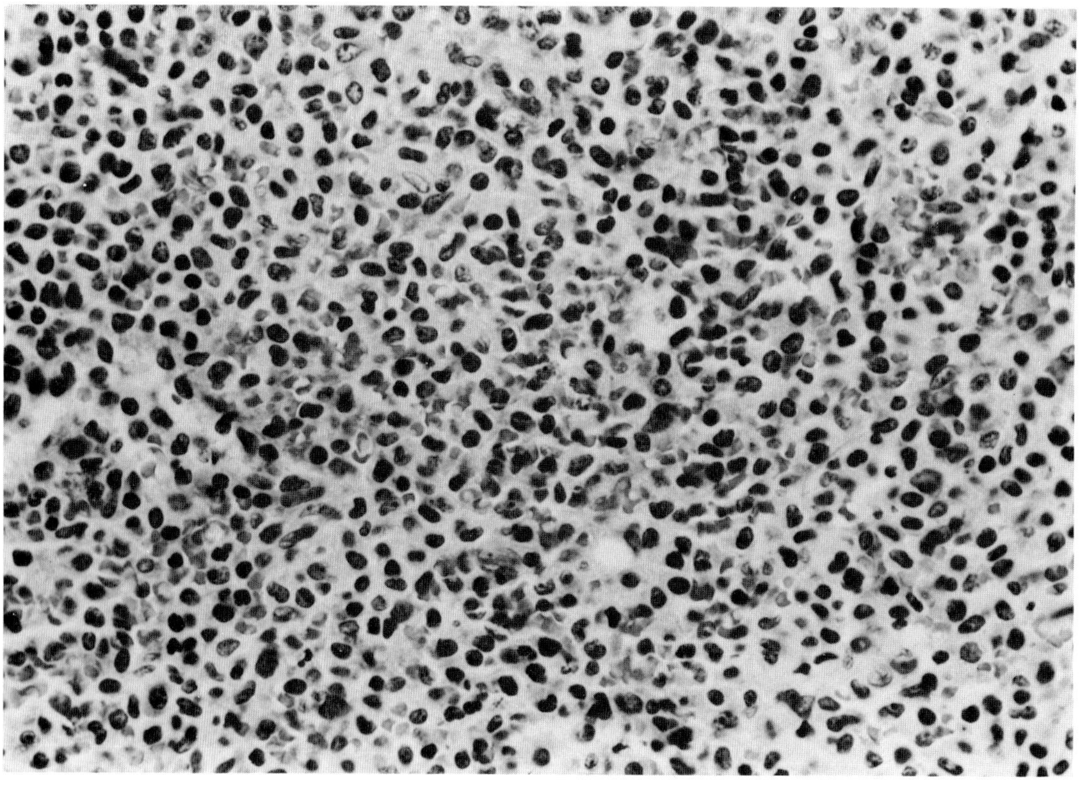

Figure 4-34. H & E section of bone marrow biopsy in hairy cell leukemia, showing mononuclear cells that appear widely separated.

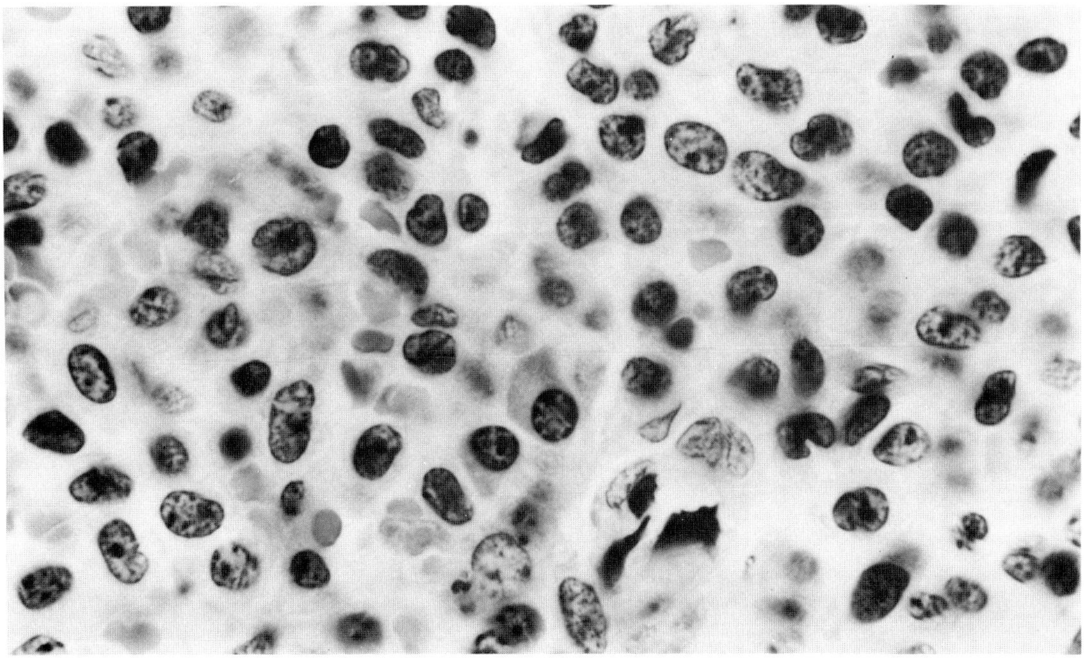

H & E section of bone marrow biopsy in hairy cell leukemia, demonstrating cells with elongated, pale-appearing nuclei. **Figure 4-35.**

H & E section of bone marrow biopsy in hairy cell leukemia, showing higher-power view of hairy cells with irregular nuclear contours. **Figure 4-36.**

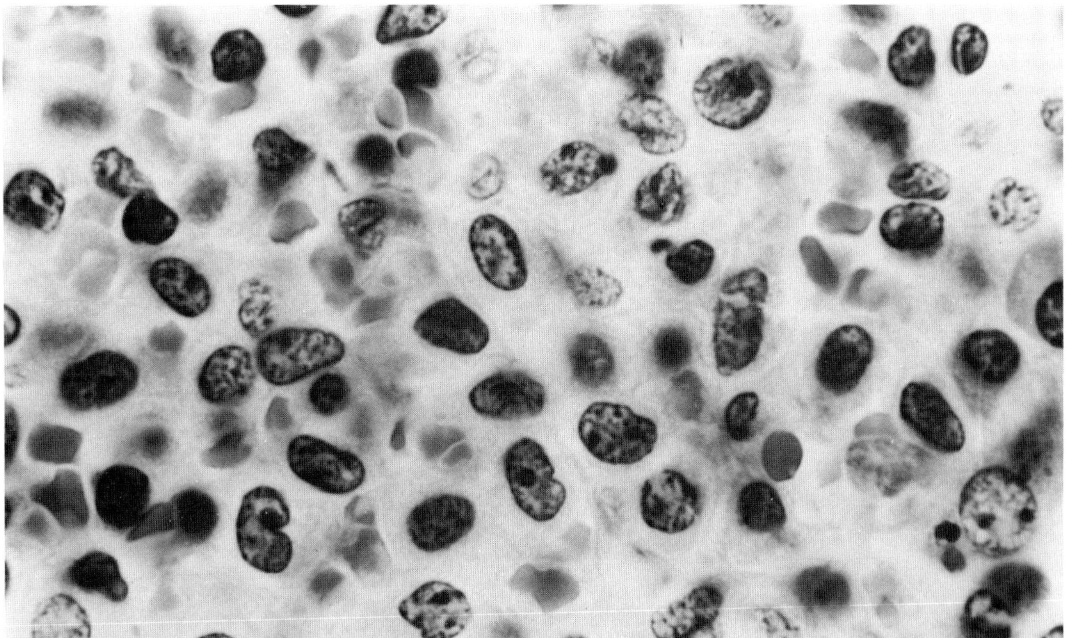

Cytochemically, hairy cells contain little if any glycogen with the PAS reagent. Activities of specific esterase and myeloperoxidase are not detectable in hairy cells. Using alpha naphthyl acetate or alpha naphthyl butyrate as the substrate,[503] typical hairy cells contain fluoride sensitive activity of nonspecific esterase (Fig. 4-31). From the cytochemical standpoint, the most useful test for hairy cells is the demonstration of acid phosphatase activity that is resistant to inhibition by L-tartrate, as seen in Figure 4-37 and Plate *6D*.[520,907,909,1109,1111] In fact, activity of this enzyme often seems more intense when tartrate is added to the incubation medium. Although it has been described in other types of pathologic blood cells, including so-called lymphosarcoma cells,[714,888] tartrate resistant acid phosphatase is probably the most distinctive cytochemical marker for hairy cells available currently. Using the pyronine stain for RNA, hairy cells contain scant amounts of cytoplasmic RNA.

At present, the cell of origin of the hairy cell is unknown. Although some studies have indicated monocytic properties such as phagocytosis and fluoride sensitive nonspecific esterase activity, other studies, including scanning electron microscopic pictures and immunologic characterization, suggest that the hairy cell exhibits properties of B-lymphocytes.

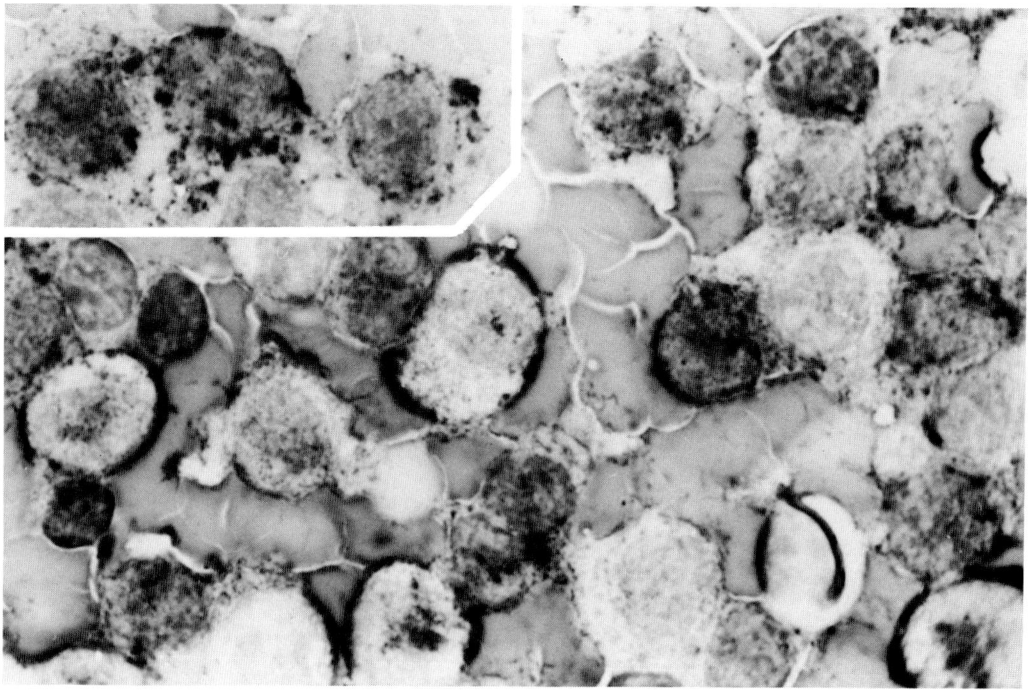

Figure 4-37. Tartrate-resistant acid phosphatase activity in hairy cells from bone marrow in hairy cell leukemia.

Leukemic phase of Burkitt's lymphoma

Within the spectrum of lymphocytic leukemia, the leukemic phase of Burkitt's lymphoma represents a rare and newly described variant, and only a few cases have been reported.[322,346,416,469,996] Clinically, it is associated with a fulminating clinical course because Burkitt's lymphoma, along with oat cell carcinoma of the lung, is one of the most rapidly proliferating neoplasms known, with a tumor doubling time of less than 30 days. Hematologically, it is characterized by the occurrence of lymphoid cells with multiple cytoplasmic vacuoles in the blood, the bone marrow, and the visceral organs. In some instances, patients with this type of leukemia have an antecedent history of Burkitt's lymphoma, usually originating in a lymph node, the gastrointestinal tract, or the central nervous system.[747] In other instances, the leukemia apparently does not have an antecedent history of Burkitt's lymphoma and seems to arise *de novo*.[183] Immunologically, this leukemia is of B-lymphocyte origin. In the FAB classification,[87] lymphoblastic leukemia with Burkitt cells is designated as L3.

Peripheral blood samples from most patients with the leukemic phase of Burkitt's lymphoma show markedly increased numbers of large, atypical-appearing lymphoid cells. In other patients, few of these cells are seen. Burkitt lymphoma cells have round to oval nuclei with large, prominent aggregates of nuclear chromatin. Usually, a nucleolus is visualized. Cytoplasm is deeply basophilic and contains numerous vacuoles of various sizes. Some vacuoles show coalescence. Platelets may be decreased in number, and erythrocytes show variability in size and shape. Neutrophils are usually reduced in number.

In most cases, aspirates of bone marrow from patients with the leukemic phase of Burkitt's lymphoma are markedly hypercellular (Fig. 4-38 and Plate 4*B*).[95,183,322,346,996,1104] Megakaryocytes are reduced in number, as are erythroblasts and granulocytic cells. Most of the cells in the bone marrow are large, mononuclear cells with abundant basophilic cytoplasm containing many vacuoles of differing sizes. Nuclei of these leukemic cells are round to oval in configuration. Chromatin aggregates appear blocklike and stripe-shaped, and they are connected to each other and to the nuclear membrane by coarse strands of chromatin. Heterochromatin appears darkly stained. In most instances, a nucleolus is visualized. Both small and large leukemic cells are present. Large macrophages are scattered at irregular intervals throughout the marrow specimen (Fig. 4-39). These macrophages have a small nucleus with an irregular configuration of the nuclear membrane, abundant cytoplasm, and frequently amorphous-appearing cytoplasmic debris. Macrophages of a similar type are seen in lymph nodes from patients with Burkitt's lymphoma, and they impart a distinctive "starry sky" appearance to the node.

H & E stained biopsies of bone marrow from patients with Burkitt's leukemia are hypercellular. Megakaryocytes, erythroblasts, and gran-

(text continued on p. 232)

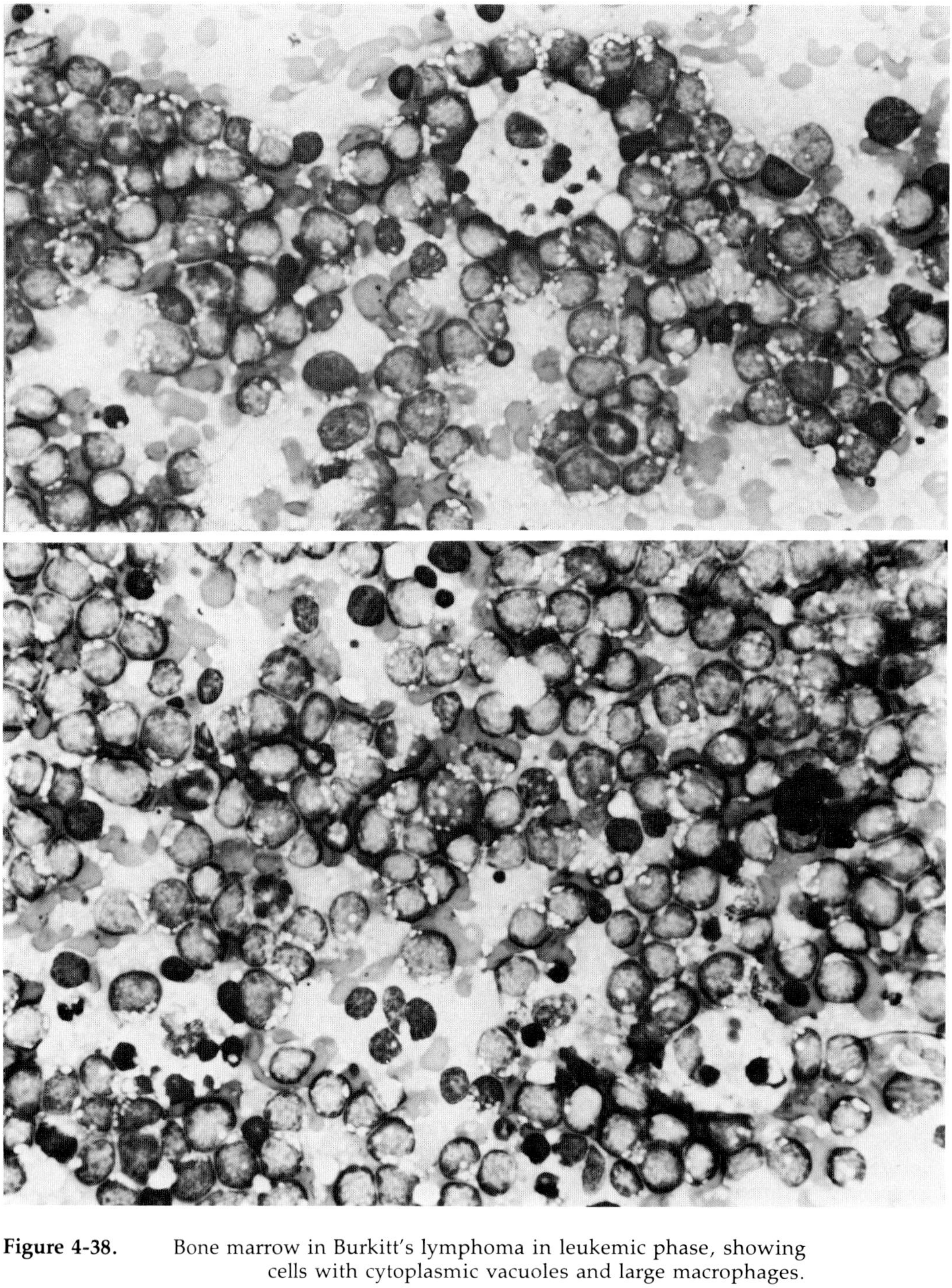

Figure 4-38. Bone marrow in Burkitt's lymphoma in leukemic phase, showing cells with cytoplasmic vacuoles and large macrophages.

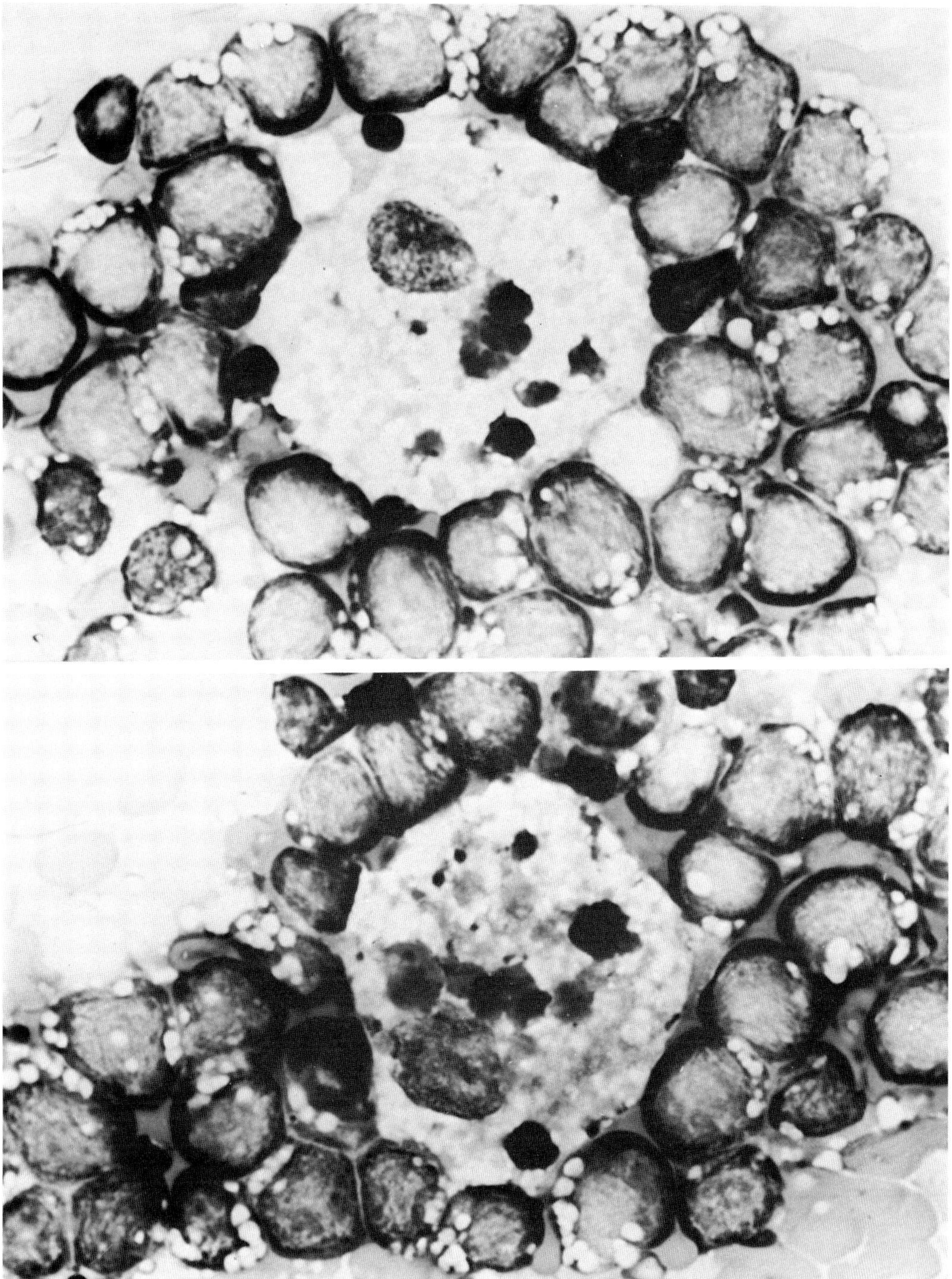

Figure 4-39. Bone marrow in Burkitt's lymphoma in leukemic phase, showing leukemic cells with prominent aggregates of nuclear chromatin, multiple cytoplasmic vacuoles, and large macrophages.

ulocytic cells are markedly reduced in number. For the most part, the marrow is replaced by large mononuclear cells with well-defined nuclear membrane, irregular nuclear shapes, and large aggregates of nuclear chromatin scattered throughout the nucleus and adherent to the nuclear membrane (Figs. 4-40, 4-41). Usually a nucleolus is visualized. As on aspirates of bone marrow, macrophages are interspersed throughout the specimen (Figs. 4-40, 4-41, 4-42).

Cytochemically, PAS stain for glycogen is weakly positive in some Burkitt's lymphoma cells. In most of the cells, however, the stain for glycogen is negative.[87] This reaction contrasts with the blocklike pattern of glycogen in many leukemic lymphoblasts. Activities of myeloperoxidase and specific esterase are not detectable in Burkitt's lymphoma cells. Similarly, these cells do not contain material that is stained by Sudan black B. Several cytochemical stains are useful in the diagnosis of Burkitt's lymphoma. These include the pyronine stain for RNA and the oil red O stain for lipids. Using the methyl green-pyronine stain, cytoplasm of Burkitt's lymphoma cells stain intensely positive, indicating a high cytoplasmic content of RNA (Fig. 4-43 and Plate 5*F*). Using the oil red O stain (Fig. 4-44 and Plate 6*F*), vacuoles in the cytoplasm of Burkitt's lymphoma cells stain intensely positive, indicating a high content of lipid. Stains for acid phosphatase activity and nonspecific esterase activity show weak positivity and are not useful diagnostically.

Plasma cell leukemia

Plasma cell leukemia may be conceptualized as the leukemic phase of multiple myeloma. It is believed to be a disorder of B-lymphocytes. In most cases, plasma cell leukemia is characterized by proliferation of malignant plasma cells in the blood, the bone marrow, and the visceral organs. Although rare, plasma cell leukemia is the most frequent leukemic manifestation of multiple myeloma. In some cases of multiple myeloma, particularly those treated with alkylating agents, acute myeloblastic or monocytic leukemia develops as a terminal event.[766,1047] As yet, it is uncertain whether acute leukemia in these patients was part of the natural evolutionary history of multiple myeloma or whether alkylating agents used to treat the myeloma acted as leukemogens.

Plasma cell leukemia occurs rarely in patients with Waldenström's macroglobulinemia.[1116] Although it can occur in patients with any of the several recognized immunoglobulin subtypes of multiple myeloma, plasma cell leukemia is particularly frequent in patients who have the IgD type of multiple myeloma.[533] As in multiple myeloma, many patients who have plasma cell leukemia have a demonstrable monoclonal gammopathy in plasma or urine. Some may have had an antecedent history of multiple myeloma or lymphoproliferative disorders such as chronic lymphocytic leukemia.[20]

(text continued on p. 238)

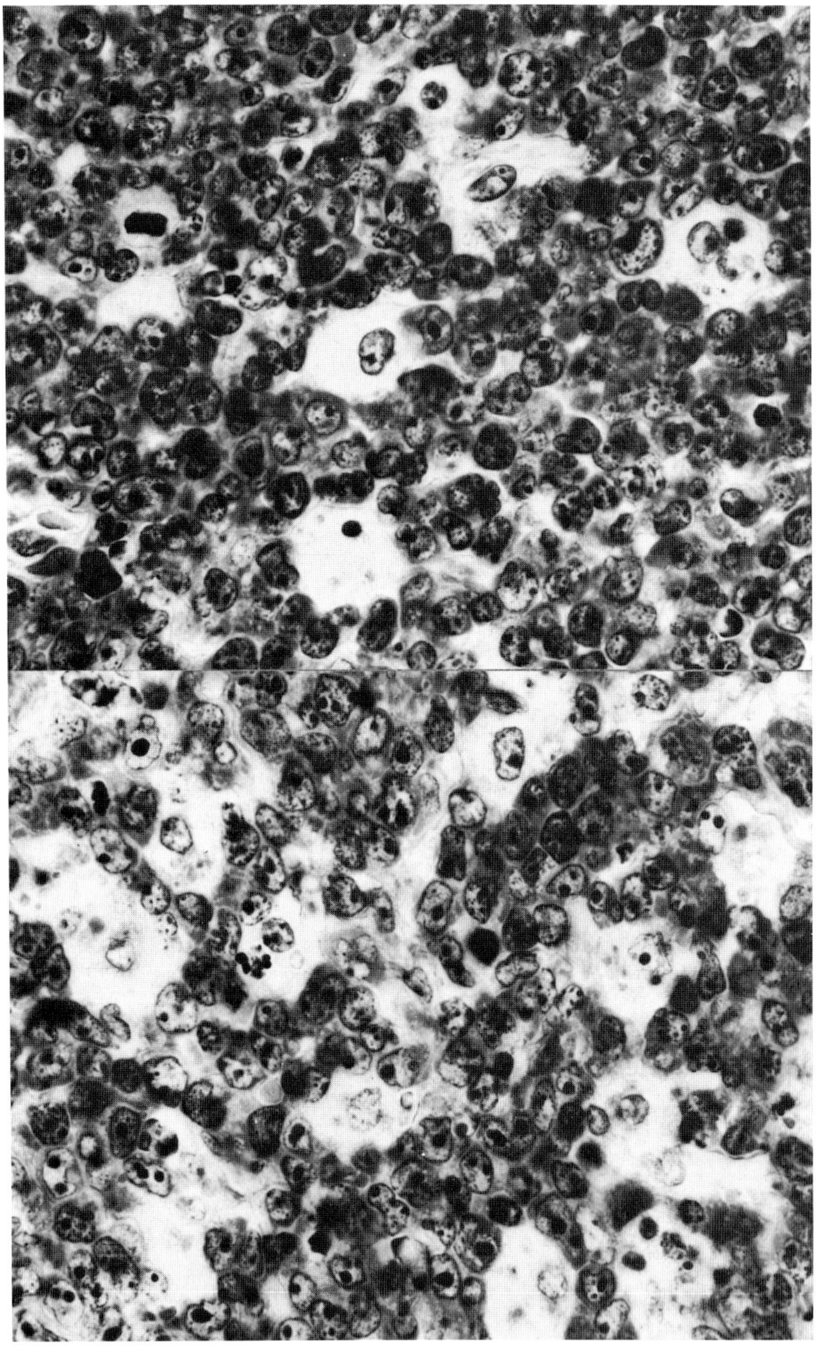

Figure 4-40. H & E section of bone marrow biopsy in Burkitt's lymphoma in leukemic phase, showing leukemic cells with large nucleoli and numerous macrophages.

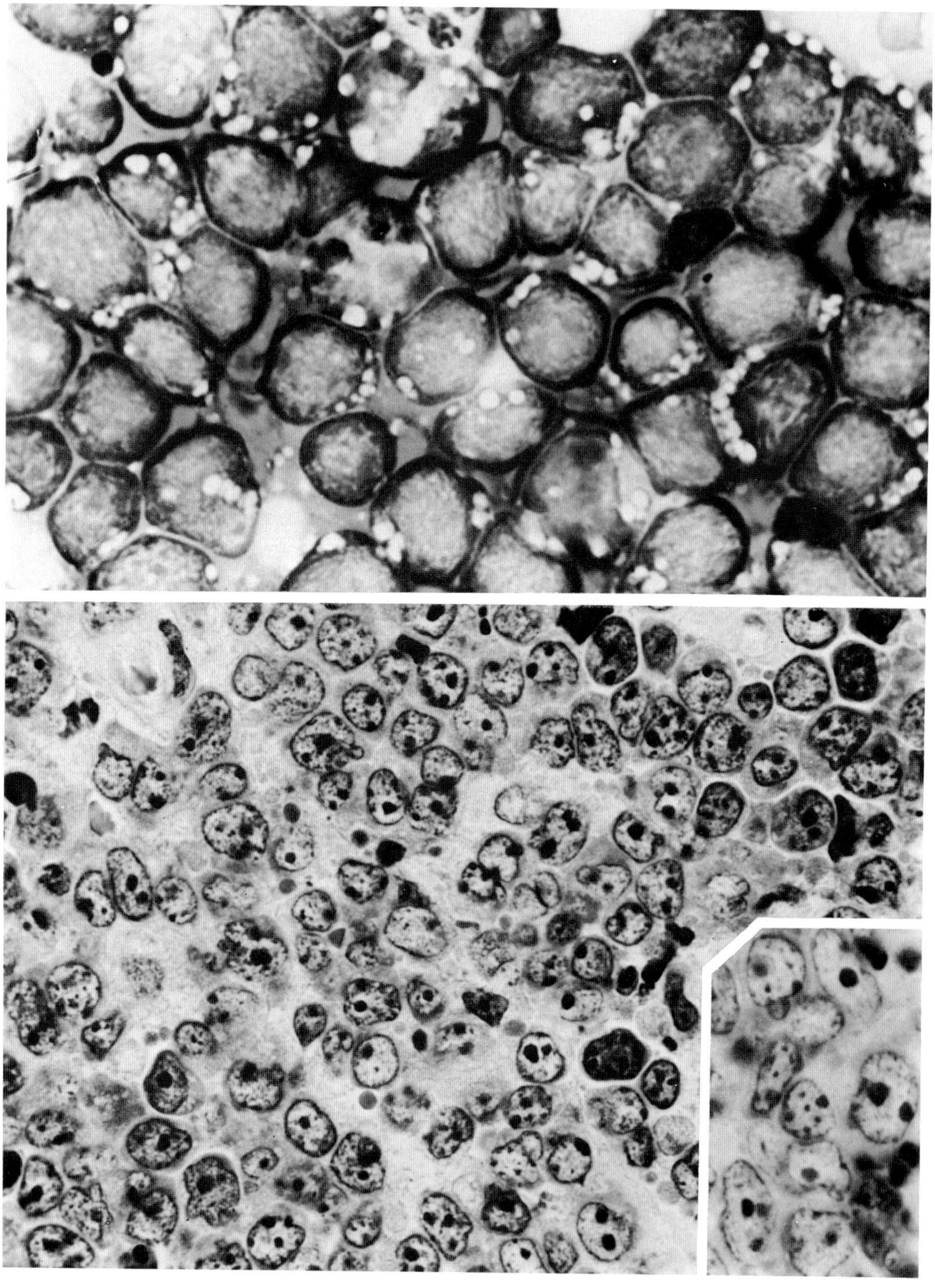

Figure 4-41. Wright's stain of Burkitt's lymphoma cells in bone marrow (**top**). Higher-power view of Burkitt's cells on H & E section of bone marrow biopsy, depicting prominent nucleoli and dispersed chromatin (**bottom, inset**).

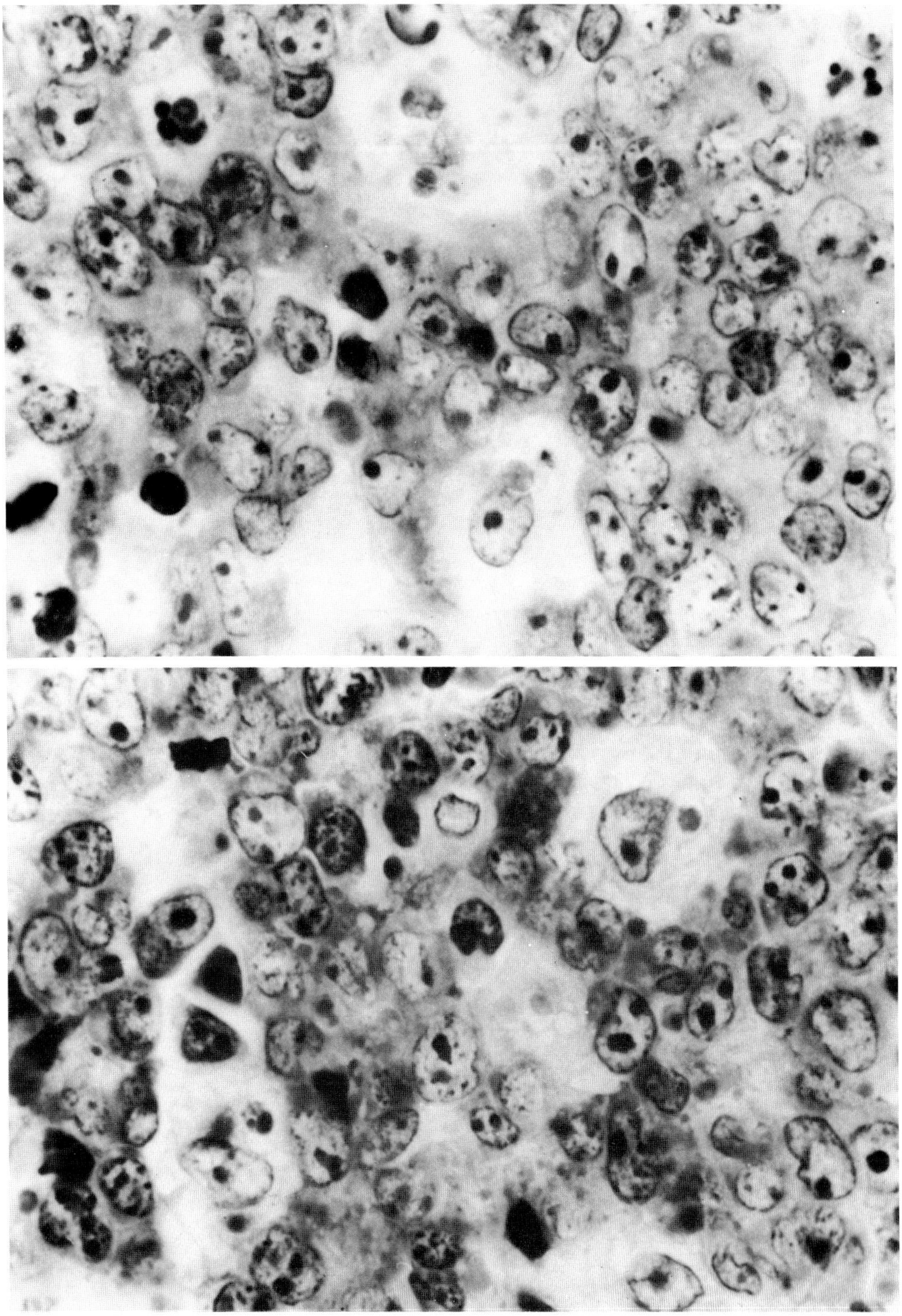

Higher-power view of Burkitt's lymphoma cells and macrophages on H & E section of bone marrow biopsy. **Figure 4-42.**

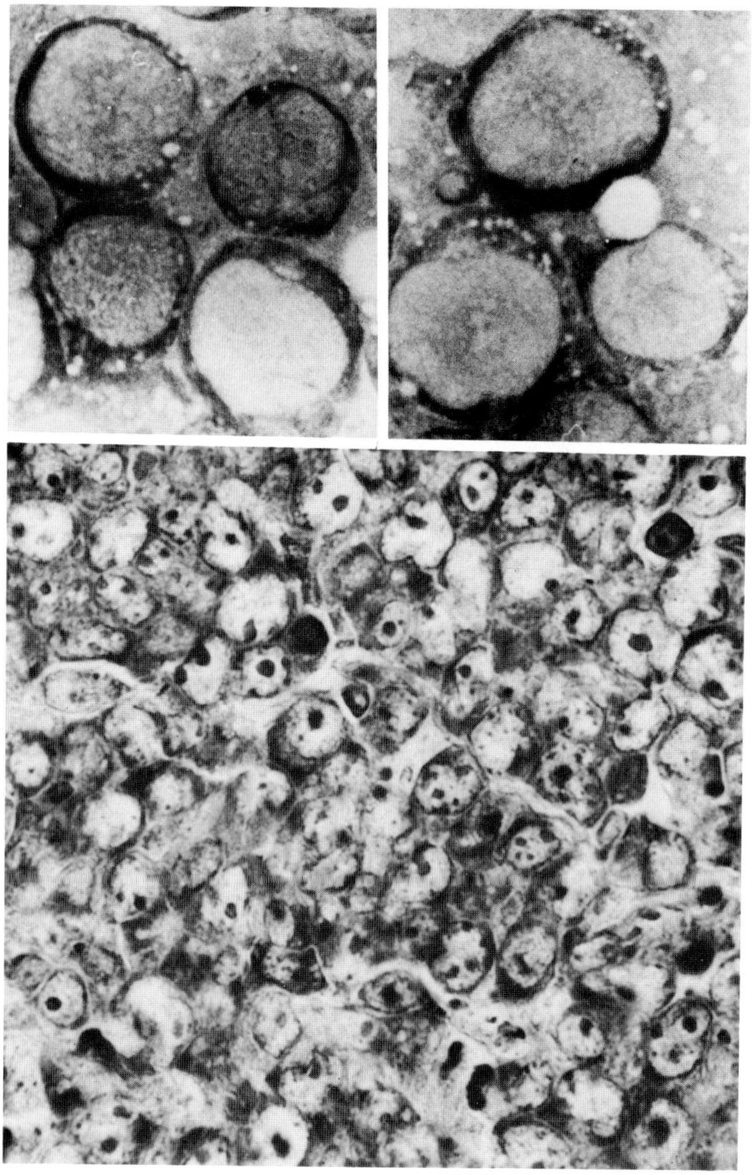

Figure 4-43. Methyl green-pyronine stain applied to Burkitt's lymphoma cells. Intense pyroninophilia on bone marrow aspirate (**top**). Intense pyroninophilia on B_5-formalin fixed biopsy specimen (**bottom**).

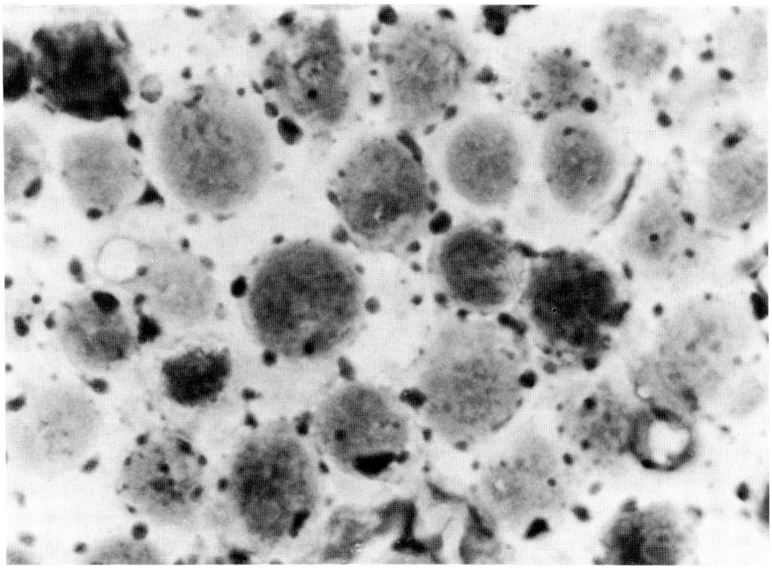

Figure 4-44. Oil red O positive material in the cytoplasm of Burkitt's lymphoma cells. Presumably, the lipid material having affinity for the oil red O stain is contained within vacuoles.

Figure 4-45. Bone marrow in plasma cell leukemia, illustrating giant plasma cells with prominent nucleoli. Using Wright stain, the cytoplasm of these cells stained orange. This color is believed to indicate a high content of glycoprotein, as found in IgA myeloma. Cells having this tinctorial property have been called flame cells.

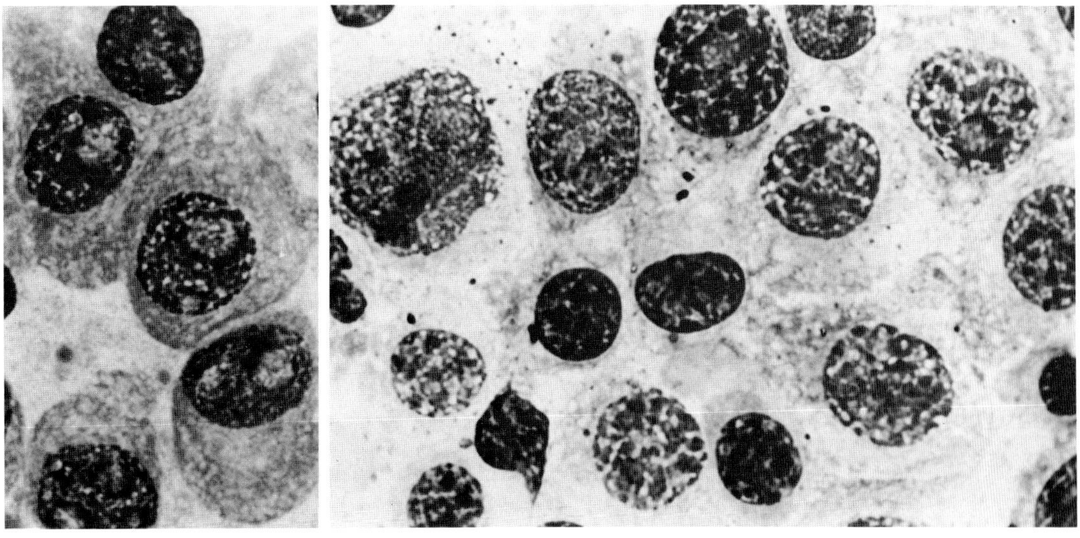

Plasma cells occur in varying numbers in the peripheral blood of patients with plasma cell leukemia.[181,202,203,215,248,284,293,295,310,320,338,343,423,428,429,454–456,466,473,487,534,582,598,600,611,632,633,715,718,720,722,737,740,754,755,781,796,798,799,807,808,810,815–818,850,896,899,902,956,983,1017,1018,1029,1031,1053,1086,1103,1107,1116,1121] They vary in morphologic appearance. Some are mature-appearing plasma cells, whereas others are large and pleomorphic and may be binuclear. Cytoplasm is deeply basophilic and may contain several vacuoles of varying sizes. Nuclei of these plasma cells have prominent aggregates of nuclear chromatin in some instances, and delicate, dispersed-appearing chromatin in others.[975] Large nucleoli with perinucleolar condensations of chromatin are found in some of these plasma cells. Rouleau formation of erythrocytes is seen frequently, and normoblastemia and thrombocytopenia occur often.

As illustrated in Figures 4-45 through 4-49, aspirates of bone marrow from patients with plasma cell leukemia show findings identical to those in multiple myeloma.[35,80,81,103,176,592] Plasmacytosis is marked, and plasma cells vary in appearance from normal and typical-appearing in some cases to giant and pleomorphic in others (Fig. 4-48). Some of the plasma cells contain cytoplasmic vacuoles, and others contain crystalloids, Russell's bodies (Fig. 4-49), or cytoplasmic fibrils.[738] Megakaryocytes, erythroblasts, and granulocytic cells are markedly reduced in number.

Biopsies of bone marrow from patients with plasma cell leukemia demonstrate marked hypercellularity. Most of the marrow is composed of plasma cells with eccentric nuclei and deeply staining cytoplasm (Fig. 4-50). Giant and multinuclear plasma cells may be found. Megakaryocytes, erythroblasts, and granulocytic cells are markedly reduced in number.

Cytochemically, plasma cells in the peripheral blood and bone marrow of patients with plasma cell leukemia show small amounts of glycogen demonstrable with the PAS reagent. In rare cases of plasma cell leukemia, large amounts of glycogen are present in plasma cells. Sometimes, PAS positive nuclear inclusions (Dutcher bodies) can be identified. Using stains for specific esterase and myeloperoxidase, activities of these enzymes cannot be demonstrated in plasma cells. Activities of nonspecific esterase (Fig. 4-51 and Plate 6C) and acid phosphatase (Fig. 4-51, *inset*) are intense and sometimes localized as a chunklike pattern of reaction product in both normal plasma cells and myeloma cells.[242,607] As in normal plasma cells, pyronine stain is intense, reflecting increased cytoplasmic content of RNA (Fig. 4-52 and Plate 5*E*).

Figure 4-46. Bone marrow in plasma cell leukemia, depicting large numbers of malignant plasma cells. Some of these cells contain multiple cytoplasmic vacuoles. ➡

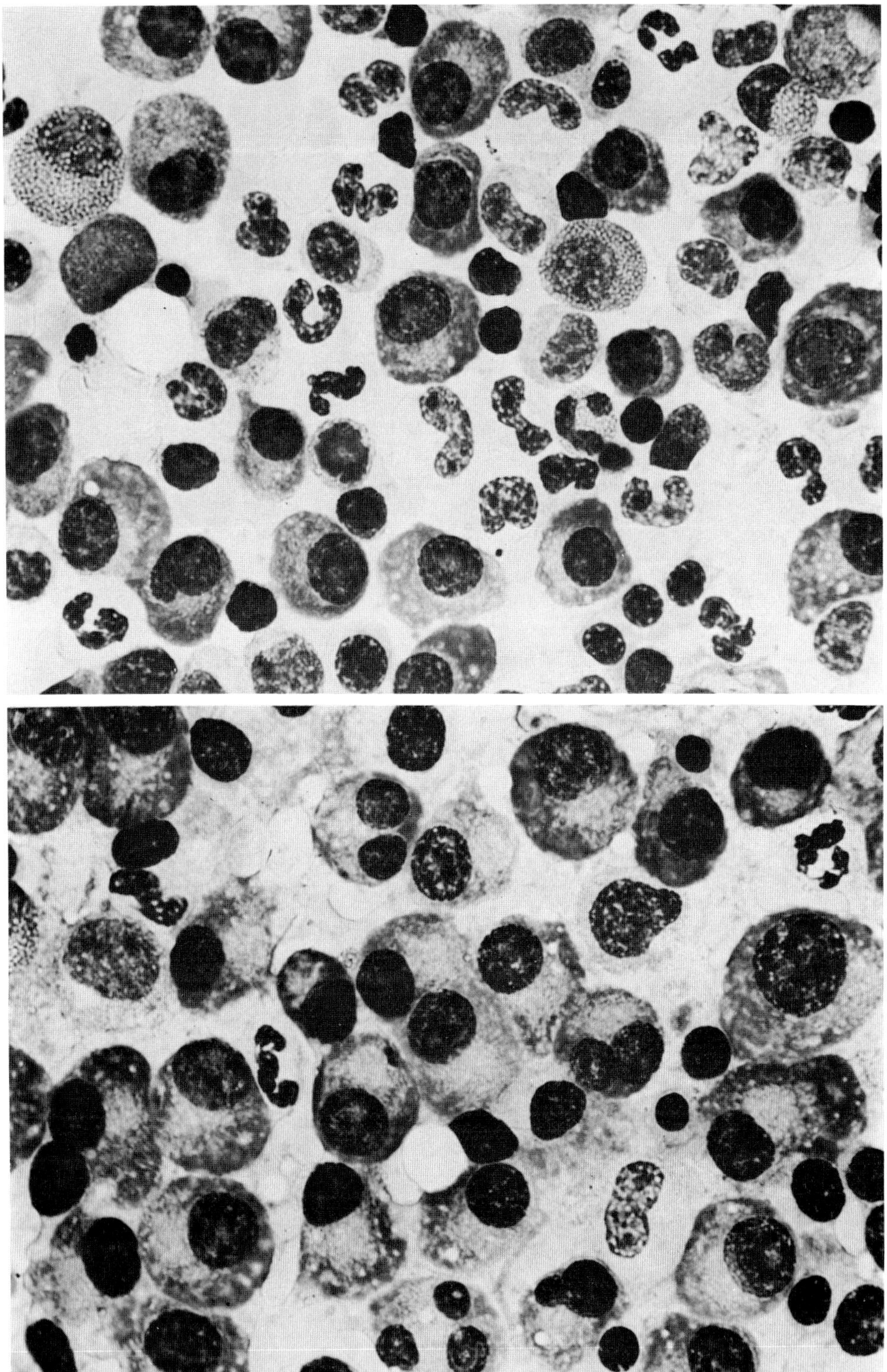

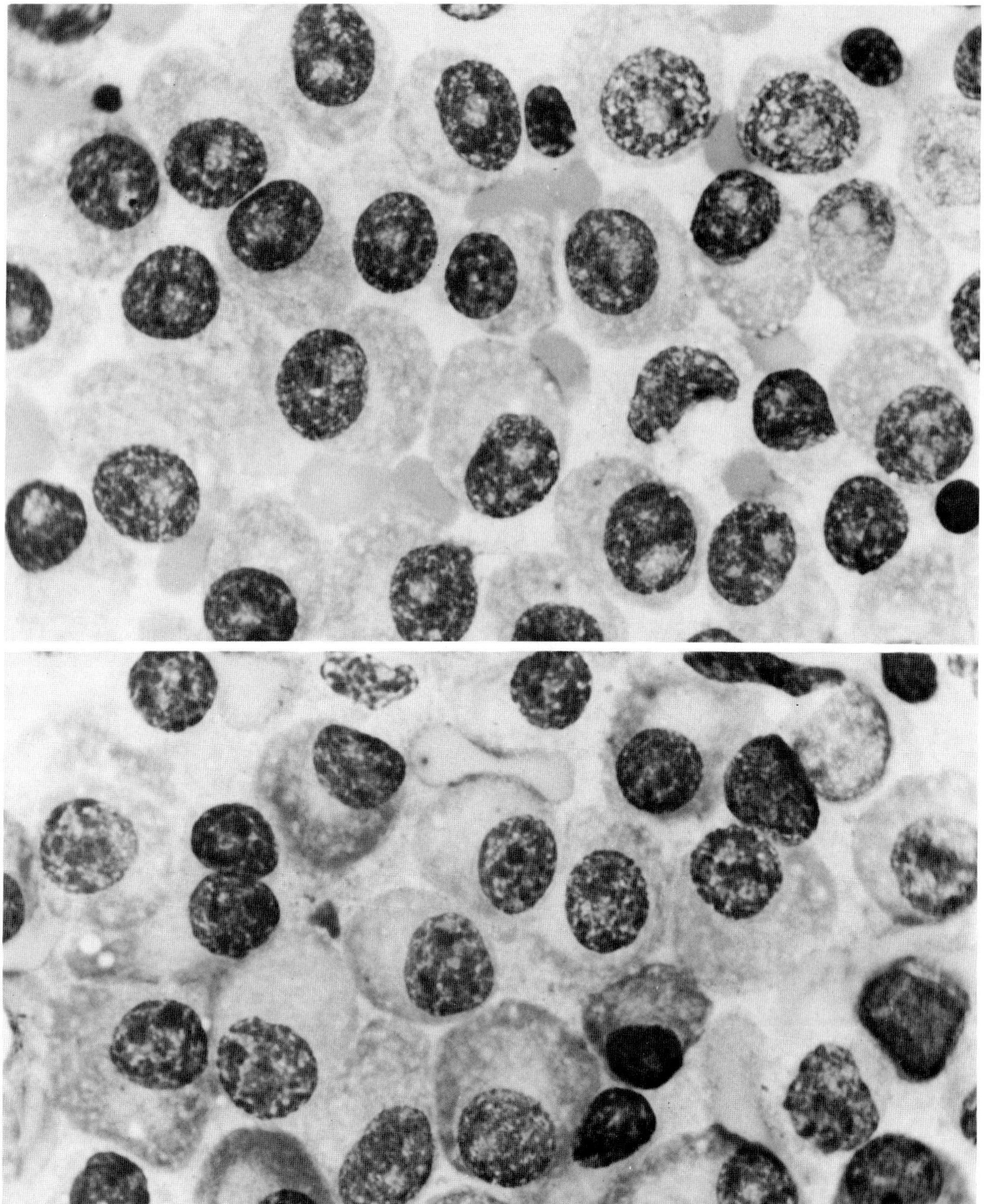

Figure 4-47. Bone marrow in two cases of plasma cell leukemia. Pleomorphism of plasma cells is shown. Many of the cells have large nucleoli with perinucleolar condensations of chromatin.

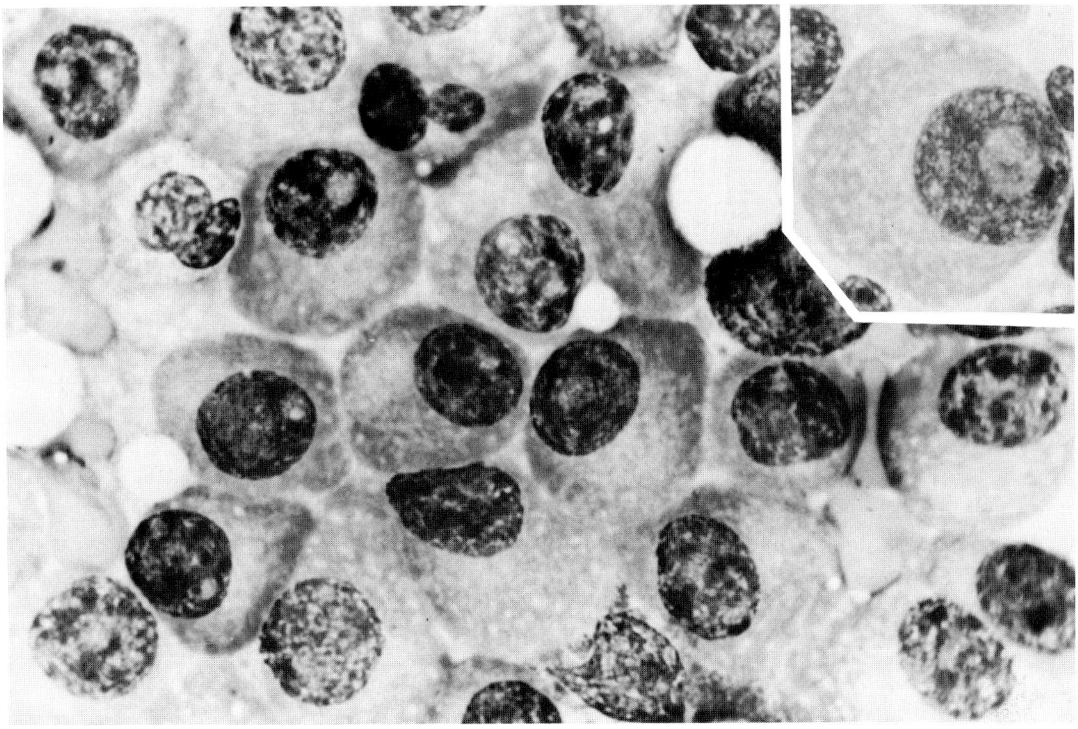

Bone marrow in plasma cell leukemia, showing malignant plasma cells with considerable pleomorphism and prominent nucleoli. Gigantic plasma cell with unusually large nucleolus (**inset**).

Figure 4-48.

Bone marrow in plasma cell leukemia, showing malignant plasma cells of various sizes and shapes. Malignant plasma cells containing Russell's bodies (**inset**). These eosinophilic-staining globular inclusions are thought to represent precipitation of immunoglobulin.

Figure 4-49.

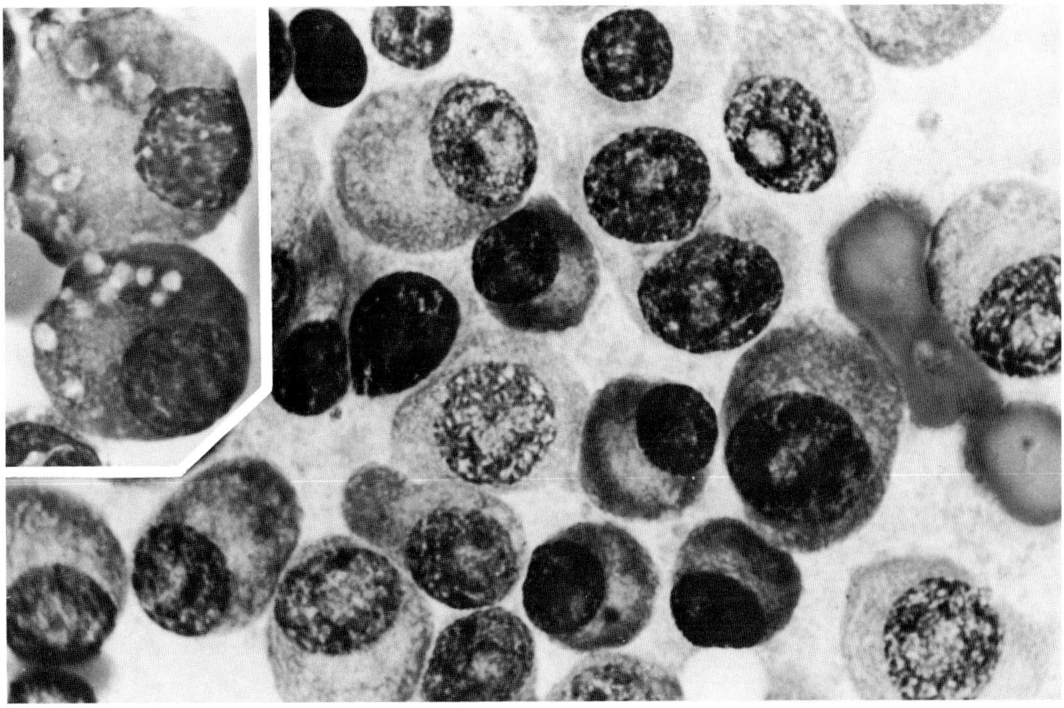

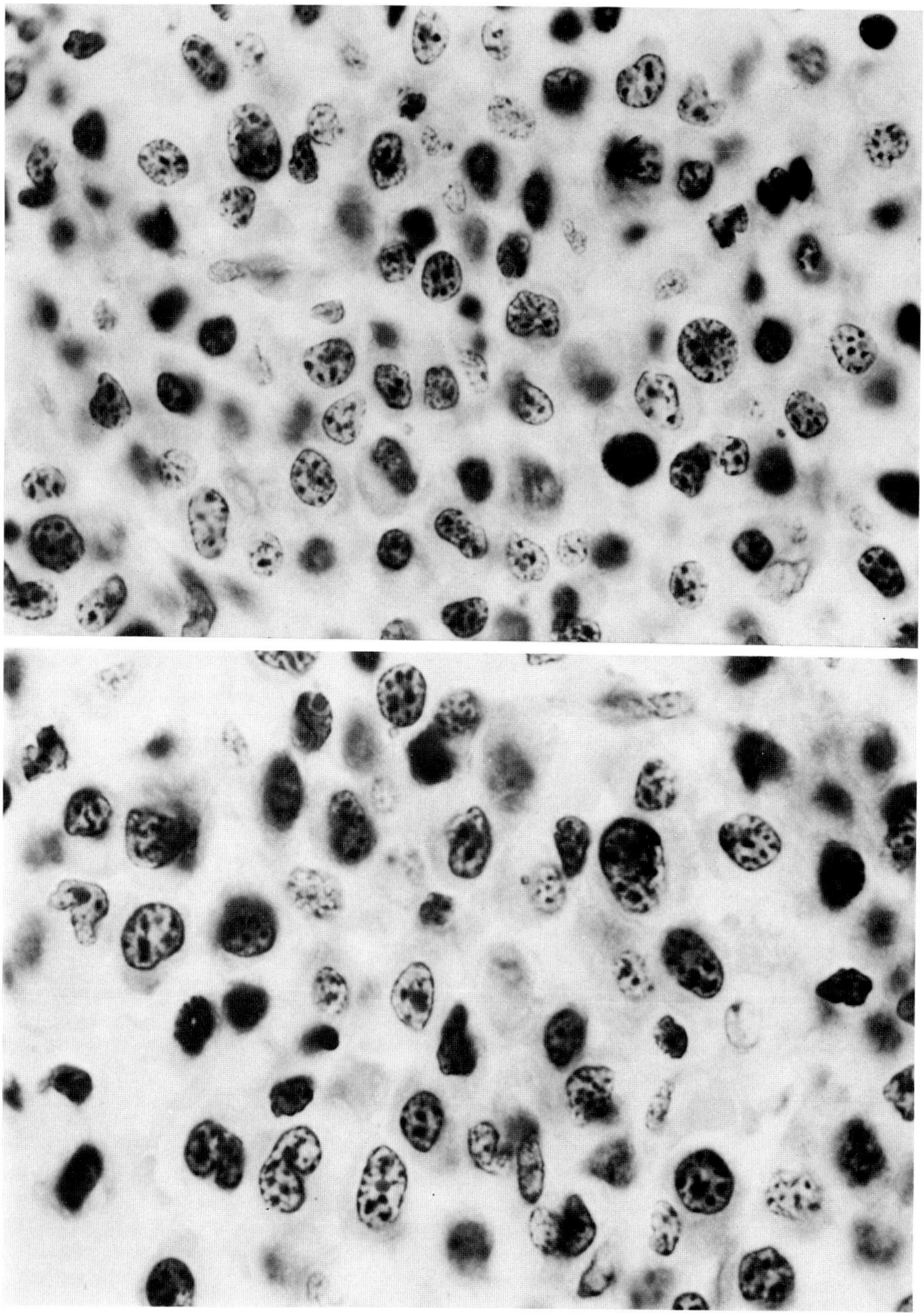

Figure 4-50. H & E section of bone marrow biopsy in plasma cell leukemia. Low-power view (**top**) and higher-power view (**bottom**) of marrow show nuclei with prominent aggregates of chromatin, sometimes arranged in "spoke of the wheel" pattern. Generally, this chromatin pattern is seen more often in normal plasma cells than in myeloma cells, in which the nuclear chromatin is more delicate and aggregates of chromatin smaller than in normal plasma cells.

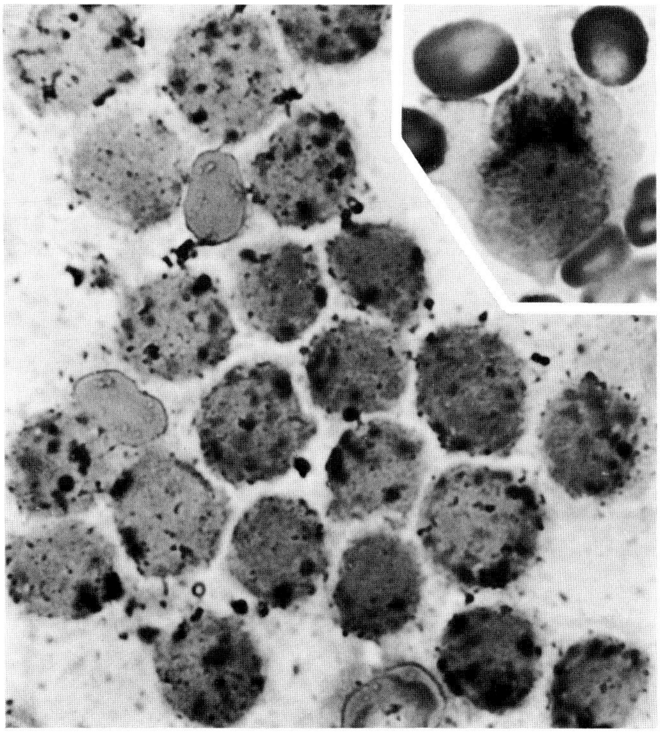

Nonspecific esterase activity in malignant plasma cells in plasma cell leukemia. Unipolar localization of acid phosphatase activity in malignant plasma cell in plasma cell leukemia (**inset**).

Figure 4-51.

Methyl green-pyronine stain in plasma cell leukemia. Cells with darkly stained cytoplasm indicate high cellular content of RNA, demonstrable with the pyronine stain.

Figure 4-52.

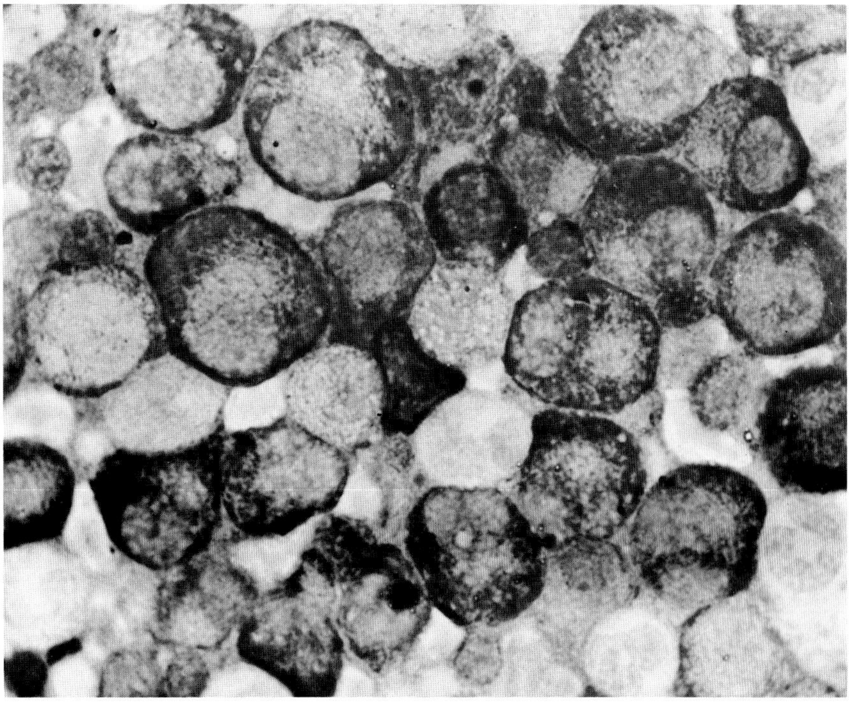

5

Cytochemistry of Leukemic Cells

Over the past century since the first description of differential staining of human blood cells by Ehrlich,[289] dyes and indicators have found increasing application in making better and more precise distinctions between the formed elements of the blood. Limitations inherent in panoptically stained specimens were appreciated early in the story of cytochemistry, and additional tests were developed in an effort to solve one of the early problems of morphologic hematology, namely, the distinction between monocytes, metamyelocytes, and lymphocytes. To some extent, the same problems exist a century later.

In a sense, routine panoptic stains such as Wright's or Giemsa's represent cytochemistry because, using a mixture of dyestuffs, they effect differences in tinctorial properties of the different blood cells contained in a mixture of these cells. Presumably, the color differences are due to more important biochemical and biophysical differences among cell types. As yet, the substance or substances in a cell that have a specific affinity for a dyestuff are not understood. For example, the cytoplasm of monocytes stains lilac using Wright's or Giemsa's stain, whereas the granules of eosinophils stain orange. To date, the substance or substances within the monocyte cytoplasm that react with the components of the panoptic stain to produce a lilac color are unknown. Similarly, although granules of eosinophils have an affinity for eosin in the dye mixture, the precise substance or substances in the granule that interact with the dye to create the orange color are unidentified.

Recognizing that staining of blood cells by panoptic stains is largely empirical, although admittedly very useful, efforts were soon directed toward identification of properties that were distinctive for one cell type compared to another. Based upon the presence or absence of these properties, distinction between cell types could be made. The myeloperoxidase stain is an example of this concept. Soon after the division of blood cells into mononuclear and polynuclear,[289,290] it became apparent that more needed to be done to identify cells that did not fit readily into either category. Originally used in an effort to identify the oxidizing ferment in vegetables, the myeloperoxidase stain was soon applied to the study of normal and leukemic blood cells.[25,371,800,933,943] It identifies a distinctive property not found in lymphocytes, and remains one of the most useful tests in the cytochemistry of normal and leukemic blood cells. Similarly, the monocyte was first identified on the basis of what seemed to be a unique supravital staining reaction, in which monocytes contained a "rosette" of vacuoles stained by neutral red.[966] Subsequently, this rosette was identified in lymphocytes, and the neutral red-Janus green stain for moncytes fell into disuse. It was, however, the first of several tests over the next half century that dealt specifically with efforts to identify blood monocytes on the basis of distinctive staining properties and to distinguish them from granulocytic cells.

Over the past several decades and with the use of more sophisticated techniques to demonstrate enzymes and other nuclear and cytoplasmic constituents, it has become appreciated that cytochemistry represents biochemistry on a microscopic and submicroscopic level. Like any other discipline, cytochemistry has limitations, including instability of reagents, importance of fixatives, regulations of pH and metallic ion content, preservation of color reaction over a period of time, and variability of dye lots. Most important, cytochemistry must deal with the problem of specificity: Is the substance reputedly being measured by the test actually the substance itself? Cytochemistry must also deal with the problem of identity: Is the substance creating a color reaction with a dye in the test tube the same as the substance that may be causing the same color reaction on a microscopic slide or in a suspension of cells? Ideally, a cytochemical test for light microscopic use should be rapid, colorful, simple, and distinctive.

In an effort to improve the specificity of cytochemical reactions and to confer diagnostic value on these reactions, several parameters have entered the field of cytochemistry of blood cells. First, the specificity of cytochemical reactions can often be improved with the use of inhibitors. In the case of cytochemical identification of enzymes, the use of inhibitors has been especially helpful with acid phosphatase and nonspecific esterase. In hairy cells from patients with hairy cell leukemia, activity of acid phosphatase is resistant to inhibition by L-tartrate.[1109] This property forms the basis for one of the few diagnostic tests for hairy cells now available because it is found rarely if at all in other types of normal and abnormal blood cells. Similarly, activity of nonspecific esterase that is sensitive to inhibition by sodium fluoride is believed to be a property of cells of monocytic origin, including macrophages.[1108] Also, myeloperoxidase in eosinophils shows a distinctive resistance to inhibition by sodium cyanide.[1108]

In addition to the use of inhibitors to enhance the diagnostic value of cytochemical tests, differences in the spatial distribution and configuration of the cytochemical reaction product have been recognized. Basically, these involve differences in the staining pattern that may show some cellular selectivity. For example, the PAS stain for glycogen demonstrates a large, chunklike pattern of diastase-sensitive material in erythroblasts from patients with erythroleukemia.[408,411,412,505,519,821,825] In megaloblasts from patients with untreated pernicious anemia, glycogen appears as an intense but diffuse cytoplasmic staining.[499] In lymphoblasts from patients with acute lymphoblastic leukemia, glycogen appears as prominent but punctate deposits in the cytoplasm.[32,90,412,447]

Similar differences in staining pattern are found in the case of several enzymes. In normal and leukemic monocytes, activity of fluoride-sensitive nonspecific esterase activity is intense and diffuse.[317,1108,1110] In malignant cells from patients with histiocytic lymphoma, activity of nonspecific esterase is often in a unipolar distribution[507] Similarly, activity of acid phosphatase is intense and diffuse in hairy cells.[1109] In leukemic lymphoblasts from patients with T-cell acute lymphoblastic leukemia, activity of acid phosphatase shows an intense clustering in a unipolar area.[188,189,192,211,212,670]

By direct staining with one or more dyes, or as the result of more complex reactions involving sensitized dye complexes, it is now possible to distinguish between the various formed elements of the blood using criteria based primarily on distinctive properties of these cells rather than on their morphology. As a tool for detecting properties of cells, cytochemistry has found increasing applications as a cellular probe and as a rapid, inexpensive way of determining the origin of cell types. In recent years, developments in cytochemistry have been rapid. New uses have been found for traditional dyes, and new dyes have been synthesized to meet specific needs. Greater understanding of the interaction of dyes with cellular constituents has been attained, and many of the limitations of cytochemistry have been recognized. With further developments in the automation of cytochemistry and in the synthetic organic chemistry of dyestuffs, further improvements in the identification of cells can be anticipated in the near future.

Of the various cytochemical stains developed in recent years, a few have proved valuable in the diagnosis of the cytologic types of leukemia. A discussion of these stains follows.

Prussian blue stain

One of the oldest cytochemical tests, Prussian blue stain is an acidified solution of sodium nitroprusside.[198,507,701,771,777,1035] It identifies iron in the ferric or plus-3 oxidation state. In keeping with the current trend to identify distinctive staining patterns and the presence or absence of a substance or substances, Prussian blue stain in chronic refractory anemia, such as Di Guglielmo syndrome, shows large agglomerates of iron, indicating slow turnover of iron. In contrast, the pattern of iron storage in autoimmune hemolytic anemia is one of small particles, reflecting increased turnover and utilization of iron.

In the cytochemistry of leukemia, the Prussian blue reaction identifies ringed sideroblasts, especially in patients with preleukemic disorders of erythropoiesis and in other preleukemic myeloblastic or monocytic disorders. In these conditions, erythroblasts often show mitochondria surrounding the nucleus and containing ferric iron within the lamellae and cristae of mitochondria. When stained with the Prussian blue reaction, these mitochondria appear as a typical blue "necklace" pattern surrounding the nucleus (Fig. 5-1). Occasionally, the mitochondria containing iron are large and form clusters within one area of the cytoplasm of the erythroblasts. Rarely, the Prussian blue stain reaction reveals a diffuse staining of the cytoplasm of erythroblasts.

PRUSSIAN BLUE*

Fixative (formalin-methanol)

37% formalin	10 ml
100% methanol	90 ml

(store in refrigerator)

Solutions

A. 2% HCl

Concentrated HCl	2 ml
Distilled H_2O	100 ml

B. 2% Potassium ferrocyanide

Potassium ferrocyanide	2 g
Distilled H_2O	100 ml

(shelf life 6 weeks)

C. 0.1% Nuclear Fast Red

Nuclear Fast Red	0.1 gm
Aluminum Sulfate	5.0 gm
Distilled H_2O	100 ml

mix and filter through Whatman #1 paper

Procedure

1. Fix coverslips in coplin jar for 30 min.
2. Rinse in distilled H_2O.
3. Mix equal parts solution A and B.
4. Pour solution (step 3) into coplin jar for 30 min.
5. Rinse in distilled H_2O.
6. Counterstain with 0.1% nuclear fast red for 1 min.
7. Wash in distilled H_2O.
8. Air dry and mount with Permount.

Results

Substances containing ferric iron, such as hemosiderin, stain blue.

* Modified from Thompson SW, Hunt RD: Selected Histochemical and Histopathological Methods. Springfield, Thomas, 1966

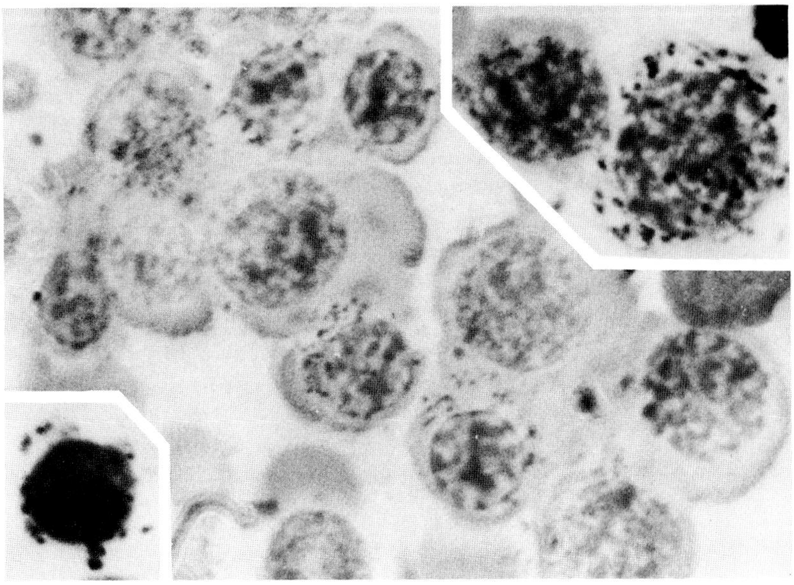

Figure 5-1. Prussian blue stain for iron, showing siderotic "granules" in the cytoplasm of erythroblasts from bone marrow in chronic erythremic myelosis.

PAS stain

Approximately four decades after the advent of the myeloperoxidase reaction, the PAS stain for glycogen was applied to the study of blood cells.[30,32] Soon thereafter, it was appreciated that erythroblasts in various types of leukemia, particularly erythroleukemia, contained substantial amounts of glycogen and that the glycogen often appeared as large, chunklike aggregates,[825] as seen in Figure 5-2. When leukemic lymphoblasts are stained with the PAS reagent, glycogen appears as large and small punctate particles surrounding the nuclei in some lymphoblasts, and large aggregates in other lymphoblasts (Fig. 5-3 and Plate 5C). In contrast, the PAS reaction in leukemic myeloblasts and monocytes shows a fine, dustlike pattern of glycogen distribution or, more often diffuse pale staining. In erythroblasts from patients with chronic erythremic myelosis (Plate 5*A*), the PAS stain is diffusely positive in many large megaloblastoid intermediate macronormoblasts, and it shows punctate positivity in proerythroblasts.[29,32,75–79,89,90,130,228,309,315,341,347,348,412,424,444,622,625,685,728,827,841,842,856,923,949,1024–1026,1072,1105]

Currently, in the cytochemical diagnosis of leukemia, the PAS reaction is most useful in detecting the large, blocklike pattern of glycogen in erythroblasts in erythroleukemia, in demonstrating the blocklike and punctate staining pattern of glycogen in lymphoblasts in acute lymphoblastic leukemia, and in determining whether erythroblasts in chronic erythremic myelosis and "preleukemia" are PAS positive or PAS negative. Although the pathophysiologic basis for these staining patterns is as yet unknown, at least they are sufficiently distinctive to make them useful diagnostically.

PAS (PERIODIC ACID-SCHIFF)*

Solutions

1. Schiff's reagent

Basic fuchsin	1.0 g
Sodium metabisulfite	2.0 g
IN HCl	10.1 ml
Distilled H_2O	200.0 ml
Activated charcoal	1.0 g

Bring H_2O to boiling, remove from heat, and very slowly add basic fuchsin. When fuchsin is dissolved, cool and filter. Add sodium metabisulfite and IN HCl. Let solution bleach for 24 hr. Add charcoal and shake for 1 min. Filter through Whatman #1 filter paper. Solution should appear colorless. Store in refrigerator. Discard when reagent becomes light pink.

* Modified from Thompson SW, Hunt RD: Selected Histochemical and Histopathological Methods. Springfield, Thomas, 1966

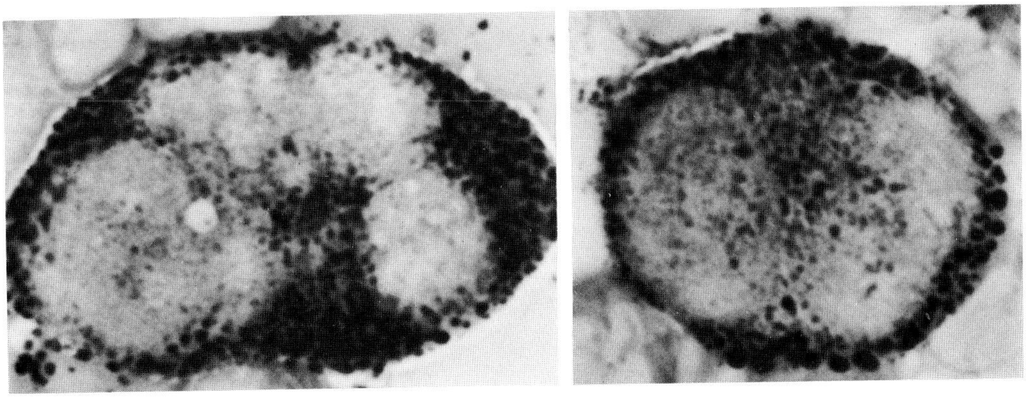

Figure 5-2. PAS stain, showing blocklike pattern of glycogen in cytoplasm of giant erythroblasts from bone marrow in erythroleukemia. (From Kass L: Enzymatic abnormalities in erythroleukemia. Acta Haematol 59:302, 1978. Reproduced with permission of S. Karger AG, Basel, Switzerland)

Figure 5-3. PAS stain, demonstrating punctate positive reaction for glycogen in the cytoplasm of leukemic lymphoblasts in acute lymphoblastic leukemia.

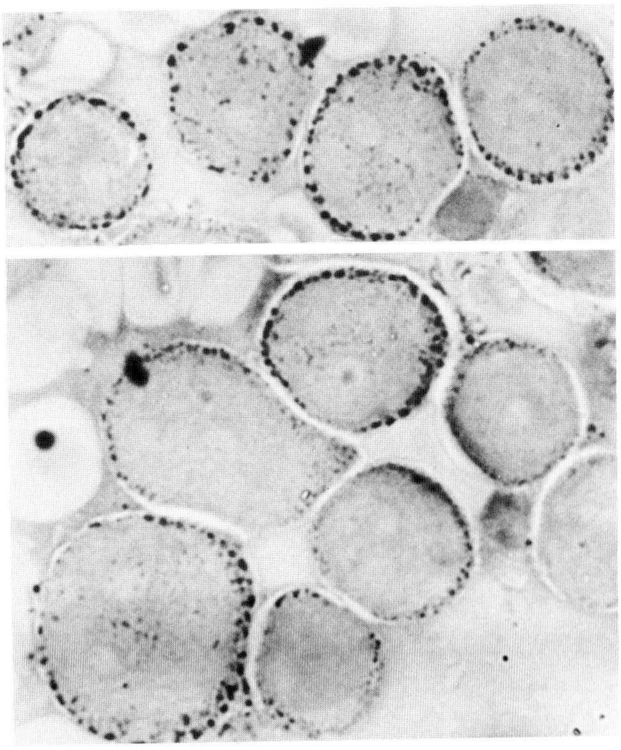

2. 1% Periodic acid

Periodic acid	1 g
Distilled H_2O	100 ml

3. Sulfurous acid rinse
   a. 10% sodium metabisulfite solution

Sodium metabisulfite	10 g
Distilled H_2O	100 ml

   b. IN HCl

HCl	83.5 ml
Distilled H_2O	916.5 ml

To make sulfurous acid rinse:
Put 6 ml of solution 3a in 100 ml flask. To this, add 5 ml solution 3b. Dilute to 100 ml with distilled H_2O (make fresh).
Solutions 3a and 3b may be kept 4 months.

Procedure

1. In staining jar, fix blood films or marrows in esterase fixative (*vide infra*) for 15 min.
2. Wash well in distilled H_2O.
3. Place in periodic acid for 10 min.
4. Wash in distilled H_2O.
5. Place in Schiff's reagent for 30 min.
6. Rinse in distilled H_2O, followed by three changes freshly prepared sulfurous acid rinse.
7. Wash in distilled H_2O.
8. Counterstain with Mayer's hematoxylin for 8 min.
9. Wash in distilled H_2O.
10. "Blue" in NH_4OH water (6 drops of 28% NH_4OH per 500 ml H_2O; 6 quick dips).
11. Wash in distilled H_2O.
12. Air dry and mount with Permount.

Diastase may be used to confirm that the staining reaction was due to glycogen.

Methyl green-pyronine stain for RNA

Combined with methyl green as the nuclear counterstain, pyronine has proven to be a valuable reagent for detecting the presence or absence of RNA, as well as increased or decreased amounts of RNA, on the basis of differences in the intensity of the staining reaction.[260,550,589] Although pyronine may identify several substances other than RNA,[1035] its staining affinity for this substance is believed to be reasonably specific. In the cytochemical study of leukemic blasts, pyroninophilia can be found in most types of blasts, and by itself it is not a useful diagnostic tool. However, pyroninophilia is unusually intense in lymphoid cells from patients who have the leukemic phase of Burkitt's lymphoma (Plate 5*F*). Combined with other diagnostic hallmarks of this disorder, pyroninophilia of Burkitt lymphoma cells is a distinctive finding.

In other types of hematopoietic cells, pyroninophilia may also be intense, and as such it helps to distinguish these cells from other cells that show little if any pyroninophilia. For example, normal proerythroblasts, early intermediate normoblasts, plasma cells, and megakaryocytes show intense pyroninophilia. Correspondingly, pathologic erythroblasts in erythroleukemia, as well as pathologic plasma cells in multiple myeloma and plasma cell leukemia show intense pyroninophilia (Plate 5*E*). Although the staining reactions are usually intense and diffuse in these cells, occasionally a striated or corrugated staining pattern of RNA can be detected.

METHYL GREEN-PYRONINE (METHOD 1)*

FAA Fixative

95% ethanol	90 ml
Glacial acetic acid	5 ml
37% formalin	5 ml
(store in refrigerator)	

Staining Solution

Methyl green	0.5 g
Pyronine Y	0.2 g
Dissolve in 100 ml	.10 M Sodium acetate (1.36 g Sodium acetate per 100 ml distilled H_2O)

Adjust to pH 4.4 with glacial acetic acid

Procedure

1. Fix in cold FAA fixative for 1 hr.
2. Rinse in distilled H_2O.
3. Stain in methyl-green pyronine for 30 min.
4. Wash in distilled H_2O air dry, and mount with Permount.

Results

DNA stains green and RNA stains red.

* Modified from Thompson SW, Hunt RD: Selected Histochemical and Histological Methods. Springfield, Thomas, 1966

METHYL GREEN-PYRONINE (METHOD 2)*

Fix coverslips or slides in Carnoy's fixative for 1 hr. Wash in distilled water and air dry.

Staining Solution

Methyl Green (Stock)

Methyl green	2.1 g
Distilled water	75.0 ml

Pyronine GS (Stock)

Pyronine GS	0.9 g
Distilled water	75.0 ml

Sodium Acetate (Stock)

Sodium acetate	1.64 g
Distilled water	100.0 ml

Acetic Acid (Stock)

Glacial acetic acid	1.2 ml
Distilled water	100.0 ml

Prepare the above solutions separately. Agitate occasionally during the 4 days following preparation.

Methyl Green-Pyronine (MGP) Working Solution

Methyl green (stock)	75.0 ml
Pyronine (stock)	75.0 ml

Cap solution tightly and place in a 60°C oven for 2 hr. Cool and add

Sodium acetate (stock)	90.0 ml
Acetic acid (stock)	60.0 ml

Allow solution to stand overnight at room temperature to reach pH 4.8.

Procedure

1. Place slides in MGP working solution for 10 min.
2. Blot slides dry with filter paper.
3. Differentiate slides in distilled water with 1 or 2 rapid dips.
4. Dip slides 3 times rapidly in each of 2 changes of cold 80% alcohol.
5. Dip slides 3 times in each of 2 changes of acetone.

Check results microscopically. If nuclei appear too bluish green, repeat the 2 rapid dips in distilled water as in step 4. If nuclei appear too red, dip

* Modified from Peter V. Emanuele, Histopathology Laboratories, Armed Forces Institute of Pathology, Washington, D.C.

slides in the cold 80% alcohol as in step 5. Keep alcohol in the refrigerator until ready to use.

6. Dip slides 2 times in a change of equal parts of acetone-xylene.
7. Clear slides in 3 changes of xylene for 1 min each.
8. Mount coverglass with Permount.

Results

Compared with Method 1, Method 2 provides more intense pyroninophilia and contrast between nucleus and cytoplasm. DNA stains blue to blue green, and RNA stains red.

Myeloperoxidase

Nearly three quarters of a century old in terms of its application to the study of blood cells, the stain for myeloperoxidase had its origin in studies of the oxidizing ferment in vegetables. A number of authors then became aware of the usefulness of reagents such as benzidine in detecting activity of this enzyme. Subsequently, the enzyme assay for peroxidase was applied to the study of leukemic blood cells.[25,371,800,933,943] Soon it became apparent that with the use of this cytochemical test, several kinds of leukemic blasts could be detected, namely, those that had myeloperoxidase activity and those that did not. Because of the occurrence of myeloperoxidase activity in mature granulocytic cells such as neutrophils, it was established that any primitive-appearing leukemic blast that contained myeloperoxidase activity was of granulocytic origin and that cells that did not contain myeloperoxidase activity were of nongranulocytic or lymphoblastic origin.

Myeloperoxidase is a lysosomal enzyme found in primary granules or lysosomes of granulocytic cells.[7,9,10,21–23,23,31,36–43,45–59,53–55,64,70,138,148,154–165,180,197,200,210,216,220,221,238,244,245,276,279,280,287,303,308,328,339,354,359,364,371,372,383,393,394,427,432,433,435,462,467,468,471,472,483,485,522,525,532,558,572,573,581,588,593,602,603,608,614,615,626,629,634,651,652,654,679,680,705,708,729,730,731,734–737,739,742–744,747,748,762,764,765,769,770,774,785,788,793,801,804,811,820,821,823,830,833,845,854,861,863,865,893,895,904,906,910,914,924,925,936,937,939,940,944,964,974,979,982,986,1003,1006,1007,1041,1050,1051,1054,1080,1082,1084,1085,1088,1094,1112,1113,1122–1124,1129] Because of their high content of primary granules, promyelocytes and myelocytes stain intensely for myeloperoxidase activity, whereas neutrophils show weaker activity. Eosinophils contain a myeloperoxidase that is resistant to inhibition by cyanide.[1108] Normal monocytes as well as some leukemic monocytes and most leukemic myeloblasts contain peroxidase activity. Leukemic myeloblasts contain weak activity of myeloperoxidase (Fig. 5-4), and Auer rods may show strong activity. In promyelocytic leukemia, activity of myeloperoxidase is unusually strong in leukemic blasts (Fig. 5-5). The myeloperoxidase stain has been complemented by staining for lipid by Sudan black B. Probably, both of these stains provide comparable information regarding the lineage of the leukemic cell. Unfortunately, not all leukemic myeloblasts contain activity of myeloperoxidase, and cells of disputed lineage like these create diagnostic problems.

Although it has undergone a variety of modifications over the years, the myeloperoxidase reaction is still a mainstay in the cytochemistry of leukemia. With the growing recognition that benzidine may be a carcinogen, substitutes for benzidine have been reported, including o-tolidine, diaminobenzidine,* and pyrocatechol-phenylenediamine.[398] These may give results equivalent to those obtained with benzidine. As alternatives to benzidine, several other tests using single reagents have been developed.† These tests provide rapid visual distinctions between granulocytic cells and other cell types and, in some cases, distinctions between mature and immature granulocytes based on color differences. In recent years, various other substances have been successfully detected using an antibody to peroxidase,[279,280,371] thereby establishing new uses for a vintage cytochemical test.

* Richard Savage, M.D., personal communication.

† Kass L: Saturn blue—a new stain for granulocytic cells. Arch Pathol Lab Med 104:551, 1980

Kass L: Niagara sky blue 6 B—a new stain for granulocytic cells. Am J Clin Pathol 74:801, 1980

Kass L: Preferential staining of granulocytic cells by sulphonaphthyl red. Stain Technol 55:137, 1980

Kass L: Staining of granulocytic cells by chlorazol black E. Am J Clin Pathol (in press)

Figure 5-4. Myeloperoxidase activity in leukemic myeloblasts in acute myeloblastic leukemia, benzidine stain. →

Figure 5-5. Intense activity of myeloperoxidase in leukemic promyelocytes in acute promyelocytic leukemia, benzidine stain. →

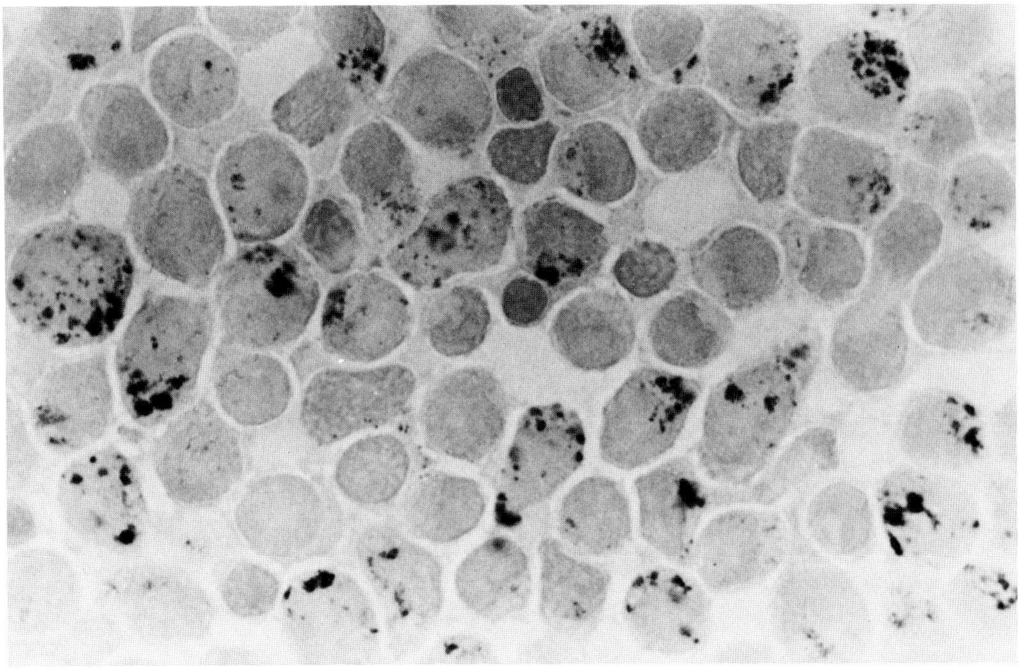

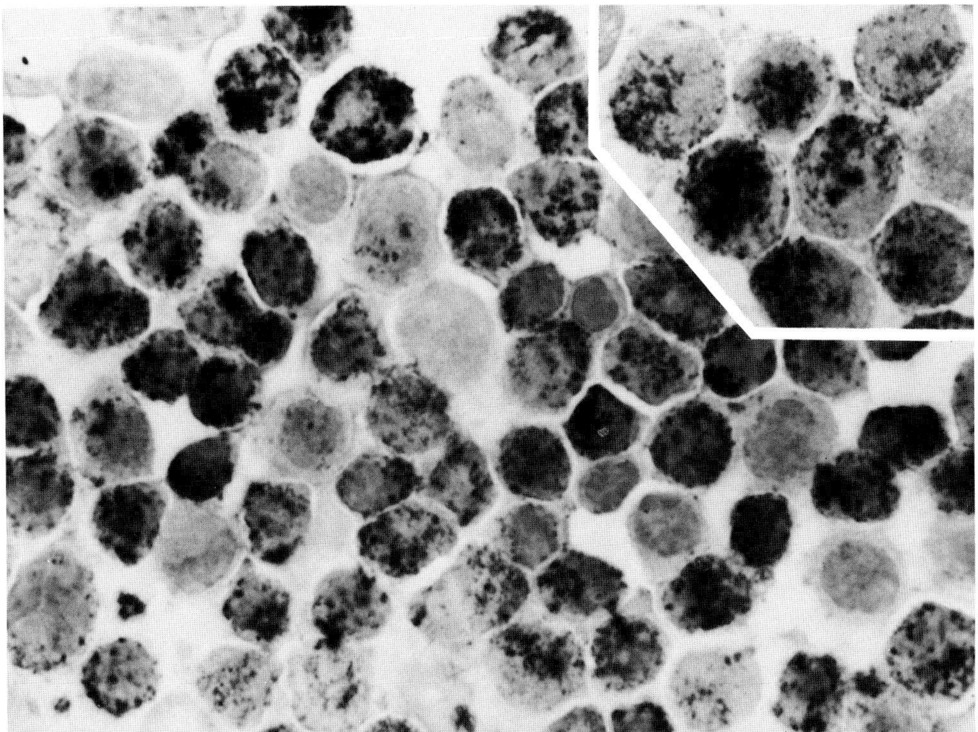

MYELOPEROXIDASE*

Solutions

1. Copper sulfate 0.5%

Copper sulfate ($CuSO_4$)	500 mg
Distilled H_2O	100 ml

2. Benzidine solution

Benzidine base	200 mg
Distilled H_2O	100 ml

Dissolve as much of the benzidine as possible in H_2O, then filter through Whatman #1 filter paper. To filtered solution, add 5 drops 3% H_2O_2. Both solutions 1 and 2 should be prepared freshly before use.

3. Safranin 1%

Safranin	1 g
Distilled H_2O	100 ml

Procedure

1. Using unfixed films of peripheral blood or bone marrow that are no more than 24 hr old, flood films with Solution 1 (copper sulfate) for 60 sec. Tip coverslip and touch edge to piece of filter paper to remove excess copper sulfate.
2. Flood coverslip with Solution 2 (benzidine solution) and pour off after 2 min.
3. Apply safranin solution to coverslip as counterstain for 2 min.
4. Wash with distilled water, air dry, and mount with Permount.

Results

Peroxidase-containing granules in normal granulocytic cells are stained bluish green. Eosinophil and basophil granules are intensely blue. The cytoplasm of monocytes is stained faintly, and their granules are blue. The nuclei of lymphocytes are red, with no blue or green cytoplasmic staining, and erythrocytes are unstained.

MYELOPEROXIDASE REACTION USING O-TOLIDINE†

1. Fix smears for 30 sec in a solution of 90 ml absolute ethanol and 10 ml 40% formalin.
2. Wash with tap water and air dry.
3. In a graduate cylinder, dissolve 250 mg o-tolidine in 6 ml absolute ethanol. Using distilled water, dilute the solution to 4 ml and add 0.02 ml 20% H_2O_2.

* Modified from Kaplow LS: Simplified myeloperoxidase stain using benzidine dihydrochloride. Blood 26:214, 1965

† Modified from Quaglino D, Flemans R: Peroxidase staining in leukocytes. Lancet 2:1020, 1958

4. Pour the unfiltered solution on coverslips of blood or bone marrow and leave for 7 min at room temperature.
5. Wash with distilled water and counterstain with 1% aqueous neutral red.

Results

Peroxidase activity appears as a yellow to yellow-brown reaction product.

Sudan black B stain

As a stain for a variety of lipids, including phospholipids, sterols, and neutral fat, Sudan black B has been used to distinguish cells of granulocytic origin, which show strong staining, from other types of cells that demonstrate little if any staining.[412,593,957] As such, staining of granulocytic cells with Sudan black B complements the reaction for myeloperoxidase, and, for the most part, the two tests show comparable sensitivity. In rare instances, however, leukemic blasts that do not demonstrate granules using conventional panoptic stains show weak sudanophilia and little if any activity of myeloperoxidase. In leukemic myeloblasts, Sudan black B shows weak staining, and in more mature granulocytic cells, such as promyelocytes, myelocytes, and neutrophils, staining is considerably stronger. In Auer rods and in granules of leukemic promyelocytes, staining by Sudan black B is unusually intense. In lymphoblasts from patients with acute lymphoblastic leukemia, little or no staining by Sudan black B is seen. Similarly, in monocytes from patients with acute monocytic leukemia, staining with Sudan black B is weak to absent.

SUDAN BLACK B STAIN*

Solutions

1. Formalin 37% aqueous solution
2. Sudan black B stock solution, made by dissolving 0.3 g Sudan black B powder in 100 ml absolute ethanol. The dye is poorly soluble in ethanol and may take up to several days to dissolve.
3. Buffer solution, made by combining 16 gm phenol with 30 ml absolute ethanol and adding this mixture to a solution of 0.3 g $Na_2PO_4 \cdot 12\ H_2O$ in 100 ml distilled water.
4. Staining solution, made by mixing 60 ml of Sudan black B stock solution with 40 ml buffer. Filter by suction, and store at room temperature for no longer than several weeks before making new solution.

* Modified from Hayhoe FGJ, Quaglino D. Doll R: The Cytology and Cytochemistry of Acute Leukaemias. MRC Special Reports Series 304. London, H.M. Stationery Office, 1964

5. Mayer's or Harris' hematoxylin can be used as a counterstain.

Procedure

1. Fix air-dried films of bone marrow or blood in formalin vapor for 10 min.
2. Immerse fixed film in Sudan black B staining solution in a coplin jar for 60 min.
3. Wash in 70% ethanol for 2 or 3 min to remove excess stain.
4. Wash in tap water and counterstain for 10 min in hematoxylin solution.
5. Wash in tap water for 5 min, air dry and mount face down with immersion oil or aqueous mounting medium, such as Plasdone for conventional light microscopy.

Results

Granules (lysosomes and specific granules) in granulocytic cells stain brown to black. A few darkly stained granules occur in monocytes. In plasma cells, erythroblasts, and megakaryocytes, brown or black granular structures are not observed.

Acid phosphatase

Using naphthol AS-TR phosphoric acid and various dyes—such as fast garnet GBC or fast corinth V—as couplers, activity of acid phosphatase can be demonstrated.[14,15,42,43,45–49,53,59,130,179,192,214,281,286,288,294,303,304,350,353,388,450,476,482,541,557,591,687,692,702,745,749,750,789,866,867,912,942,959,960,1016,1055,1056,1069,1087,1089,1095]

Acid phosphatase is a lysosomal enzyme and is believed to be a reliable marker for lysosomes. In granulocytes, acid phosphatase activity is contained within primary granules. Accordingly, activity of acid phosphatase is more intense in immature granulocytic cells—such as promyelocytes and myelocytes—that have a high content of primary granules or lysosomes than in neutrophils that have a lower content of lysosomes.

For the cytochemical study of leukemic blood cells, the presence or absence of acid phosphatase activity, the pattern of the staining reaction, and the response of the reaction to various types of enzyme inhibitors are important variables in the interpretation of the staining reaction. For example, myeloma cells and plasma cell leukemia cells show intense, chunk-like aggregates of reaction product, representing acid phosphatase activity in a perinuclear or unipolar distribution. In erythroblasts from patients with erythroleukemia, acid phosphatase activity is intense and localized in a unipolar area. In normal T-lymphocytes, acid phosphatase activity may be intense and sharply localized in a unipolar area, whereas in lymphocytes from patients with chronic lymphocytic leukemia, acid phosphatase activity is often undetectable.[264]

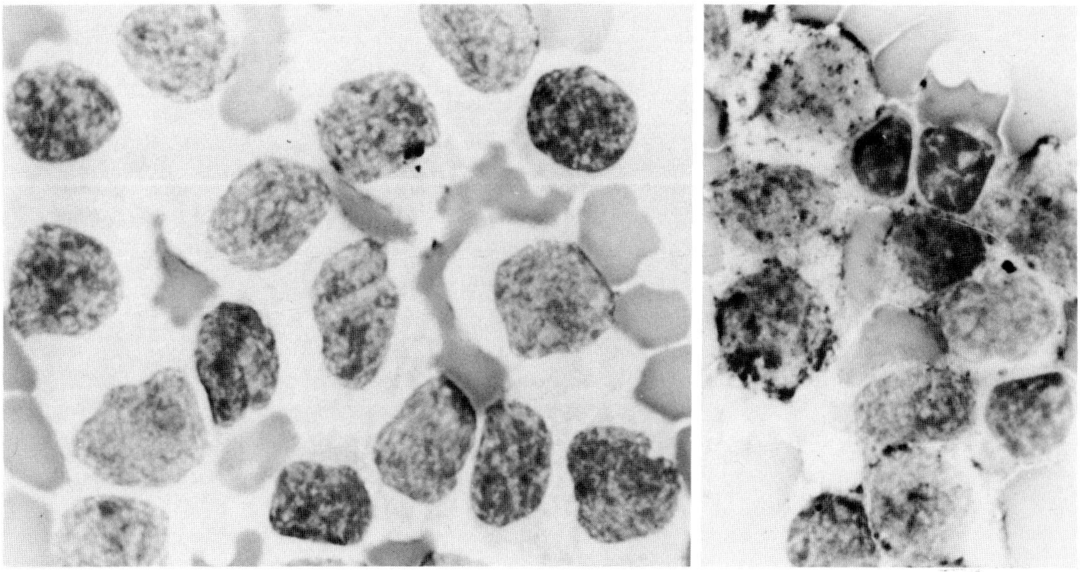

Figure 5-6. Hairy cells in hairy cell leukemia, Wright stain (**left**). Acid phosphatase activity, tartrate resistant, in hairy cells in hairy cell leukemia (**right**).

Figure 5-7. Lymphoblasts in T-cell convoluted type of lymphoblastic leukemia. Acid phosphatase activity in unipolar distribution in T-cell convoluted lymphoblasts (**inset**). In myeloma cells, similar unipolar localization of acid phosphatase activity can be found.

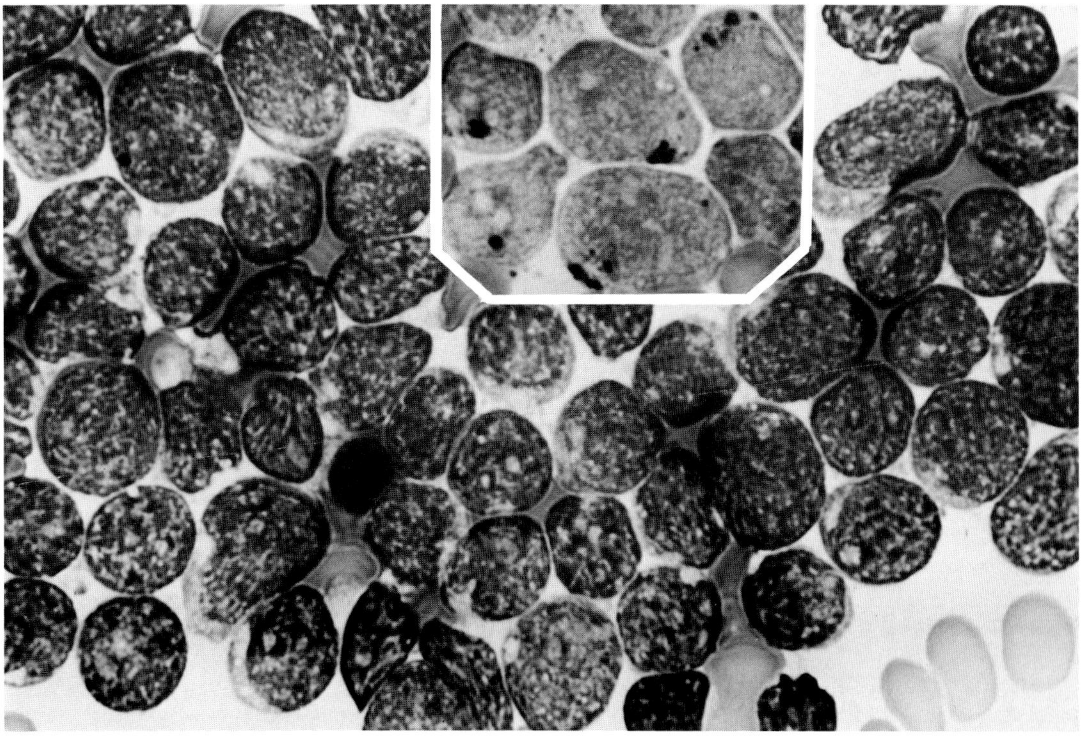

In hairy cell leukemia, the use of inhibitors has assumed diagnostic importance. Hairy cells contain activity of acid phosphatase that is resistant to inhibition by L-tartrate,[112,1109] as seen in Figure 5-6 and Plate 6*D*. This tartrate resistance is found almost exclusively in hairy cells and rarely in other atypical lymphoid cells, such as so-called lymphosarcoma cells and Sézary cells. Accordingly, the enzymatic reaction in the presence of the inhibitor has diagnostic value and it may be distinctive for hairy cells. Paradoxically, the enzymatic reaction for acid phosphatase may actually be more intense in the presence of the inhibitor than without it.

Detection of acid phosphatase activity also has diagnostic value in distinguishing between the various subtypes of lymphoblasts in acute lymphoblastic leukemia. In normal T-lymphocytes and in so-called T-cell convoluted lymphoblasts from patients with T-cell lymphoblastic leukemia, acid phosphatase activity shows a unique unipolar localization,[188,189] as seen in Figure 5-7. Similarly, in some leukemic monocytes, activity of acid phosphatase is intense and localized in a unipolar area (Fig. 5-8 and Plate 6*E*).

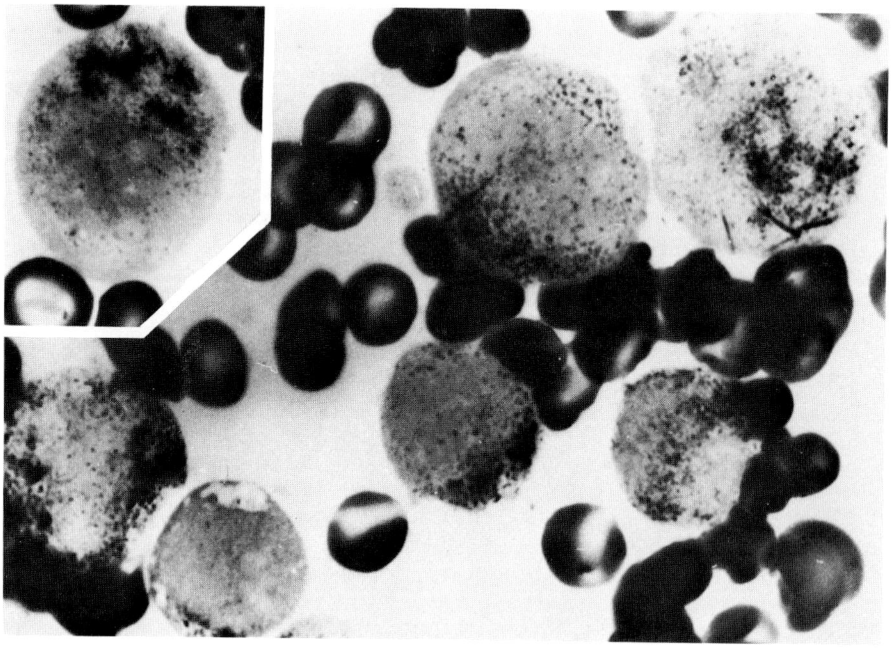

Figure 5-8. Acid phosphatase activity in leukemic monocytes in acute monocytic leukemia, showing intense activity localized in unipolar area in some cells.

ACID PHOSPHATASE*

Fixative (3% glutaraldehyde in 60% acetone)

70% glutaraldehyde	4.3 ml
Acetone	60.0 ml
Distilled H_2O	40.0 ml
(If 25% glutaraldehyde is used)	
25% glutaraldehyde	12 ml
Acetone	60.0 ml
Distilled H_2O	32.3 ml

Keep fixative in freezer at −4°C.

Solutions

1. 0.2 M acetate buffer pH 5.2

solution A		
	2.4 ml	glacial acetic acid
	197.6 ml	distilled H_2O
solution B		
	5.4 gm	sodium acetate
	200 ml	distilled H_2O
mix	42 ml	solution A with
	158 ml	solution B

2. 0.2 M L-tartrate pH 4.9
   Dissolve 3 g L-tartaric acid in 50 ml distilled H_2O. Add 35 ml IN NaOH and adjust to pH 4.9 by addition of IN NaOH. Bring volume to 100 ml with distilled H_2O.
3. 10% Manganese Chloride (Mn Cl_2)

Manganese Chloride	1 g
Distilled H_2O	10 ml

Preweighed Chemicals

Naphthol AS-TR phosphoric acid, free acid (Sigma)	15 mg
Fast corinth V (Chroma)	35 mg

Incubation Mixtures

A. Acid phosphatase without tartrate
   Make fresh each time:
   In a 50 ml beaker, dissolve the naphthol AS-TR phosphoric acid with a few drops dimethylformamide (DMF). Add the following:

Preweighed fast corinth V	35 mg
0.2 M acetate buffer pH 5.2	25 ml
Distilled H_2O	25 ml

* Modified from a technique developed by Thomas Dutcher, M.D.

Filter through Whatman #1 filter paper into another beaker, and to this add 5 drops 10% Mn Cl_2.

B. Acid phosphatase with tartrate
Make fresh each time:
To 36 ml of incubation mixture A above, add 4 ml of 0.2 M L-tartrate.

Procedure

1. Fix coverslips in fixative for 2 min.
2. Place one coverslip in incubation medium A and the other in incubation medium B.
3. Incubate for 2 hr at 37°C.
4. Wash several times in distilled H_2O.
5. Counterstain in Mayer's hematoxylin for 5 min.
6. Wash well in distilled H_2O.
7. "Blue" in NH_4OH water (6 drops 28% NH_4OH in 500 ml water; 5 quick dips).
8. Wash well in cold running H_2O.
9. Air dry and mount with Permount.

Results

Nearly all nucleated cells and platelets will have varying degrees of red reaction product on the film incubated in mixture A.

All nucleated cells and platelets will either be completely devoid of red reaction or will contain only minute amounts on the film incubated in mixture B.

On coverslips incubated in both mixtures A and B, hairy cells in hairy cell leukemia will show strong activity of acid phosphatase.

Esterases

With the use of naphthol AS-D chloroacetate as substrate and various azo dyes as couplers, a "specific" esterase can be demonstrated in the granules of granulocytic cells.[141,222,262,325,515–517,575,697,698,773,775,858,859,883] Using present techniques, activity of this enzyme is intense in both mature and immature granulocytic cells, and it can be demonstrated in tissue sections of bone marrow and other hematopoietic tissues. Activity of specific esterase is weak in leukemic myeloblasts (Fig. 5-9), and it may show weak activity in some leukemic monocytes. The cytochemical test for specific esterase is disadvantageous because there is often considerable nonspecific precipitation of dye coupler throughout the slide preparation, making it difficult to distinguish between weak enzymatic activity and stain artifact.

Nonspecific esterase is a ubiquitous enzyme found in the cytoplasm of many kinds of cells. Using alpha naphthyl acetate or alpha

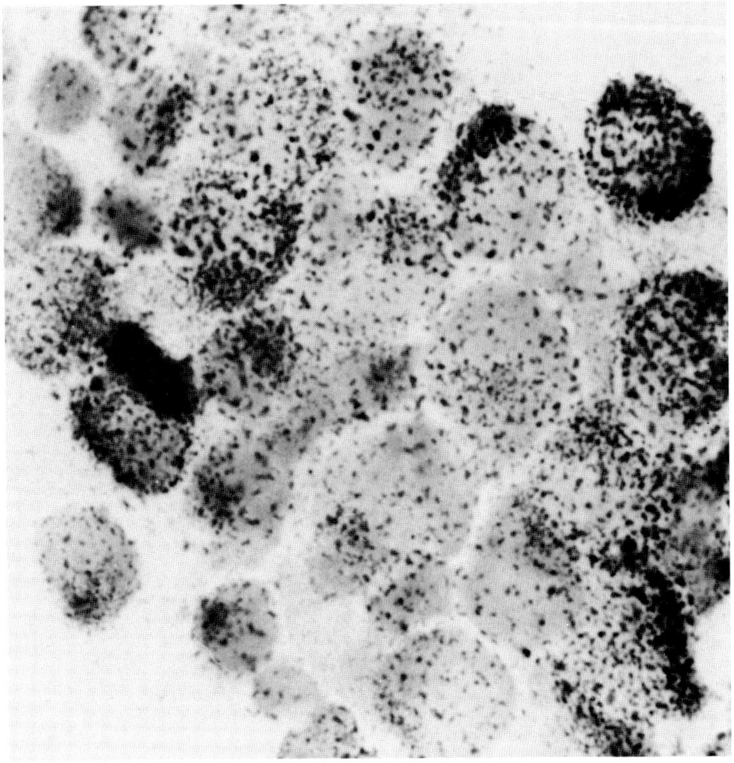

Figure 5-9. Activity of specific esterase in leukemic myeloblasts in acute myeloblastic leukemia, using naphthol AS-D chloroacetate and fast blue BBN.

naphthyl butyrate as substrate and several dyes as couplers and indicators, activity of nonspecific esterase can be detected.[51,71,141,144,171,180,189,231,241,297,298,301,306,317,323,329,355,357,358,421,422,425,430,513,514,516,517,526,540,542,590,599,605,614,703,710,828,834,898,903,914,915,919,920,948,961,962,1030,1039,1073,1088,1115] Chemically, the reaction for nonspecific esterase activity involves liberation of the highly reactive alpha naphthol. This substance couples with a sensitized dye coupler such as hexazotized new fuchsin or pararosanaline and forms a colored insoluble complex at or near the site or sites of enzymatic activity. With the use of sodium fluoride as an enzyme inhibitor, activity of nonspecific esterase can be further refined. In cells of monocytic origin, including normal and leukemic monocytes, nonspecific esterase is sensitive to inhibition by sodium fluoride, and it may in fact be obliterated by the fluoride.[317] In contrast, nonspecific esterase activity in granulocytic cells is relatively resistant to inhibition by sodium fluoride.

As with other types of cytochemical stains, the pattern of the staining reaction may have diagnostic value. For example, normal and leukemic monocytes show an intense but diffuse cytoplasmic staining, indicating activity of nonspecific esterase (Fig. 5-10 and Plate 6*A*). In contrast, leukemic cells in leukemic histiocytic lymphoma and leukemic lymphoblasts in T-cell convoluted acute lymphoblastic leukemia show a unique unipolar localization of the enzyme,[44] as seen in Figure 5-11 and Plate 6*B*. Similar unipolar localization of nonspecific esterase activity can be found in myeloma cells and in leukemic plasma cells. In other types of hematopoietic cells, such as erythroblasts in erythroleukemia, nonspecific esterase activity is intense and diffuse, but as yet it does not have diagnostic value.

To date, the most widely accepted use for this cytochemical test in a diagnostic setting is demonstrating the intense, diffuse fluoride sensitive activity found in normal and leukemic monocytes. Of the few tests available for distinguishing monocytes from other kinds of cells, the reaction for nonspecific esterase activity is probably the most reliable and consistent. For optimal results, reagents should be made freshly before use each use, and fresh substrate directly from the reagent bottle should be used. At temperatures around 4°C, substrate deterioration occurs in approximately 3 to 4 weeks.

SPECIFIC ESTERASE AND NONSPECIFIC ESTERASE*

Chemicals

1. 1 mg naphthol AS-D chloroacetate
2. 10 mg alpha naphthyl acetate
3. 4 mg fast blue BBN

* Modified from Yam LT, Li CY, Crosby WJ: Cytochemical identification of monocytes and granulocytes. Am J Clin Pathol 55:283, 1971

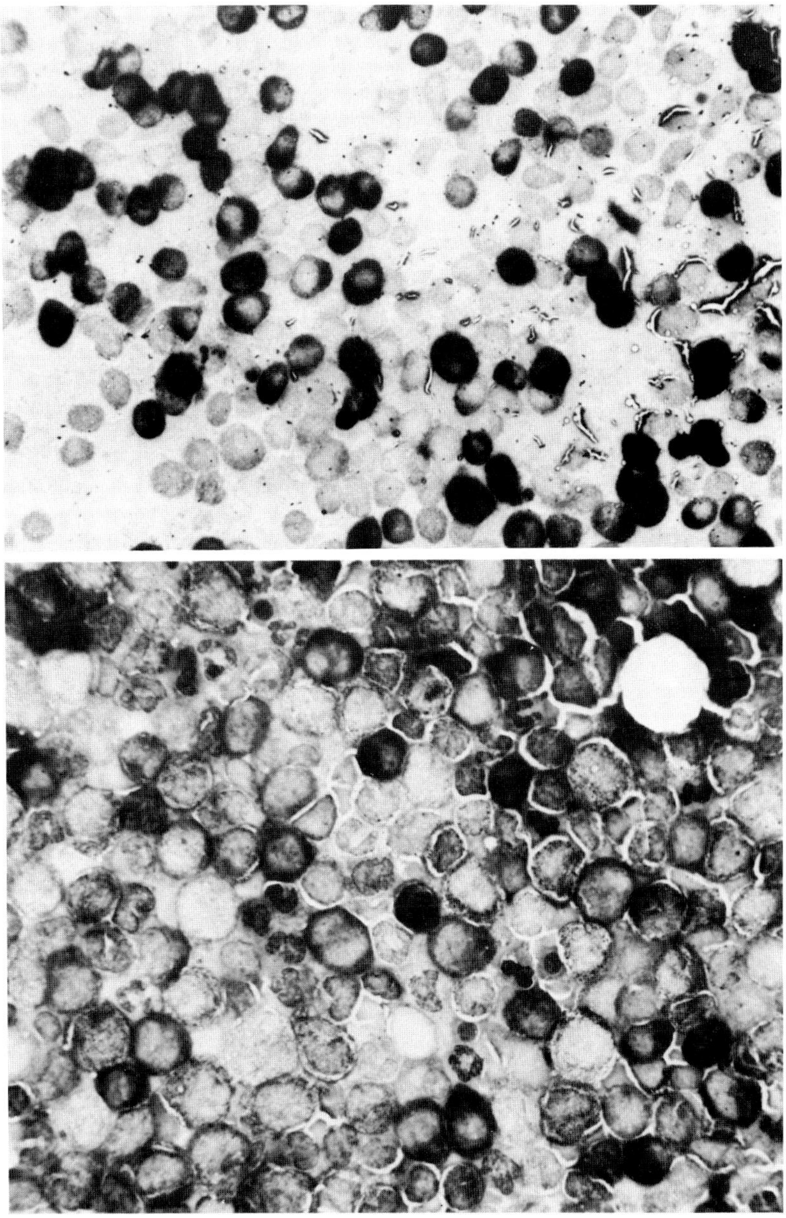

Nonspecific esterase reaction, alpha naphthyl acetate and pararosaniline, depicting intense and diffuse activity in leukemic monocytes from two cases of acute monocytic leukemia. **Figure 5-10.**

Solutions

1. Buffered formalin acetone (esterase fixative)
   - 20 mg $Na_2\ HPO_4$
   - 100 mg $K\ H_2\ PO_4$
   - 30 ml Distilled H_2O
   - 45 ml Acetone
   - 25 ml 37% formalin

   Adjust to pH 6.6
2. Phosphate buffer (0.15 M; pH 7.6)
   a. 20.7 g $Na\ H_2\ PO_4$ (monobasic) per 1000 ml distilled H_2O
   b. 21.3 g $Na_2\ HPO_4$ (dibasic) per 1000 ml distilled H_2O
   c. Mix 130 ml of *a* and 870 ml of *b*, and adjust to pH 7.6.
3. 4% Sodium Nitrite—Must be made freshly before use with fresh sodium nitrite.

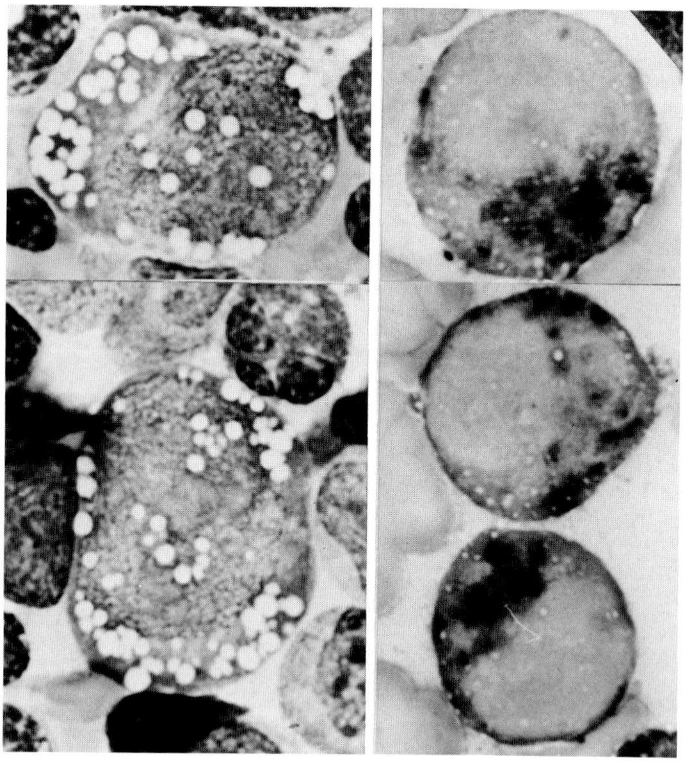

Figure 5-11. Histiocytic lymphoma cells, Wright stain (**left**). Nonspecific esterase reaction, alpha naphthyl acetate and pararosaniline, showing unipolar localization of enzymatic activity in leukemic histiocytic lymphoma cells (**right**). Similar localization of this enzyme is seen in myeloma cells and in T-cell convoluted leukemic lymphoblasts.

.2 g	Sodium nitrite
5 ml	distilled H_2O

4. 1% methyl green

1.4 g	sodium acetate
1.4 g	methyl green

Mix with 100 ml distilled H_2O, and adjust to pH 4.2 with acetic acid.

5. Pararosaniline

1 g	pararosaniline hydrochloride
20 ml	distilled H_2O
2 ml	concentrated HCl

Heat solution to dissolve dye, cool and filter. Store in refrigerator no longer than 2 months.

6. New fuchsin

1 g	new fuchsin
20 ml	distilled H_2O
2 ml	concentrated HCl

Heat solution to dissolve dye, cool and filter. Store in refrigerator. Either pararosaniline [step 5] or new fuchsin [step 6] can be used.

Procedure for Specific Esterase

1. Fix films in esterase fixative for 30–45 sec.
2. Add a few drops dimethylformamide to 1 mg naphthol AS-D chloroacetate. Use 15 ml beaker.
3. Add 9.5 ml phosphate buffer.
4. Add 5 mg fast blue BBN.
5. Mix well and filter directly onto films.
6. Incubate at room temperature for 40 min.
7. Wash with distilled H_2O.
8. Counterstain with 1% methyl green for 2 min.
9. Wash in distilled H_2O, air dry, and mount with Permount on cleaned glass slides.

Procedure for Nonspecific Esterase

1. In staining jars, fix blood and bone marrow films in esterase fixative for 30–45 sec.
2. Wash thoroughly in distilled water.
3. Make fresh 4% sodium nitrite solution.
4. Mix 0.5 ml of pararosaniline or new fuchsin and 0.5 ml sodium nitrite. Let react for 1 min only to effect hexazotization.
5. Add a few drops of acetone to 10 mg alpha naphthyl acetate. Use 15 ml beaker.
6. Add 0.6 ml of step 4 solution to step 5.
7. Add 8.9 ml phosphate buffer to step 6. Using concentrated

HCl or NaOH adjust to pH 6.1. Filter and pour into coplin jar containing bone marrow films.

8. Incubate for 1½ hr at room temperature.
   For fluoride inhibition, dissolve 75 mg sodium fluoride in 50 ml filtered incubation mixture, and incubate coverslips for 90 minutes at room temperature in small coplin jar.
9. Wash coverslips in distilled H_2O.
10. Counterstain in 1% methyl green for 2 min.
11. Wash in distilled H_2O.
12. Air dry and mount with Permount.

Note: Although it is best to store films in freezer, they may be kept at room temperature for 2 weeks without substantial change in enzyme activity.

Results

Specific esterase activity appears as blue stained particles corresponding in location to cytoplasmic granules. Both primary and secondary granules seem to contain activity of specific esterase. Accordingly, specific esterase activity is intense in promyelocytes as well as in neutrophils. In leukemic myeloblasts, specific esterase activity is present but weak. In Auer rods, specific esterase activity is strong. In erythroblasts from patients with erythroleukemia, specific esterase activity can be demonstrated, and is usually weak. Lymphoid cells including leukemic lymphocytes and leukemic lymphoblasts do not possess activity of specific esterase.

Nonspecific esterase activity appears as an orange to orange-brown cytoplasmic color. Depending on the type of cell, the intensity of staining can be weak or strong, and the pattern of staining can be diffuse or localized. Increased nonspecific esterase activity is found in a variety of cells including erythroblasts in chronic erythemic myelosis and erythroleukemia, normal and leukemic promyelocytes, normal and pathological megakaryocytes, macrophages, and hairy cells.

Diagnostically, demonstration of nonspecific esterase activity is useful in several instances. In normal and leukemic monocytes, activity of nonspecific esterase is intense. By adding sodium fluoride to the incubation medium, nonspecific esterase activity in monocytes and cells that derive from them, like macrophages, can be obliterated. In granulocytic cells and erythroblasts, nonspecific esterase activity is resistant to inhibition by sodium fluoride.

Localization of nonspecific esterase activity can have diagnostic value. In histiocytic lymphoma cells, T-cell convoluted leukemic lymphoblasts, normal T lymphocytes, normal plasma cells, and myeloma cells, the reaction product appears chunklike rather than diffuse. In histiocytic lymphoma cells, the reaction product can appear in a unipolar distribution.

Alkaline phosphatase

Using naphthol AS-MX phosphoric acid as substrate, activity of alkaline phosphatase can be demonstrated.[6,45–48,52,53,93,94,107,209,362,363,404–408,410,438,491,492,587,609,627,719,784,822,869,887,997,1000,1051,1118] In blood cells, alkaline phosphatase activity occurs in secondary or specific granules of granulocytes. Compared with myeloperoxidase activity, which is regarded as a marker for lysosomes or primary granules, alkaline phosphatase is believed to be a marker for secondary granules. Because of its localization in secondary granules, alkaline phosphatase activity is highest in granulocytic cells containing many secondary granules, such as neutrophils. In contrast, activity of alkaline phosphatase is weak or absent in less mature granulocytes, such as promyelocytes and myelocytes, that contain few secondary granules. In monocytes, activity of alkaline phosphatase is weak to absent.

To date, the most widely accepted use for the alkaline phosphatase reaction in the study of leukemic blood cells has been the demonstration of reduced to absent activity of alkaline phosphatase in granulocytes from patients with chronic granulocytic leukemia. In the majority of patients with this disorder, activity of alkaline phosphatase is virtually absent in granulocytes at all stages of maturation. Demonstration of alkaline phosphatase activity may be a useful diagnostic adjunct in distinguishing chronic granulocytic leukemia from other conditions that can resemble it, such as leukocytosis or leukemoid reaction. In the latter two conditions, as well as in agnogenic myeloid metaplasia, activity of alkaline phosphatase is normal or elevated. In patients with chronic granulocytic leukemia who show terminal evolution into acute leukemia, leukemic blasts often show high alkaline phosphatase activity.

LEUKOCYTE ALKALINE PHOSPHATASE (LAP)*

Chemicals and Solutions

1. Fast blue RR salt capsule
2. Naphthol AS-MX phosphate alkaline solution (obtained commercially)
3. Mayer's hematoxylin solution

Procedure

Fixing

1. Make films and air dry.

* Modified from Kaplow LS: A histochemical procedure for localizing and evaluating leukocyte alkaline phosphatase activity in smears of blood and marrow. Blood 10:1023, 1955

2. Using a coplin jar, fix blood films in cold 10% formalin 90% methanol for 30 sec.
3. Rinse thoroughly but gently with tap water.
4. Air dry.
5. Store in freezer.

Staining

1. Pipet 2 ml of naphthol AS-MX phosphate alkaline solution into an Erlenmeyer flask.
2. Add 48 ml of distilled water.
3. Add the contents of a fast blue RR salt capsule and mix thoroughly by shaking 15 to 30 sec.
4. Filter through Whatman #2 filter paper onto blood films in coplin jar.
5. Incubate for 30 min at room temperature and away from direct light.
6. Wash gently with tap water.
7. Counterstain with Mayer's hematoxylin solution for 3 min.
8. Wash with tap water for 20 sec, dry, and mount on glass slide.

PRECIPITATED AZO DYE IN CYTOPLASM

Cell Ratings	Size of Granule	Intensity of Stain
0	No precipitated dye	
1+	small	faint to moderate
2+	small to medium	moderate to strong
3+	medium to large	strong
4+	medium and large	brilliant

Multiply the number of cells counted in each rating by that rating. Add the totals together to obtain the total number of units. Normal is 11 to 95 units. In granulocytes from patients with chronic granulocytic leukemia, activity of alkaline phosphatase is weak to absent.

References

1. Abbrederis K, Schmalzl F, Braunsteiner H: Differential diagnosis of acute leukemia by means of cytochemical methods. Schweiz Med Wochenschr 99:1425, 1969
2. Abbrederis K, Schmalzl F, Braunsteiner H: Zur typenverteilung, Alters- und Geschlechtsdisposition der akuten Leukämien des Erwachsenen. Blut 27:302, 1973
3. Ackerman G: Microscopic and histochemical studies on the Auer bodies in leukemic cells. Blood 5:847, 1950
4. Ackerman GA: Morphology and cytochemistry of the granule–vacuole body of leukemic cells. Blood 12:443, 1957
5. Ackerman GA: Histochemical demonstration of dehydrogenase activity in the cells of normal human blood and bone marrow. J Biophys Biochem Cytol 8:61, 1960
6. Ackerman GA: Substituted naphthol AS phosphate derivatives for the localization of leukocyte alkaline phosphatase activity. Lab Invest 11:563, 1962
7. Ackerman GA: Histochemical differentiation during neutrophil development and maturation. Ann NY Acad Sci 113:537, 1964
8. Ackerman GA: Ultrastructural localization of glycogen in erythrocytes and developing erythrocytic cells in normal human bone marrow. Z Zellforsch Mikrosk Anat 140:433, 1973
9. Ackerman GA, Clark MA: Ultrastructural localization of peroxidase activity in normal human bone marrow cells. Z Zellforsch Mikrosk Anat 117:463, 1971
10. Ackerman GA, Clark MA: Ultrastructural localization of peroxidase activity in human basophil leukocytes. Acta Haematol (Basel) 45:280, 1971
11. Aksoy M, Erdem S, Din Col G: Leukemia in shoe-workers exposed chronically to benzene. Blood 44:837, 1974
12. Albrecht M, Fülle HH: Extremely small megakaryocytes in chronic myelocytic leukemia, acute leukemia, and erythroleukemia. Klin Wochenschr 52:649, 1974
13. Allegra SR, Broderick PA: Acute aleukemic megakaryocytic leukemia. Am J Clin Pathol 55:197, 1971
14. Allison AC: Lysosomes in cancer cells. J Clin Pathol (Suppl) 7:43, 1974
15. Allison AC, Mallucci L: Lysosomes in dividing cells, with special reference to lymphocytes. Lancet 2:1371, 1964
16. Altman AJ, Palmer CG, Baehner RL: Juvenile "chronic granulocytic" leukemia: a panmyelopathy with prominent monocytic involvement and circulating monocyte colony-forming cells. Blood 43:341, 1974
17. Anderson RC: Familial leukemia. Am J Dis Child 81:313, 1951
18. Andreewa P, Huhn D, Thiel E, et al: Comparison of enzyme-cytochemical findings and immunological marker investigations in acute lymphatic leukemia [ALL]. Blut 36:299, 1978

19. Anteunis A, Audebert AA, Krulik M, et al: Acute eosinophilic leukemia: An ultrastructural study. Virchows Arch 27:237, 1978
20. Appay MD, Bariety J, Bretton R: Characterization by immunoperoxidase of the lymphocytes with bundle-shaped tubular [BST] inclusions in human normal peripheral blood. Blood 55:67, 1980
21. Archer GT, Air G, Jackas M et al: Studies on rat eosinophil peroxidase. Biochim Biophys Acta 99:96, 1965
22. Archer GT, Hirsch JG: Isolation of granules from eosinophil leucocytes and study of their enzyme content. J Exp Med 118:277, 1963
23. Archer RK, Broome J: Studies on the peroxidase reaction of living eosinophils and other leucocytes. Acta Haematol (Basel) 29:147, 1963
24. Arlin ZA, Chiao JW, Fried J et al: Relationship of a T-lymphocyte marker to phase of cell cycle and morphology of leukemic cells. Cancer 39:1101, 1977
25. Armitage FL: Modified peroxidase stain for blood and bone marrow films. J Pathol Bacteriol 49:579, 1939
26. Asamer H, Schmalzl F, Braunsteiner H: Der immunzytologische Lysozymnachweis in menschlichen Blutzellen. Acta Haematol (Basel) 41:49, 1969
27. Asamer H, Schmalzl F, Braunsteiner H: Immunocytological demonstration of lysozyme (muramidase) in human leukaemic cells. Br J Haematol 20:571, 1971
28. Ashburn P, Cooper MR, McCall CE et al: Nitroblue tetrazolium reduction: false positive and false negative results. Blood 41:921, 1973
29. Astaldi G, Mauri C: Recherches sur l'activité proliférative de l'hémocytoblaste de la leucémie aiguë. Rev Belge Pathol et Med exper 23:69, 1953
30. Astaldi G, Rondanelli EG, Bernardelli E et al: An abnormal substance present in the erythroblasts of thalassaemia major; cytochemical investigations. Acta Haematol (Basel) 12:145, 1954
31. Astaldi G, Strosselli E: Histochemistry of the giant granulation of polynuclear cells in Chediak's anomaly. Nouv Rev Fr Hematol 1:626, 1961
32. Astaldi G, Verga L: The glycogen content of the cells of lymphatic leukaemia. Acta Haematol (Basel) 17:129, 1957
33. Audebert AA, Zittoun R, Cadiou M et al: Clinical conference: Modes of blast transformation in chronic myeloid leukemia. Nouv Rev Fr Hematol 15:270, 1975
34. Auer J: Some hitherto undescribed structures found in the large lymphocytes of a case of acute leukaemia. Am J Med Sci 131:1002, 1906
35. Azar HA: Plasma cell myelomatosis and other monoclonal gammopathies. Pathol Annu 7:1, 1972
36. Baggiolini M: The enzymes of the granules of polymorphonuclear leukocytes and their functions. Enzyme 13:132, 1972
37. Baggiolini M, Bretz U, Gusus B: Biochemical characterization of azurophil and specific granules from human and rabbit polymorphonuclear leukocytes. Schweiz Med Wochenschr 104:129, 1974
38. Baggiolini M, De Duve C, Masson PL et al: Association of lactoferrin with specific granules in rabbit heterophil leukocytes. J Exp Med 131:559, 1970
39. Baggiolini M, Hirsch JG, De Duve AC: Resolution of granules from rabbit heterophil leukocytes into distinct populations by zonal sedimentation. J Cell Biol 40:529, 1969
40. Baggiolini M, Hirsch JG, De Duve AC: Further biochemical and morphological studies of granule fractions from rabbit heterophil leukocytes. J Cell Biol 45:586, 1970
41. Bainton DF: Origin, content, and fate of PMN granules. In Williams RC Jr., Fudenberg HH (eds): Phagocytic Mechanisms in Health and Disease, p 123. New York, Intercontinental Medical Book Corporation, 1972
42. Bainton DF: Sequential degranulation of the two types of polymorphonuclear leukocyte granules during phagocytosis of microorganisms. J Cell Biol 58:249, 1973

43. Bainton DF: Neutrophil granules. Br J Haematol 29:17, 1975
44. Bainton DF, Bozdech MJ: Fine-structure localization of alpha–naphthyl esterases in human lymphocytes. Clin Res 27:634, 1979
45. Bainton DF, Farquhar MG: Origin and nature of polymorphonuclear leukocyte granules. J Cell Biol 27:6A, 1965
46. Bainton DF, Farquhar MG: Enzyme composition of polymorphonuclear leukocyte granules. J Cell Biol (Suppl) 31:8A, 1966
47. Bainton DF, Farquhar MG: Origin of granules in polymorphonuclear leukocytes. J Cell Biol 28:277, 1966
48. Bainton DF, Farquhar MG: Differences in enzyme content of azurophil and specific granules of polymorphonuclear leukocytes: I. Histochemical staining of bone marrow smears. J Cell Biol 39:286, 1968
49. Bainton DF, Farquhar MG: Segregation and packaging of granule enzymes in eosinophilic leukocytes. J Cell Biol 45:54, 1970
50. Bainton DF, Friedlander LM, Shohet SB: Abnormalities in granule formation in acute myelogenous leukemia. Blood 49:693, 1977
51. Bainton DF, Golde DW: Differentiation of macrophages from normal human bone marrow in liquid culture: Electron microscopy and cytochemistry. J Clin Invest 61:1555, 1978
52. Bainton DF, Ullyot JL: Azurophil and specific granules of blood neutrophils in chronic myelogenous leukaemia: an ultrastructural and cytochemical analysis. Br J Haematol 29:149, 1974
53. Bainton DF, Ullyot JL, Farquhar MG: The development of neutrophilic polymorphonuclear leukocytes in human bone marrow. J Exp Med 134:907, 1971
54. Balazs A: Enzyme histochemical differentiation of bone marrow and blood myeloid cells in rats. Haematologia 2:197, 1968
55. Balazs A: Granule formation in rat myeloid cells. An electron microscopic study. Z Zellforsch 99:286, 1969
56. Balazs A, Blazsek I, Rappay G et al: Electroncytochemical investigation of activity and localization of several enzymes in differentiating myeloid cells. Acta Biol Acad Sci Hung 23:389, 1972
57. Balogh K, Cohen RB: Histochemical demonstration of diaphorases and dehydrogenases in normal human leukocytes and platelets. Blood 17:491, 1961
58. Bank A, Larsen PR, Anderson HM: Di Guglielmo's syndrome after polycythemia. N Engl J Med 275:489, 1966
59. Barka T, Anderson PJ: Histochemical methods for acid phosphatase using hexazonium pararosanalin as coupler. J Histochem Cytochem 10:741, 1962
60. Barnard DL, Burns GF, Gordon J et al: Chronic myelomonocytic leukemia with paraproteinemia but no detectable plasmacytosis. Cancer 44:927, 1979
61. Barrett S, Garratty E, Garratty G: Cell surface heterogeneity of human blood neutrophils and monocytes. Br J Haematol 43:575, 1979
62. Barry WE, Day HJ: Refractory sideroblastic anemia: Clinical and hematologic study of ten cases. Ann Intern Med 61:1029, 1964
63. Bastos AL, Marques D, Campos EM et al: Cytophotochemistry: A cytochemical principle for dye-inducible fluorescence (DIF) reactions of tumor cells. J Pathol 107:xiv, 1972
64. Baudhuin P: Peroxisomes (microbodies, glyoxysomes). In Lima-de-Faria A (ed): A Handbook of Molecular Cytology. Amsterdam, North-Holland Pub. Co., 1969
65. Bean RHD: Phenylbutazone and leukaemia. Br Med J 2:1552, 1960
66. Beard MEJ, Fairley GH: Acute leukemia in adults. Semin Hematol 11:5, 1974
67. Bearman RM, Pangalis GA, Rappaport H: Prolymphocytic leukemia: Clinical, histopathological, and cytochemical observations. Cancer 42:2360, 1978

68. Bearman RM, Pangalis GA, Rappaport H: Acute ("malignant") myelosclerosis. Cancer 43:279, 1979
69. Beattie JW, Seal RME, Crowther KV: Chronic monocytic leukaemia. Q J Med 20:131, 1951
70. Becker RP, De Bruyn PP: Backscattered electron imaging of endogenous and exogenous peroxidase activity in rat bone marrow, p 171. In Johari O, Becker RP (ed): Proceedings of the Scanning Electron Microscopy Symposium, 9th, V.2. Chicago, IIT Research Institute, 1976
71. Beckman H, Neth R, Landbeck G et al: Cytochemical studies of blood monocytes in reactive and leukemic monocytosis. Klin Wochenschr 46:888, 1968
72. Beckmann H, Neth R, Landbeck G et al: Cytochemical differentiation and course of disease in acute childhood leukemias. Deutsch Med Wochenschr 94:2320, 1969
73. Beckstead JH, Bainton DF: Enzyme histochemistry on bone marrow biopsies: Reactions useful in the differential diagnosis of leukemia and lymphoma applied to 2-micron plastic sections. Blood 55:386, 1980
74. Behnke O: An electron microscope study of the megakaryocyte of the rat bone marrow: I. The development of the demarcation membrane system and the platelet surface coat. J Ultrastruct Res 24:412, 1968
75. Belpomme D, Dantchev D, Du Rusquec E et al: T and B lymphocyte markers on the neoplastic cell of 20 patients with acute and 10 patients with chronic lymphoid leukemia. Biomedicine 20:109, 1974
76. Belpomme D, Dantchev D, Du Rusquec E et al: The T or B nature of neoplastic cells in lymphatic leukemia. Bull Cancer (Paris) 61:387, 1976
77. Belpomme D, Dantchev D, Joseph R, et al: Further studies of acute and chronic leukemias: T and B cell membrane markers and scanning electron microscopy. In Astaldi G et al (eds): Current Studies on Standardization Problems in Clinical Pathology, Haematology, and Radiotherapy in Hodgkin's Disease, pp. 143–55. Amsterdam: Excerpta Medica, 1975
78. Belpomme D, Dantchev D, Karima AM et al: Search for correlations between immunological and morphological criteria used to classify lymphoid leukemias and non-Hodgkin's hematosarcomas with special reference to scanning electron microscopy and T and B membrane markers. Recent Results Cancer Res 56:131, 1976
79. Belpomme D, Lelarge N, Mathé G et al: Etiological, clinical and prognostic significance of the T-B immunological classification of primary acute lymphoid leukemias and non Hodgkin's lymphomas. Haematol Blood Transfusion 20:33, 1977
80. Beltran G, Stuckey WJ: Nuclear lobulation and cytoplasmic fibrils in leukemic plasma cells. Am J Clin Pathol 58:159, 1972
81. Ben-Bassat H, Frand UI, Isersky C et al: Plasma cell leukemia with IgD paraprotein. Arch Intern Med 121:361, 1968
82. Ben-Bassat H, Polliack A, Inbar M et al: Membrane properties and surface morphology of cell lines derived from patients with lymphoma, leukemia and infectious mononucleosis. Isr J Med Sci 14:1221, 1978
83. Ben-Bassat H, Polliack A, Mitrani-Rosenbaum S, et al: A comparative study of human cell lines derived from patients with lymphoma, leukemia and infectious mononucleosis: Membrane properties, ultrastructure, and surface morphology. Cancer 40:1481, 1977
84. Benedetti G: L'eritremia cronica (considerazioni su di un caso tipico di mielosi aneritremica cronica). Hematologica 19:229, 1939
85. Benestad HB: A modified technique for combination of cytochemistry and radioautography of haemic cells. Scand J Haematol 18:54, 1977
86. Bennett JM: Myelomonocytic leukemias: A historical review and perspectives. Cancer 27:1218, 1971

87. Bennett JM, Catovsky D, Daniel MT et al: Proposals for the classification of the acute leukaemias. French–American–British (FAB) Co-operative Group. Br J Haematol 33:451, 1976
88. Bennett JM, Catovsky D, Daniel MT, et al: Letter: A variant form of hypergranular promyelocytic leukemia (M3). Ann Intern Med 92:261, 1980
89. Bennett JM, Dutcher TF: The cytochemistry of acute leukemia: Observations on glycogen and neutral fat in bone marrow aspirates. Blood 33:341, 1969
90. Bennett JM, Henderson ES: Lymphoblastic leukaemia. Br Med J 2:513, 1969
91. Bennett JM, Klemperer MR, Segel GB: Survival prediction based on morphology of lymphoblasts. Recent Results Cancer Res 43:23, 1973
92. Bennett JM, Reed CE: Acute leukemia cytochemical profile: Diagnostic and clinical implications. Blood Cells 1:101, 1975
93. Bennett JM, Rutenburg AM: Diagnostic value of leukocyte alkaline phosphatase: A three year experience. Ann Histochim 11:173, 1966
94. Bentfeld ME, Nichols BA, Bainton DF: Ultrastructural localization of peroxidase in leukocytes of rat bone marrow and blood. Anat Rec 187:219, 1977
95. Berard C et al: Histopathological definition of Burkitt's tumor. Bull WHO 40:601, 1969
96. Berceanu S, Papazian R: Some biologic regulating factors of proliferation and differentiation in acute leukemias. Rev Rhum Med Intern 11:163, 1974
97. Berkowitz LR, Ross DW, Orringer EP: Hairy cell leukemia with acquired dyserythropoiesis. Arch Intern Med 140:554, 1980
98. Bernard J: Preleukemic states. Nouv Rev Fr Hematol Blood Cells 17:2, 1976
99. Bernard J, Boiron M, Lortholary P et al: The very acute leukemias. Cancer Res 25:1675, 1965
100. Bernard J, Jacquillat C, Chavelet F et al: Leucémie aiguë d'une enfant de 5 mois née d'une mère atteinte de leucémie aiguë au moment de l'accouchement. Nouv Rev Fr Hematol 4:140, 1964
101. Bernard J, Weil M, Boiron M et al: Acute promyelocytic leukemia: Results of treatment by daunorubicin. Blood 41:489, 1973
102. Bernard J, Weil M, Flandrin G et al: Clinical study of acute monoblastic leukemia (abstr) Proc Am Assoc Cancer Res 16:201, 1975
103. Bernier GM, Berman JH, Fanger MW: Plasma cell leukemia with excretion of half-molecules of immunoglobulin A (alpha-1 lambda-1). Ann Intern Med 86:572, 1977
104. Bernstein R, Katz J, Bezwoda W et al: Acute blastic transformation in patients with chronic granulocytic leukaemia and myelofibrosis. S Afr Med J 50:616, 1975
105. Bernstein RE: A rapid screening dye test for the detection of glucose-6-phosphate dehydrogenase deficiency in red cells. Nature 194:192, 1962
106. Berretsi A, Malaskova V, Oberling F et al: Application des techniques cytochimiques à 47 cas de leucémies aiguës. Modifications cytochimiques au cours de l'évolution. Semin Hop Paris 49:633, 1973
107. Berzy I, Solti V: Occurrence of alkaline phosphatase reaction in the immature myeloid and reticular cells of the peripheral blood. Folia Haematol (Leipz) 97:168, 1972
108. Bessis M: Living Blood Cells and Their Ultrastructure. Weed RI (transed): New York, Springer-Verlag, 1973
109. Bessis M: Blood Smears Reinterpreted. New York, Springer International, 1977
110. Bessis M, Breton-Gorius J: Pathologie et asynchronisme de développement des organelles cellulaires au cours des leucémies aiguës granulocytaires. Etude au microscope électronique. Nouv Rev Fr Hematol 9:245, 1969

111. Bessis M, Dreyfus B, Breton-Gorius J et al: Etude au microscope électronique de onze cas d'anémies réfractaires avec enzymopathies multiples. Nouv Rev Fr Hematol 9:87, 1969
112. Bezwoda WR, Derman DP, Bothwell TH et al: The diagnosis and management of hairy-cell leukaemia. S Afr Med J 55:577, 1979
113. Biberfeld P: Morphogenesis in blood lymphocytes stimulated with phytohaemagglutinin (PHA). Acta Pathol Microbiol Scand [A] (Suppl) 223:1, 1971
114. Biberfeld P, Hedfors E.: Atypical blood lymphocytes in sarcoidosis: Morphology, cytochemistry, and membrane properties. Scand J Immunol 3:615, 1974
115. Biermer A: Halt Zunachst einen Vortrag über eine von ihm ofters beobachtete eigenthumliche Form von progressiver pernizioser Anamie, welche mit capillaren Blutungen der Haut, Retina, des Gehirns, etc. Corr Blf schweiz Aerzte 2:15, 1872
116. Bizzozero OJ, Johnson KG, Ciocco A: Radiation-related leukemia in Hiroshima and Nagasaki, 1946–1964. I. Distribution, incidence, and appearance time. N Engl J Med 274:1095, 1966
117. Blair TR, Bayrd ED, Pease GL: Atypical leukemia. JAMA 198:21, 1966
118. Blankenship RM, Greenburg BR, Lucas RN et al: Familial sea-blue histiocytes with acid phosphatemia. A syndrome resembling Gaucher disease: The Lewis variant. JAMA 225:54, 1973
119. Block M, Jacobson LO, Bethard WF: Preleukemic acute human leukemia. JAMA 152:1018, 1953
120. Bloom GE, Warner S, Gerald PS et al: Chromosome abnormalities in constitutional aplastic anemia. N Engl J Med 174:8, 1966
121. Bloomfield CD Brunning RD: Prognostic implications of cytology in acute leukemia in the adult: The case for subacute leukemia. Hum Pathol 5:641, 1974
122. Bloomfield CD, Brunning RD: Acute leukemia as a terminal event in nonleukemic hematopoietic disorders. Semin Oncol 3:297, 1976
123. Boggs DR: Hematopoietic stem cell theory in relation to possible lymphoblastic conversion of chronic myeloid leukemia. Blood 44:449, 1974
124. Boggs DR: The pathogenesis and clinical patterns of blastic crisis of chronic myeloid leukemia. Semin Oncol 3:289, 1976
125. Boggs DR, Wintrobe MM, Cartwright GE: The acute leukemias. Analysis of 322 cases and review of the literature. Medicine 41:163, 1962
126. Bohn L: Histochemical demonstration of lactate dehydrogenase in cells in blood films. Scand J Haematol 2:108, 1965
127. Bohnel J, Stacher A: Zur Frage der "Präleukämien". Verh Dtsch Ges Inn Med 79:415, 1973
128. Boll I: Leukocyte morphology and its diagnostic importance. Dtsch Med J, 22:760, 1971
129. Boll I: Morphological phase–contrast cinematographic studies on the behavior of bone marrow cells in vitro. X. Myeloblasts and monoblasts. Blut 32:115, 1976
130. Boll I, Heyer B, Düring G et al: Quantitative determination of the enzyme contents and the PAS-positive substances of bone marrow cells in vivo and in vitro. Acid and alkaline phosphatase, alpha-naphthylacetate-esterase, peroxidase and PAS-reaction in 3-day bone marrow clot culture. Klin Wochenschr 48:1233, 1970
131. Boll I, Waechter R von, Meyer-Burg J: Die akute Erythroblastose des Erwachsenen (Morbus DiGuglielmo). Morphologische, Cytochemische und zellkinetische Beobachtungen. Klin Wochenschr 50:517, 1972
132. Bondare B, Barawika I, Kolodjasnaja I: The karyotype of bone marrow cells in the preleukemic phase of chronic myelocytic leukemia. Folia Haematol (Leipz) 105:58, 1978

133. Boni P: Acute total hyperplastic myelosis. Haematologica 24:69, 1942
134. Borges WH, Nicklas JW, Hamm CW: Prezygotic determinants in acute leukemia. J Pediatr 70:180, 1967
135. Borkowski W, Armata, J.: Morphology of paralymphoblasts and the clinical course of acute lymphoblastic leukemia in children. Pediatr Pol 52:999, 1977
136. Bouroncle BA, Wiseman BK, Doan CA: Leukemic reticuloendotheliosis. Blood 13:609, 1958
137. Boysen G: An evaluation of aryl sulphatase activity in leukaemic cells. Scand J Haematol 6:246, 1969
138. Boyum A: Separation of leukocytes from blood and bone marrow. Scand J Clin Lab Invest (Suppl) 21:97, 1, 1968
139. Branehog I, Ridell B, Swolin B et al: Megakaryocyte quantifications in relation to thrombokinetics in primary thrombocythaemia and allied diseases. Scand J Haematol 15:321, 1975
140. Brauer MJ, Dameshek W: Hypoplastic anemia and myeloblastic leukemia following chloramphenicol therapy. Report of three cases. N Engl J Med 277:1003, 1967
141. Braunsteiner H: Esterases in leukocytes. J Histochem Cytochem 7:202, 1959
142. Braunsteiner H: Use of cytochemical methods in the study of acute leukemia. Acta Haematol Jap 35:594, 1972
143. Braunsteiner H: Prognostic significance of cytochemical findings in differentiated myelogenous leukaemias of adults (author's trans). Wien Klin Wochenschr (Suppl) 89:1, 1977
144. Braunsteiner H, Schmalzl F: Cytochemical studies of monocytes. Demonstration of a characteristic esterase. Nouv Rev Fr Hematol 8:289, 1968
145. Braunsteiner H, Schmalzl F: Cytochemistry of monocytes and macrophages. In Van Furth R (ed): Mononuclear Phagocytes, p 62. Philadelphia, F.A. Davis, 1970
146. Brausil B: Diagnostic differences between atypical myeloblasts and lymphoblasts in the phase contrast microscope. Acta Haematol (Basel) 12:276, 1954
147. Brenner JF, Necheles TF, Bonacossa IA et al: Scene segmentation techniques for the analysis of routine bone marrow smears from acute lymphoblastic leukemia patients. J Histochem Cytochem 25:601, 1977
148. Breton-Gorius J: Utilisation de la diaminobenzidine pour la mise en évidence, au microscope électronique, de l'hémoglobine intracellulaire. La réactivité des différents organelles des érythroblastes. Nouv Rev Fr Hematol 10:243, 1970
149. Breton-Gorius J: Aspects ultrastructuraux de la maturation des mégakaryocytes humains normaux. Nouv Rev Fr Hematol 13:504, 1973
150. Breton-Gorius J: Development of two distinct membrane systems associated in giant complexes in pathological megakaryocytes. Ser Haematol 8:49, 1975
151. Breton-Gorius J, Coquin Y, Guichard J: Activités peroxydasiques de certaines granulations des neutrophiles dans deux cas de déficit congénital en myéloperoxydase. CR Acad Sci [D], Paris 280:1753, 1975
152. Breton-Gorius J, Coquin Y, Guichard J: Cytochemical distinction between azurophils and catalase-containing granules in leukocytes. I. Studies in developing neutrophils and monocytes from patients with myeloperoxidase deficiency: Comparison with peroxidase-deficient chicken heterophils. Lab Invest 38:21, 1978
153. Breton-Gorius J, Coquin Y, Vilde JL et al: Cytochemical and ultrastructural studies of aberrant granules in the neutrophils of two patients with myeloperoxidase deficiency during a preleukemic state: Relationship to abnormal bactericidal activity. Nouv Rev Fr Hematol Blood Cells 17:187, 1976
154. Breton-Gorius J, Dreyfus B, Sultan C: Identification of circulating micromegakaryocytes in a case of refractory anemia: An electron microscopic–cytochemical study. Blood 40:453, 1972

155. BRETON-GORIUS J, GUICHARD J: Electron microscopic study of the localization of peroxidases in human bone marrow cells. Nouv Rev Fr Hematol 9:678, 1969
156. BRETON-GORIUS J, GUICHARD J: Ultrastructural localization of peroxidase activity in human platelets and megakaryocytes. Am J Pathol 66:277, 1972
157. BRETON-GORIUS J, GUICHARD J: Fine structural and cytochemical identification of microperoxisomes in developing human erythrocytic cells. Am J Pathol 79:523, 1975
158. BRETON-GORIUS J, GUICHARD J: Two different types of granules in megakaryocytes and platelets as revealed by the diaminobenzidine method. J Microsc Biol Cell 23:197, 1975
159. BRETON-GORIUS J, GUICHARD J: Améliorations techniques permettant de révéler la peroxydase plaquettaire. Nouv Rev Fr Hematol 16:381, 1976
160. BRETON-GORIUS J, GUICHARD J: Cytochemical distinction between azurophils and catalase-containing granules in leukocytes: Distribution in human promyelocytes and promonocytes. J Reticuloendothel Soc 24:637, 1978
161. BRETON-GORIUS J, HOUSSAY D: Auer bodies in acute promyelocytic leukemia. Demonstration of their fine structure and peroxidase localization. Lab Invest 28:135, 1973
162. BRETON-GORIUS J, HOUSSAY D, DREYFUS B: Partial myeloperoxidase deficiency in a case of preleukaemia. I. Studies of fine structure and peroxidase synthesis of promyelocytes. Br J Haematol 30:273, 1975
163. BRETON-GORIUS J, HOUSSAY D, VILDE JL et al: Partial myeloperoxidase deficiency in a case of preleukaemia. II. Defects of degranulation and abnormal bactericidal activity of blood neutrophils. Br J Haematol 30:279, 1975
164. BRETON-GORIUS J, REYES F, DUHAMEL G: Megakaryocytes and platelet peroxidase. Identification by the ultrastructural demonstration of platelet peroxidase. Blood 51:45, 1978
165. BRETON-GORIUS J, REYES F, VERNANT JP et al: The blast crisis of chronic granulocytic leukaemia: Megakaryoblastic nature of cells as revealed by the presence of platelet-peroxidase. A cytochemical ultrastructural study Brit J Haematol 39:295, 1978
166. BRIÈRE J, CASTRO-MALASPINA H, TANZER J et al: Subacute myelogenous leukemia with Philadelphia chromosome. Nouv Rev Fr Hematol 15:407, 1975
167. BRIGGS RS, PERILLIE PE, FINCH SC: Lysozyme in bone marrow and peripheral blood cells. J Histochem Cytochem 14:167, 1966
168. BROUET JC, SELIGMAN M: The immunological classification of acute lymphoblastic leukemia. Cancer 42:817, 1978
169. BROUN GO: Chronic erythromonocytic leukemia. Am J Med 47:785, 1969
170. BRUCHER H: The monocytes. Haematol Bluttransfus 7:10, 1969
171. BRUCHER H, DILL A, GRÄBER M: Studies on blood and tissue macrophages. Acta Haematol (Basel) 41:76, 1969
172. BRUN B, VERNANT JP, REYES F et al: Leucémie aiguë monoblastique. Aspects cliniques et thérapeutiques de 20 cas. Ann Med Intern 127:807, 1976
173. BRYNES RK, GOLOMB HM, DESSER RK, et al: Acute monocytic leukemia. Cytologic, histologic, cytochemical, ultrastructural and cytogenic observations. Am J Clin Pathol 65:471, 1976
174. BUDMAN DR, KOZINER B, ARLIN Z: Massive lymphadenopathy mimicking lymphoma in leukemic reticuloendotheliosis. Am J Med 66:160, 1979
175. BUNTING CH: Leukocytes. Physiol Rev 2:505, 1922
176. BURK KH, DREWINKO B, TRUJILLO JM: Establishment of a human plasma cell line in vitro. Cancer Res 38:2508, 1978
177. BURKE JS, BYRNE GE, JR, RAPPAPORT H: Hairy cell leukemia (leukemic reticuloendotheliosis). I. A clinical pathologic study of 21 patients. Cancer 33:1399, 1974

178. BURKE JS, MACKAY B, RAPPAPORT H: Hairy cell leukemia (leukemic reticuloendotheliosis) II. Ultrastructure of the spleen. Cancer 37:2267, 1976
179. BURSTONE MS: Histochemical demonstration of acid phosphatase with naphthol AS-phosphates. J Natl Can Inst 21:523, 1958
180. BURSUKER I, GOLDMAN R: Quantitative and functional differences induced in bone marrow-derived mononuclear phagocytes by inflammatory stimuli. J Reticuloendothel Soc 26:533, 1979
181. BUSKARD NA, BOYES DA, GROSSMAN L: Plasma cell leukemia following treatment with radiotherapy and melphalan. Can Med Assoc J 117:788, 1977
182. CADMAN EC, CAPIZZI RL, BERTINO JR: Acute nonlymphocytic leukemia. A delayed complication of Hodgkin's disease therapy. Analysis of 109 cases. Cancer 40:1280, 1977
183. CARBONE PP, BERARD CW, BENNETT JM et al: Burkitt's tumor. Ann Intern Med 70:817, 1969.
184. CARDAMONE JM, KIMMERLE RI, MARSHALL EY: Development of acute erythroleukemia in B-cell immunoproliferative disorders after prolonged therapy with alkylating drugs. Am J Med 57:836, 1974
185. CAREY RW, TAFT PD, BENNETT JM et al: Carcinocythemia (carcinoma cell leukemia). An acute leukemia-like picture due to metastatic carcinoma cells. Am J Med 60:273, 1976
186. CARMEL R, COLTMAN CA, YATTEAU RF et al: Association of paroxysmal nocturnal hemoglobinuria with erythroleukemia. N Engl J Med 283:1329, 1970
187. CASTOLDI GL, GRUSOVIN GD, SCAPOLI GL: Consecutive cytochemical staining for the analysis of the blastic population in the acute phase of chronic myeloid leukemia. Biomedicine 23:12, 1975
188. CATOVSKY D: Letter: T-cell origin of acid-phosphate-positive lymphoblasts. Lancet 2:327, 1975
189. CATOVSKY D, GALETTO J, OKOS A et al: Cytochemical profile of B and T leukaemic lymphocytes with special reference to acute lymphoblastic leukaemia. J Clin Pathol 27:767, 1974
190. CATOVSKY D, GALTON DAG: Lysozyme activity and nitroblue-tetrazolium reduction in leukaemic cells. J Clin Pathol 26:60, 1973
191. CATOVSKY D, GALTON DA, ROBINSON J: Myeloperoxidase-deficient neutrophils in acute myeloid leukaemia. Scand J Haematol 9:142, 1972
192. CATOVSKY D, GREAVES MF, PAIN C et al: Acid–phosphatase reaction in acute lymphoblastic leukaemia. Lancet 1:749, 1978
193. CATOVSKY D, SHAW MT, HOFFBRAND AV et al: Sideroblastic anaemia and its association with leukaemia and myelomatosis: A report of five cases. Br J Haematol 20:385, 1971
194. CAWLEY JC: The microtubules of leukaemic Rieder cells: An ultrastructural study. Scand J Haematol 9:417, 1972
195. CAWLEY JC, HAYHOE FGJ: Ultrastructure of Haemic Cells. Philadelphia, WB Saunders, 1973
196. CHAN BWB, FLEMANS RJ, ZBINDEN G: Acute leukemia with megakaryocytic predominance. Cancer 28:1343, 1971
197. CHANG YT, ANDERSEN RN: Cultivation of mouse bone marrow cells: Cytochemical studies of the granulocyte colony feeder cells. J Reticuloendothel Soc 18:34, 1975
198. CHAYEN J, BITENSKY L, BUTCHER RG: Practical histochemistry. New York, John Wiley & Sons, 1973
199. CHELLOUL N, DANIEL MT, FLANDRIN G: Adult lymphohistiocytosis of "hairy leukocyte leukemia": A histologic, cytologic, and ultrastructural study. Pathol Eur 8:235, 1973
200. CHESSELLS JM, HARDISTY RM, RAPSON NT: Acute lymphoblastic leukemia in children: Classification and prognosis. Lancet 2:1307, 1977

201. CHIYODA S, MIURA Y: Alkaline phosphatase activity in chronic myelogenous leukemia cells in cultures (author's trans). Acta Haematol Jpn 40:172, 1977
202. CHIZHOVA ZP, VOLODIN NN, REZNIKOV IUP: Plasmoblastic leukemia in children. Vopr Okhr Materin Det 16:77, 1971
203. CHOW C, SPITZER RW: Laboratory diagnosis of Bence Jones proteinuria in a patient with plasma cell leukemia. Clin Chem 21:1683, 1975
204. CLEIN GP: The classification of acute leukaemia. In Clefton FJ, Crowther D, Malpas JS: Advances in acute leukaemia, p. 51. Amsterdam, American Elsevier, 1974
205. CLINE MJ: The White Cell. Cambridge, Harvard University Press, 1975
206. COHEN HJ: Human lymphocyte surface immunoglobulin capping. Normal characteristics and anomalous behavior of chronic lymphocytic leukemic lymphocytes. J Clin Invest 55:84, 1975
207. COHEN JR, CREGER WP, GREENBERG PL et al: Subacute myeloid leukemia. A clinical review. Am J Med 66:959, 1979
208. COHEN MH, BENNETT JM, BERARD CW et al: Burkitt's tumor in the United States. Cancer 23:1259, 1969
209. COHEN T, CREGER WP: Acute myeloid leukemia following seven years of aplastic anemia induced by chloramphenicol. Am J Med 43:762, 1967
210. COHN ZA, HIRSCH JG: The isolation and properties of the specific cytoplasmic granules of rabbit polymorphonuclear leukocytes. J Exp Med 112:983, 1960
211. COLEMAN CN, WILLIAMS CJ, FLINT A et al: Hematologic neoplasia in patients treated for Hodgkin's disease. N Engl J Med 297:1249, 1977
212. COLLINS RD, SMITH JL, CLEIN GP et al: Absence of B- and T-cell markers on acute lymphoblastic leukaemic cells and persistence of the T-cell marker on mitogen-transformed T-lymphocytes. Br J Haematol 26:615, 1974
213. COPELLI M: Di una emopatia sistemizzata rappresentata da una iperplasia eritoblastica (eritromatosis). Pathologica 4:460, 1912
214. COPPOLA A: Acid phosphatase of azurophilic granules of C3HST4 lymphoma. Acta Neuropathol (Suppl) 6:53, 1975
215. CORDIER JF, CREYSSEL R, COEUR P et al: Plasmocyte leukemia. Incidence of Bence Jones proteinuria and of lambda isotypia. Nouv Presse Med 4:3145, 1975
216. COTRAN RS, LITT M: The entry of granule-associated peroxidase into the phagocytic vacuoles of eosinophils. J Exp Med 129:1291, 1969
217. COURT BROWN WMC, DOLL R: Leukaemia in childhood and young adult life. Trends in mortality in relation to etiology. Br Med J 1:981, 1961
218. COURT BROWN WM, DOLL R: Mortality from cancer and other causes after radiotherapy for ankylosing spondylitis. Br Med J 2:1327, 1965
219. COWAN HB, PARMLEY RT, CRIST WM et al: Crystalloid-containing giant granules in leukemic monoblasts. Blood 55:946, 1980
220. CRADDOCK CG: Production and distribution of granulocytes and the control of granulocyte release. In Wolstenholme GEW, O'Connor M (eds): Ciba Foundation Symposium on Haemopoiesis, p 237. London, J & A Churchill, 1960.
221. CRONKITE EP, VINCENT PC: Granulocytopoiesis, myeloproliferative disorders of animals and man, p 3. Atomic Energy Commission Symposium Series, 1970
222. CROSS JP, HADAYAH N: Chloroacetate esterase staining in autoradiography of bone marrow. J Anat 111:513, 1972
223. DAMESHEK W: Pernicious anemia, megaloblastosis, and the DiGuglielmo syndrome. Blood 13:1085, 1958
224. DAMESHEK W: Sideroblastic anaemia: Is this a malignancy? Br J Haematol 11:52, 1965
225. DAMESHEK W: The DiGuglielmo syndrome revisited. Blood 34:567, 1969

226. Dameshek W, Baldini M: The DiGuglielmo syndrome. Blood 13:192, 1958
227. Dameshek W, Gunz F: Leukemia, 2nd ed. New York, Grune & Stratton, 1965
228. Dancey JT, Deubelbeiss KA, Harker LA et al: Section preparation of human marrow for light microscopy. J Clin Pathol 29:704, 1976
229. Daniel MT, Flandrin G: Fine structure of abnormal cells in hairy cell (tricholeukocytic) leukemia, with special reference to their *in vitro* phagocytic capacity. Lab Invest 30:1, 1974
230. Daniel MT, Flandrin G, Bernard J: Acute mast-cell leukemia. Cytochemical and ultrastructural study, about a particular case (author's trans) Nouv Rev Fr Hematol 15:319, 1975
231. Daniel MT, Flandrin G, Lejeune F et al: I. Les esterases spécifiques monocytaires. Utilisation dans la classification des leucémies aiguës. Nouv Rev Fr Hematol 11:233, 1971
232. Dantchev D: Scanning electron microscopy morphology of mononuclear leukocytes in normal subjects and in patients with lymphoid and monocytoid neoplasias. Recent Results Cancer Res 64:94, 1978
233. Dantchev D, Belpomme D: Critical study of the mononuclear leukocyte morphology based on scanning electron microscopy in normal subjects and in patients with lymphoid or monocytoid proliferative disorders. Comparison with the T, B or null cell membrane phenotypes. Biomedicine 26:202, 1977
234. Dausset J: Essai de différentiation des diverses variétés de leucémies aiguës par l'examen du sang et de la moelle à l'état frais. Sang 21:610, 1950
235. Davies AJ, Belpomme D, Mathé G: Clinical significance and prognostic value of the T-B immunological classification of human primary acute lymphoid leukaemias. Lancet 1:555, 1977
236. Davis AT, Brunning RD, Quie PG: Polymorphonuclear leukocyte myeloperoxidase deficiency in a patient with myelomonocytic leukemia. N Engl J Med 285:789, 1971
237. Davis WC, Douglas SD: Defective granule formation and function in the Chediak–Higashi syndrome in man and animals. Semin Hematol 9:431, 1972
238. De Duve C, Wattiaux R: Functions of lysosomes. Annu Rev Physiol 28:435, 1966
239. Deegan ML, Cossman J, Chosney BT et al: Hairy cell leukemia. An immunologic and ultrastructural study. Cancer 38:1952, 1976
240. DeGowin RL: Benzene exposure and aplastic anemia followed by leukemia 15 years later. JAMA 185:748, 1963
241. DeJong MC: Esterase staining of human peripheral blood mononuclear cells. J Immunol Methods 26:193, 1979
242. De Pascale A, Sannito F: Histochemical behavior of intralysosome acid phosphatase in the elements of the plasmacytic series. Boll Soc Ital Biol Sper 44:125, 1968
243. DeSouza E, Kothare SN: A method for the cytochemical demonstration of succinic dehydrogenase in human leucocytes. J Histochem Cytochem 7:77, 1959
244. Desser RK, Himmelhoch SR, Evans WH et al: Guinea pig heterophil and eosinophil peroxidase. Arch Biochem Biophys 148:452, 1972
245. DeWald B, Rindler-Ludwig R, Bretz U et al: Subcellular localization and heterogeneity of neutral proteases in neutrophilic polymorphonuclear leukocytes. J Exp Med 141:709, 1975
246. Dicke KA, Spitzer G, Cork A et al: In vitro colony growth of acute myelogenous leukemia. Nouv Rev Fr Hematol Blood Cells 17:125, 1976
247. Dicke KA, Spitzer G, Scheffers PH et al: In vitro colony growth of acute myelogenous leukemia. Haematol Bluttransfus 19:63, 1976
248. Diehl V, Schaadt M, Kirchner H et al: Long-term cultivation of plasma cell leukemia cells and autologous lymphoblasts (LCL) in vitro: A comparative study. Blut 36:331, 1978

249. Di Guglielmo G: Un caso di eritroleucemia. Folia Med 3:386, 1917
250. Di Guglielmo G: Eritroleucemia e piastrinemia. Folia Med 6:3, 1920
251. Di Guglielmo G: L'eritremie acuta. Communicazione al XXIX Congresso di Medicina Interna tenuto in Roma dal 24 al 27 Ottobre del 1923. Napoli. Stabilimento Tipografico N. Jovene
252. Di Guglielmo G: Eritremie acute. Boll Soc Med Chir Pavia 1:665, 1926
253. Di Guglielmo G: Le eritremie. Haematologica 9:301, 1928
254. Di Guglielmo G: Les maladies érythrémiques. Rev Hematol 1:355, 1946
255. Di Guglielmo G: Acute erythremic disease, p 33. Proc Sixth Int Congr Internat Soc Hematol New York, Grune & Stratton, 1958
256. Di Guglielmo G: Le Malattie Eritremiche ed Eritroleucemiche. Rome, II Pensiero Scientifico Editore, 1962
257. Di Guglielmo G, Ferrara A; Contributo alla diagnostica delle malattie eritremiche ed eritroleucemiche. Haematologica 41:605, 1956
258. Di guglielmo G, Quattrin N: Mielosi eritremica cronica. Haematologica 24:1, 1942
259. Di Majorica G, Lanzavecchia G, LeCoultre A: Studio sulla morfologia dei leucociti umani normali e leucemia col metodo della sezioni sottili al microscopio elettronico. Rendeconti del r. Instituto lombardo di scienze e lettere 90:559, 1956
260. Dimmock VM: A rapid method for methyl green-pyronin staining of blood and bone marrow smears. Med Lab Sci 34:173, 1977
261. Doan CA, Wiseman BK: The monocyte, monocytosis, and monocytic leukosis: A clinical and pathological study. Ann Intern Med 8:383, 1934
262. Domanksy VY, Belchenko DI: The activity and isoenzyme spectrum of esterases in erythroid and myeloid cells of rabbit bone marrow. Vopr Med Khim 21:489, 1975
263. Dorr AD, Moloney WC: Acquired pseudo-Pelger anomaly of granulocytic leukocytes. N Engl J Med 261:742, 1959
264. Douglas SD, Cohnen G, König E et al: Lymphocyte lysosomes and lysosomal enzymes in chronic lymphocytic leukemia. Blood 41:511, 1973
265. Downey H: The myeloblast—its occurence under normal and pathological conditions, and its relations to lymphocytes and other blood cells. Folia Haematol 34:65, 1927
266. Downey H: Monocytic leukemia and leukemic reticuloendotheliosis, p. 1273. In Downey H (ed): Handbook of Hematology, Vol 2. New York, Paul B Hoeber, 1938
267. Downey H: The myeloblast. In Downey H (ed): Handbook of Hematology, Vol 3, p 1962. New York, Paul B Hoeber, 1938
268. Drescher J, Müller M: Morphological, cytochemical and radioautographic studies on infectious mononucleosis cells in vitro. Blut 21:315, 1970
269. Dreyfus B: Granulocytopathies acquises au cours des leucémies et des anémies réfractaires. Nouv Rev Fr Hematol 13:243, 1973
270. Dreyfus B: Preleukemic states. I. Definition and classification. II. Refractory anemia with an excess of myeloblasts in the bone marrow (smoldering acute leukemia). Nouv Rev Fr Hematol Blood Cells 17:33, 1976
271. Dreyfus B, Rochant H, Sultan C et al: Les anémies réfractaires avec excès de myéloblastes dans la moelle. Etude de onze observations. Presse Med 78:359, 1970
272. Duhamel G, Gorin NC, Najaman A et al: Myeloid leukemia with myelofibrosis (author's trans). Nouv Rev Fr Hematol 15:99, 1975
273. Duhamel G, Muratore R, Byron PA: Anémies réfractaires avec myéloblastose partielle. Analyse d'un protocole portant sur 77 cas. Nouv Rev Fr Hematol 16:81, 1976
274. Duhamel G, Najman A, Gorin N et al: Clinical conference: Acute transformation in 45 cases of chronic myeloid leukemia. Nouv Ref Fr Hematol 15:301, 1975

275. DULTSINA SM, POTAPOVA SG, GERASIMOVA NH: Cytological characteristics of the monocytes under normal conditions (review of the literature and personal data). Probl Gematol Pereliv Krovi 23:47, 1978
276. DUNN WB, HARDIN JH, SPICER SS: Ultrastructural localization of myeloperoxidase in human neutrophil and rabbit heterophil and eosinophil leukocytes. Blood 32:935, 1968
277. DURANT JR, TASSONI EM: Coexistent DiGuglielmo's leukemia and Hodgkin's disease. AM J Med Sci 254: 824, 1967
278. DUS D, HARLOZIŃSKA A, KOTLAREK-HAUS S et al: Human CGL bone marrow cells grown in vitro, their immunological, cytological and cytochemical characteristics and attempts of their xenotransplantation. Folia Haematol 101:576, 1974
279. DVORAK AM, DVORAK HF, KARNOVSKY MJ: Cytochemical localization of peroxidase activity in the developing erythrocyte. Am J Pathol 67:303, 1972
280. DVORAK AM, DVORAK HF, KARNOVSKY MJ: Uptake of horseradish peroxidase by guinea pig basophilic leukocytes. Lab Invest 26:27, 1972
281. EANES RZ, SCOTT RB, SWENSON B et al: Characterization of marrow granulocyte development: Enzyme-specific activity profiles in response to inflammatory reactions. Br J Haematol 38:463, 1978
282. EASTMAN P, SCHWARTZ R, SCHRIER SL: Distinctions between idiopathic ineffective erythropoiesis and DiGuglielmo's disease: Clinical and biochemical differences. Blood 40:487, 1972
283. EASTMAN P, WALLERSTEIN RO, SCHRIER SL: Conversion of polycythemia vera to chronic DiGuglielmo's syndrome. JAMA 204:1141, 1968
284. EBBESEN P, RASK-NIELSEN R, MCINTIRE KR: Plasma cell leukemia in BALB c mice inoculated with subcellular material. I. Incidence and morphology. J Natl Cancer Inst 41:473, 1968
285. EBSTEIN W: Ueber die acute Leukamie und Pseudoleukamie. Dtsch Arch klin Med 44:343, 1889
286. ECONOMOPOULOS P: Acid phosphatase activity in basophil cells of bone marrow. Mikroskopie 28:285, 1972
287. ECONOMOPOULOS P: The formation and degradation of granules in eosinophilic cells. Acta Anat 86:53, 1973
288. EDELSON PJ, STITES DP, GOLD S et al: Disorders of neutrophil function: Defects in the early stages of the phagocytic process. Clin Exp Immunol 13:21, 1973
289. EHRLICH P: Uber die specifischen Granulationen des Blutes. Arch Anat Physiol, Physiologische Abteilung (Suppl) 571, 1879
290. EHRLICH P, LAZARUS A: Histology of the blood. Normal and pathological. In Nothnagel H (ed.): Diseases of the Blood, p 17. New York, WB Saunders, 1905
291. ELIZALDE A, MILLER SP: Cytologic studies of enzymes in malignant tissues. Cancer 20:423, 1967
292. EMBERGER JM, NAVARRO M, RIZKALLA N et al: Histological and cytological study of bone marrow and spleen during primary medullary and splenic histiolymphocytosis. Nouv Rev Fr Hematol 13:365, 1973
293. EMMERICH B, MAURER R, SCHMIDT G et al: Effect of several chemotherapeutic agents on protein synthesis in leukemic plasma cells (author's trans). Blut 38:155, 1979
294. ENOMOTO T, KITANI T: Electron microscopic studies on peroxidase and acid phosphatase reaction in human leukocytes (in normal and leukemic cells and on phagocytosis). Acta Haematol Jpn 29:554, 1966
295. ERDMANN H, KUBANEK B, HEIMPEL H: Plasma cell leukemia Ig-A paraproteinemia. Helv Med Acta (Suppl) 49:154, 1969
296. ERF LA, RHOADS CP: The hematological effects of benzene (benzol) poisoning. J Indust Hygiene 21:421, 1939

297. Erokhina VI, Letskiy VB, Abdulkadyrov KM et al: Cytochemical characteristics of bone marrow reticular cells in healthy people and in patients with various forms of leukoses-reticuloses. Probl Gematol Pereliv Krovi 16:7, 1971
298. Evans WH, Wilson S, Mage MG et al: Biochemical characterization of isolated immature granulocytes from guinea pig bone marrow. J Reticuloendothel Soc 9:209, 1971
299. Ewald O: Die leukaemische Reticuloendotheliose. Deutsche Arch Klin Med 142:222, 1923
300. Fairbanks VF, Lampe LT: A tetrazolium-linked cytochemical method for estimation of glucose-6-phosphate dehydrogenase activity in individual erythrocytes: Applications in the study of heterozygotes for glucose-6-phosphate dehydrogenase deficiency. Blood 31:589, 1968
301. Farid NR, Noel EP, Kutas C: A population of human lymphocytes staining for esterases. Experientia 32:340, 1976
302. Farnes P: Histochemical approaches to cell characterization in vitro. Nat Cancer Inst Monogr 26:199, 1967
303. Farnes P, Barker BE: Tissue culture studies of human bone marrow. I. Maturation of granulocytes in chemically defined medium. J Natl Cancer Inst 39:1045, 1967
304. Farquhar MG, Bainton DF, Baggiolini M et al: Cytochemical localization of acid phosphatase activity in granule fractions from rabbit polymorphonuclear leukocytes. J Cell Biol 54:141, 1972
305. Favara BE: Diagnostic methods in pediatric oncology. Pediatr Clin North Am 23:55, 1976
306. Fedorko ME: Morphologic and functional characteristics of bone marrow macrophages from Imferon-treated mice. Blood 45:435, 1975
307. Fedorko M, Hirsch JG: Structure of monocytes and macrophages. Semin Hematol 7:109, 1970
308. Feinstein G, Janoff A: Variations in elastase-like esterase activities in human leukocytes during cell maturation. Proc Soc Exp Biol Med 152:36, 1976
309. Feldges AJ, Aur RJ, Verzosa MS et al: Periodic acid–Schiff reaction, a useful index of duration of complete remission in acute childhood lymphocytic leukemia. Acta Haematol 52:8, 1974
310. Fernández Rañada JM, Ribera C, López Pascual J et al: Plasma cell leukaemia. Haematological features and response to treatment (author's trans) Sangre 20:498, 1975
311. Ferrata A: Le Emopatie. Soc Edit Lib, 2, Parte spec., 1923
312. Fieschi A, Bianchini E, Cambiaggi G et al: Studies on the biological behavior of the cells of acute leukaemia. Acta Haematol 16:126, 1956
313. Finkel HE, Brauer MJ: Giant eosinophilic granules in the chronic DiGuglielmo syndrome. N Engl J Med 274:209, 1966
314. Fischer M, Mitrou PS: Panmyelopathie mit hyperplastischem Knochenmark und Präleukämie. Med Klin 71:2127, 1976
315. Fischer R, Gropp F: Ergebnisse cytologischer und cytochemischer Untersuchungen an Lymphozyten in vitro. Blut 15:129, 1967
316. Fischer R, Henneheuser HH, Kaufer C: Der cytochemische Nachweis von Naphtol-AS-D-Chloroacetat-Esterase in Auer Stäbchen. Klin Wochenschr 44:1401, 1966
317. Fischer R, Schmalzl F: Uber die Hemmbarkeit der Esteraseaktivität in Blutmonocyten durch Natriumfluorid. Klin Wochenschr 42:751, 1964
318. Fisher WB, Armentrout SA, Weisman R Jr et al: "Preleukemia." A myelodysplastic syndrome often terminating in acute leukemia. Arch Intern Med 132:226, 1973
319. Fishman SA, Ritz ND: Erythroleukemia following melphalan therapy for multiple myeloma. NY State J Med 75:2402, 1975

320. FITZGERALD PH, RASTRICK JM, HAMER JW: Acute plasma cell leukaemia following chronic lymphatic leukaemia: Transformation or two separate diseases? Br J Haematol 25:171, 1973
321. FLANDRIN G, BERNARD J: Cytological classification of acute leukemias: A survey of 1400 cases. Blood 1:7, 1975
322. FLANDRIN G, BROUET JC, DANIEL MT et al: Acute leukemia with Burkitt's tumor cells: A study of six cases with special reference to lymphocyte surface markers. Blood 45:183, 1975
323. FLANDRIN G, DANIEL MT: Cytochemistry of esterases in hematology. Pathol Biol 19:547, 1971
324. FLANDRIN G, DANIEL MT: Practical value of cytochemical studies for the classification of acute leukemias. In Mathé G, Pouillart P, Schwarzenberg L (eds): Nomenclature, Methodology and Results of Clinical Trials in Acute Leukemias, pp 43–56. London, William Heinemann, 1973
325. FLANDRIN G, DANIEL MT, BLANCHET P et al: II. La leucémie aiguë monocytaire. Situation clinique et pronostique actuelle à la lumière des techniques de détermination des estérases spécifiques. Nouv Rev Fr Hematol 11:241, 1971
326. FLANDRIN G, DANIEL MT, COUDERC O: Classification cytologique des leucémies aiguës lymphoblastiques: Incidences cliniques et pronostiques. Actualités Hématol 7:25, 1973
327. FLANNERY EP, DILLON DE, FREEMAN MV et al: Eosinophilic leukemia with fibrosing endocarditis and short Y chromosome. Ann Intern Med 77:223, 1972
328. FOA R, FOA C, MURATORE R et al: Ultrastructural localization of peroxidases in myeloid cells. CR Soc Biol 164:591, 1970
329. FORBES IJ, ZALEWSKI PD: A subpopulation of human B lymphocytes that rosette with mouse erythrocytes. Clin Exp Immunol 26:99, 1976
330. FORNI A, MOREO L: Chromosome studies in a case of benzene-induced erythroleukaemia. Eur J Cancer 5:459, 1969
331. FOUCAR K, MCKENNA RW, BLOOMFIELD CD et al: Therapy-related leukemia: A panmyelosis. Cancer 43:1285, 1979
332. FRAENKEL A: Ueber acute Leukamie. Dtsch Med Wschr 21:639, 1895
333. FRANK GA, BERGOL'TS VM: Morphology and morphogenesis of viral reticulosis in mice of the CC 57 BR line. Probl Gematol Pereliv Krovi 15:27, 1970
334. FRAUMENI JF: Bone marrow depression induced by chloramphenicol or phenylbutazone. JAMA 201:828, 1967
335. FRAUMENI JF, MANNING MD: Acondroplasia and leukaemia. Br Med J 3:680, 1967
336. FREEMAN JA: The ultrastructure and genesis of Auer bodies. Blood 15:449, 1960
337. FREEMAN JA: Origin of Auer bodies. Blood 27:499, 1966
338. FULLE HH, PRIBILLA W: Diagnosis and treatment of plasma cell leukemia. Dtsch Med Wochenschr 98:874, 1973
339. FURTH R VAN, HIRSCH JG, FEDORKO ME: Morphology and peroxidase cytochemistry of mouse promonocytes, monocytes, and macrophages. J Exp Med 132:794, 1970
340. GAHMBERG CG, NILSSON K, ANDERSSON LC: Specific changes in the surface glycoprotein pattern of human promyelocytic leukemic cell line HL-60 during morphologic and functional differentiation. Proc Natl Acad Sci USA 76:4087, 1979
341. GAHRTON G: A quantified periodic acid–Schiff reaction in normal and leukemic leukocytes. Eur J Cancer 1:299, 1965
342. GAHRTON G, YATAGANAS X: Quantitative cytochemistry of glycogen in blood cells. Methods and clinical application. Prog Histochem Cytochem 9:1, 1976
343. GAILANI S, SEON BK, HENDERSON ES: Plasma cell leukemia: Response to conventional myeloma therapy. J Med 8:403, 1977

344. Gall JA, Boggs DR, Chervenick PA et al: Discordant patterns of chromosome changes and myeloblast proliferation during terminal phase of chronic myeloid leukemia. Blood 47:347, 1976
345. Galton DAG, Goldman JM, Wiltshaw E et al: Prolymphocytic leukemia. Br J Haematol 27:7, 1974
346. Ganick DJ, Finlay JL: Acute lymphoblastic leukemia with Burkitt cell morphology and cytoplasmic immunoglobulin. Blood 56:311, 1980
347. Gavosto F, Maraini G, Pileri A; Nucleic acids and protein metabolism in acute leukaemia cells. Blood 16:1555, 1960
348. Gavosto F, Pileri A, Maraini G: Incorporazione di timidina marcata con tritio negli elementi del middollo osseo normale c leucemico-indagini autoradiografiche. Haematologica 44:977, 1959
349. Geary CG, Catovsky D, Wiltshaw E et al: Chronic myelomonocytic leukaemia. Br J Haematol 30:289, 1975
350. Geyer G, Schaaf P, Linss W et al: Enzyme distribution patterns in the granules of eosinophil leukocytes of the mouse. Acta Histochem 38:189, 1970
351. Glick AD, Horn RG: Identification of promonocytes and monocytoid precursors in acute leukaemia of adults: Ultrastructural and cytochemical observations. Br J Haematol 26:395, 1974
352. Gmyrek D, Witkowski R, Syllm-Rapoport L et al: Chromosomal aberrations and abnormalities of red-cell metabolism in a case of Fanconi's anemia before and after development of leukaemia. German Med Monthly 13:105, 1968
353. Goldberg AF, Barka T: Acid phosphatase activity in human blood cells. Nature 195:297, 1962
354. Goldberg ED, Stepanova EI, Fedorovich TV et al: Activity of peroxidase and lipid and glycogen content of the hematopoietic elements in healthy persons. Lab Delo 5:25960, 1976
355. Goldberg ED, Stepanova EI, Titov SP et al: Cytochemical study of hydrolytic enzymes in the hematopoietic cells of healthy persons. Probl Gematol Pereliv Krovi 21:39, 1976
356. Goldman JM: Acute promyelocytic leukaemia. Br Med J 1:380, 1974
357. Goldman JM, Catovsky D: The function of the phagocytic leukocytes in leukaemia. Br J Haematol (Suppl) 23:223, 1972
358. Goldman JM, Th'ng KH: Phagocytic function of leukocytes from patients with acute myeloid and chronic granulocytic leukaemia. Br J Haematol 25:299, 1973
359. Goldstein DJ: Selective staining of eosinophil granules in sections by alkaline orcein in contracted urea solution. Stain Tech 38:49, 1963
360. Goldstone AH, McVerry BA, Janossy G: Clonal identification in acute lymphoblastic leukemia. Blood 53:892, 1979
361. Golomb HM, Rowley JD, Vardiman JW et al: "Microgranular" acute promyelocytic leukemia: A distinct clinical, ultrastructural, and cytogenetic entity. Blood 55:253, 1980
362. Gomori G: Histochemical staining methods. In Visschir MB (ed): Methods in Medical Research, Vol 4 p 14. Chicago, Year Book Medical Publishers, 1951
363. Gomori G: Chloroacyl esters as histochemical substrates. J Histochem Cytochem 1:469, 1953
364. Goodpasture EW: A peroxidase reaction with sodium nitroprusside and benzidine in blood smears and tissues. J Lab Clin Med 4:442, 1919
365. Goodwin AF: Some new observations on Auer bodies in acute myelogenous leukemia. Folia Haematol 51:359, 1934
366. Gordon PA, Stuart J: Red cell cytochemistry in glucose-6-phosphate dehydrogenase deficiency. Br J Haematol 27:358, 1974
367. Gorman AM, O'Connel LG: Pseudo-Chediak-Higashi anomaly in acute leukemia. Am J Clin Pathol 65:1030, 1976

368. Goud TJ, Schotte C, van Furth R: Identification and characterization of the monoblast in mononuclear phagocyte colonies grown in vitro. J Exp Med 142:1180, 1975
369. Gourdin MF, Reyes F, Lejonc JL et al: Surface features of cells in human lymphoproliferative disorders. An immunoelectron microscopy study. Haematol Bluttransfus 19:207, 1976
370. Gracheva AI: Electron microscope technic for studying blood cells and several problems in interpreting the structure of these cells. Tsitologiia 16:511, 1974
371. Graham GS: The oxidizing ferment of the myelocyte series of cells and its demonstration by an alphanaphtol–pyronin method. J Med Res 35:231, 1916
372. Graham RC, Jr, Karnovsky MJ: The early stages of absorption of injected horse-radish peroxidase in the proximal tubules of mouse kidney: Ultrastructural cytochemistry by a new technique. J Histochem Cytochem 14:291, 1966
373. Gralnick HR: Classification of acute leukemia. Ann Intern Med 87:740, 1977
374. Gralnik HR, Harbor J, Vogel C: Myelofibrosis in chronic granulocytic leukemia. Blood 37:152, 1971
375. Gralnick HR, Sultan C: Acute promyelocytic leukaemia: Haemorrhagic manifestation and morphologic criteria. Br J Haematol 29:373, 1975
376. Gralnik HR, Tan HK: Acute promyelocytic leukemia: A model for understanding the role of the malignant cell in hemostasis. Hum Pathol 5:661, 1974
377. Grande M, Liso V: Cytochemical study of the cell colonies from agar culture of normal bone marrow and of peripheral blood from patients with chronic myeloid leukemia. Nouv Rev Fr Hematol Blood Cells 18:563, 1977
378. Grant MD, Coleman M: Hodgkin's disease with acute lymphoblastic leukemia. JAMA 231:623, 1975
379. Greenberg PL, Nichols WC, Schrier SL: Granulopoiesis in acute myeloid leukemia and preleukemia. N Engl J Med 284:1225, 1971
380. Gregory L, William R, Thompson E: Leucocyte function in Down's syndrome and acute leukaemia. Lancet 1:1359, 1972
381. Grozdea J, Columbies P, Ducos J: A study of the leukocyte peroxidases in the peripheral blood and bone marrow of twenty patients with acute and chronic granulocytic leukemia. Rev Fr Etud Clin Biol 12:180, 1967
382. Grusovin GD, Castoldi GL: Characterization of blast cells in acute nonlymphoid leukemias by consecutive cytochemical reactions. Acta Haematol 55:338, 1976
383. Gunz FW: Studies of leukaemic blood 'in vitro' with special reference to the effect of some therapeutic agents, Ph.D. thesis, Cambridge University, 1949
384. Gunz FW: Erythroleukemia. In Williams, WJ et al (eds.): Hematology, p 699. New York, McGraw-Hill, 1972
385. Gunz FW, Baikie AG: Leukemia, 3rd ed. New York, Grune & Stratton, 1974
386. Gunz FW, Burry AF: Cellular types in acute leukaemia: Diagnosis and significance. J Clin Pathol 16:325, 1963
387. Gunz FW, Fitzgerald PH, Crossen PE et al: Multiple cases of leukemia in a sibship. Blood 27:482, 1966
388. Gupta S, Good RA, Siegal FP: Rosette-formation with mouse erythrocytes. II. A marker for human B and non-T lymphocytes. Clin Exp Immunol 25:319, 1976
389. Haegert DG, Cawley JC, Collins RD et al: Leukaemic reticuloendotheliosis: A morphological and immunological study of four cases. J Clin Pathol 27:967, 1974
390. Haghbin M, Murphy ML, Tan CT: Treatment of acute nonlymphoblastic leukemia in children with a multiple-drug protocol. Cancer 40:1417, 1977
391. Haglid K, Carlsson CA, Thulin CA: Lactate dehydrogenase isoenzymes and proteins in human gliomas. Neurochirurgia 13:19, 1970

392. HAKIM J, BOIVIN P, TROUBE H et al: Activitiés fonctionnelles et enzymatiques des granulocytes du sang de malades ayant une anémie réfractaire acquise. Nouv Rev Fr Hematol 14:397, 1974
393. HALE AJ, WILSON SJ: The deoxyribonucleic acid content of leukocytes in normal and in leukaemic human blood. J Pathol Bacteriol 77:605, 1959
394. HALE AJ, WILSON SJ: The deoxyribonucleic acid content of the leukocytes in human blood, bone marrow and lymph glands. J Pathol Bacteriol 82:483, 1961
395. HALLBERG T, GURNER BW, COOMBS RRA: Opsonic adherence of sensitized ox red cells to human lymphocytes as measured by rosette formation. Int Arch Allergy Appl Immunol 44:500, 1973
396. HAMILTON-PATERSON JL: Preleukaemic anaemia. Acta Haematol 2:309, 1949
397. HAMMOUDA F, QUAGLINO D, HAYHOE FGJ: Blastic crisis in chronic granulocytic leukaemia. Cytochemical, cytogenetic, and autoradiographic studies in four cases. Br Med J 1:1275, 1964
398. HANKER JS, YATES PE, METZ CB, et al: A new specific noncarcinogenic reagent for demonstration of horseradish peroxidase. Histochem J 9:789, 1977
399. HARLEY RD, BAIRD HW, CRAVEN EM: Ataxia-telangiectasia. Arch Ophthalmol 77:582, 1967
400. HASSELBACK R, CURTIS J, SOOTS M et al: The influence of morphology on prognosis in acute leukemia. Can Med Assoc J 96:1610, 1967
401. HAWKSLEY JC: A note on the occurrence of Auer's bodies in monocytic leukaemia. J Pathol Bacteriol 40:365, 1935
402. HAYASHI M, NAKAJIMA Y, FISHMAN WH: The cytologic demonstration of β-glucuronidase employing naphthol AS-BI glucuronide and hexazonium pararosanilin: A preliminary report. J Histochem Cytochem 12:293, 1964
403. HAYHOE FGJ: The cytochemical demonstration of lipids in blood and bone marrow cells. J Pathol Bacteriol 65:413, 1953
404. HAYHOE FGJ: Clinical and cytological recognition and differentiation of the leukaemias. In Hayhoe FGJ (ed): Lectures on Haematology, p 113. Cambridge, University Press, 1960
405. HAYHOE FGJ: Leukaemia: Research and clinical practice. London, J & A Churchill, 1960
406. HAYHOE FGJ: Cytochemical techniques in diagnostic haematological cytology. In Europ Soc Hematol Proc, Part II, 7th Congress, London, 1959, p 331. Basel, S Karger, 1960
407. HAYHOE FGJ: Clinical and cytological recognition and differentiation of leukemias. In Zarafonetic CJD (ed): Proceedings of the International Conference on Leukemia and Lymphoma, Ann Arbor 1967, p 307. Philadelphia, Lea & Febiger, 1968
408. HAYHOE FGJ, CAWLEY JC: Acute leukemia: Cellular morphology, cytochemistry and fine structure. Clin Hematol 1:49, 1972
409. HAYHOE FGJ, FLEMANS RJ: An Atlas of Haematological Cytology. London, Wolfe Medical Books, 1969
410. HAYHOE FGJ, QUAGLINO D: Cytochemical demonstration and measurement of leukocyte alkaline phosphatase activity in normal and pathological states by a modified azo-dye coupling technique. Br J Haematol 4:375, 1958
411. HAYHOE FGJ, QUAGLINO D: Refractory sideroblastic anaemia and erythraemic myelosis: Possible relationship and cytochemical observations. Br J Haematol 6:381, 1960
412. HAYHOE FGJ, QUAGLINO D, DOLL R: The Cytology and Cytochemistry of Acute Leukaemias, MRC Special Report Series No 304. London, HMSO, 1960
413. HAYNES CR, DORNER I, LEONARD GL et al: Persistent polyagglutinability in vivo unrelated to T-antigen activation. Transfusion 10:43, 1970
414. HEATH CW, JR, BENNETT JM, WHANG-PENG J et al: Cytogenic findings in erythroleukemia. Blood 33:453, 1969
415. HECHT F, KOLER RD, RIGAS DA et al: Leukaemia and lymphocytes in ataxia-telangiectasia. Lancet 2:1193, 1966

416. HEIDEMAN RL, FALLETTA JM, STENBER CP et al: Acute leukemia in Burkitt lymphoma. Med Pediatr Oncol 2:215, 1976
417. HEIMPEL H, BAUKE J: Präleukämien. Med Klin 67:997, 1972
418. HELLER A: Cytochemical findings in panmyelopathy. Internist 12:161, 1971
419. HELLER A, GROSS R: Histochemical findings on preleukemic states. Blut 28:452, 1974
420. HELLRIEGEL K-P, SCHMALZL F: Conventional morphological examination of blood and bone marrow cells in the diagnosis of preleukemic syndromes. In Schmalzl F, Hellriegel K-P, (eds): Preleukemia. New York, Springer, 1980
421. HERMANSKÝ F, POCH T, PÖSSNEROVA V: Cytochemistry of reticular cells and myelomonocytes in pathological conditions. Folia Haematol 98:180, 1972
422. HERTL M: Morphologische und zytochemische Untersuchungen am Zellen der akuten leukose. Folia Haematol Neue Folge 3:296, 1959
423. HESSE R, HEGEWALD G: Plasmacytic leukemia. Z Aerztl Fortbild 66:1201, 1972
424. HEYDEN HW VON, HEYDEN D VON: Cytological characteristics of normal human peripheral blood cells in culture before establishment into lymphocytoid cell lines. Cell Tissue Kinet 6:271, 1973
425. HEYDEN HW VON, HEYDEN D VON: Characteristics of macrophages in vitro derived from peripheral blood cells. Blut 29:37, 1974
426. HICKLING RA: Chronic nonleukaemic myelosis. QJ Med 6:253, 1937
427. HICKS RG, BENNETT JM: An improved cytochemical method for nitroblue tetrazolium reduction by neutrophils. Am J Med Technol 37:226, 1971
428. HIDDEMANN W, KNOLL O, ZUMKLEY H et al: A case of plasma cell leukemia with terminal kidney failure following intravenous urography. Verh Dtsch Ges Inn Med 83:1110, 1977
429. HIDDEMANN W, KNOLL O, ZUMKLEY H et al: Plasma-cell leukemia with kidney failure after intravenous urography. Med Welt 28:1282, 1977
430. HIGGY KE, BURNS GF, HAYHOE FGJ: Discrimination of B, T, and null lymphocytes by esterase cytochemistry. Scand J Haematol 18:437, 1977
431. HILLESTAD LK: Acute promyelocytic leukaemia. Acta Med Scand 159:189, 1957
432. HIMMELHOCH SR, EVANS WH, MAGE MG et al: Purification of myeloperoxidases from the bone marrow of the guinea pig. Biochemistry 8:914, 1969
433. HIRSCH JG, COHN ZA: Degranulation of polymorphonuclear leukocytes following phagocytosis of microorganisms. J Exp Med 112:1005, 1960
434. HOELZER D, KURRLE E, SCHMÜCKER H et al: Evidence for differentiation of human leukemic blood cells in diffusion chamber culture. Blood 49:729, 1977
435. HOLOWIECKI J, SZERUDA-RUDZKA E, STELLA B: Dobór metod cytochemicznych w diagnostyce ostrych białaczek. Acta Haematol Pol 4:111, 1973
436. HOLZHEY P, KOLARZ G, PIETSCHMANN H et al: Acute lymphadenosis with an unusual course. Wien Z Inn Med 52:419, 1971
437. HOPFNER C, POTRON G, ADNET JJ et al: Blue histiocytes and Gaucher cells in the spleen and lymph nodes in chronic myelogenous leukemia (author's trans). Nouv Rev Fr Hematol 14:607, 1974
438. HORN RG, SPICER SS, WETZEL BK: Phagocytosis of bacteria by heterophil leukocytes: Acid and alkaline phosphatase cytochemistry. Am J Pathol 45:327, 1964
439. HOSSFELD DK, TORMEY D, ELLISON RR: Ph^1-positive megakaryoblastic leukemia. Cancer 36:576, 1975
440. HUHN D, KUBOTH W, SCHMALZL F: DiGuglielmo syndrome. Dtsch Med Wochenschr 98:855, 1973
441. HUHN D, RODT H, THIEL E et al: T-cell leukemias of adulthood. Blut 33:141, 1976
442. HUHN D, SCHMALZL F: Electron microscopic and cytochemic findings in DiGuglielmo's syndrome and in other forms of leukemia. In Neth R, Gallo RC et al (eds): Workshop—Modern Trends in Human Leukemia, Wilsede 1973, p 36. New York, Grune & Stratton, 1974

443. Huhn D, Schmalzl F, Demmler K: Monocytic leukemia. Light and electron microscopic morphology and cytochemistry. Dtsch Med Wochenschr 96:1594, 1971
444. Hulea SA, Arnstein HR: Intracellular distribution of ribonuclease activity during erythroid cell development. Biochim Biophys Acta 476:131, 1977
445. Humbert JR, Hathaway WE, Robinson A et al: Preleukaemia in children with a missing bone marrow C chromosome and a myeloproliferative disorder. Br J Haematol 21:705, 1971
446. Humble JG: The early diagnosis of monocytic leukaemia based on a study of 91 cases. J Clin Pathol 26:389, 1973
447. Humphrey GB, Nesbit ME, Brunning RD: Prognostic value of the periodic acid–Schiff (PAS) reaction in acute lymphoblastic leukemia. Am J Clin Pathol 61:393, 1974
448. Hurdle AD, Garson OM, Buist DG: Clinical and cytogenetic studies in chronic myelomonocytic leukaemia. Br J Haematol 22:773, 1972
449. Hutton JJ, Coleman MS: Terminal deoxynucleotidyl transferase measurements in the differential diagnosis of adult leukaemias. Br J Haematol 34:447, 1976
450. Imanaka T: Electron microscopic study on the localization of acid phosphatase in normal human leukocytes. I. Acid phosphatase activity in peripheral leukocytes and developing neutrophils in bone marrow. Acta Haematol Jpn 35:677, 1972
451. Inbar M, Larnicol N, Jasmin C et al: A method for the quantitative detection of human acute lymphatic leukemia. Eur J Cancer 13:1231, 1977
452. Insel RA, Melewicz FM, La Via MF et al: Morphology, surface markers, and in vitro responses of a human leukemic T cell. Clin Immunol Immunopathol 4:382, 1975
453. Isaacs R: Lymphosarcoma cell leukemia. Ann Intern Med 11:657, 1937
454. Ishige K, Takagi T, Oguro M et al: A case of plasma cell leukemia (author's trans). Rinsho Ketsueki 19:272, 1978
455. Isobe T, Ikeda Y, Imura H et al: Plasma cell leukemia. A clinical study of 13 cases, with a demonstration of small-sized plasma cells. Nippon Ketsueki Gakkai Zasshi 40:529, 1977
456. Isobe T, Ikeda Y, Ohta H: Comparison of sizes and shapes of tumor cells in plasma cell leukemia and plasma cell myeloma. Blood 53:1028, 1979
457. Israels MCG: Immature cell erythraemia in an adult. J Pathol Bacteriol 481:299, 1939
458. Israels MCG, Wilkinson JF: Achrestic anaemia. QJ Med 5:69, 1936
459. Izrael V, Jacquillat C, Chastang C et al: Les myéloblastoses partielles. Bull Cancer 61:341, 1974
460. Jackson CW: Cholinesterase as a possible marker for early cells of the megakaryocytic series. Blood 40:586, 1972
461. Jackson H, Parker F, Jr, Lemon HM: Agnogenic myeloid metaplasia of the spleen. A syndrome simulating other more definite hematologic disorder. N Engl J Med 222:985, 1940
462. Jacob J: The practice and application of electron microscope autoradiography. Int Rev Cytol 30:91, 1971
463. Jacobsen KM: Monozytenleukose. Acta Med Scand 111:30, 1942
464. Jacquillat C, Chastang C, Tanzer J et al: Clinical conference: Prognostic factors in chronic myeloid leukemia. 798 Observations. Nouv Rev Fr Hematol 15:229, 1975
465. Jacquillat C, Weil M, Auclerc G et al: Cytologie et pronostic des leucémies aiguës lymphoblastiques. Actualités Hématol 7:42, 1973
466. Jagwe JG, Serwadda AM: Plasma cell leukaemia in Mulago hospital: A case history and review of the literature. East Afr Med J 49:537, 1972
467. Jain NC: Peroxidase activity in leukocytes of some animal species. Folia Haematol 88:297, 1967

468. Jain NC: A comparative cytochemical study of leukocytes of some animal species. Folia Haematol 94:49, 1970
469. Jaiyesimi F, Oluboyede O, Taylor D et al: Burkitt's lymphoma presenting as acute leukemia. Acta Haematol 54:115, 1975
470. Jansa P: On the argument about the origin of monocytes. Vnitr Lek 15:646, 1969
471. Jansa P: Experience with cytochemical examination of blood and bone marrow in clinical practice. Vnitr Lek 18:367, 1972
472. Jansa P, Papousek F: Cytochemistry of leukocytes in malignancy. Folia Haematol 96:34, 1971
473. Jean G, Lambertenghi-Deliliers G, Ranzi T et al: Ultrastructural aspects of bone marrow and peripheral blood cells in a case of plasma cell leukemia. Acta Haematol 45:36, 1971
474. Jehn V, Thiel E, Baumgart R: Auer-bodies in acute lymphocytic leukemia (ALL). Blood 55:167, 1980
475. Jensen MK, Roll K: Phenylbutazone and leukaemia. Acta Med Scand 178:505, 1965
476. Jerie M: Cytochemical intracellular localization of the activity of acid phosphatase in normal cells of the blood and bone marrow. Cas Lek Cesk 108:168, 1969
477. Jerusalem C: On the morphology, origin and function of various types of monocytes. Verh Anat Ges 115:233, 1965
478. Johnstone JM: The appearance and significance of tissue mast cells in human bone marrow. J Clin Pathol 7:275, 1954
479. Jolly J: Traité technique d'hématologie, morphologie, histogenèse, histophysiologie, histopathologie, 2 vols. Paris, Maloine, 1923
480. Jones ME, Saleem A: Acute promyelocytic leukemia. A review of literature. Am J Med 65:673, 1978
481. Joseph RR, Belpomme D, Mathé G: Increase in "Null" cells in acute lymphocytic leukaemia in remission on long-term immunotherapy. Br J Cancer 33:567, 1976
482. Joshi KC, Andleigh HS: Cytochemical demonstration of acid phosphatase activity in blood and bone-marrow cells in health and various hematological disorders. Indian J Med Res 55:1184, 1967
483. Joshi KC, Joshi KR: Cytochemical study of granulopoiesis with special reference to ribonucleoprotein. Indian J Med Res 55:421, 1967
484. Judkiewicz L, Pluzanska A, Polkowska-Kulesza E et al: Morphology of cell nucleoli, fluorescent microscopy findings and degree of blastic lymphocyte transformation in chronic lymphocytic leukemia, Hodgkin's disease and aplastic anemia. Acta Haematol Pol 2:327, 1971
485. Kageoka T, Nakashima K, Miwa S: Simultaneous demonstration of peroxidase and lysozyme activities in leukemic cells. Am J Clin Pathol 67:481, 1977
486. Kahn P: Comparative cytochemical studies of "Gaucher's-Cells" and phagocytizing reticulum cells in the bone marrow of healthy subjects. Wien Klin Wochenschr 81:416, 1969
487. Kaiser R, Spöttl F: Acute primary plasma cell leukemia (author's trans). Wien Klin Wochenschr 89:95, 1977
488. Kamada N, Okada K, Ito T et al: Chromosomes 21-22 and neutrophil alkaline phosphatase in leukaemia. Lancet 1:364, 1968
489. Kamada N, Okada K, Oguma N et al: C-G translocation in acute myelocytic leukemia with low neutrophil alkaline phosphatase activity. Cancer 37:2380, 1976
490. Kaplan ME, Clark C: An improved rosetting assay for detection of human T lymphocytes. J Immunol Methods 5:131, 1974
491. Kaplow LS: A histochemical procedure for localizing and evaluating leukocyte alkaline phosphatase activity in smears of blood and marrow. Blood 10:1023, 1955

492. Kaplow LS: Simplified myeloperoxidase stain using benzedine dihydrochloride. Blood 26:215, 1965
493. Kaplow LS: Cytochemical heterogeneity of human circulating monocytes. Acta Cytol 19:358, 1975
494. Karanas A, Silver RT: Characteristics of the terminal phase of chronic granulocytic leukemia. Blood 32:445, 1968
495. Karle H, Hansen HE, Killmann SA: Intracellular lysozyme in mature neutrophils and blast cells in acute leukemia. Blood 44:247, 1974
496. Kass L: Demonstration of histones in proerythroblasts in pernicious anemia and the DiGuglielmo syndrome. J Histochem Cytochem 20:817, 1972
497. Kass L: Metachromatic staining of basis nucleoproteins in chronic erythremic myelosis. Am J Clin Pathol 62:21, 1974
498. Kass L: Rapid detection of ringed sideroblasts in erythremic myelosis. Arch Pathol 99:225, 1975
499. Kass L: Pernicious Anemia. Philadephia, WB Saunders, 1976
500. Kass L: Cytochemical detection of homocysteine in pernicious anemia and in chronic erythremic myelosis. Am J Clin Pathol 67:53, 1977
501. Kass L: Esterase activity in erythroleukemia. Am J Clin Pathol 67:368, 1977
502. Kass L: Esterase reactions in acute myelomonocytic leukemia. Am J Clin Pathol 67:485, 1977
503. Kass L: Nonspecific esterase activity in 'hairy cells.' Acta Haematol 58:103, 1977
504. Kass L: Enzymatic abnormalities in erythroleukemia. Acta Haematol 59:302, 1978
505. Kass L: Preleukemic Disorders. Springfield, Charles C Thomas, 1979
506. Kass L: Bone Marrow Interpretation. Philadelphia, JB Lippincott, 1979
507. Kass L: Unipolar nonspecific esterase activity in leukemic histiocytic lymphoma. Arch Pathol Lab Med 103:366, 1979
508. Kass L: New aspects of preleukemic disorders. CRC Crit Rev Clin Lab Sci 10:329, 1979
509. Kass L: Affinity of chronic erythremic myelosis erythroblast nuclei for gold chloride. Stain Technol 54:17, 1979
510. Kass L, Eickholt M: Rapid detection of ringed sideroblasts with bromchlorphenol blue. Am J Clin Pathol 70:738, 1978
511. Kass L, Gray R: Ultrastructural localization of histones in chronic erythremic myelosis. Am J Pathol 81:493, 1975
512. Kass L, Hadi MZ: Phosphorylase activity in chronic erythremic myelosis. Am J Clin Pathol 64:503, 1975
513. Kass L, Munster D: Nonspecific esterases in normal human blood cells: Characterization by isoelectric focusing in polyacrylamide gel. Exp Hematol 7:272, 1979
514. Kass L, Munster D: Nonspecific esterases in leukemic blasts: Characterization by isoelectric focusing in polyacrylamide gel. Am J Clin Pathol 72:611, 1979
515. Kass L, Peters CL: Nonspecific esterase activity in pernicious anemia and chronic erythremic myelosis: A cytochemical and electrophoretic study. Am J Clin Pathol 68:273, 1977
516. Kass L, Peters CL: Esterases in acute leukemias: A cytochemical and electrophoretic study. Am J Clin Pathol 69:57, 1978
517. Kass L, Peters CL, Vugrin D: Electrophoretic properties of esterases in chronic granulocytic leukemia. Am J Clin Pathol 69:329, 1978
518. Kass L, Schnitzer B: Monocytes, Monocytosis, and Monocytic Leukemia. Springfield, Charles C Thomas, 1973
519. Kass L, Schnitzer B: Refractory Anemia. Springfield, Charles C Thomas, 1975
520. Katayama I, Yang JPS: Reassessment of a cytochemical test for differential diagnosis of leukemic reticuloendotheliosis. Am J Clin Pathol 68:268, 1977

521. Kelemen E: Granulocyte alkaline phosphatase activity: A measure of the emergence time of mature marrow neutrophils? Acta Haematol 50:19, 1973
522. Kelenyi G, Németh A: Comparative histochemistry and electron microscopy of the eosinophil leucocytes of vertebrates. I. A study of avian, reptile, amphibian and fish leukocytes. Acta Biol. Acad Sci Hung 20:405, 1969
523. Keller DF: Glucose-6-phosphate dehydrogenase deficiency. CRC Monoscience Series. London, Butterworth & Co, 1971
524. Keusch F, Ruttner JR, Pedio G et al: Occurrence of plasmacytoid cells in long-term blood-cell cultures from a patient with 'hairy cell' leukemia. Exp Cell Biol 44:20, 1976
525. Keyserlingk DG: On the disintegration of cells in the human and rat bone marrow. Acta Haematol 40:252, 1968
526. Khitsov GA, Kuzmin VI, Svetlichnyi IS: Histochemical methods of detecting the activity of nonspecific esterases by analyzing punctates in hematology. Probl Gematol Pereliv Krovi 18:62, 1973
527. Killmann SA: Preleukemia: Does it exist? Nouv Rev Fr Hematol Blood Cells 17:81, 1976
528. Kinet-Denoël C, Breton-Gorius J: Teneurs en ADN, ultrastructure et activité peroxydasique des mégacaryocytes médullaires dans un cas d'anémie réfractaire. Nouv Rev Fr Hematol 13:661, 1973
529. Kirchen ME, Marshall GJ: Marrow storage cells: An ultrastructural study. J Reticuloendothel Soc 19:109, 1976
530. Kjeldsberg CR, Bearman RM, Rappaport H: Prolymphocytic leukemia. An ultrastructural study. Am J Clin Pathol 73:150, 1980
531. Kjeldsberg CR, Nathwani BN, Rappaport H: Acute myeloblastic leukemia developing in patients with mediastinal lymphoblastic lymphoma. Cancer 44:2316, 1979
532. Klebanoff SJ: Antimicrobial mechanisms in neutrophilic polymorphonuclear leukocytes. Semin Hematol 12:117, 1975
533. Knolle J, Meyer KH: IgD-multiple myeloma. Review of 102 cases reported in the literature (author's trans). Immun Infekt 31:125, 1975
534. Knospe WH, Gregory SA: Smouldering acute leukemia. Clinical and cytogenetic studies in six patients. Arch Intern Med 127:910, 1971
535. Knospe WH, Gregory SA, Trobaugh FE, Jr et al: Chronic lymphocytic leukemia: Correlation of clinical course and therapeutic response with in vitro testing and morphology of lymphocytes. Am J Hematol 2:73, 1977
536. Knowles DM, Hoffman HT, Ferrarini M et al: The demonstration of acid α-naphthyl acetate esterase activity in human lymphocytes: Usefulness as a T-cell marker. Cell Immunol 35:112, 1978
537. Kohli P, Brunner HE, Siegenthaler W: Erythroleukämie nach chronischer Benzolintoxikation. Untersuchung der Ferro-und Erythrocytenkinetik mit radioaktivem Eisen und Chrom. Schweiz Med Wochenschr 97:368, 1967
538. Kohn G, Manny N, Eldor A et al: De Novo appearance of the pH-1 chromosome in a previously monosomic bone marrow (45,XX,-6): conversion of a myeloproliferative disorder to acute myelogenous leukemia. Blood 45:653, 1975
539. Kolmeire KH, Bayrd ED: Familial leukemia: Report of instance and review of the literature. Proc Mayo Clin 38:523, 1963
540. Komatsu H, Furusawa S: Staining for esterases, dehydrogenases and phosphorylases. Rinsho Byori (34 Spec No):67, 1978
541. Komiyama A, Spicer SS: Ultrastructural localization of a characteristic acid phosphatase in granules of rabbit basophils. J Histochem Cytochem 22:1092, 1974
542. Korenevskaia MI, Alpidovskii VK, Lenskaia RV et al: Cytogenetic and histocytochemical study of the reticular cells in patients with acute reticulosis. Probl Gematol Pereliv Krovi 15:27, 1970

543. KOROSTOFF NR, SUN NCJ, OKUN DB: Atypical blast crisis in chronic myelogenous leukemia. JAMA 245:1245, 1981
544. KOZINER B, FILIPPA DA, MERTELSMANN R et al: Characterization of malignant lymphomas in leukemic phase by multiple differentiation markers of mononuclear cells. Correlations with clinical features and conventional morphology. Am J Med 63:556, 1977
545. KOZINER B, MCKENZIE S, STRAUS D et al: Cell marker analysis in acute monocytic leukemias. Blood 49:895, 1977
546. KOZINETS GI, BYKOVA IA, SUKIASCOVA TG et al: Use of the PAS reaction for assessing ineffective erythropoiesis. Lab Delo 9:518, 1978
547. KRAUSE JR, SRODES C, LEE RE: Use of the bone marrow imprint in the diagnosis of leukemic reticuloendotheliosis ("hairy cell leukemia"). Am J Clin Pathol 68:368, 1977
548. KRUSE H: Zytochemie der Leukozyten im Kindesalter. I. Zytochemische Reaktionen der Leukozyten bei Infektionen—Neutrophile Granulozyten—Eosinophile–Monozyten. Folia Haematol 101:904, 1974
549. KRUSE H: Cytochemistry of leukcocytes in children. IV. The use of cytochemical tests in the differentiation and prognostic evaluation of acute leukemias. Folia Haematol 102:389, 1975
550. KURNICK NB: Histological staining with methyl-green-pyronin. Stain Technol 27:233, 1952
551. KURZ R, HAAS H: Value of the cytological and cytochemical classification in the management of acute childhood leukemia. Acta Haematol 52:1, 1974
552. KUSHNER JP, LEE GR, WINTROBE MM et al: Idiopathic refractory sideroblastic anemia. Clinical and laboratory investigation of 17 patients and review of the literature. Medicine 50:139, 1971
553. KYLE RA, BIEGER RC, GLEICH GJ: Diagnosis of syndromes associated with hyperglobulinemia. Med Clin North Am 54:917, 1970
554. KYLE RA, GREIPP PR: The laboratory investigation of monoclonal gammopathies. Mayo Clin Proc 53:719, 1978
555. KYLE RA, PIERRE RV, BAYRD ED: Multiple myeloma and acute leukemia associated with alkylating agents. Arch Intern Med 135:185, 1975
556. LABEDZKI L, GRIPS KH: Chronic monocytic leukemia. Dtsch Med Wochenschr 99:690, 1974
557. LAHE BD: The histochemistry of phosphatases: The use of lead acetate instead of lead nitrate. J R Microsp Soc 85:73, 1966
558. LAMBIN P, LETO A: La recherche des peroxydases leucocytaires par la méthode au cuivre de Sato et son application au diagnostic différentiel des leucémies aiguës. Rev Belge Sci Med 2:236, 1930
559. LARSEN TE: Emperipolesis of granular leukocytes within megakaryocytes in human hemopoietic bone marrow. Am J Clin Pathol 53:485, 1970
560. LASZLO J, HUANG AT: Diagnosis and management of myeloproliferative disorders. Curr Probl Cancer 2:1, 1977
561. LATNER AL: Wellcome award lecture: Biochemistry of Cancer. Ann Clin Biochem 13:458, 1976
562. LAWKOWICZ W, KRZEMINSKA-LAWKOWICZOWA I, KNAP J: Neoplastic variant of acute erythroblastosis—proerythroblastoma. Acta Haematol Pol 7:65, 1976
563. LAWRINSON W, GROSS S: Nuclear arylsulfatase activity in primitive hemic cells. Lab Invest 13:1612, 1964
564. LAWS WC, BOHANNON RA, ROBINSON AJ et al: Acute promyelocytic leukemia with hypofibrinogenemia. Calif Med 109:219, 1968
565. LEAVELL BS, TWOMEY J: Possible leukemoid reaction in disseminated tuberculosis: Report of a case with Auer rods. Trans Am Clin Climatol Assoc 75:166, 1963
566. LEDER LD: Enzyme histochemical findings in chronic erythroblastosis and acute erythremia. Klin Wochenschr 43:795, 1965

567. LEDER LD: Die fermentcytochemische Erkennung normaler und neoplasticher Erythropoesezellen in Schnitt und Ausstrich. Blut 15:289, 1967
568. LEDER LD: The origin of blood monocytes and macrophages. A review. Blut 16:86, 1967
569. LEDER LD: A case of acute leukemia with pseudo-Pelger cells containing Auer bodies. Acta Haematol 42:58, 1969
570. LEDER LD: The origin of the blood monocytes and their relationship to the so-called histiocytes. Haematol Bluttransfus 7:19, 1969
571. LEDER LD: Acute myelo-monocytic leukemia with atypical naphthol-as-d-chloracetate esterase positive eosinophils. Acta Haematol 44:52, 1970
572. LEDER LD: A particularly sensitive modification of the peroxidase reaction. Acta Haematol 44:292, 1970
573. LEDER LD, NICHOLAS R: Long-term investigations of the cytologic and enzyme cytochemical behavior of skin window macrophages. Klin Wochenschr 43:684, 1965
574. LEDER LD, PAPE B: Cytochemical demonstration of transitional forms between promyelocytes and eosinophil myelocytes. Klin Wochenschr 48:1009, 1970
575. LEDER LD, PAPE B: Detection of naphthol-as-d-chloracetate esterase in the normal eosinophilic preliminary stage. Klin Wochenschr 48:885, 1970
576. LEE JAH: Acute myeloid leukaemia in adolescents. Br Med J 1:988, 1961
577. LEE SL, LIVINGS D, JAMES GW et al: Morphological classification of acute leukemias. Cancer Chemother Rep 16:151, 1962
578. LEE SL, KOPEL S, GLIDEWELL O: Cytomorphological determinants of prognosis in acute lymphoblastic leukemia of children. Semin Oncol 3:209, 1976
579. LEE SL, ROSNER F, ROSENTHAL RL: Reticulum cell leukemia. Clinical and hematologic entity. NY J Med 69:422, 1969
580. LEHRER RI, GOLDBERG LS, APPLE MA et al: Refractory megaloblastic anemia with myeloperoxidase deficient neutrophils. Ann Intern Med 76:447, 1972
581. LEHRER RI, HANIFIN J, CLINE MJ: Defective bactericidal activity in myeloperoxidase-deficient human neutrophils. Nature 223;78, 1969
582. LEIBETSEDER F, SAMITZ H, SORGO G: Plasma cell leukemia. Blut 27:351, 1973
583. LEINONEN EA: Cytochemical studies of acute leukemias. Acta Haematol 43:219, 1970
584. LEJEUNE F, TURPIN F, LORTHOLARY P: Unusual giant inclusions in blast cells during acute transformation of chronic myeloid leukemia. Cytochemical and ultrastructural study (author's trans). Pathol Biol 26:7, 1978
585. LEJONC RJ, GOURDIN MF, MANNONI P et al: The surface morphology of human B lymphocytes as revealed by immunoelectron microscopy. J Exp Med 141:392, 1975
586. LENNERT K, STEIN H, KAISERLING E: Cytological and functional criteria for the classification of malignant lymphomata. Br J Cancer (Suppl) 2:29, 1975
587. LEONARD BJ, ISRAELS MCG, WILKINSON JF: Alkaline phosphatase in the white cells in leukaemia and leukaemoid reactions. Lancet 1:289, 1958
588. LETZKY WB, PAPISOWA DG: Cytochemical studies on blood and bone marrow leukocytes in various age groups. Folia Haematol 102:530, 1975
589. LEVY SB: A histological stain for detecting friend leukemic cells in fixed hemopoietic tissue. Arch Pathol Lab Med 100:463, 1976
590. LI CY, LAM KW, YAM LT: Esterases in human leukocytes. J Histochem Cytochem 21:1, 1973
591. LI CY, YAM LT, LAM KW: Studies of acid phosphatase isoenzymes in human leukocytes: Demonstration of isoenzyme cell specificity. J Histochem Cytochem 18:901, 1970
592. LIANG W, ROWLEY JD: Letter: 14q+ marker chromosomes in multiple myeloma and plasma-cell leukaemia. Lancet 1:96, 1978
593. LILLIE RD, BURTNER HJ: Stable sudanophilia of human neutrophil leukocytes in relation to peroxidase and oxidase. J Histochem Cytochem 1:8, 1953

594. Limarzi LR, Levinson SA: An undescribed type of erythropoiesis observed in human sternal marrow. Arch Pathol 36:127, 1943
595. Lin PS, Tsai S, Wallach DF et al: Visualization of surface topology of human lymphoid cells by scanning electron microscopy. Bibl Haematol 40:263, 1975
596. Linman JW, Bagby CG Jr: The preleukemic syndrome: Clinical and laboratory features, natural course, and management. Nouv Rev Fr Hematol Blood Cells 17:11, 1976
597. Linman JW, Saarni MI: The preleukemic syndrome. Semin Hematol 11:93, 1974
598. Lisiewicz J: Mechanism and frequency of hemorrhagic complications in patients with plasma cell myeloma (author's trans). Przegl Lek 34:495, 1977
599. Liso V, Lortholary P, Lejeune F et al: Nonspecific esterase activity of polynuclears in Chediak's Disease. Pathol Biol 19:507, 1971
600. Liso V, Troccoli G, Grande M: Cytological and clinical aspects of plasmacell leukemia. Minerva Med 66:2526, 1975
601. Liso V, Troccoli G, Grande M: Cytochemical study of acute promyelocytic leukaemia. Blut 30:261, 1975
602. Liso V, Troccoli G, Specchia G et al: Cytochemical "normal" and "abnormal" eosinophils in acute leukemias. Am J Hematol 2:123, 1977
603. Lison L: Sur la recherche histochimique des oxydases par la réaction du bleu d'indophenol. Le cas des lipides. Bull Soc Chim Biol 18:185, 1936
604. Litwin J, Stacher A: Therapiebedingte Veränderungen zytochemischer Befunde akuter Leukämien. Blut 28:161, 1974
605. Litwin J, Szmigielski S, Zupánska B: Histochemical studies of the activity of nonspecific esterases in blood and bone marrow cells. Pol Arch Med Wewn 35:647, 1965
606. Locquin M, Bessis M: Sur quelques caractères des corps d'Auer et de Charcot. CR Soc Biol 143:609, 1949
607. Löffler H: Unspecific esterases and acid phosphatase in plasmacytomas. Blut 15:330, 1967
608. Löffler H: Cytochemical changes in cytoplasm during hemoglobin synthesis. Haematol Bluttransfus 10:43, 1972
609. Löffler H, Pralle H: Cytochemical and biochemical demonstration of an alkaline phosphatase in human red cells. Klin Wochenschr 48:763, 1970
610. Löffler H, Pralle H, Lück R et al: Der cytochemische ermittelte Leukosetypals prognostischer Parameter bei unreifzelligen Leukosen. Klin Wochenschr 52:134, 1974
611. Lorie IuI, Relizinger MA, Frenkel MA et al: Myeloma disease developing into acute plasma cell leukemia. Probl Gematol Pereliv Krovi 14:22, 1969
612. Lorie IuI, Soloveva EA, Streliukhina NV et al: Clinico-morphological variants of acute leukemia. Probl Gematol Pereliv Krovi 14:22, 1969
613. Lortholary P, Lejeune F, Tanzer J et al: Etude cytochimique des leucoblastes au cours de la transformation aiguë de la leucémie myéloide chronique. Presse Médicale 74:1237, 1966
614. Lotem J, Sachs L: Control of normal differentiation of myeloid leukemic cells. VI. Inhibition of cell multiplication and the formation of macrophages. J Cell Physiol 85:587, 1975
615. Lotem J, Sachs L: Control of normal differentiation of myeloid leukemic cells. XII. Isolation of normal myeloid colony-forming cells from bone marrow and the sequence of differentiation to mature granulocytes in normal and D+ myeloid leukemic cells. J Cell Physiol 92:97, 1977
616. Loukas DF: Leukemic reticuloendotheliosis: A medical technologist's diagnosis. Am J Med Technol 42:367, 1976
617. Love R, Studzinski GP, Walsh RJ: Nuclear, nucleolinar and cytoplasmic acid phosphatases in cultured mammalian cells. Exp Cell Res 58:62, 1969

618. Love R, Walsh RJ: Nucleolinar morphology in normal diploid, neoplastic, and aneuploid cells in vitro. Cancer Res 30:990, 1970
619. Low F, Freeman J: Electron Microscopic Atlas of Normal and Leukemic Human Blood. New York, McGraw-Hill, 1958
620. Lukes RJ, Collins RD: A functional approach to the classification of malignant lymphoma. Recent Results Cancer Res 46:18, 1974
621. Luzzatto AM: Sull'anemia grave megaloblastica senza reporto ematologico corrispondente. Riv Ven 47:193, 1907
622. Mabry EW, Carubelli R: Sialic acid in human cancer. Experientia 28:182, 1972
623. Macdiarmid WW: Chromosomal changes following treatment of polycythaemia with radioactive phosphorus. QJ Med 34:133, 1965
624. MacMahon B, Levy MA: Prenatal origin of childhood leukemia. N Engl J Med 270:1082, 1964
625. MacPherson GG: Synthesis and localization of sulphated mucopolysaccharide in megakaryocytes and platelets of the rat, an analysis by electron-microscope autoradiography. J Cell Sci 10:705, 1972
626. Mage MG, Evans WH, Himmelhoch SR et al: Immunological identification and quantifications of guinea pig heterophil and eosinophil peroxidases. J Reticuloendothel Soc 9:201, 1971
627. Makita T, Sandborn EB: The ultrastructural localization of adenosinetriphosphatase and alkaline phosphatase activity in eosinophil leukocytes. Histochemie 24:99, 1970
628. Malpuech G, Perrimond H, Benzou MJ et al: Erythroleucémie révélée par un diabète insipide. Pédiatrie 31:618, 1976
629. Mann JR, Simpson JS, Munkley RM et al: Lysosomal enzyme cytochemistry in acute leukaemia. J Clin Pathol 24:831, 1971
630. Marchal G, Lemoin J, Bloch-Michel R: Leucémie chronique à monocytes, sans splénomégalie, ni adénomégalie. Sang 8:694, 1934
631. Martin WJ, Bayrd ED: Erythroleukemia, with special emphasis on the acute or incomplete variety. Report of five cases. Blood 9:321, 1954
632. Martinez-Cairo S, Gutierrez Lizardi P, Ramos Torres J: Leukemia of plasma cells. Prensa Med Mex 41:400, 1976
633. Masera P, Benzio G, Pagliardi GL et al: Physiopathological and cytogenetic findings in 2 cases of plasma cell leukemia. Minerva Med 61:2005, 1970
634. Mason DY, Farrell C, Taylor CR: The detection of intracellular antigens in human leucocytes by immunoperoxidase staining. Br J Haematol 31:361, 1975
635. Mason JE Jr, DeVita VT, Canellos GP: Thrombocytosis in chronic granulocytic leukemia: Incidence and clinical significance. Blood 44:483, 1974
636. Mathé G: Editorial: Towards dismembering the so-called acute lymphoid leukemia. Nouv Presse Med 6:2399, 1977
637. Mathé G, Belpomme D: T and B lymphocytic nature of leukemias and lymphosarcomas: a new but still uncertain parameter for their classification. Biomedicine 20:81, 1974
638. Mathé G, Belpomme D, Dantchev D et al: Recent progress made in the classification of lymphoid and monocytoid leukemias and of lympho- and reticulosarcomas. Ann Med Interne 126:523, 1975
639. Mathé G, Belpomme D, Dantchev D et al: Progress in the classification of lymphoid and/or monocytoid leukemias and of lympho- and reticulosarcomas (non-Hodgkin's lymphomas). Biomedicine 22:177, 1975
640. Mathé G, Belpomme D, Dantchev D et al: Immunoblastic acute lymphoid leukaemia. Biomedicine 20:333, 1974
641. Mathé G, De Vassal F, Schwarzenberg L et al: Results in children of acute lymphoid leukaemia protocol ICIG-ALL 9 consisting of chemotherapy for only nine months followed by active immunotherapy. Comparison with the results of more prolonged chemotherapy protocols. Recognition of two

groups of acute lymphoid leukaemias from prognostic parameters. In Salmon SE, Jones SE (eds): Adjuvant Therapy of Cancer, pp 357–72. Amsterdam, North-Holland Publishers, 1977

642. MATHÉ G, DE VASSAL F, SCHWARZENBERG L et al: Preliminary results of three protocols for the treatment of acute lymphoid leukaemia of children: Distinction of two groups of patients according to predictable prognosis. Med Pediatr Oncol 4:17, 1978
643. MATHÉ G, DE VASSAL F, SCHWARZENBERG L et al: Results in children of acute lymphoid leukemia: Protocol ICIG-ALL 9 consisting of chemotherapy for only 9 months followed by active immunotherapy: Comparison with the results of more prolonged chemotherapy protocols. Recognition of two groups of acute lymphoid leukemias from prognostic parameters. Recent Results Cancer Res 62:67, 1977
644. MATHÉ G, POUILLART P, AMIEL JL et al: Cytological classification of 100 cases of acute lymphoid leukemia: Factor of prognosis after immunotherapy. Bull Cancer 61:309, 1974
645. MATHÉ G, POUILLART P, HAYAT M et al: Essential factor to therapeutic indications and prognosis of acute leukemias: Their cytological subclassification. Rev Prat II 23:151, 1973
646. MATHÉ G, POUILLART P, SCHLUMBERGER JR et al: Cytology in classification of non-leukemic diffuse cancerous lymphomas (lympho- and reticulosarcomas). Bull Cancer 61:23, 1974
647. MATHÉ G, POUILLART P, SCHLUMBERGER JR et al: Cytology in the classification of diffuse non-leukaemic malignant lymphomata (lympho- and reticulosarcomata). Br J Cancer 31 (Suppl) 2:53, 1975
648. MATHÉ G, POUILLART P, SCHWARZENBERG L et al: Leukaemic conversion of non-Hodgkin's malignant lymphomata. Br J Cancer 31 (Suppl) 2:96, 1975
649. MATHÉ G, POUILLART P, STERESCU M et al: Subdivision of classical varieties of acute leukemia: Correlation with prognosis and cure expectancy. Eur J Clin Biol Res 16:554, 1971
650. MATHÉ G, POUILLART P, WEINER R et al: Classification and subclassification of acute leukemias correlated with clinical expression, therapeutic sensitivity and prognosis. In Mathé G et al (eds): Nomenclature, Methodology and Results of Clinical Trials in Acute Leukemias: Recent Results in Cancer Research, vol 43, p. 6. New York, Springer-Verlag, 1973
651. MATTA MC, BASTOS AL, MARQUES D et al: Photoreversibility of fluorescence in nile blue stain granules of tumor cells. J Pathol 107:xiv, 1972
652. MAUPIN B, SCHLUMBERGER JR, GASC JP et al: Contribution of cytochemistry to comparative hematology: Reptile blood. Nouv Rev Fr Hematol 9:663, 1969
653. MAXIMOW AA: The lymphocytes and plasma cells. In Cowdry E: Special Cytology, 2nd ed, Vol II, p 601. New York, Paul B Hoeber, 1932
654. MAXWELL MH: Electron cytochemistry of developing and mature eosinophils in the bone marrow of the fowl and the duck. Histochem J 10:63, 1978
655. MAYER EP, BRATESCU A, DRAY S et al: Enumeration of human B lymphocytes in stained blood smears by their binding of bacteria. Clin Immunol Immunopathol 9:37, 1978
656. MCCAFFREY R, HARRISON TA, BALTIMORE D: Terminal deoxynucleotidyl transferase in human bone marrow. Blood 44:931, 1974
657. MCCAFFREY R, HARRISON TA, PARKMAN R et al: Terminal deoxynucleotidyl transferase activity in human leukemic cells and in normal human thymocytes. N Engl J Med 292:775, 1975
658. MCCONNELL I, HURD CM: Lymphocyte receptors. 1. Receptors for Fc of IgG and complement (C3b) on immunoglobulin-bearing, antigen-binding and antibody-secreting cells. Immunology 30:825, 1976
659. MCDONALD CR, PAINE CJ: Extranodal lymphoma with an unusual constellation of features. South Med J 68:1177, 1975

660. McDuffie NG: Crystalline patterns in Auer bodies and specific granules of human leukocytes. J Microsc 6:321, 1967
661. McDuffie NG: Nuclear blebs in human leukaemic cells. Nature 214:1341, 1967
662. McKenna RW, Bloomfield CD, Dick F et al: Acute monoblastic leukemia: Diagnosis and treatment of ten cases. Blood 46:481, 1975
663. McKenna RW, Parkin J, Brunning RD: Morphologic and ultrastructural characteristics of T-cell acute lymphoblastic leukemia. Cancer 44:1290, 1979.
664. Mdzewski B, Swierkowska K, Kling E et al: The patterns and distribution of euchrysine-binding grains in nonstimulated human peripheral T and non-T lymphocytes. Cell Immunol 42:170, 1979
665. Meacham GC, Weisberger AS: Early atypical manifestations of leukemia. Ann Intern Med 41:780, 1954
666. Meister H: Cytochemistry of monocytic leukemia. Haematologia 4:311, 1970
667. Meister H: Applied cytochemistry in hematology. Folia Haematol 100:186, 1973
668. Meister H: Chronic myelosis. Z Gesamte Inn Med 31:203, 1976
669. Meister H, Fuchs R: Cytochemistry of malignant erythroblastosis. Folia Haematol 98:71, 1972
670. Melvin SL: Comparison of techniques for detecting T-cell acute lymphocytic leukemia. Blood 54:210, 1979
671. Merker H: The differentiation between anaemia sideroblastica refractoria and erythremic myelosis by cytochemical methods. Proc 9th Congr Eur Soc Haematol (Lisbon) 2:290, 1963
672. Meuret G: Human monocytopoiesis. Exp Hematol 2:238, 1974
673. Meyers B: Erythroleukaemia. Proc R Soc Med 21:741, 1928
674. Meytes D, Katz D, Ramot B: Preleukemia and leukemia in polycythemia vera. Blood 47:237, 1976
675. Micu D: Metabolic disturbances of the lymphoid cells and their significance for clinical pathology. Med Interne 15:3, 1977
676. Miescher PA, Farguet JJ: Chronic myelomonocytic leukemia in adults. Semin Hematol 11:129, 1974
677. Miller A, DeLuca D, Decker J et al: Specific binding of antigen to lymphocytes. Evidence for lack of unispecificity in antigen-binding cells. Am J Pathol 65:451, 1971
678. Miller DR, Newstead GJ, Young LW: Perinatal leukemia with a possible variant of the Ellis-van Creveld syndrome. J Pediatr 74:300, 1969
679. Miller F: Electron microscopic cytochemistry of leukocyte granules. In Uyeda R (ed): International Congress for Electron Microscopy, 6th, Kyota, 1966, Tokyo, Maruzen, 1966
680. Miller F, Herzog V: Die Lokalisation von Peroxydase und saurer Phosphatase in eosinophilen Leukocyten während der Reifung. Elektronenmikroskopisch-cytochemische Untersuchungen am Knochenmark von Ratte and Kaninchen. Z Zellforsch 97:84, 1969
681. Miller RW: Radiation, chromosomes, and viruses in the etiology of leukemia. N Engl J Med 271:30, 1964
682. Miller RW: Relation between cancer and congenital defects in man. N Engl J Med 275:87, 1966
683. Miller RW, Todaro GJ: Viral transformation of cells from persons at high risk of cancer. Lancet 1:81, 1969
684. Milner G: Preleukaemia and smouldering leukaemia. J Clin Pathol 29:88, 1976
685. Minata F, Hardin JH, Spicer SS: Coexistence of acid Q and acid mucosubstance in the nucleoid of human blood platelet granules. Histochemie 35:319, 1973

686. Mintz U, Djaldetti M, Rozenszajn L et al: Giant lysosome-like structures in promyelocytic leukemia. Ultrastructural and cytochemical observations. Biomedicine (Express) 19:426, 1973
687. Mirecka J, Cichocki T: The simultaneous detection of acid and alkaline phosphatase in neutrophil granulocytes. Folia Histochem Cytochem 7:431, 1969
688. Mitus WJ, Bergna LJ, Mednicoff IB et al: Cytochemical studies of glycogen content of lymphocytes in lymphocytic proliferations. Blood 13:748, 1958
689. Mitus WJ, Mednicoff IB, Vitrels B et al: Neoplastic lymphoid reticulum cells in the peripheral blood: A histochemical study. Blood 17:206, 1961
690. Miura AB: Multiple vacuoles formation in erythroblasts in an erythroleukaemic patient. Scand J Haematol 16:183, 1976
691. Miura AB: Hairy cell leukemia: Diagnostic criteria from the clinical point of view (author's trans). Acta Haematol Jpn 39:705, 1976
692. Miura H: Acid phosphatase in blood corpuscles and the hematopoetic organs. Jpn J Clin Pathol 14:404, 1966
693. Moeschlin S: Erythroblastosen, Erythroleukämien und Erythroblastämien. Folia Haematol 64:262, 1940
694. Moeschlin S: Phase-contrast microscopy of leukocytes. In The Leukaemias: Etiology, Pathophysiology and Treatment. Henry Ford Hosp Int Symposium, p 37. New York, Academic Press, 1957
695. Mollin DL: A symposium on sideroblastic anaemia. Introduction: Sideroblasts and sideroblastic anaemia. Br J Haematol 11:41, 1965
696. Mollo F, Stramignoni A: Nuclear projections in blood and lymph node cells of human leukaemias and Hodgkin's disease and in lymphocytes cultured with phytohaemagglutinin. Br J Cancer 21:519, 1967
697. Moloney WC, Fleigelman L, McPherson K: The use of naphthol AS-D chloroacetate as a substrate for the study of leukocyte esterase activity. J Histochem Cytochem 7:306, 1959
698. Moloney WC, McPherson K, Fliegelman L: Esterase activity in leukocytes demonstrated by the use of naphthol AS-D chloroacetate substrate. J Histochem Cytochem 8:200, 1960
699. Moore GE, Kitamura H, Toshima S: Morphology of cultured hematopoietic cells. Cancer 22:245, 1968
700. Morrison M, Samwick AA, Loewe L: Dwarf cell (atypical myelogenous) leukemia. Acta Haematol 2:125, 1949
701. Morse WI: Stainable ferric iron particles in erythroid marrow cells and erythrocytes. Can Med Assoc J 72:418, 1955
702. Mover S, Li CY, Yam LT: Semiquantitative evaluation of tartrate-resistant acid phosphatase activity in human blood cells. J Lab Clin Med 80:711, 1972
703. Mueller J, Brun del Re G, Buerki H et al: Nonspecific acid esterase activity: A criterion for differentiation of T and B lymphocytes in mouse lymph nodes. Eur J Immunol 5:270, 1975
704. Mukherji PS: Acute myeloblastic leukaemia following chloramphenicol treatment. Br Med J 1:1286, 1957
705. Müller D: Preservation of blood smears for cytochemical enzyme determination. Klin Wochenschr 44:1316, 1966
706. Müller-Hermelink U, Müller-Hermelink HK: Scanning electron microscopic investigations of acute leukemia. Virchows Arch (Cell Pathol) 23:227, 1977
707. Mund J, Jacobi H, Künzer W: Cell morphology and prognosis of paraleukoblast leukemia. Dtsch Med Wochenschr 94:2550, 1969
708. Murata F, Spicer SS: Morphologic and cytochemical studies of rabbit heterophilic leukocytes. Evidence for tertiary granules. Lab Invest 29:65, 1973
709. Murphy SB, Borella L, Sen L et al: Lack of correlation of lymphoblast cell size with presence of T-cell markers or with outcome in childhood acute lymphoblastic leukaemia. Br J Haematol 31:95, 1975

710. NACHLAS MM, SELIGMAN AM: Evidence for the specificity of esterase and lipase by the use of three chromogenic substrates. J Biol Chem 181:343, 1949
711. NACHLAS MM, TSOU KC, DE SOUZA E et al: Cytochemical demonstration of succinic dehydrogenase by the use of a new p-nitrophenyl substituted ditetrazole. J Histochem Cytochem 5:420, 1957
712. NAEGELI O: Ueber rothes Knockenmark und Myeloblasten. Dtsch Med Wochenschr 26:287, 1900
713. NAEGELI O: Blutkrankkeiten and Blutdiagnostik, 5th ed Berlin, Springer-Verlag 1931
714. NAEIM F, CAPOSTAGNO VJ, JOHNSON CE JR et al: Sézary syndrome: Tartrate-resistant acid phosphatase in the neoplastic cells. Am J Clin Pathol 71:528, 1979
715. NAGARATNAM N, TALWAITE SN, FERNANDO DJ et al: Uncommon manifestations of myelomatosis and plasma cell leukaemia. Ceylon Med J 22:167, 1977
716. NAJEAN Y, PECKING A, BROQUET M: Anémies réfractaires avec myéloblastose partielle. (1) Analyse d'un protocole, groupant 79 cas I.—Caractères cliniques et évolution sous androgénothérapie. Nouv Rev Fr Hematol 16:67, 1976
717. NAKAJIMA E: Shuju naru hoyotai o hogan seru kotsuzui bosaiho hakketsubyo no ichi rei. Chiba Igaku Semmon Gakko Koyukai Zasshi, 68, 1914
718. NAKATA K, AOKI S, FUJIMOTO K et al: Case of plasma cell leukemia following prolonged cholangitis. Jpn J Clin Med 29:2318, 1971
719. NANBA K, JAFFE ES, BRAYLAN RC et al: Alkaline phosphatase-positive malignant lymphoma. A subtype of B-cell lymphoma. Am J Clin Pathol 68:535, 1977
720. NASU H, WARABI H: Immunochemical properties of IgG pyroglobulin in a patient with plasma cell leukemia. Rinsho Ketsueki 20:485, 1979
721. NATHWANI BN, KIM H, RAPPAPORT H: Malignant lymphoma, lymphoblastic. Cancer 38:964, 1976
722. NAZARLI AG, KULIEV SA, PETERBURSKAIA RS: Case of plasmacytic variant of acute leukemia in a 13 year old boy. Pediatriia 8:84, 1975
723. NECHELES TF, BRENNER JF, FRISTENSKY R et al: The computer-assisted morphological classification of acute leukemia. I. Preliminary results. Biomedicine 25:241, 1976
724. NELSON DA: Cytomorphological diagnosis of the acute leukemias. Semin Oncol 3:201, 1976
725. NICHOLS BA, BAINTON DF: Differentiation of human monocytes in bone marrow and blood. Sequential formation of two granule populations. Lab Invest 29:27, 1973
726. NICHOLS BA, BAINTON DF, FARQUHAR MG: Differentiation of monocytes. Origin, nature, and fate of their azurophil granules. J Cell Biol 50:498, 1971
727. NICHOLS TM: An electron microscope investigation of Auer bodies in myeloblastic leukemia. Can Med Assoc J 102:865, 1970
728. NIGAM VN, CANTERO A: Polysaccharides in cancer. Adv Cancer Res 16:1, 1972
729. NIKULINA LD: Significance of the determination of the activity of leukocytes enzymes in the prognosis of acute leukemia in children. Probl Gematol Pereliv Krovi 20:37, 1975
730. NILSSON K, PONTEN J: Classification and biological nature of established human hematopoietic cell lines. Int J Cancer 15:321, 1975
731. NOWELL PC: Differentiation of human leukemic leukocytes in tissue culture. Exp Cell Res 19:267, 1960
732. NOWELL PC, HUNGERFORD DA: A minute chromosome in human chronic granulocytic leukemia. Science 132:1497, 1960
733. NUNNENSIEK C, RÖSSLER R, NEUMANN G: Chronische Erythroleukämie bei Di-Guglielmo-Syndrom. Med Welt 27:1154, 1976
734. OCHS HD, SLICHTER SJ, HARKER LA et al: The Wiskott-Aldrich Syndrome: Studies of lymphocytes, granulocytes, and platelets. Blood 55:243, 1980

735. ODEBERG H, OLOFSSON T, OLSSON I: Granulocyte function in chronic granulocytic leukaemia. I. Bactericidal and metabolic capabilities during phagocytosis in isolated granulocytes. Br J Haematol 29:427, 1975
736. OEHMICHEN M, SAEBISCH R: Histochemical staining autoradiographic technic. Acta Histochem 41:353, 1971
737. OGAWA M, KOCHWA S, SMITH C et al: Clinical aspects of IgE myeloma. N Engl J Med 281:1217, 1969
738. OGAWA M, PRESTON RA, CARSON MD et al: Cytoplasmic fibrils in plasma cell leukemia. Acta Haematol 56:123, 1976
739. OGAWA T, KOERTEN HK, DAEMS WT: Peroxidase activity in monocytes and tissue macrophages of mice. Cell Tissue Res 188:361, 1978
740. OKADA H, HATTORI Y, KANOH T et al: An autopsied case of plasma cell leukemia with long-term survival. Jpn J Clin Hematol 18:145, 1977
741. OKADA H, HOSHINO T, YAMAMOTO T: Acute granulocytic leukemia terminating in profound megaloblastoid cell proliferation: A case report. Acta Haematol Jpn 34:205, 1971
742. OKUN MR, DONNELLAN B, PEARSON SH et al: Melanin: A normal component of human eosinophils. Lab Invest 30:681, 1974
743. OKUN MR, EDELSTEIN L, NIEBAUER G et al: The histochemical tyrosine-dopa reaction for tyrosinase and its use in localizing tyrosinase activity in mast cells. J Invest Dermatol 53:39, 1969
744. OLIVER C, ESSNER E: Distribution of anomalous lysosomes in the beige mouse: a homologue of Chediak–Higashi syndrome. J Histochem Cytochem 21:218, 1973
745. OLIVER C, ESSNER E: Formation of anomalous lysosomes in monocytes, neutrophils, and eosinophils from bone marrow of mice with Chediak–Higashi syndrome. Lab Invest 32:17, 1975
746. OLSEN T: Histiocytic medullary reticulosis. A post-mortem finding in a patient with erythroleukaemia and skin carcinomata. Med J Aust 1:242, 1973
747. OLUMIDE AA, ADELOYE A: Burkitt's lymphoma of the spinal cord. Trop Geogr Med 30:215, 1978
748. ORNSTEIN L, ANSLEY H, SAUNDERS A: Improving manual differential white cell counts with cytochemistry. Blood Cells 2:557, 1976
749. ORZECHOWSKI-JUZWENKO K, KOTLAREK-HAUS S, BRODZKA W: Activity of acid phosphatase in blood cells and bone marrow cells in some internal diseases. Acta Haematol Pol 3:229, 1972
750. OSCULATI F: Fine structural localization of acid phosphatase and arylsulfatase in the chick heterophile leukocytes. Z Zellforsch Mikrosk Anat 109:398, 1970
751. OSGOOD EE: Monocytic leukemia. Report of six cases and review of one hundred and twenty-seven cases. Arch Intern Med 59:931, 1937
752. OSGOOD EE: A unifying concept of the etiology of the leukemias, lymphomas, and cancers. J Natl Cancer Inst 18:155, 1957
753. OSGOOD EE: Acute monocytic leukemia as an explanation for "hiatus leukemicus" and "myelo-monocytic leukemia." Blood 33:268, 1969
754. OTA H, IKEDA Y, KAKIZAWA H: Three cases of plasma cell leukemia—therapeutic effect and cytologic study (author's trans). Jpn J Clin Hematol 15:201, 1974
755. PAINE CJ, RICHARDSON JV, EICHNER ER: Diverse clinical expression of multiple myeloma—atypical presentations. Am J Med Sci 267:99, 1974
756. PAINTRAND M, DANTCHEV D, MATHÉ G: Electron microscopic aspects of cells in the four subvarieties of acute lymphoid leukemia. Recent Results Cancer Res 43:71, 1973
757. PALAKAVONGS P, TEICHBERG S, VINCIGUERRA V et al: Ultrastructural cytochemical analysis of blastic transformation of chronic myelocytic leukemia. Blood 49:535, 1977
758. PALÚ G, POWLES R, SELBY P et al: Patterns of maturation in short-term culture of human acute myeloid leukaemic cells. Br J Cancer 40:719, 1979

759. PANGALIS GA, NATHWANI BN, RAPPAPORT H: Malignant lymphoma, well-differentiated lymphocytic: Its relationship with chronic lymphocytic leukemia and macroglobulinemia of Waldenström. Cancer 39:999, 1977
760. PANGALIS GA, NATHWANI BN, RAPPAPORT H et al: Acute lymphoblastic leukemia: The significance of nuclear convolutions. Cancer 43:551, 1979
761. PANGALIS GA, WALDMAN SR, RAPPAPORT H: Cytochemical findings in human non-neoplastic blood and tonsillar B and T lymphocytes. Am J Clin Pathol 69:314, 1978
762. PANIZON F, CRAMER R, SORANZO MR et al: Leucocyte disfunction in Chediak–Higashi syndrome (abstr). 4th meeting, European Society for Paediatric Haematology and Immunology, pp 316–17. Berlin, Springer-Verlag, 1974
763. PANTAZOPOULOUS N, SINKS LF: Morphological criteria for prognostication of acute lymphoblastic leukaemia. Br J Haematol 27:25, 1974
764. PAPADIMITRIOU JM, VAN DUIJN P, BREDEROO P et al: A new method for the cytochemical demonstration of peroxidase for light, fluorescence and electron microscopy. J Histochem Cytochem 24:82, 1976
765. PAPPENHEIM A: Ueber verschiedene typen von Lymphozyten und Monozyten, zum Teil im scheinbar normalen Blut. Folia Haematol 12:26, 1911
766. PARKER AC: A case of acute myelomonocytic leukemia associated with myelomatosis. Scand J Haematol 11:257, 1973
767. PARKS-WEBER F: A case of leukanaemia. Trans Pathol Soc Lond 55:288, 1904
768. PARMENTIER C, MICHEAU C, MORARDET N et al: Granulocyte stem cells in a case of Chediak–Higashi–Steinbrinck syndrome. Biomedicine (Express) 23:59, 1975
769. PARMLEY RT, OGAWA M, DARBY CP JR et al: Congenital neutropenia: Neutrophil proliferation with abnormal maturation. Blood 46:723, 1975
770. PARMLEY RT, SPICER SS: Cytochemical and ultrastructural identification of a small type granule in human late eosinophils. Lab Invest 30:557, 1974
771. PARMLEY RT, SPICER SS, ALVAREZ CJ: Ultrastructural localization of nonheme cellular iron with ferrocyanide. J Histochem Cytochem 26:729, 1978
772. PARMLEY RT, SPICER SS, WRIGHT NJ: Diaminobenzidine activity in the platelet dense body. J Histochem Cytochem 22:1063, 1974
773. PARWARESCH MR, DANNENBERG K, LEDER LD: The demonstration of naphthol-AS-D-chloroacetate-esterase in normal basophil granulopoietic cells. Klin Wochenschr 48:1008, 1970
774. PARWARESCH MR, LEDER LD, DANNENBERG KE: On the origin of human basophilic granulocytes. Acta Haematol 45:273, 1971
775. PARWARESCH MR, SADIGHI R: Cytophotometric measurements of the activity of naphthol-AS-D-chloroacetate-esterase in the basophil leukocytes of normal human marrow. Virchows Arch (Zellpathol) 6:88, 1970
776. PAUL J: Metabolic processes in normal and cancer cells. In: Ambrose EJ, Roe FJC (eds): The Biology of Cancer, pp 52–62. New York, Van Nostrand, 1966
777. PEARSE AGE: Histochemistry, Theoretical and Applied, 3rd ed, p 896. London, Churchill, 1968–72
778. PEDERSEN B: The blastic crisis of chronic myeloid leukaemia: Acute transformation of a preleukaemic condition? Br J Haematol 25:141, 1973
779. PEDERSEN B, HAYHOE FG: Cellular changes in chronic myeloid leukaemia. Br J Haematol 21:251, 1971
780. PEDERSEN B, HAYHOE FGJ: Relation between phagocytic activity and alkaline phosphatase content of neutrophils in chronic myeloid leukaemia. Br J Haematol 21:257, 1971
781. PEDRAZA MA: Plasma-cell leukemia with unusual immunoglobulin abnormalities. Am J Clin Pathol 64:410, 1975
782. PEPYS MB, SATEGNA-GUIDETTI C, MIRJAH DD et al: Enumeration of immunoglobulin-bearing lymphocytes in whole peripheral blood. Clin Exp Immunol 26:91, 1976

783. Perillie PE: Studies of the changes in leukocyte alkaline phosphatase following pyrogen stimulation in chronic granulocytic leukemia. Blood 29:401, 1967
784. Perillie PE, Finch SC: Alkaline phosphatase activity of exudative leukocytes in acute leukemia. Blood 18:572, 1961
785. Persidsky MD, Ellett MH: Lysosomes and cell cryoinjury. Cryobiology 8:345, 1971
786. Perugini S: Sul significato dell affinita dei granuli neutrofili dei leukociti per i coloranti tipo Sudan. Experientia 10:381, 1954
787. Petersen HS: Erythroleukaemia in a melphalan-treated patient with primary macroglobulinaemia. Scand J Haematol 10:5, 1973
788. Petersen HS: Nuclear arylsulphatase activity in leukaemia and some other haematological disorders. Scand J Haematol 10:35, 1973
789. Peterson IS: Differentiation of bone marrow cells by examination of acid phosphatase in children with diseases of the blood system. Probl Gematol Pereliv Krovi 18:15, 1973
790. Peterson LC, Bloomfield CD, Brunning RD: Blast crisis as an initial or terminal manifestation of chronic myeloid leukemia. A study of 28 patients. Am J Med 60:209, 1976
791. Peterson LC, Bloomfield CD, Sundberg RD et al: Morphology of chronic lymphocytic leukemia and its relationship to survival. Am J Med 59:316, 1975
792. Petrov MN, Vashkinel VK: Morphology and function of thrombocytes in patients with lymphocytic leukemia. Ter Arkh 47:33, 1975
793. Petrova AS, Kazanova LI, Solovieva EA: Cytochemical methods in diagnosis of malignant neoplasms. Neoplasma 19:327, 1972
794. Pezzimenti JF, Kim HC, Lidenbaum J: Erythroleukemia-like syndrome due to busulfan toxicity in polycythemia vera. Cancer 38:2242, 1976
795. Phillips BA, Yarbro JW, Konior GS: Acquired Pelger–Huet anomaly in erythroleukemia. J Surg Oncol 2:407, 1970
796. Piaserico PL, Dalla Pria AF: Plasmacellular leukemia. Anatomoclinical study of a case. Arch De Vecchi Anat Patol 61:503, 1976
797. Pierre RV: Preleukemic states. Semin Hematol 11:73, 1974
798. Pietschmann H, Stockinger L: Clinical and electron microscopic studies in lymphoid plasma cellular reticulosis. Blut 20:76, 1970
799. Pietschmann H, Uhlrich D, Kolarz G et al: Cytochemical studies on myeloma cells. Folia Haematol 99:131, 1973
800. Piney A: The oxydase reaction in acute myeloid leukaemia. J Pathol Bacteriol 28:97, 1925
801. Pluss HJ: Cytochemical methods in hematology. Schweiz Med Wochenschr 101:1500, 1971
802. Pluss H, Hitzig WH: Cytochemical studies in acute leukemia of childhood. Schweiz Med Wochenschr 96:1212, 1966
803. Poch J, Lodrová V: Cytochemistry reactions for the demonstration of myeloperoxidase. Their relation to sudanophilia and stainability of hemoglobin. Cas Lek Cesk 115:381, 1976
804. Pokrovskii AA, Kravchenko LV, Tutelian VA et al: Lysosomal component in the mechanism of the toxic effect of sporofusarin. Farmakol Toksikol 39:93, 1976
805. Polliack A: Surface morphology of leukemic cells: Application of scanning electron microscopy to the study of human leukemias. Isr J Med Sci 13:701, 1977
806. Polliack A: Scanning electron microscopy and the surface morphology of human leukocytes: Current status. Isr J Med Sci 15:629, 1979
807. Polliack A, Gamliel H, Froimovici M et al: Myeloma cells: Surface morphology as seen by scanning and transmission electron microscopy. Isr J Med Sci 14:1252, 1978
808. Polliack A, Rachmilewitz D, Zlotnick A: Plasma cell leukemia. Unassembled light and heavy chains in the urine. Arch Intern Med 134:131, 1974

809. Polliack A, Siegal FP, Clarkson BD et al: A scanning electron microscopy and immunological study of 84 cases of lymphocytic leukaemia and related lymphoproliferative disorders. Scand J Haematol 15:359, 1975
810. Poulet J, Cocheton JJ, Combrisson A et al: Sigmoid localization of Kahler's disease. Acute terminal plasmacytic leukemia with gastric and duodenal localizations. Semin Hop Paris 50:2923, 1974
811. Pozo-Olano JD: On the myeloperoxidase reaction of leukocytes. Enzymol Biol Clin 9:234, 1968
812. Prchal JT, Throckmorton DW, Carroll AJ III et al: A common progenitor for human myeloid and lymphoid cells. Nature 274:590, 1978
813. Presentey B, Jerushalmy Z, Mintz V: Eosinophilic leukaemia: Morphological, cytochemical, and electron microscopic studies. J Clin Pathol 32:261, 1979
814. Pretlow TG II: Chronic monocytic dyscrasia culminating in acute leukemia. Am J Med 46:130, 1969
815. Proto C, Grasso Capriloli G, Levi GF: Leukemic myelosis with plasma cells. Anatomo-clinical facts, differentiating the plasmacytoma. Haematologica 52:1, 1967
816. Pruzanski W, Platts ME, Ogryzlo MA: Leukemic form of immunocytic dyscrasia (plasma cell leukemia). A study of ten cases and a review of the literature. Am J Med 47:60, 1969
817. Pruzanski W, Ogryzlo MA: The changing pattern of diseases associated with M components. Med Clin North Am 56:371, 1972
818. Pruzanski W, Rother I: IgD plasma cell neoplasia: Clinical manifestations and characteristic features. Can Med Assoc J 102:1061, 1970
819. Pulvertaft RJV, Humble JC: The bone marrow in leukemia. Acta Haematol 24:68, 1960
820. Quaglino D: Cytochemical techniques in acute leukaemias. Proceedings of the 7th Congress of the European Society of Hematology, London, Part II, 1959, p 386. Basel, Karger, 1960
821. Quaglino D: Cytochemistry of blood and bone marrow cells with particular reference to leukaemia and allied diseases. PhD Thesis, Cambridge University, 1961
822. Quaglino D: Alkaline phosphatase positivity in erythroblasts. Haematol Lat 7:263, 1964
823. Quaglino D, Flemans R: Peroxidase staining in leukocytes. Lancet 2:1020, 1958
824. Quaglino D, Hayhoe FGJ: Observations on the periodic acid–Schiff reaction in lymphoproliferative diseases. J Pathol Bacteriol 78:521, 1959
825. Quaglino D, Hayhoe FGJ: Periodic acid–Schiff positivity in erythroblasts with special reference to DiGuglielmo's disease. Br J Haematol 6:26, 1960
826. Quaglino D, Hayhoe FGJ: Acetone fixation for the cytochemical demonstration of dehydrogenases in blood and bone marrow cells. Nature 187:85, 1960
827. Quaglino D, Hayhoe FGJ: Phosphorylase activity in haemic cells. Nature 194:929, 1962
828. Queisser W, Neu G: Proliferation of an unspecific esterase positive (monocytopoietic) cell system in human bone marrow as observed from nucleic acid synthesis. Klin Wochenschr 48:1229, 1970
829. Queisser W, Queisser U, Ansmann M et al: Megakaryocyte polyploidization in acute leukaemia and preleukaemia. Br J Haematol 28:261, 1974
830. Quie PG, White JG, Holmes B et al: In vitro bactericidal capacity of human polymorphonuclear leukocytes: Diminished activity in chronic granulomatous disease of childhood. J Clin Invest 46:668, 1967
831. Radaszkiewicz T, Denk H: Rosette formation of guinea pig lymphoid cells with rabbit erythrocytes—differences between thymocytes and peripheral blood lymphocytes. Cell Immunol 16:374, 1975
832. Radnay J, Goldman I, Rozenszajn LA: Growth of human B-lymphocyte colonies in vitro. Nature 278:351, 1979

833. Radwanska ATV: Undifferentiated and low differentiated cells in normal and leukemic bone marrow. Ann Med Sect Pol Acad Sci 19:167, 1974
834. Radzun HJ, Parwaresch MR, Kulenkampff Ch et al: Lysosomal acid esterase: Activity and isoenzymes in separated normal human blood cells. Blood 55:891, 1980
835. Ramasamy R: A fluorescent stain for viable rosette-forming cells. J Immunol Methods 5:305, 1974
836. Rappaport AH, Cohen RJ, Castro JR: Erythroleukemia following total-body irradiation for advanced lymphocytic lymphoma. Radiology 115:179, 1975
837. Rappaport H: Tumors of the hematopoietic system. In Atlas of Tumor Pathology, Sect 3, Fasc 8, p 37. Washington, Armed Forces Institute of Pathology, 1966
838. Rappaport H: Histologic criteria for diagnosis and classification of acute leukemias. Recent Results Cancer Res 43:35, 1973
839. Rappoport AE, Kugel VH: Monocytic leukemia. A case report illustrating variations in the clinical picture. Blood 2:332, 1947
840. Rastetter J: Classification of leukemias: Their morphology and relations to prognosis. Verh Dtsch Ges Inn Med 79:261, 1973
841. Reed CE, Bennett JM: N-acetyl-beta-glucosaminidase activity in normal and malignant leukocytes. J Histochem Cytochem 23:752, 1975
842. Reed RC, Gutterman JU, Mavligit GM et al: Sialic acid on leukemia cells: Relation to morphology and tumor immunity. Proc Soc Exp Biol Med 145:790, 1974
843. Reisner EH: The nature and significance of megaloblastic blood formation. Blood 13:313, 1958
844. Reizenstein P, Langerlof B: Aregenerative anemia with hypercellular sideroblastic marrow. A preleukemic condition. Acta Haematol 47:1, 1972
845. Repine JE, Clawson CC, Brunning RD: Abnormal pattern of bactericidal activity of neutrophils deficient in granules, myeloperoxidase, and alkaline phosphatase. J Lab Clin Med 88:788, 1976
846. Reschad H, Schilling-Torgau V: Ueber eine neue Leukämie durch echte Uebergangsformen (Splenozytenleukämie) und ihre Bedeutung für die Selbständigkeit diesen Zellen. Münch Med Wochenschr 60:1981, 1913
847. Reyes F, Gourdin MF, Farcet JP et al: Morphologic criteria of surface studies by electron microscopy and classification of lymphoproliferative syndromes. Critical study. Ann Med Interne 127:865, 1976
848. Rheingold JJ: Acute leukemia. Its smouldering phase, or leukemia never starts on Thursday. JAMA 230:985, 1974
849. Rheingold JJ, Kaufman R, Adelson E et al: Smouldering acute leukemia. N Engl J Med 268:812, 1963
850. Riccardi A, Martinotti A, Perugini S: Cytokinetic studies in two cases of plasma cell leukemia. Haematologica 62:581, 1977
851. Richter HJ: Morphology of prolymphocytic leukemia. Verh Dtsch Ges Pathol 62:511, 1978
852. Richter MN: The presence of Auer rods in leukemic tissues. Arch Intern Med 31:677, 1923
853. Richter W, Fleischer J, Irmscher J: Zytochemische Befunde bei chronischen monozytären Leukosen. Folia Haematol 102:524, 1975
854. Rickard KA, Dunleavy L, Brown R et al: Fractionation of normal human bone marrow on albumin gradients. Aust J Exp Biol Med Sci 52:767, 1974
855. Riddell EM, Davidson RJL: Coexistence of pernicious anaemia and acute erythraemic myelosis. J Clin Pathol 21:590, 1968
856. Riesco A, Leyton C: Mucopolysaccharides in peripheral leukocytes of cancer patients. Br J Cancer 25:284, 1971
857. Rigby PG, Pratt PT, Rosenlof RC et al: Genetic relationships in familial leukemia and lymphoma. Arch Intern Med 121:67, 1968

858. RINDLER R, HÖRTNAGL H, SCHMALZL F et al: Hydrolysis of a chymotrypsin substrate and of naphthol-AS-D-chloroacetate by human leukocyte granules. Blut 26:239, 1973
859. RINDLER R, SCHMALZL F, HÖRTNAGEL M et al: Naphthol-AS-D-chloroacetate esterases in granule extracts from human neutrophil leukocytes. Blut 23:223, 1971
860. ROBERTS BE, ABBOTT CR, FORTT RW et al: Preleukemia. A report of four cases. Acta Haematol 39:20, 1968
861. ROBERTSON PB, RYEL RB, TAYLOR RE et al: Collagenase: Localization in polymorphonuclear leukocyte granules in the rabbit. Science 177:64, 1972
862. ROCHANT H, DREYFUS B, BOUGUERRA M et al: Refractory anemias, preleukemic conditions, and fetal erythropoiesis. Blood 39:721, 1972
863. ROELS F, WISSE E, DEPREST B et al: Cytochemical discrimination between catalases and peroxidases using diaminobenzidine. Histochemistry 41:281, 1975
864. ROHKRÄMER H: Zur Klinik und Pathologie der aleukämischen Systemreticulose. Zugleich ein Beitrag zur Urethanbehandlung. Ztschr Klin Med 146:310, 1950
865. ROOT RK, ROSENTHAL AS, BALESTRA DJ: Abnormal bactericidal, metabolic, and lysosomal functions of Chediak–Higashi syndrome leukocytes. J Clin Invest 51:649, 1972
866. ROSALES CL, BENNETT JM, RUTENBERG AM: An azo dye method for the histochemical demonstration of leukocyte acid phosphatase. J Histochem Cytochem 13:714, 1965
867. ROSALES CL, BENNETT JM, RUTENBUG AM: Histochemical demonstration of leukocyte acid phosphatase in health and in disease. Br J Haematol 12:172, 1966
868. ROSAS-URIBE A, RAPPAPORT H: Malignant lymphoma, histiocytic type with sclerosis (sclerosing reticulum cell sarcoma). Cancer 29:946, 1972
869. ROSENBLUM D, HIMMELHOCH SR, PETERSON EA et al: Purification of alkaline phosphatase from guinea pig bone marrow. Arch Biochem Biophys 141:303, 1970
870. ROSENTHAL S, CANELLOS GP, GRALNICK HR: Erythroblastic transformation of chronic granulocytic leukemia. Am J Med 63:116, 1977
871. ROSENTHAL S, CANELLOS GP, WHANG-PENG J et al: Blast crisis of chronic granulocytic leukemia. Morphologic variants and therapeutic implications. Am J Med 63:542, 1977
872. ROSNER F: Acute leukemia as a delayed consequence of cancer chemotherapy. Cancer 37:1033, 1976
873. ROWLEY JD: A new consistent chromosome abnormality in chronic myelogenous leukaemia identified by quenacrine fluorescence and Giemsa staining. Nature 243:290, 1973
874. ROZENSZAJN LA, GUTMAN A, RADNAY J et al: A study of the nature of "hairy" cells, with emphasis on enzymatic markers. Am J Clin Pathol 66:432, 1976
875. ROZENSZAJN LA, JOSEPH D: The isozyme patterns of glucose-6-phosphate dehydrogenase in blood cells, bone marrow and other human tissues. Biochim Biophys Acta 364:38, 1974
876. ROZENSZAJN LA, KOLMAN S, SHOHAM D et al: Glucose-6-phosphate dehydrogenase isoenzymes in blood cells. Nature 226:862, 1970
877. ROZENSZAJN LA, LEIBOVICH M, SHOHAM D et al: The esterase activity in megaloblasts, leukaemic and normal haemopoietic cells. Br J Haematol 14:605, 1968
878. ROZENSZAJN LA, SHOHAM D: The demonstration of dehydrogenases and diaphorases in cells of peripheral blood and bone marrow. Blood 29:737, 1967
879. ROZMAN C, WOESSNER S, SAEZ-SERRANIA J: Acute erythromyelosis after benzene poisoning. Acta Haematol 40:234, 1968
880. RUBIN AD, DOUGLAS SD, CHESSIN LN et al: Chronic reticulolymphocytic leukemia: Reclassification of "leukemic reticuloendotheliosis" through functional characterization of the circulating mononuclear cells. Am J Med 47:149, 1969

881. RUBNITZ AS: Etiology of acute leukemias. J Lab Clin Med 14:497, 1929
882. RUBNITZ AS: Classification of leukemias. Am J Clin Pathol 2:243, 1932
883. RUDENS IuF, BUĬKIS IM: Cytochemical analysis of naphthol-AS-D-chloroacetate esterases. Lab Delo 10:592, 1971
884. RUNDLES RW: Monocytic leukemia. In Williams WJ et al (eds): Hematology, p 896. New York, McGraw-Hill, 1972
885. RUSSELL NH, KEENAN JP, BELLINGHAM AJ: Thrombocytopathy in preleukemia: Association with a defect of Thromboxane A_2 activity. Br J Haematol 41:417, 1979
886. RUSSELL WO: Tissue diagnosis of cancer: Changed concepts, technical advances and future applications. Proc Natl Cancer Conf 5:673, 1964
887. RUTENBURG AM, ROSALES CL, BENNETT JM: An improved histochemical method for the demonstration of leukocyte alkaline phosphatase activity: Clinical Application. J Lab Clin Med 65:698, 1965
888. RUTTNER JR, PEDIO G: "Sezary-like" cells in malignant and non-malignant diseases. Exp Cell Biol 45:141, 1977
889. RYDER RJW: Chronic monocytic leukemia. Blut 14:47, 1966
890. SAARNI MI, LINMAN JW: Myelomonocytic leukemia: Disorderly proliferation of all marrow cells. Cancer 27:1221, 1971
891. SAARNI MI, LINMAN JW: Preleukemia: The hematologic syndrome preceding acute leukemia. Am J Med 55:38, 1973
892. SABIN FR: On the origin of the cells of the blood. Physiol Rev 2:38, 1922
893. SAILLEN R: Biosynthesis of the porphyrin system in the myeloblasts. Helv Med Acta 30:208, 1963
894. SAKURAI T, KUDO H, NAGASAWA T et al: Correlation of marrow morphology and cytogenetics after blastic transformation in a case with chronic myelocytic leukemia (author's trans). Jpn J Clin Hematol 16:636, 1975
895. SALMON SE, CLINE MJ, SCHULTZ J et al: Myeloperoxidase deficiency. Immunologic study of a genetic leukocyte defect. N Engl J Med 282:250, 1970
896. SALMON SE, MCINTYRE OR, OGAWA M: IgE myeloma: Total body tumor cell number and synthesis of IgE and DNA. Blood 37:696, 1971
897. SALOMON H, TATARSKY I: Preleukemic leukemia. A report of four cases Isr J Med Sci 5:1178, 1969
898. SALVIDIO E: Determination of enzymatic activity of the cells of acute and chronic leukaemia by the Carlsberg micromethods. Proc 6th Congr Eur Soc Haematol, p 319. Basel, Karger, 1957
899. SANDRI R, CONTE L, PIASERICO PL: Acute plasma-cell leukemia. Report of a case. Minerva Med 67:2598, 1976
900. SARIN PS, ANDERSON PN, GALLO RC: Terminal deoxynucleotidyl transferase activities in human blood leukocytes and lymphoblast cell lines and in blast cells of some patients with chronic myelogenous leukemia in acute phase. Blood 47:11, 1976
901. SAVAGE RA, HOFFMAN GC, LUCAS FV: Morphology and cytochemistry of "microgranular" acute promyelocytic leukemia. Am J Clin Pathol 75:548, 1981
902. SAWKAR LA: Hyperviscosity syndrome in acute plasma cell leukaemia. J Assoc Physicians India 20:397, 1972
903. SBARRA AJ, PAUL BB, JACOBS AA et al: Role of the phagocyte in host–parasite interactions. 38. Metabolic activities of the phagocyte as related to antimicrobial action. J Reticuloendothel Soc 12:109, 1972
904. SCARPELLI DG, PEARSE AGE: Physical and chemical protection of cell constituents and the precise localization of enzymes. J Histochem Cytochem 6:369, 1958
905. SCHADE H, SCHOELLER L, SCHULTZE KW: D-trisomie (Pätausyndrom) mit kongenitaler myeloischer Leukämie. Med Welt 50:2690, 1962
906. SCHAEFER HE, FISCHER R: Peroxidase activity as a cause of stabile granulocyte sudanophilia. Klin Wochenschr 48:1424, 1970

907. Schaefer HE, Hellriegel KP, Fischer R: Tartrate-resistant acid phosphatase in various cell types of lymphoreticular and hematopoietic system. Blut 34:393, 1977
908. Schaefer HE, Hellriegel KP, Hennekeuser HH et al: Eosinophilic leukemia, an immature cell myelosis with chloroacetate esterase-positive eosinophilia. A morphological and cytochemical study on the problem of monophylic myeloses. Blut 26:7, 1973
909. Schaefer HE, Hellriegel KP, Zach J et al: Cytochemical polymorphism of acid phosphatase in hairy cell leukemia. Blut 31:365, 1975
910. Schaefer HE, Kaufer C, Fischer R: Comparative enzyme-cytochemical investigations in blood and bone marrow cells of laboratory animals. Virchows Archiv (Zellpathol) 4:310, 1970
911. Schaeffer HE, Krüger GR, Fischer R: Morphological classification of malignant lymphomas: Ultrastructural, cytochemical and immunological results. Acta Neuropathol (Suppl) 6:21, 1975
912. Schafer AI, Rosenthal DS, Moloney WC: Hairy cell leukemia: Diagnosis and management. Am J Med Sci 275:297, 1978
913. Schäfer KH, Beckmann H, Neth R: The application of cytochemical testing procedures in the diagnosis of hematologic diseases in children. Mschr Kinderheilk 113:428, 1965
914. Schiff R, Lewis JP, Cross CE: Esterase markers in mouse radiation chimeras. J Histochem Cytochem 20:472, 1972
915. Schiff R, Sonnenschien C: Somatic cell hybridization. Cellular interaction upon esterase isozymes. Exp Cell Res 70:269, 1972
916. Schilling V: Der Monozyt in trialistischer Auffassung und seine Bedeutung im Krankheitsbilde. Med Klin 22:536, 1926
917. Schleicher EM: Giant orthochromatic erythroblasts. Their importance for the promegaloblast and pronormoblast problem. J Lab Clin Med 29:127, 1944
918. Schmalzl F, Braunsteiner H: On the origin of monocytes. Acta Haematol 39:177, 1968
919. Schmalzl F, Braunsteiner H: Cytochemical demonstration of esterase activity in blood and bone marrow cells. Klin Wochenschr 46:642, 1968
920. Schmalzl F, Braunsteiner H: The cytochemistry of monocytes and macrophages. Ser Haematol 3:93, 1970
921. Schmalzl F, Braunsteiner H: The application of cytochemical methods to the study of acute leukemia. A review. Acta Haematol 45:209, 1971
922. Schmalzl F, Huber C, Braunsteiner H: Demonstration of proliferating monocytic precursors by the combination of cytochemical and autoradiographic methods. Klin Woschenschr 47:887, 1969
923. Schmalzl F, Huhn D, Abbrederis K et al: Acute lymphocytic leukemia. Cytochemistry and ultrastructure. Blut 29:87, 1974
924. Schmalzl F, Huhn D, Asamer H et al: Cytochemistry and ultrastructure of pathologic granulation in myelogenous leukemia. Blut 27:243, 1973
925. Schmalzl F, Kurz R, Glatzl J: Congenital agranulocytosis. II. Cytochemical and experimental studies using Rebuck's skin window technic. Paediatr Paedol 6:84, 1971
926. Schmalzl F, Lederer B, Braunsteiner H: Atypical myeloblastic leukemia with differentiation into "paraneutrophils." Blut 20:337, 1970
927. Schnitzer B, Kass L: Leukemic phase of reticulum cell sarcoma (histiocytic lymphoma). A clinopathologic and ultrastructural study. Cancer 31:547, 1973
928. Schnitzer B, Kass L: Refractory sideroblastic anemia (chronic erythremic myelosis): An ultrastructural study of the bone marrow. Am J Clin Pathol 60:343, 1973
929. Scholnik AP, Kass L: A direct cytochemical method for the identification of lysozyme in various tissues. J Histochem Cytochem 21:65, 1973

930. Schrek R, Donnelly WJ: "Hairy" cells in blood in lymphoreticular neoplastic disease and flagellated cells in normal lymph nodes. Blood 27:199, 1966
931. Schrek R, Rappaport H, Rubnitz ME et al: Cytology and reactions of viable cells from malignant lymphoma. Cancer 23:1061, 1969
932. Schubert JC, Martin H: Observations on 7 patients with "blood mast cell leukemia." Klin Wochenschr 46:929, 1968
933. Schultze WH: Die Oxydasereaktion an Gewebsschnitten und ihre Bedeutung für die Pathologie. Zugleich ein Beitrag zur Differentialdiagnose der Leukämien. Beitr Z Pathol Anat v.z.alg. Path 45:127, 1909
934. Schwartz E: Cellular gigantism and pluripolar mitosis in human hematopoiesis. Am J Anat 79:75, 1946
935. Schwartz SO, Critchlow J: Erythremic myelosis (DiGuglielmo's disease). Critical review with report of four cases, and comments on erythroleukemia. Blood 7:765, 1952
936. Scolari L, Masala C: The immunoperoxidase method. Application to the study of immunoglobulin producing cells in the bone marrow of patients with monoclonal gammopathies. Boll Soc Ital Biol Sper 50:549, 1974
937. Scott RB, Eanes RZ, Cooper LW et al: Characterization of marrow granulocyte development: Changes in response to inflammatory reactions. Br J Haematol 37:503, 1977
938. Scott RB, Ellison RR, Ley AB: A clinical study of twenty cases of erythroleukemia (DiGuglielmo's syndrome). Am J Med 37:162, 1964
939. Scott RB, Grogan WM, Collins JM: Separation of rabbit marrow precursor cells by combined isopycnic sedimentation and electronic cell sorting. Blood 51:1137, 1978
940. Scott RE, Horn RG: Fine structural features of eosinophile granulocyte development in human bone marrow. Evidence for granule secretion. J Ultrastr Res 33:16, 1970
941. Sebahoun G, Sultan C, Dreyfus B: Quelques remarques sur les érythroleucémies. A propos de 13 observations et revue de la littérature. Ann Med Interne 125:633, 1974
942. Seeman PM, Palade GE: Acid phosphatase localization in rabbit eosinophils. J Cell Biol 34:745, 1967
943. Sehrt E: Die histologische Darstellung der Lipoide der weissen Blutzellen (neutro-und eosinophile Leukozyten, Mastzellen, Uebergangszellan und Mononukleäre) und die Beziehung dieser Lipoide zur Oxydasereaktion. München med Wchnschr 74:139, 1927
944. Seligsohn U, Ramot B: Chronic monocytic leukemia: A case with an eight-year survival. Isr J Med Sci 3:868, 1967
945. Senn JS, Pinkerton PH: Defective in vitro colony formation of human bone marrow preceding overt leukemia. Br J Haematol 23:277, 1972
946. Sexauer J, Kass L, Schnitzer B: Subacute myelomonocytic leukemia. Clinical, morphologic and ultrastructural studies of 10 cases. Am J Med 57:853, 1974
947. Shaklai M, Nir M, Feuerman E et al: Cutaneous involvement in erythroleukemia. Report of a case. Dermatologica 149:385, 1974
948. Shamoto M, Suchi T: Malignant histiocytosis with chronic course. Ultrastructural and ultrastructural cytochemical studies. Virchows Arch (Pathol Anat) 376:95, 1977
949. Sharp JA: The association of lymphocytes with larger motile cells in cultures of mammalian thymus. J Pathol 103:87, 1971
950. Shaw MT: The importance of cytochemical differentiation of acute leukemias. Proc Am Assoc Cancer Res (Abstr) 15:40, 1974
951. Shaw MT: The cytochemistry of acute leukemia: A diagnostic and prognostic evaluation. Semin Oncol 3:219, 1976

952. Shaw MT, Bottomley RH, Grozea PN et al: Heterogeneity of morphological, cytochemical and cytogenetic features in the blastic phase of chronic granulocytic leukemia. Cancer 35:199, 1975
953. Shaw MT, Bottomley SS, Bottomley RH et al: The relationship of erythromonocytic leukemia to other myeloproliferative disorders. Am J Med 55:542, 1973
954. Shaw MT, Klemp VA: Cytochemical staining with oil red O and β-glucuronidase in the diagnosis of acute leukemia. Am J Clin Pathol 61:169, 1974
955. Shaw MT, Nordquist RE: "Pure" monocytic or histiomonocytic leukemia: A revised concept. Cancer 35:208, 1975
956. Shaw MT, Twele TW, Nordquist RE: Plasma cell leukemia: Detailed studies and response to therapy. Cancer 33:619, 1974
957. Sheehan HL, Storey GW: An improved method for staining leukocyte granules with Sudan black B. J Pathol Bacteriol 59:336, 1947
958. Sheets R, Drevets CC, Hamilton HE: Erythroleukemia (DiGuglielmo syndrome). A report of clinical observations and experimental studies in seven patients. Arch Intern Med 111:295, 1963
959. Shibata A, Sakaki H: Phosphatase staining. Rinsho Byori (34 Spec No):33, 1978
960. Shimada Y: Ultrastructural cytochemical studies on human blood cells. Peroxidase and acid phosphatase reaction. Nagoya Med J 15:239, 1969
961. Shnitka TK, Seligmann AM: Role of esterase inhibition on localization of esterase and the simultaneous cytochemical demonstration of inhibitor sensitive and resistant enzyme species. J Histochem Cytochem 9:504, 1961
962. Shoham D, David EB, Rozenszajn LA: Cytochemical and morphologic identification of macrophages and eosinophils in tissue cultures of normal human bone marrow. Blood 44:221, 1974
963. Shoham D, Steiner Z, Bar-Sela S et al: Chronic myeloid leukemia with transition to basophilic leukemia. Harefuah 94:111, 1978
964. Shrader RE, Hurley LS: Enzyme histochemistry of peripheral blood and bone marrow in zinc-deficient rats. Lab Invest 26:566, 1972
965. Silverstein MN, Ellefson RD, Ahern EJ: The syndrome of the sea-blue histiocyte. N Engl J Med 282:1, 1970
966. Simpson ME: The experimental production of macrophages in the circulating blood. J Med Res 43:77, 1922
967. Sinn CM, Dick FW: Monocytic leukemia. Am J Med 20:588, 1956
968. Sjogren U, Brandt L: Differences in morphology and mitotic activity between intra- and extra-medullary erythropoietic tissue in chronic myeloid leukaemia. Scand J Haematol 13:116, 1974
969. Skeel RT, Bennett JM, Henderson ES: Bone marrow lymphocytosis in acute leukemia: A second look. Blood 35:356, 1970
970. Skeel RT, Henderson ES, Bennett JM: The significance of bone marrow lymphocytosis of acute leukemia patients in remission. Blood 32:767, 1968
971. Skinnider L: Differences in Cytochalasin D-induced surface alterations between chronic lymphocytic leukaemic and normal lymphocytes. Experientia 33:1652, 1977
972. Skinnider LF, Card RT, Padmanabh S: Chronic myelomonocytic leukaemia. An ultrastructural study by transmission and scanning electron microscopy. Am J Clin Pathol 67:339, 1977
973. Skinnider LF, Ghadially FN: Glycogen in erythroid cells. Arch Pathol 95:139, 1973
974. Skinnider LF, Ghadially FN: Ultrastructure and cell surface abnormalities in neoplastic histiocytes. Br J Cancer 35:657, 1977
975. Smith H: Lymphoid nuclei within nuclei of myeloma cells. Lancet 2:1059, 1978

976. Smith J, Newell DG, Gordon J et al: B-cell acute lymphatic leukaemia: immunoglobulin synthesis, morphology and clinical features. Scand J Haematol 22:246, 1979
977. Smith JL, Barker CR, Clein GP et al: Characterization of malignant mediastinal lymphoid neoplasm (Sternberg sarcoma) as thymic in origin. Lancet 1:74, 1973
978. Smith WB, Ablin A, Goodman JR et al: Atypical megakaryocytes in preleukemic phase of acute leukemia. Blood 42:535, 1973
979. Smutka P, Brunning RD: An evaluation of nuclear arylsulfatase activity in acute leukemias. Acta Haematol 41:290, 1969
980. Sobel HJ, Marquet E: Granular cells and granular cell lesions. Pathol Annu 9:43, 1974
981. Sohn KY, Boggs DR: Klinefelter's syndrome, LSD usage and acute lymphoblastic leukemia. Clin Genet 6:20, 1974
982. Spicer SS, Horn RG, Wetzel BK: Ultrastructural and cytochemical characteristics of leukocytes in various stages of development. Biochem Pharmacol (Suppl):143, 1968
983. Spiegelberg HL, Heath VC, Lang JE: IgG half-molecules: Clinical and immunologic features in a patient with plasma cell leukemia. Blood 45:305, 1975
984. Spiers ASD, Liew A, Baikie AG: Neutrophil alkaline phosphatase score in chronic granulocytic leukaemia: Effects of splenectomy and antileukaemic drugs. J Clin Pathol 28:517, 1975
985. Spitzer G, Dicke KA, Gehan EA et al: A simplified in vitro classification for prognosis in adult acute leukemia: The application of in vitro results in remission-predictive models. Blood 48:795, 1976
986. Spitznagel JK, Dalldorf FG, Leffell MS et al: Character of azurophil and specific granules purified from human polymorphonuclear leukocytes. Lab Invest 30:774, 1974
987. Srivastava BIS, Khan SA, Henderson ES: High terminal deoxynucleotidyl transferase activity in acute myelogenous leukemia. Cancer Res 36:3847, 1976
988. Srivastava BIS, Khan SA, Minowada J et al: High terminal transferase activity in pediatric patients with acute lymphocytic and acute myelocytic leukemias. Int J Cancer 22:4, 1978
989. Stahl A, Papy MC, Muratore R et al: Erythromyélose avec pseudo-syndrome de Pelger-Huët et anomalies chromosomiques complexes. Nouv Rev Fr Hematol 5:879, 1965
990. Stass SA, Schumacher HR, Keneklis TP et al: Terminal deoxynucleotidyl transferase immunofluorescence of bone marrow smears. Am J Clin Pathol 72:898, 1979
991. Stavem P, Harboe M: Acute erythroleukaemia in a patient treated with melphalan for the cold agglutinin syndrome. Scand J Haematol 8:375, 1971
992. Stavem P, Hjort PF, Jeremic M et al: Numerous lipid-containing vacuoles in the myeloblasts of two patients with acute leukaemia. Scand J Haematol 6:39, 1969
993. Stavem P, Hjort PF, Vogt E et al: Ring-shaped nuclei of granulocytes in a patient with acute erythroleukaemia. Scand J Haematol 6:31, 1969
994. Stavem P, Ly B, Blichfeldt P et al: Acute eosinophilo-myelomonocytic leukaemia, one of the "in between leukaemias." Scand J Haematol 21:355, 1978
995. Stern R, Widirstky ST, Wurster-Hill DH et al: Hand-mirror cell leukemia associated with mental retardation: Immunologic, chromosome, and morphologic studies. Blood 54:703, 1979
996. Stevens DA, O'Connor GT, Levine PH et al: Acute leukemia with 'Burkitt's lymphoma cells' and Burkitt's lymphoma. Simultaneous onset in American siblings; description of a new entity. Ann Intern Med 76:967, 1972
997. Stewart CA: Leukocyte alkaline phosphatase in myeloid maturation. Pathology 6:287, 1974

998. Stodtmeister R: Akute Erythroblastose und erythroblastiche Reaktion. Klin Wochenschr 20:444, 1941
999. Stodtmeister R, Fliedner TM: Morphological aspects of myelofibrosis, observed in rats following sublethal whole body irradiation and subsequent allogeneic bone marrow cell transfusion. Folia Haematol 100:23, 1973
1000. Stolbach LL, Krant MJ, Fishman WH: Ectopic production of an alkaline phosphatase isoenzyme in patients with cancer. N Engl J Med 281:757, 1969
1001. Storti E, Perugini S: Cytochemical researches on the lipids of the hematic cells with particular attention to those of acute leukosis. Acta Haematol 5:321, 1951
1002. Storti E, Perugini S, Soldati M: Quadro citotopochimico dei polisaccaridi nelle cellule del sangue e degli organi emopoietici dell'uomo normali. Medicina (Parma) 3:145, 1953
1003. Strauss RG, Bove KE, Jones JF et al: An anomaly of neutrophil morphology with impaired function. N Engl J Med 290:478, 1974
1004. Stryckmans PA: Current concepts in chronic myelogenous leukemia. Semin Hematol 11:101, 1974
1005. Stuart AE, Young GA, Grant PF: Identification of human mononuclear cells by anti-monocyte serum. Br J Haematol 34:457, 1976
1006. Stuart J, Gordon PA, Lee TR: Polymorph cytochemistry in systemic infection. Br J Haematol 30:126, 1975
1007. Stuart J, Gordon PA, Lee TR: Enzyme cytochemistry of blood and marrow cells. Histochem J 7:471, 1975
1008. Stuart J, Simpson JS: Dehydrogenase enzyme cytochemistry of unfixed leukocytes. J Clin Pathol 23:517, 1970
1009. Stuart J, Simpson JS, Mann JR: Intracellular hydrogen transport systems in acute leukaemia. Br J Haematol 19:739, 1970
1010. Stuart J, Simpson JS, Wallin JM: Glycolytic and respiratory enzyme activity in relation to leukocyte maturation. Br J Haematol 23:79, 1972
1011. Stuart J, Skowron PN: A cytochemical study of marrow enzymes in megaloblastic anaemia. Br J Haematol 15:443, 1968
1012. Sullivan AL, Weintraub LR: Sideroblastic anemias. An approach to diagnosis and management. Med Clin North Am 57:335, 1973
1013. Sultan C, Heilmann-Gouault M, Tulliez M: Relationship between blast-cell morphology and occurrence of a syndrome of disseminated intravascular coagulation. Br J Haematol 24:255, 1973
1014. Sultan C, Imbert M, Ricard MF et al: Pure acute monocytic leukemia. A study of 12 cases. Am J Clin Pathol 68:752, 1977
1015. Sultan C, Marquet M, Joffroy Y: Etude de certaines dysmyélopoïèses acquises idiopathiques et secondaires par culture de moelle in vitro. Ann Med Interne 125:599, 1974
1016. Suzuki A: Histochemical demonstration of acid phosphatase activity in various blood cells. Tohoku J Exp Med 89:27, 1966
1017. Suzuki A, Shibata A, Onodera S et al: Histochemical study on plasma cells. Tohoku J Exp Med 97:1, 1969
1018. Suzuki I, Takahashi M, Itoh S: Ultrastructural study of human myeloma cells in relation to its function. J Clin Pathol 23:339, 1970
1019. Sweet DL, Jr, Golomb HM, Ultmann JE: Chronic lymphocytic leukaemia and its relationship to other lymphoproliferative disorders. Clin Haematol 6:141, 1977
1020. Sydenstricker VP, Phinizy TB: Acute monocytic leukemia. A case with partial autopsy. Am J Med Sci 184:770, 1932
1021. Syrén E, Raeste AM: Identification of blood monocytes by demonstration of lysozyme and peroxidase activity. Acta Haematol 45:29, 1971
1022. Takahashi I, Nakanishi T, Sakato J et al: Preleukemia: Hematological disorder prior to onset of leukemia. Acta Med Okayama 29:437, 1975

1023. TAKAHASHI K, TERASHIMA K, KOJIMA M et al: Pathological, histochemical and ultrastructural studies on sea-blue histiocytes and Gaucher-like cells in acquired lipidosis occurring in leukemia. Acta Pathol Jpn 27:775, 1977

1024. TAKEUCHI T: Histochemical demonstration of branching enzyme (amylo-1, 4→1,6-transglucosidase) in animal tissues. J Histochem Cytochem 6:208, 1958

1025. TAKEUCHI T, KINOSHITA K: Histochemical demonstration of phosphorylase in blood and bone marrow cells. Blood 11:375, 1956

1026. TAKEUCHI T, KURIAKI H: Histochemical detection of phosphorylase in animal tissues. J Histochem Cytochem 3:153, 1955

1027. TAMAOKI N, ESSNER E: Distribution of acid phosphatase, (β-glucuronidase) and N-aceytl-β-glucosaminidase activities in lymphocytes of lymphatic tissues of man and rodents. J Histochem Cytochem 17:238, 1969

1028. TAN HK, LAMBERG JD: Diagnosis of acute leukemia. Variability of morphologic criteria. Am J Clin Pathol 68:440, 1977

1029. TANGÜN Y, SARAÇBAŞI Z, INCEMAN S: Clinical and hematological features in 2 cases of acute plasma cell leukemia. Turk Tip Cem Mec 34:543, 1968

1030. TAVASSOLI M: Cytochemistry of marrow and extramedullary adipocytes in monolayer cultures. Scand J Haematol 20:330, 1978

1031. TAYLOR CR, MASON DY: The immunohistological detection of intracellular immunoglobulin in formalin–paraffin sections from multiple myeloma and related conditions using the immunoperoxidase technique. Clin Exp Immunol 18:417, 1974

1032. TELDGES AJ, ANN RJA, NERZOSA MS et al: Periodic acid–Schiff reaction, a useful index of duration of complete remission in acute childhood lymphocytic leukemia. Acta Haematol 52:8, 1974

1033. TESTA JR, GOLOMB HM, ROWLEY JD et al: Hypergranular promelocytic leukemia (APL): Cytogenetic and ultrastructural specificity. Blood 52:272, 1978

1034. THEOLOGIDES A: Unfavorable signs in patients with acute myelogenous leukemia. Ann Interne Med 76:95, 1972

1035. THOMPSON SW, HUNT RD: Selected Histochemical and Histopathological Methods. Springfield, Charles C Thomas, 1966

1036. TRANCHIDA L, PALUTKE M, POULIK MD et al: Primary acquired sideroblastic anemia preceding monoclonal gammopathy and malignant lymphoma. Am J Med 55:559, 1973

1037. TREADGOLD CH: Myeloid leukaemia in a child with blood picture of so-called megaloblastic degeneration. Lancet 1:94, 1913

1038. TREBICHAVSKY I, BONA C, ANTEUNIS A et al: Structure of the lymphocyte membrane. I. The distribution of immunoglobulins on the surface of rabbit lymphocytes of different origin. (Electron-microscopy-peroxidase-labelling-endocytosis-IgG-IgM-IgA-thymocytes). Folia Biol 18:30, 1972

1039. TRUBOWITZ S, MASEK B: A histochemical study of the reticuloendothelial system of human marrow—its possible transport role. Blood 32:610, 1968

1040. TRUBOWITZ S, MASEK B, FRASKA JM: Leukemic reticuloendotheliosis. Blood 38:288, 1971

1041. TSCHUDI P, ARCHER RK, GERBER H: Cytochemical staining of equine blood and bone marrow cells. Equine Vet J 9:205, 1977

1042. TSUKIMOTO I, WONG KY, LAMPKIN BC: Surface markers and prognostic factors in acute lymphoblastic leukemia. N Engl J Med 294:245, 1976

1043. TULLIEZ M, RICARD MF, JAN F et al: Preleukaemic abnormal myelopoiesis induced by chlorambucil: A case study. Scand J Haematol 13:179, 1974

1044. TULLIEZ M, VERNANT JP, BRETON-GORIUS J et al: Pseudo–Chediak–Higashi anomaly in a case of acute myeloid leukemia: Electron microscopic studies. Blood 54:863, 1979

1045. TURNER A, KJELDSBERG CR: Hairy cell leukemia: A review. Medicine 57:477, 1978

1046. Turpin F, Lejeune F, Vilde JL et al: Peroxidase deficiency in neutrophils in systemic mastocytosis (author's trans). Nouv Rev Fr Hematol Blood Cells 20:77, 1978
1047. Tursz T, Flandrin G, Brouet JC et al: Coexistence of a myeloma and a granulocytic leukemia in the absence of any treatment. Study of 4 cases. Nouv Rev Fr Hematol 14:693, 1974
1048. Tzortzatou F, Hayhoe FGJ: Cytochemical demonstration of dihydrofolate reductase activity. Br J Haematol 28:199, 1974
1049. Uchiyama T, Junj Y, Sagawa K et al: Adult T-cell leukemia: Clinical and hematologic features of 16 cases. Blood 50:481, 1977
1050. Undritz E: Die Modification 3 der Peroxydasereaktion nach Graham-Knoll zum Nachweis von Auer-Stäbchen und Nucleolen. Schweiz Med Wschr 95:1479, 1965
1051. Undritz E: The Alius–Grignaschi anomaly: The hereditary constitutional peroxidase defect of the neutrophils and monocytes. Blut 14:129, 1966
1052. VanDen Bogaert P, Von Hove W: Polycythemia vera suivie d'une erythroleucémie aiguë. Nouv Presse Med 79:1685, 1971
1053. Van der Burg ME, Mulder H: Plasma cell leukemia with plasma cells in the urine. Ned Tijdschr Geneeskd 122:305, 1978
1054. Van der Meer JWM, Beelen RHJ, Fluitsma DM et al: Ultrastructure of mononuclear phagocytes developing in liquid bone marrow cultures. A study on peroxidatic activity. J Exp Med 149:17, 1979
1055. Van Furth R: An approach to the characterization of mononuclear phagocytes involved in pathological processes, Vol 2, p 91. In Giroud JP et al (eds): Future Trends in Inflammation II. Basel, Brikhäuser, 1975
1056. Van Furth R, Fedorko ME: Ultrastructure of mouse mononuclear phagocytes in bone marrow colonies grown in vitro. Lab Invest 34:440, 1976
1057. Van Hegan RI: Distribution of malignant lymphoid cells. J Clin Pathol (Suppl) 13:70, 1979
1058. Van Hove W, Hamers J: Erythroleukemia. A clinical study of 11 cases and a review of the literature. Acta Clin Belg 27:405, 1972
1059. Van Slyck EJ, Rebuck JW: Pseudo–Chediak–Higashi anomaly in acute leukemia. Am J Clin Pathol 62:673, 1974
1060. Vardiman JW, Golomb HM, Rowley JD et al: Acute nonlymphocytic leukemia in malignant lymphoma. Cancer 42:229, 1978
1061. Vaughan JM, Harrison CV: Leucoerythroblastic anaemia and myelosclerosis. J Pathol Bacteriol 48:339, 1939
1062. Verloop MC, Bos CC: Differential diagnosis between Björkman's anemia refractoria sideroblastica and erythraemic myelosis (DiGuglielmo's disease). Proc 9th Congr Eur Soc Hematol 2:964, 1963
1063. Verloop MC, Punt K, Hellemann P et al: Sideroachrestiche anemie bij kinderen. Nederl T Geneesk 108:434, 1964
1064. Vigliani EC, Saita G: Benzene and leukemia. N Engl J Med 271:872, 1964
1065. Vilter RW, Jarrold T, Will JJ et al: Refractory anemia with hyperplastic bone marrow. Blood 15:1, 1960
1066. Vilter RW, Will JJ, Jarrold T: Refractory anemia with hyperplastic bone marrow (aregenerative anemia). Semin Hematol 4:175, 1967
1067. Viswanathan B: Chronic myelomonocytic leukemia. A review. J Kans Med Soc 77:59, 1976
1068. Voisin GA: Cytomorphologic distinction of lymphocytes of thymic origin and medullary origin. Pathol Biol 23:482, 1975
1069. Volante G, Bussolati G, Stramignoni A: Research on mononuclear cells in human blood in cultures. I. Morphological and histochemical data with and without treatment with phytohemagglutinin. Pathologica 58:1, 1966

1070. Von Leube W: Ueber einen Fall von rapid verlaufender schwerer Anämie mit gleichzeitiger leukämischer Beschaffenheit des Blutes. Sitzungs-Ber d physikmed Ges zu Würzburg 3:46, 4:49, 1900
1071. Vowels MR, Willoughby MLN: Cyclic chemotherapy in acute lymphoblastic leukaemia of childhood. 5-year survivals. Arch Dis Child 48:436, 1973
1072. Wachstein M: The distribution of histochemically demonstrable glycogen in human blood and bone marrow cells. Blood 4:54, 1949.
1073. Wachstein M, Wolf G: The histochemical demonstration of esterase activity in human blood and bone marrow smears. J Histochem Cytochem 6:457, 1958
1074. Wagner VM, Baehner RL: Lack of correlation between blast cell size and length of first remission in acute lymphocytic leukemia in childhood. Med Pediatr Oncol 3:373, 1977
1075. Warburg O (ed): The Metabolism of Tumors (Dickens F trans). London, Constable, 1930
1076. Warburg O: On the origin of cancer cells. Science 123:309, 1956
1077. Ward HP, Block MH: The natural history of agnogenic myeloid metaplasia (AMM) and a critical evaluation of its relationship with the myeloproliferative syndrome. Medicine 50:357, 1971
1078. Warter J, Mantz JM, Grappe JM et al: Plasmacytic leukemia. Apropos of 3 cases. Semin Hop Paris 43:503, 1967
1079. Watanabe S, Shimosato Y, Kameya T, et al: Leukemic distribution of a human acute lymphocytic leukemia cell line (Ichikawa strain) in nude mice conditioned with whole-body irradiation. Cancer Res 38:3494, 1978
1080. Wattiaux R: Biochemistry and function of lysosomes. In Lima-de-Faria A (ed): Handbook of Molecular Cytology. Amsterdam, North-Holland, 1969
1081. Weber TH, Wegelius R, Borgström GH et al: A case of pure monocytic leukaemia in a child—characterization of cellular morphology, membrane markers, surface glycoproteins and karyotype. Scand J Haematol 22:47, 1979
1082. Wegmann R: Enzymatic hematology and its functional correlations. Gegenbaurs Morphol Jahrb 116:401, 1971
1083. Weil M, Sebaoun G, Semama A et al: Leucémie aiguë monoblastique. Acta Haematol 10:156, 1976
1084. Weisenburger DD: Acute myelofibrosis terminating as acute myeloblastic leukemia. Am J Clin Pathol 73:128, 1980
1085. West BC, Rosenthal AS, Gelb NA et al: Separation and characterization of human neutrophil granules. Am J Pathol 77:41, 1974
1086. Wetter O, van de Weert M, Reis HE et al: Mitogenic stimulation of human leukemic plasma cells. Exp Hematol 2:73, 1974
1087. Wetzel BK, Horn RG, Spicer SS: Letter: Cytochemical localization of nonspecific phosphatase activity in rabbit myeloid elements. J Histochem Cytochem 11:812, 1963
1088. Wetzel BK, Horn RG, Spicer SS: Fine structural studies on the development of heterophil, eosinophil and basophil granulocytes in rabbits. Lab Invest 16:349, 1967
1089. Wetzel BK, Spicer SS, Horn RG: Fine structural localization of acid and alkaline phosphatases in cells of rabbit blood and bone marrow. J Histochem Cytochem 15:311, 1967
1090. Whang-Peng J, Gralnick HR, Johnson RE et al: Chronic granulocytic leukemia (CGL) during the course of chronic lymphocytic leukemia (CLL): Correlation of blood, marrow and spleen morphology and cytogenetics. Blood 43:333, 1974
1091. Whitby LEH, Christie JM: Monocytic leukaemia. Lancet 1:80, 1935
1092. White JG: Fine structural demonstration of acid phosphatase activity in Auer bodies. Blood 29:667, 1967
1093. White JG: Serotonin storage organelles in human megakaryocytes. Am J Pathol 63:403, 1971

1094. WHITE JG, ESTENSEN RD: Selective labilization of specific granules in polymorphonuclear leukocytes by phorbol myristate acetate. Am J Pathol 75:45, 1975

1095. WILLCOX MB, GOLDE DW, CLINE MJ: Cytochemical reactions of human hematopoietic cells in liquid culture. J Histochem Cytochem 24:979, 1976

1096. WILLIAMS MJ: Myeloblastic leukemia preceded by prolonged hematologic disorder. Blood 10:502, 1955

1097. WILLIGHAGEN RG: Enzyme histochemistry of human tumours. Beitr Pathol 141:280, 1970

1098. WINCHESTER RJ, ROSS G: Methods of enumerating lymphocyte populations. In Rose and Friedman (eds): Manual of Clinical Immunology, p 64, Washington, DC. Am Soc for Microbiol, 1976

1099. WINKELSTEIN A, SPARKES RS, CRADDOCK CG: Trisomy of group C in a myeloproliferative disorder. Report of case. Blood 27:722, 1966

1100. WINKLER K, STÜHRK H, GAEDICKE G et al: Clinical significance of cytochemical findings of childhood acute lymphoblastic leukaemia (author's trans). Klin Paediatr 189:22, 1977

1101. WINTROBE MM, MITCHELL DM: Atypical manifestations of leukemia. QJ Med 33:67, 1940

1102. WOLFF LJ, RICHARDSON ST, NEIBURGER JR et al: Poor prognosis of children with acute lymphocytic leukemia and increased B cell markers. J Pediatr 89:956, 1976

1103. WOODRUFF RK, MALPAS JS, PAXTON AM et al: Plasma cell leukemia (PCL): A report on 15 patients. Blood 52:839, 1978

1104. WRIGHT D: "Burkitt's Lymphoma: A review of the pathology, immunology and possible etiologic factors." Sommers S (ed): Pathology Annual. New York, Appleton-Century-Crofts, 1971

1105. WULFF HR, SORENSEN M: A new technique for the histochemical demonstration of phosphorylase in blood films. Acta Haematol 35:304, 1966

1106. WULFHEKEL U, DÜLLMANN J, BARTELS H et al: On the ultrastructure and cytochemistry of eosinophil-myelomonocytic leukemias. Virchows Arch (Pathol Anat) 365:289, 1975

1107. WURSTER-HILL DH, MCINTYRE OR, CORNWELL GG III: Chromosome studies in myelomatosis. Virchows Archiv (Cell Pathol) 29:93, 1978

1108. YAM LT, LI CY, CROSBY WJ: Cytochemical identification of monocytes and granulocytes. Am J Clin Pathol 55:283, 1971

1109. YAM LT, LI CY, LAM KW: Tartrate-resistant acid phosphatase isoenzyme in the reticulum cells of leukemic reticuloendotheliosis. N Engl J Med 284:357, 1971

1110. YAM LT, LI CY, WOLFE HJ et al: Histochemical study of acute leukemia. Arch Pathol 97:129, 1974

1111. YAM LT, TAVASSOLI M, JACOBS P: Differential characterization of the "reticulum cell' in lymphoreticular neoplasms. Am J Clin Pathol 64:171, 1975

1112. YAMADA E, YAMAUCHI R: Some observations on the cytochemistry and morphogenesis of the granulocytes in the rat bone marrow as revealed by electron microscopy. Acta Haematol Jpn, 29:530, 1966

1113. YAMASHITA K, IWAMOTO T, IIJIMA S: Immunohistochemical observation of lysozyme in macrophages and giant cells in human granulomas. Acta Pathol Jpn 28:689, 1978

1114. YOFFEY JM: Quantitative Cellular Haematology. Springfield, Charles C Thomas, 1960

1115. YOSHIMURA Y, MORI M: Isoenzymes and histochemistry of non-specific esterases in normal and cancerous skin. Histochem J 2:275, 1970

1116. YOSHITOSHI Y, SUZUKI H, OASHI T et al: Case of familial Waldenström's croglobulinemia with plasma cell leukemia and collection of 38 cases reported in Japan. Naika 25:955, 1970

1117. Zagury D, Avrameas S, Zeitoun P: Demonstration of IgG on the lymphocyte surface and separation of lymphocyte populations from antiserums. CR Acad Sci (D) 273:719, 1971
1118. Zaitseva KK: Methods of bone decalcification offering a possibility of determining alkaline phosphatase in the bone marrow cells. Arkh Patol 27:78, 1965
1119. Zanger B, Dorsey HN: Fever—a manifestation of preleukemia. JAMA 236:1266, 1976
1120. Zarrabi MH, Rosner F, Bennett JM: Letter: Acute leukemia and non-Hodgkin's lymphoma. N Engl J Med 298:280, 1978
1121. Zawadzki ZA, Kapadia S, Barnes AE: Leukemic myelomatosis (plasma cell leukemia). Am J Clin Pathol 70:605, 1978
1122. Zeya HI, Laszlo J: Granule assembly in precursors of human leukemia granulocytes. Am J Pathol 71:467, 1973
1123. Zeya HI, Spitznagel JK: Characterization of cationic protein-bearing granules of polymorphonuclear leukocytes. Lab Invest 24:229, 1971
1124. Zeya HI, Spitznagel JK: Isolation of polymorphonuclear leukocyte granules from rabbit bone marrow. Lab Invest 24:237, 1971
1125. Zittoun R: Annotation: Subacute and chronic myelomonocytic leukaemia: A distinct haematological entity. Br J Haematol 32:1, 1976
1126. Zittoun R, Bernardou A, Bilski-Pasquier G et al: Les leucémies myélomonocytaires subaiguës. Etude de 27 cas et revue de la littérature. Semin Hop Paris 48:1943, 1972
1127. Zittoun R, Certin M, Andebert A et al: Evolution des anémies réfractaires. Rapports avec les leucémies myélo-monocytaires. Ann Med Interne 125:593, 1974
1128. Zittoun R, Diebold J, Bernardou A et al: Diagnostic des états préleucémiques et des leucémies myélo-monocytaires et atypiques. Nouv Rev Fr Hematol 13:380, 1973
1129. Zucker-Franklin D, Hirsch JG: Electron microscope studies on the degranulation of rabbit peritoneal leukocytes during phagocytosis. J Exp Med 120:569, 1964

Index

Page number followed by f indicates figure; page number followed by t indicates table